Statistical Inference for Diffusion Type Processes

KENDALL'S LIBRARY OF STATISTICS

Published titles:

1. Multivariate Analysis
   Part 1: Distributions, Ordination and Inference
   WJ Krzanowski (University of Exeter) and FHC Marriott (University of Oxford)
   1994,
2. Multivariate Analysis
   Part 2: Classification, Covariance Structures and Repeated Measurements
   WJ Krzanowski (University of Exeter) and FHC Marriott (University of Oxford)
   1995,
3. Multilevel Statistical Models
   Second edition
   H Goldstein (University of London)
   1995,
4. The Analysis of Proximity Data
   BS Everitt (Institute of Psychiatry) and S Rabe-Hesketh (Institute of Psychiatry)
   1997,
5. Robust Nonparametic Statistical Methods
   Thomas P Hettmansperger (Penn State University) and Joseph W McKean (Western Michigan University)
   1998,
6. Statistical Regression with Measurement Error
   Chi-Lun Cheng (Academia Sinica, Republic of China) and John Van Ness (University of Texas at Dallas)
   1999,
7. Latent Variable Models and Factor Analysis
   D.J. Bartholemew (London School of Economics) and M. Knott (London School of Economics)
   1999,
8. Statistical Inference for Diffusion Type Processes
   B.L.S. Prakasa Rao (Indian Statistical Institute)
   1999,

Statistical Inference for Diffusion Type Processes

B.L.S. Prakasa Rao
Indian Statistical Institute, New Delhi

John Wiley & Sons, Ltd

Co-published in the USA by
Oxford University Press Inc., New York

First published in Great Britain in 1999 by
Arnold, a member of the Hodder Headline Group,
338 Euston Road, London NW1 3BH

Co-published in the USA by
Oxford University Press Inc.,
198 Madison Avenue, New York, NY 10016

John Wiley & Sons Ltd, The Atrium, Southern Gate, Chichester, West Sussex, PO19 8SQ, United Kingdom

For details of our global editorial offices, for customer services and for information about how to apply for permission to reuse the copyright material in this book please see our website at www.wiley.com.

British Library Cataloguing in Publication Data
A catalogue record for this book is available from the British Library

Library of Congress Cataloging-in-Publication Data
A catalog record for this book is available from the Library of Congress

ISBN 978-0-470-71112-5

1 2 3 4 5 6 7 8 9 10

Typeset in 9/11 pt Times by Focal Image Ltd, Torquay

To the memory of my father-in-law, K. Siva Rama Sastri
and to my mother-in-law, K.S. Subhadra Devi
for their love and affection for an academician

Contents

Preface

Decision making in all spheres of activity involves uncertainty. If rational decisions have to be made, they have to be based on past observations of the phenomenon. Data collection, model building and inference from the data collected, validation of the model and refinement of the model are the key steps or building blocks involved in any rational decision making process. Stochastic processes are widely used for model building in social, physical, engineering and life sciences. A recent application is in the area of financial economics. Statistical inference for stochastic processes is of great importance from the theoretical as well as from an applications point of view in model building. During the past 20 years, much progress has been made in the study of inferential aspects for continuous- as well as discrete-time stochastic processes. Diffusion type processes are a large class of continuous-time processes which are widely used for stochastic modeling. Our aim in this book is to bring together several methods of estimation of parameters involved in such processes when the process is observed continuously over a period of time or when sampled data are available, as is generally feasible. We discuss parametric as well as nonparametric aspects of the problem. Diffusion type processes form a large subclass of processes known as semimartingales. The general asymptotic theory of semimartingales and their statistical inference is discussed in *Semimartingales and their Statistical Inference*.

It is a great pleasure for me to thank the Indian Statistical Institute for its support. Thanks are due to several colleagues who have sent preprints and reprints to me over the years which have been extensively used in the preparation of this book. Mr V.P. Sharma and Ms Simmi Marwah of the Indian Statistical Institute have ably assisted me in preparing the manuscript. Thanks are also due to Ms Nicki Dennis, Director, Applied Science and Technology Publishing, Arnold, for her interest in the project right from the outset. Finally, it is a great pleasure for me to thank my wife Vasanta for her endurance and patience during endless hours I was away in the office working on this book; she is familiar with the idiosyncrasies of an academician and a workaholic.

B.L.S. Prakasa Rao
New Delhi, July 1998

Introductory Notes

1. Diffusion Type Processes

The concept of a semimartingale is of major interest in stochastic modeling as it includes several types of processes such as point and diffusion processes. A semimartingale essentially is the sum of a local martingale and a process which is of bounded variation. We give a short review of the theory of diffusion and diffusion type processes which form a major class of semimartingales. A brief introduction to the theory of semimartingales has been given by Prakasa Rao (1987). The discussion here is based on Basawa and Prakasa Rao (1980), Karlin and Taylor (1981), Liptser and Shiryayev (1977; 1978) and other related books.

2. Parametric Inference for Diffusion Type Processes from Continuous Paths

Diffusion processes can be used for stochastic modeling with applications to physical, biological and medical sciences, and more recently to economic and social sciences. However, all these models involve unknown parameters or unknown functions which need to be estimated from observations on the processes. The parameters in the model may be finite-dimensional or infinite-dimensional. They have to be estimated either from one or many continuous realizations of the process if a continuous realization is possible, or from a discrete sampled data set for the process if the process is not continuously observable. In this chapter we discuss different methods of estimation of the drift parameter for a diffusion type process when a continuous sample path of the process is available. Earlier work on the estimation of the drift parameter for a linear stochastic differential equation is presented in Basawa and Prakasa Rao (1980). Results discussed in Section 2.2 on maximum likelihood estimation in the scalar parameter case for a nonlinear stochastic differential equation are due to Mishra and Prakasa Rao (1985). The discussion in the vector parameter case follows Basu (1983a), which generalizes and uses the techniques in Prakasa Rao and Rubin (1981). The material on the maximum likelihood estimation using the log-likelihood ratio process follows Kutoyants (1984a). The discussion on local asymptotic mixed normality is based on the work of Dietz (1989). Results on large deviations for maximum likelihood estimation are due to Levanony (1994). Section 2.3 contains results due to Lanska (1979) on the method of minimum contrast for the estimation of the drift parameter. Section 2.4 contains results due to Prakasa Rao (1982) on maximum probability estimation. Section 2.5 follows the work of Basu (1983b) on the Bernstein–von Mises theorem and its application to Bayes estimation for diffusion processes generalizing earlier work in

Borwanker *et al.*(1971), Prakasa Rao (1981) and Prakasa Rao and Rubin (1981). The work in Section 2.6 on the minimum distance method for the estimation of the drift parameter is due to Kutoyants (1991) and Dietz and Kutoyants (1997). The discussion on M-estimation method in Section 2.7 follows Yoshida (1988, 1990). Results on recursive estimation in Section 2.8 are due to Levanony *et al.*(1994). The material in Section 2.9 on sequential estimation follows Sørensen (1986). The work in Section 2.10 on estimation from incomplete observations is due to Larédo (1990) and Section 2.11 contains results due to Larédo (1990) and Genon-Catalot and Larédo (1987) on estimation from first hitting times.

3. Parametric Inference for Diffusion Type Processes from Sampled Data

We consider parametric inference for diffusion type processes when the process is observed at discrete time instants. It is important to know in this context whether a sampled process inherits the properties of the original process such as mixing if the original process is mixing. Section 3.1 contains results due to Prakasa Rao (1990b). Earlier work on maximum likelihood estimation from discrete data for linear stochastic differential equations is due to Le Breton (1976), Dorogovcev (1976), Prakasa Rao and Rubin (1981) and Prakasa Rao (1983b). Section 3.2 contains results due to Penev (1985) extending the work of Prakasa Rao (1983b) to the multidimensional case. The material in Section 3.3 is based on the work of Yoshida (1992) and Pedersen (1995a; 1995b). The discussion in Section 3.4 follows Shoji (1995) and Shoji and Ozaki (1997). The work on estimation by the method of martingale estimating functions in Section 3.5 follows Bibby and Sørensen (1995a; 1995b; 1995c) and Kessler and Sørensen (1995). This Section also contains discussion on estimation by approximation of the transition function by a suitable Gaussian density due to Kessler (1997), and recent results on estimation by approximation of the transition function by Gaussian and close to Gaussian densities due to Ait-Sahalia (1997). The general results on estimation through deterministic or random sampling in Section 3.6 are based on the work of Genon-Catalot and Jacod (1993; 1994). The discussion on simulation-based estimation methods in Section 3.7 is based on Broze *et al.* (1995), Clement (1997) and Gallant and Long (1997). Results on Bayesian inference in Section 3.8 are a variation of work of Bishwal (1996) following Yoshida (1992).

4. Nonparametric Inference for Diffusion Type Processes from Continuous Sample Paths

Several methods for the estimation of functionals of a probability density function and estimation of a regression function by nonparametric techniques are extensively discussed in Prakasa Rao (1983a). A general review of the parametric and nonparametric methods of statistical inference from sampled data for stochastic processes is given in Prakasa Rao (1988b). Theorems 4.2.1–4.2.3 concerning estimation of the marginal density and its derivative are due to Prakasa Rao (1979). Pham Dinh Tuan (1981) studied the estimation of the drift coefficient by using nonparametric methods for the estimation of a regression function. Theorems 4.2.4–4.2.7 discuss estimation of the drift by the kernel method and are due to Kutoyants (1985a). The method of sieves was introduced by Grenander (1981). Theorem 4.2.8, dealing with the consistency of an estimator of the drift obtained using the method of sieves, is due to Geman and Hwang (1982). Genon-Catalot and Larédo (1986) considered the problem of nonparametric estimation of the drift from observations related to hitting times; Theorem 4.2.9 is due to them. Estimation of the drift in a nonstationary difusion model by the kernel method was investigated by Kutoyants (1985b). The discussion on efficient density estimation for the stationary distribution of an ergodic process follows Kutoyants (1997a; 1997b). Theorem 4.3.1 is due to

Kutoyants (1985b). Nguyen and Pham Dinh Tuan (1982) discussed the same problem using the method of sieves. Theorem 4.3.2 is from Nguyen and Pham Dinh Tuan (1982). Leskow and Rozanski (1989) considered the problem of drift estimation for a slightly more general class of stochastic differential equations and suggested an estimator using the method of sieves. Theorem 4.3.3 is due to them. All the results in Section 4.4 are due to McKeague and Tofoni (1991).

5. Nonparametric Inference for Diffusion Type Processes from Sampled Data

A general review of the methods of statistical inference from sampled data for stochastic processes was presented in Prakasa Rao (1988b). The special case of the nonparametric density estimation for stochastic processes from sampled data was discussed in Prakasa Rao (1990a), on which results in Section 5.2 are based. The discussion on the estimation of the drift coefficient for the one-dimensional case in Section 5.3 is based on Nguyen and Pham Dinh Tuan (1982), Geman (1979), Pham Dinh Tuan (1981) and Banon and Nguyen (1981a; 1981b). The discussion in the multidimensional case, is from Eplett (1987). The results on the estimation of the diffusion coefficient in the one-dimensional case by the method of nearest neighbors is due to Florens-Zmirou (1993). Theorem 5.4.6, dealing with the multidimensional case, is from Brugière (1993). The material on the estimation of the diffusion coefficient by the method of wavelets is from Genon-Catalot *et al.*(1992). Ait-Sahalia (1996b) discussed the problem of estimation by the method of matching the drift and diffusion to the one-dimensional marginal density; Theorem 5.4.10 is taken from that work. Nonparametric estimation of the diffusion coefficient from equally spaced sample data under L_2-loss function are from Maiboroda (1995). Results on the rates of minimax risk under L_p-loss and on the construction of optimal estimators are due to Hoffmann (1997).

6. Applications to Stochastic Modeling

A large number of applications of diffusion processes for stochastic modeling are discussed in Karlin and Taylor (1981). Some recent applications to mathematical finance, forest management and stochastic hydrology are discussed in this chapter. Sections 6.2–6.4 are based on the work of Shoji (1995) and Nagahara (1996).

7. Numerical Approximation Methods for Stochastic Differential Equations

Simulation of the solution of a stochastic differential equation has been extensively discussed in Kloeden and Platen (1992) and Kloeden, Platen and Schurz (1994). The discussion in Sections 7.1 and 7.2 is based on Kloeden and Platen (1989). The material in Section 7.3 on the effects of discretization in estimation for diffusion processes is from Kloeden *et al.*(1996). The discussion on the uniform approximation for stochastic integrals of functions of a solution of a stochastic differential equation is based on Prakasa Rao and Rubin (1998).

Appendices

The discussion of the uniform ergodic theorem in Appendix A and the uniform ergodic theorem for stochastic integrals in Appendix B is due to Prakasa Rao (1982). The discussion of the conditions for the interchangeability of stochastic integration and ordinary differentiation in

Appendix B is due to Karandikar (1983). The brief introduction to wavelets in Appendix C follows Genon-Catalot *et al.*(1992). The Gronwall–Bellman type lemma discussed in Appendix D is from Liptser and Shiryayev (1977).

1

Diffusion Type Processes

1.1 Introduction

Decision making in all spheres of activity involves uncertainity. If rational decisions have to be made, they have to be based on the past observations of the phenomenon. Data collection, model building and inference from the data collected, validation of the model and refinement of the model are the key steps involved in any rational decision making process. Stochastic processes are widely used for model building, and the subject of inference for stochastic processes is of importance both from the theoretical and from the application points of view. Several books have been published during the last 20 years, including Basawa and Prakasa Rao (1980), Grenander (1981), Prakasa Rao (1983a; 1987) and Kutoyants (1984a). More recently, Karr (1991) and Fleming and Harrington (1991) are concerned with inference for stochastic processes. Prabhu (1988), Prabhu and Basawa (1991), Basawa and Prabhu (1994), Prakasa Rao (1995), and Prakasa Rao and Bhat (1996) contain papers giving comprehensive accounts of recent developments in the area of inference for stochastic processes.

The notion of a semimartingale has been found to be of major interest in stochastic modeling as it includes several types of processes such as point processes and diffusion processes which are widely used for model building. A semimartingale is essentially the sum of a local martingale and a process which is of bounded variation. A brief introduction to such processes is given in Prakasa Rao (1987). Liptser and Shiryayev (1977; 1978), Kallianpur (1980), Bremaud (1981) and Elliott (1982) deal extensively with this topic.

The asymptotic theory of semimartingales and their statistical inference are discussed in detail by Prakasa Rao (1999). We concentrate our discussion here on parametric and nonparametric inference for diffusion type processes which constitute a major class of semimartingales and which are widely used for stochastic modeling purposes.

We will discuss briefly the properties of diffusion type processes along with some examples, following Karlin and Taylor (1981). Diffusion type processes constitute a large class of semimartingales, and statistical inference for such processes will be the subject of the following chapters.

1.2 Diffusion Type Processes

1.2.1 Diffusion Processes

Let X_t denote the coordinates of a sufficiently small particle suspended in a liquid at an instant t. Suppose the velocity of the motion of the liquid at the point x and the instant t is equal to $g(t, x)$. Further suppose that the fluctuational component of the displacement is a random variable whose distribution depends on the position x of the particle, the instant t at which the displacement is observed and the quantity Δt which is the length of the interval of the time during which its displacement is observed. Suppose the average value of this displacement is zero independent of t, X_t and Δt. Thus the displacement of the particle can be written in the form

$$X_{t+\Delta t} - X_t = g(t, X_t)\,dt + \varepsilon_{t,X_t,\Delta t}, \tag{1.2.1}$$

where $E\varepsilon_{t,X_t,\Delta t} = 0$. If $g(t, x) = 0$ and the distribution of $\varepsilon_{t,x_t,\Delta t}$ is independent of x and t as in the case of the Wiener process, then $E\varepsilon^2_{t,X_t,\Delta t} = \ell\Delta t$ for some constant $\ell > 0$. It is natural to assume that the properties of the medium change slightly for small changes in t and x. This leads to a homogeneous process and it may be assumed that

$$\varepsilon_{t,X_t,\Delta t} = \sigma(t, X_t)\varepsilon_{t,\Delta t}, \tag{1.2.2}$$

where $\sigma(t, x)$ characterizes the properties of the medium at the point x at the instant t and $\varepsilon_{t,\Delta t}$ is the value of the incrment that is obtained in the homogeneous case under the condition $\sigma(t, x) = 1$. In other words, $\varepsilon_{t,\Delta t}$ must be distributed like the increment of a Wiener process $W(t)$, namely $W(t + \Delta t) - W(t)$. Hence

$$X_{t+\Delta t} - X_t \simeq g(t, X_t)\Delta t + \sigma(t, X_t)[W(t + \Delta t) - W(t)]. \tag{1.2.3}$$

This leads to the stochastic differential equation

$$dX_t = g(t, X_t)\,dt + \sigma(t, X_t)\,dW(t), \qquad t \geq 0, \tag{1.2.4}$$

which can be taken as the starting point of a diffusion process. The stochastic differential equation is interpreted in the form

$$X_t - X_0 = \int_0^t g(s, X_s)\,ds + \int_0^t \sigma(s, X_s)\,dW(s), \qquad t \geq 0, \tag{1.2.5}$$

and the precise meaning in which the stochastic integral

$$\int_0^t \sigma(s, X_s)\,dW(s)$$

is defined and the properties of such integrals are discussed, for instance, in Basawa and Prakasa Rao (1980). For a brief introduction, see Appendix B.

Itô stochastic differential equation

Consider the stochastic differential equation

$$X_t = X_s + \int_s^t \sigma(u, X_u)\,dW_u + \int_s^t g(u, X_u)\,du, \qquad 0 \leq s \leq t, t \geq 0 \tag{1.2.6}$$

where X_s is an $\mathcal{F}_s$-measurable random variable independent of $\{W_u - W_v, u \geq v \geq s\}$. Let $\mathcal{F}_{s,t}$ be the σ-algebra generated by X_s and $\mathcal{F}^W_{s,t}$, where $\mathcal{F}^W_{s,t}$ is the completion of the σ-algebra generated by $\{W_u - W_v, t \geq u \geq v \geq s\}$. Let $\mathcal{F}_t \equiv \mathcal{F}_{0,t}$. Suppose that $g : R_+ \times R^d \to R^d$ is measurable and σ is a $d \times m$ ordered measurable matrix with $\sigma_{ij} : R_+ \times R^d \to R$ such that there exists a constant $K > 0$ satisfying

$$\begin{aligned} |g(u,x) - g(u,y)| &\leq K|x - y|, \\ |\sigma(u,x) - \sigma(u,y)| &\leq K|x - y|, \\ |g(u,x)|^2 + |\sigma(u,x)|^2 &\leq K(1 + |x|^2). \end{aligned} \tag{1.2.7}$$

Under the conditions stated above, the equation

$$dX_t = g(t, X_t)\,dt + \sigma(x, X_t)\,dW_t, \qquad X_s = x \in R^d,\ t \geq s, \tag{1.2.8}$$

has a unique, continuous, strong solution $X_t = X_{s,x}(t, \omega)$ and, for each $t \geq s$, X_t is $(\mathcal{B}(R^d) \times \mathcal{F}^W_{s,t})$-measurable.

Let $s = 0$ and $X_0 = x_0 \in R^d$ where x_0 is independent of $\mathcal{F}^W_{0,T}$. Then the equation

$$dX_t = g(t, X_t)\,dt + \sigma(t, X_t)\,dW_t, \qquad 0 \leq t \leq T,\ X_0 = x_0 \in R^d, \tag{1.2.9}$$

has a unique solution $\{X_t\}$ which is a continuous Markov process relative to $\{\mathcal{F}_t\}$. If g and σ depend on X_t but not on t, then the process is a homogeneous Markov process.

In our discussion later, we will come across instances where we need to build martingales out of diffusion processes. Next we discuss a method which will lead to appropriate function $Y_t = f(X_t)$ to be martingales, using eigenfunctions.

Eigenfunctions and martingales

The generator L of a Markov process $\{X_t\}$ is defined by

$$Lf(x) = \lim_{\Delta \to 0} \frac{1}{\Delta}[(\pi_\Delta f)(x) - f(x)] \tag{1.2.10}$$

whenever it exists, where $(\pi_\Delta f)(x) = E[f(X_\Delta)|X_0 = x]$. It is known that the domain $\mathcal{D}$ of the generator L of a diffusion contains the class of bounded twice continuously differentiable functions with bounded derivatives. For an eigenfunction $\phi \in \mathcal{D}$, with eigenvalue λ, one can prove, using the Markov property, that

$$\frac{\partial}{\partial \Delta}\pi_\Delta(\phi)(x) = L\pi_\Delta(\phi)(x) = -\lambda\pi_\Delta(\phi)(x) \tag{1.2.11}$$

and hence

$$(\pi_\Delta \phi)(x) = e^{-\lambda\Delta}\phi(x). \tag{1.2.12}$$

The domain $\mathcal{D}$ can be extended to $\mathcal{D}^*$ consisting of the class of twice continuously differentiable functions f for which the process

$$N_t = f(X_t) - f(X_0) - \int_0^t Lf(X_s)\,ds \tag{1.2.13}$$

is a martingale. If

$$dX_t = a(X_t)\,dt + \sigma(X_t)\,dW_t, \tag{1.2.14}$$

then, by Itô's lemma (see Appendix B),

$$N_t = \int_0^t f'(X_s)\sigma(X_s)\,dW_s, \tag{1.2.15}$$

and hence a sufficient condition for $f \in \mathcal{D}^*$ is that

$$\int_0^t E[f'(X_s)^2 \sigma^2(X_s)]\, ds < \infty. \tag{1.2.16}$$

Let $\mathcal{D}^{**}$ be the class of twice continuously differentiable functions satisfying (1.2.16). Let

$$Y_t = e^{\lambda t} \phi(X_t) \tag{1.2.17}$$

where ϕ is an eigenfunction corresponding to the eigenvalue λ. By Itô's lemma,

$$\begin{aligned} Y_t &= Y_0 + \int_0^t e^{\lambda s}[L\phi(X_s) + \lambda\phi(X_s)]\, ds + \int_0^t e^{\lambda s}\phi'(X_s)\sigma(X_s)\, dW_s \\ &= Y_0 + \int_0^t e^{\lambda s}\phi'(X_s)\sigma(X_s)\, dW_s. \end{aligned} \tag{1.2.18}$$

Hence, if $\phi \in \mathcal{D}^{**}$, then Y is a martingale.

Remark. There are many physical, biological, economic and social phenomena which can reasonably be modeled by diffusion processes. Examples are molecular motions of particles subject to interaction; share price fluctuations in a perfect market; communications systems with noise; neurophysiological activity with disturbances; community relationships; and gene substitution in evolutionary development (cf. Karlin and Taylor 1981).

It is sometimes convenient to consider a diffusion process from a different viewpoint for stochastic modeling purposes. A diffusion process can also be defined as a continuous-time parameter process with almost surely continuous paths and with a strong Markov property. Every diffusion process has the property that, for every $\varepsilon > 0$,

$$\lim_{h\downarrow 0} \frac{1}{h} P(|X(t+h) - X(t)| > \varepsilon | X(t) = x) = 0$$

for all x in the state space. Suppose that

$$\mu(x,t) = \lim_{h\downarrow 0} \frac{1}{h} E[(X(t+h) - X(t)) | X(t) = x]$$

and

$$\sigma^2(x,t) = \lim_{h\downarrow 0} \frac{1}{h} E[(X(t+h) - X(t))^2 | X(t) = x]$$

exist for all x and t. The coefficient $\mu(x,t)$ is called the *drift* and $\sigma(x,t)$ is the *diffusion*.

Let $X(t)$ be a time-homogeneous diffusion with drift $\mu(x)$ and diffusion $\sigma(x)$. Consider

$$Y^\lambda(t) = \exp\left[\lambda X(t) - \lambda \int_0^t \mu(X(s))\, ds - \tfrac{1}{2}\lambda^2 \int_0^t \sigma^2(X(s))\, ds\right] \tag{1.2.19}$$

for $t > 0$. Then, under some regularity conditions, $\{Y^\lambda(t), \mathcal{F}_t, t > 0\}$ is a martingale, where $\mathcal{F}_t = \sigma\{X(s) : s \le t\}$. If $X(t)$ is a standard Wiener process, then $\mu(x) \equiv 0$ and $\sigma^2(x) \equiv 1$ and

$$Y^\lambda(t) = \exp[\lambda X(t) - \lambda^2 t/2]. \tag{1.2.20}$$

Conversely if Y^{λ} is a martingale for every real λ, then X is a diffusion. In fact $\{X(t), t \geq 0\}$ is a diffusion process (under some regularity conditions) with drift $\mu(x)$ and diffusion $\sigma(x)$ if and only if for every bounded and twice continuously differentiable f, the process

$$Z_f(t) = f(X(t)) - f(X(0)) - \int_0^t \left[\tfrac{1}{2}\sigma^2(X(s))f''(X(s)) + \mu(X(s))f'(X(s))\right] ds \quad (1.2.21)$$

is a martingale. A similar result also holds for nonhomogeneous diffusion processes (see Karlin and Taylor 1981, p. 377).

Stochastic Modeling

Following Karlin and Taylor (1981), we can think of stochastic models evolving in discrete time to have the structure

$$Z_{k+1} = f(Z_k, s_k) + \varepsilon_k$$

or, in general,

$$Z_{k+1} = f(Z_k, s_k, \varepsilon_k),$$

where Z_{k+1} is the characteristic at the $(k+1)$th generation, with Z_{k+1} depending on (i) Z_k, the state at the kth generation, (ii) s_k, another randomly varying or fixed parameter at kth generation, and (iii) ε_k, 'noise'. Here $\{s_k\}$ and $\{\varepsilon_k\}$ can be interpreted as, for instance, respectively the stochastic or deterministic effect and, the demographic or sampling effect in the biological context.

A continuous-time version of such a process (as an extension of the above discrete set-up) can be modeled by a diffusion process.

The following discussion on examples of diffusion processes is based on Karlin and Taylor (1981).

Example 1.2.1 (*Wiener process (Brownian motion) W with drift*)
Here $W(0) = 0$ and $W(t) - W(s)$ is normal with independent increments with mean $E[W(t) - W(s)] = \mu|s - t|$ and $\text{var}[W(t) - W(s)] = \sigma^2|t - s|$.

Example 1.2.2 (*Ornstein–Uhlenbeck process V*)
The process V is a solution of the stochastic differential equation

$$dV(t) = -\alpha V(t)\,dt + \sigma\,dW(t), \qquad t \geq 0,\ \alpha > 0. \quad (1.2.22)$$

A Wiener process models the position of a particle in a liquid, whereas the Ornstein–Uhlenbeck process models the velocity of the particle. Two factors affect the velocity in a short time: (i) the functional resistance of the surrounding medium which is assumed to reduce the magnitude of velocity by a proportional amount; and (ii) the change in velocity due to random collisions with neighboring particles.

One can build from one diffusion process several other diffusions by using Itô's lemma (see Appendix B).

(i) Geometric Wiener process Y. Suppose $\{W(t), t \geq 0\}$ is a Wiener process with drift μ and diffusion σ. Then the process $Y = e^W$ is called the *geometric Wiener process (geometric Brownian motion)*. Applying Itô's formula, we obtain

$$dY(t) = (\mu + \tfrac{1}{2}\sigma^2)\,Y(t)\,dt + \sigma Y(t)\,dW^*(t), \quad (1.2.23)$$

where W^* is a standard Wiener process. This is used to model share prices in a perfect market. Note that prices are nonnegative and exhibit long-run exponential growth (or decay). This process is also used for modeling population growth. Note that

$$dW = \mu\, dt + \sigma\, dW^*(t).$$

Since $Y = e^W \equiv F(W)$, it follows that

$$\begin{aligned} dY &= F_w(W)\, dW + \tfrac{1}{2} F_{ww}(W)\sigma^2\, dt \\ &= e^W\, dW + \tfrac{1}{2} e^W \sigma^2\, dt \\ &= Y[\mu\, dt + \sigma\, dW^*(t)] + \tfrac{1}{2} Y\sigma^2\, dt \\ &= (\mu Y + \tfrac{1}{2}\sigma^2 Y)\, dt + \sigma Y\, dW^*(t). \end{aligned}$$

(ii) Bessel process Y. Let $Y(t) = \sqrt{Z(t)}, t \geq 0$, where $Z(t) = W_1^2(t) + \cdots + W_n^2(t)$ and $W_i, 1 \leq i \leq n$, are independent standard Wiener processes. Then $\{Y(t), t \geq 0\}$ is a diffusion process with drift coefficient $\frac{n-1}{2Y(t)}$ and diffusion coefficient unity.

Example 1.2.3 (*Wright–Fisher genetic model with mutation effects only*)
Consider a population of constant size N of individuals consisting of two types, A and a. Suppose that initially there are i individuals of type A and $N - i$ individuals of type a. The next generaion is produced subject to the influences of mutation. Suppose that, at birth, the probability of conversion from type A to type a is α and from type a to type A is β. Since the parental population consists of i type-A and $N-i$ type-a individuals, the expected proportion of type A after mutation is $(i/N)(1-\alpha)+(1-i/N)\beta$ and of type a is $(i/N)\alpha+(1-i/N)(1-\beta)$. Hence the probability of an offspring of type A is

$$\begin{aligned} p_i &= \frac{i(1-\alpha) + (N-i)\beta}{\{i(1-\alpha) + (N-i)\beta\} + \{i\alpha + (N-i)(1-\beta)\}} \\ &= \frac{i(1-\alpha) + (N-i)\beta}{N}. \end{aligned}$$

Suppose $\alpha = r_1/N$ and $\beta = r_2/N$, with $r_1 > 0$ and $r_2 > 0$. Let $X(t)$ be the number of type-A individuals in the tth generation (under the Wright–Fisher model, the composition of next generation is determined through N binomial trials with p_i as the probability of success, when i is the number of type A in the previous generation).

For large N, the process X can be approximated by a diffusion process. In fact, let

$$Y_n(\tau) = \frac{X([N\tau])}{N}, \qquad 0 \leq \tau \leq 1.$$

Then $Y_n(\tau)$ represents the fraction of type-A individuals in the population at generation time $[N\tau]$. It can be shown that

$$Y_N(\tau) = \frac{X([N\tau])}{N} \to Y(\tau)$$

as $i/N \to \zeta$ and $N \to \infty$, where Y is a diffusion process with drift coefficient $\mu = -r_1\zeta + (1-\zeta)r_2$ and diffusion coefficient $\zeta(1-\zeta)$.

Example 1.2.4 (*Model for population growth in a random environment*)
The population growth model is formulated generally as

$$\frac{dN}{dt} = f(N, t) + g(N, t)B(t), \tag{1.2.24}$$

where $N(t)$ is the population size at time t and $B(t)$ is Gaussian white noise.

(a) A special case of this model is the 'exponential' growth model

$$\frac{dN}{dt} = r(t)N(t)$$

where $r(t)$ is the instantaneous rate of growth at time t. Suppose

$$r(t) = \alpha + \gamma B(t),$$

where α and γ are constants and $B(t)$ is white noise. Hence

$$\frac{dN}{dt} = (\alpha + \gamma B(t))N(t) = \alpha N(t) + \gamma N(t)B(t).$$

and the stochastic differential equation version of this model is

$$dN(t) = \alpha N(t)\,dt + rN(t)\,dW(t), \qquad (1.2.25)$$

where $\{W(t), t \geq 0\}$ is a standard Wiener process. The solution of this stochastic differential equation is the geometric Brownian motion $N(t) = \exp[(\alpha - \frac{1}{2}\gamma^2)t + \gamma W(t)]$. Note that if $0 < \alpha < \frac{1}{2}\gamma^2$,

$$N(t) \to 0 \text{ as } t \to \infty \text{ a.s.}$$

(b) Another example of a growth model is the stochastic version of the logistic equation

$$\frac{dN(t)}{dt} = N(t)[k(t) - N(t)], \qquad k(t) = \alpha + \gamma B(t),$$

namely

$$\begin{aligned} dN(t) &= N(t)[\alpha\,dt + \gamma\,dW(t) - N(t)\,dt] \\ &= N(t)(\alpha - N(t))\,dt + N(t)\gamma\,dW(t). \end{aligned} \qquad (1.2.26)$$

(Here $\alpha + \gamma B(t) = k(t)$ represents a stochastic analogue of the carrying capacity of an ecological habitat.)

(c) A third example is a stochastic analogue of

$$\frac{dN(t)}{dt} = k(t)N(t) - \beta(t)N^2(t),$$

where $\beta(t)$ measures the individual effects on reproduction by survival of others. If $k(t) = \alpha + \gamma B(t)$, the corresponding stochastic differential equation is

$$dN(t) = (\alpha N(t) - \beta(t)N^2(t))\,dt + \gamma N(t)\,dW(t). \qquad (1.2.27)$$

Example 1.2.5 (*Models for economic processes*)
Stochastic differential equations provide a mechanism to incorporate inference associated with randomness, uncertainity and risk factors connected with various economic factors such as stock prices, the labor force and technological variables. Modeling of stochastic fluctuations by stochastic differential equations leads to easier interpretation of parameter changes over time.

(a) Model for production and consumption variation in a one-sector economy. Let $K(t)$ be capital assets at time t; $L(t)$ the labor force available at time t, assumed to be proportional to the population size at time t; and $C(t)$ the consumption rate at time t. The production function $F(K, L)$ is assumed to be concave and homogeneous of order 1 with respect to K and L. An example of such a function is the Cobb–Douglas production function,

$$F(K, L) = K^{\alpha} L^{1-\alpha}, \qquad 0 < \alpha < 1. \tag{1.2.28}$$

The capital goods accumulation equation is formulated as

$$\frac{dK(t)}{dt} = F(K(t), L(t)) - \lambda K(t) - C(t), \tag{1.2.29}$$

where λ is the rate of depreciation of capital (λ nonnegative and constant) and $C(t)$ is the consumption rate.

The labor force, L is assumed to satisfy the stochastic differential equation (observe that $L(t)$ is subject to a random pertubation)

$$dL(t) = L(t)(\alpha\, dt + \sigma\, dW(t)). \tag{1.2.30}$$

Note that $L(t) = L(0) \exp\{(\alpha - \sigma^2/2)t + \sigma W(t)\}$. Let

$$k(t) = \frac{K(t)}{L(t)}$$

be the capital–labor ratio, and let

$$C(t) = \frac{C(t)}{L(t)}$$

be per-capita consumption. Note that $F(\lambda K, \lambda L) = \lambda F(K, L)$ since F is homogeneous of order 1. Define

$$f(k) = F(k, 1) = \frac{F(K, L)}{L},$$

where $k = K/L$; $f(k)$ is the per-capita output function. Define

$$s(t) = 1 - \frac{c(t)}{f(k(t))}.$$

Then $s(t)$ is the savings per unit output at time t. Furthermore, $s(t) \geq 0$ for real savings and $s(t) < 0$ for debts. Suppose $s(t) \equiv s$ is a constant independent of the rate of output, that is, the gross consumption rate per capita is a fixed fraction of gross capital output. Let

$$G = G(L, K) = \frac{K}{L} = k,$$

where K satisfies the deterministic differential equation (1.2.29) and L satisfies the stochastic differential equation (1.2.30). Applying the Itô formula, we have

$$dG(t) = (G_t + G_L L\alpha + \tfrac{1}{2} G_{LL} L^2 \sigma^2)\, dt + G_L L\sigma\, dW(t), \tag{1.2.31}$$

where

$$G_t = \frac{\partial G}{\partial t}, \qquad G_L = \frac{\partial G}{\partial L}, \qquad G_{LL} = \frac{\partial^2 G}{\partial L^2}. \tag{1.2.32}$$

It can be shown that the above stochastic differential equation reduces to

$$dk(t) = [sf(k) - \lambda k]\, dt - \alpha k\, dt + k\sigma^2\, dt - \sigma k\, dW(t), \tag{1.2.33}$$

observing that $G_t = sf(k(t)) - \lambda k(t)$, $G_L L = -k(t)$ and $\frac{1}{2}G_{LL} = k(t)/L^2$.

(b) Model for option prices. Suppose the stock prices vary in accordance with the stochastic differential equation

$$dS(t) = S(t)[\mu\, dt + \sigma\, dW(t)]. \tag{1.2.34}$$

Here $S(t)$ denotes the stock price at time t. The option price (the price of the right to buy a stock at a given time) is assumed to be a function $Z = F(S, \tau)$, where S is the current stock price and τ is the time currently left in which to exercise the option. The problem is to determine the function $F(S, \tau)$. Suppose the total investment I is allocated among options and stocks as

$$I = F(S, \tau) + \alpha S = Z + \alpha S.$$

By Itô's lemma,

$$dZ(t) = (F_S \mu S - F_t + \tfrac{1}{2}\sigma^2 S^2 F_{SS})\, dt + F_S \sigma S\, dW(t) \tag{1.2.35}$$

and

$$dI(t) = [(\alpha + F_S)\mu S - F_t + \tfrac{1}{2}\sigma^2 S^2 F_{SS}]\, dt + (\alpha + F_S)\sigma S\, dW(t), \tag{1.2.36}$$

where F_S, F_t, F_{SS} denote the partial derivatives of appropriate order. Here the risk term is $(\alpha + F_S)\sigma S$. The risk is zero if $\alpha = -F_S$. Then the expected income per unit time is

$$\tfrac{1}{2}F_{SS}\sigma^2 S^2 - F_t.$$

The market is stable if the expected income from secure investments is equal to rI. Hence

$$\tfrac{1}{2}F_{SS}\sigma^2 S^2 - F_t = rI = r(F + \alpha s) = r(F - F_S S)$$

and therefore

$$F_t = \tfrac{1}{2}S^2\sigma^2 F_{SS} + rSF_S - rF.$$

Solving this differential equation, subject to the condition $F(S, 0) = \max(0, S - a)$, where a is the fixed cost of exercising the purchase of an option, gives the function F.

Example 1.2.6 (*Model for signal detection*)
The signal process X (which is not observable) is assumed to evolve according to a linear stochastic differential equation

$$dX(t) = F(t)X(t)\, dt + G(t)\, dW_1(t) \tag{1.2.37}$$

and the observable process Z is governed by the stochastic differential equation

$$dZ(t) = H(t)X(t)\, dt + R(t)\, dW_2(t), \tag{1.2.38}$$

where W_1 and W_2 are independent Wiener processes. The problem is to study the characteristic of the process X based on observation of the process Z.

Example 1.2.7 (*Model for the position of the pole during the rotation of the earth (Arato 1982)*)
The earth's rotation is separated into three parts: the rotational motion of the earth in space due to lunar and solar gravitational attraction on the earth's equatorial bulge; polar motion or wobble, the motion of the rotational axis with respect to the earth's crust; and the change in the length of the day due to the variable speed of rotation about the instantaneous pole.

The position of the pole of rotation at time t is described by the complex-valued stochastic process

$$Z(t) = X(t) + i\,Y(t).$$

Suppose

$$Z(t) = me^{2\pi i t} + \zeta(t),$$

where the first term is a periodical component. From the equations of motion with respect to the rotating reference axis, it was shown (cf. Arato 1982) that $\{\zeta(t)\}$ can be modeled by the stochastic differential equation

$$d\zeta(t) = -\gamma\zeta(t) + d\chi(t), \tag{1.2.39}$$

with $\gamma = \lambda - i\nu, \lambda > 0, \zeta(t) = \zeta_1(t) + i\zeta_2(t), \chi(t) = W_1(t) + i\,W_2(t)$ is a complex Wiener process, $E[W_i(t)] = 0$, and $E[W_i(t)]^2 = at$.

Example 1.2.8 (*Modeling sunspot activity (Arato 1982)*)
After subtracting the mean, the number of sunspots $\zeta(t)$ can be modeled by the stochastic differential equation

$$d\zeta'(t) + (a_1\zeta'(t) + a_2\zeta(t))\,dt = dW(t), \tag{1.2.40}$$

where $\zeta'(t)$ denotes the derivative of $\zeta(t)$ in an appropriate sense.

Remark. In all the models described above, one of the basic problems of interest is the estimation of the parameters involved in the stochastic differential equation on the basis of a realization of the process satisfying the stochastic differential equation.

1.2.2 Diffusion Type Processes

Let $(\Omega, \mathcal{F}, P)$ be a probability space and $\{\mathcal{F}_t\,, 0 \le t \le T\}$ be a right continuous complete fitration. Let $W = \{W_t, \mathcal{F}_t\}$ be a Wiener process defined on $(\Omega, \mathcal{F}, P)$. Let $Z = \{Z_t, \mathcal{F}_t, 0 \le t \le T\}$ be a supermartingale of the form

$$Z_t = \exp\left(\int_0^t \beta_s\,dW_s - \tfrac{1}{2}\int_0^t \beta_s^2\,ds\right) \tag{1.2.41}$$

where $\{\beta_t, \mathcal{F}_t, t \ge 0\}$ is nonanticipative and

$$P\left(\int_0^T \beta_s^2\,ds < \infty\right) = 1. \tag{1.2.42}$$

Suppose that

$$P\left(\int_0^t \gamma^2\,ds < \infty\right) = 1, \qquad V_s = Z_s\rho_s.$$

Then, by Itô's lemma,

$$Z_t = 1 + \int_0^t \gamma_s\,dW_s, \qquad \gamma_s = Z_s\beta_s. \tag{1.2.43}$$

Theorem 1.2.1 (*Girsanov*) *If $EZ_T = 1$, then the process*

$$\tilde{W}_t = W_t - \int_0^t \beta_s\, ds, \qquad 0 \le t \le T, \tag{1.2.44}$$

is a Wiener process on $(\Omega, \mathcal{F}, \tilde{P})$ with respect to the filtration $\{\mathcal{F}_t, 0 \le t \le T\}$, and the probability measure $\tilde{P}$ is given by

$$(d\tilde{P}/dP)(\omega) = Z_T(\omega). \tag{1.2.45}$$

A proof of Theorem 1.2.1 is given in Liptser and Shiryayev (1977, p. 232). A generalization of the theorem is as follows.

Theorem 1.2.2 (*Generalization of Girsanov's theorem*) *Let $(\Omega, \mathcal{F}, P)$ be a probability space and $\{\mathcal{F}_t, 0 \le t \le T\}$ be a right continuous complete filtration. Let $W = \{W_t, \mathcal{F}_t, 0 \le t \le T\}$ be a Wiener process defined on $(\Omega, \mathcal{F}, P)$. Let $Z = \{Z_t, \mathcal{F}_t, 0 \le t \le T\}$ be a nonnegative continuous super martingale with*

$$Z_t = 1 + \int_0^t \gamma_s\, dW_s, \tag{1.2.46}$$

where $\gamma = \{\gamma_t, \mathcal{F}_t, 0 \le t \le T\}$ is such that

$$P\left(\int_0^T \gamma_s^2\, ds < \infty\right) = 1. \tag{1.2.47}$$

Suppose $EZ_T = 1$. Then $Z = \{Z_t, \mathcal{F}_t, 0 \le t \le T\}$ is a martingale. Define $d\tilde{P} = Z_T(\omega)\, dP$. Then, on $(\Omega, \mathcal{F}, \tilde{P})$, the process $\tilde{W} = \{\tilde{W}_t, \mathcal{F}_t, 0 \le t \le T\}$ defined by

$$\tilde{W}_t = W_t - \int_0^t Z_s^* \gamma_s\, ds \tag{1.2.48}$$

is a Wiener process (here $Z_s^* = Z_s^{-1}$ if $Z_s > 0$ and $Z_s^* = 0$ if $Z_s = 0$).

Remark. Let $\gamma_i = \{\gamma_i(t), \mathcal{F}_t, 0 \le t \le T\}, 1 \le i \le n$, be random processes with $P(\int_0^T \gamma_i^2(t)\, dt < \infty) = 1,\ 1 \le i \le n$, and let $W = \{(W_1(t), \ldots, W_n(t)), \mathcal{F}_t, 0 \le t \le T\}$ be an n-dimensional Wiener process. Define

$$Z_t = 1 + \int_0^t \sum_{i=1}^{n-1} \gamma_i(s)\, dW_i(s), \qquad 0 \le t \le T. \tag{1.2.49}$$

If $EZ_T = 1$, then the process

$$\tilde{W}_t = W_t - \int_0^t Z_s^* \gamma_s\, ds \tag{1.2.50}$$

is a Wiener process with respect to the measure $d\tilde{P} = Z_T\, dP$ and the filtration $\{\mathcal{F}_t\}$.

Definition. Let $(\Omega, \mathcal{F}, P)$ be a probability space and $\{\mathcal{F}_t, 0 \leq t \leq T\}$ be a filtration satisfying the usual conditions. Let $W = \{W_t, \mathcal{F}_t\}$, $0 \leq t \leq T$, be a Wiener process. A continuous random process $\zeta = \{\zeta_t, \mathcal{F}_t, 0 \leq t \leq T\}$ is called an *Itô process* if there exist nonanticipative processes $\alpha = \{\alpha_t, \mathcal{F}_t\}$ and $\beta = \{\beta_t, \mathcal{F}_t\}$, $0 \leq t \leq T$, such that $P(\int_0^T |\alpha_t|\, dt < \infty) = 1$, $P(\int_0^T \beta_t^2\, dt < \infty = 1)$ and

$$\zeta_t = \zeta_0 + \int_0^t \alpha(s, \omega)\, ds + \int_0^t \beta(s, \omega)\, dW_s \text{ a.s.} \tag{1.2.51}$$

The process $\{\zeta_t\}$ is said to have the stochastic differential

$$d\zeta_t = \alpha(t, \omega)\, dt + \beta(t, \omega)\, dW_t. \tag{1.2.52}$$

Definition. An Itô process $\zeta = \{\zeta_t, \mathcal{F}_t, 0 \leq t \leq T\}$ is called a *process of diffusion type* if $\alpha(t, \omega)$ and $\beta(t, \omega)$ are $\mathcal{F}_t^\zeta$-measurable for almost all t, $0 \leq t \leq T$, where $\mathcal{F}_t^\zeta = \sigma\{\zeta(s), 0 \leq s \leq t\}$.

Theorem 1.2.3 *Let $\zeta = (\zeta_t, \mathcal{F}_t\}$ be an Itô process with the stochastic differential*

$$d\zeta_t = A_t(\omega)\, dt + b_t(\zeta)\, dW_t, \tag{1.2.53}$$

and $\eta = (\eta_t, \mathcal{F}_t)$ be a process of diffusion type with

$$d\eta_t = a_t(\eta)\, dt + b_t(\eta)\, dW_t, \qquad \eta_0 = \zeta_0, \tag{1.2.54}$$

where ζ_0 is an $\mathcal{F}_0$-measurable random variable with $P(|\zeta_0| < \infty) = 1$. Suppose the following conditions hold:

(i) $a_t(x)$ and $b_t(x)$ satisfy the following conditions for $t \in [0, T]$, $x \in C_T = C[0, T]$:

$$|a_t(x)) - a_t(y)|^2 + |b_t(x) - b_t(y)|^2 \leq L_1 \int_0^t |x_s - y_s|^2\, dK(s) + L_2 |x_t - y_t|^2, \tag{1.2.55a}$$

$$a_t^2(x) + b_t^2(x) \leq L_1 \int_0^t (1 + x_s^2)\, dK(s) + L_2(1 + x_t^2) \tag{1.2.55b}$$

where $L_1 > 0$, $L_2 > 0$ and $0 \leq K(s) \leq 1$, with $K(\cdot)$ nondecreasing and right continuous;

(ii) for any $0 \leq t \leq T$, the equation

$$b_t(\zeta)\alpha_t(\omega) = A_t(\omega) - a_t(\zeta) \tag{1.2.56}$$

has (P-a.s.) bounded solution;

(iii) $P(\int_0^T \alpha_t^2(\omega)\, dt < \infty) = 1$; and

(iv) $E \exp(-\int_0^T \alpha_t(\omega)\, dW_t - \frac{1}{2} \int_0^T \alpha_t^2(\omega)\, dt) = 1$.

Then μ_ζ is equivalent to μ_η and (P-a.s.)

$$\frac{d\mu_\eta}{d\mu_\zeta}(\zeta) = E\left\{\exp\left(-\int_0^T \alpha_t(\omega)\, dW_t - \tfrac{1}{2}\int_0^T \alpha_t^2(\omega)\, dt\right)\Big|\mathcal{F}_T^\zeta\right\} \tag{1.2.57}$$

(here μ_ζ and μ_η are the probability measures generated by the processes ζ and η, respectively on $C[0, T]$ endowed with the supremum norm).

Proof (*Sketch of proof*). Note that

$$\alpha_t(\omega) = b_t^*(\zeta)[A_t(\omega) - a_t(\zeta)], \tag{1.2.58}$$

where $b_t^* = b_t^{-1}$ if $b_t \neq 0$ and $b_t^* = 0$ if $b_t = 0$.

Let

$$Z_t = \exp\left(-\int_0^t \alpha_s(\omega)\, dW_s - \tfrac{1}{2}\int_0^t \alpha_s^2(\omega)\, ds\right), \tag{1.2.59}$$

and

$$d\tilde{P} = Z_t\, dP. \tag{1.2.60}$$

Then, by Girsanov's theorem,

$$\tilde{W}_t = W_t + \int_0^t \alpha_s(\omega)\, ds \tag{1.2.61}$$

is a Wiener process with respect to $\{\mathcal{F}_t, 0 \leq t \leq T\}$ under $\tilde{P}$. But

$$\begin{aligned}
&\eta_0 + \int_0^t a_s(\zeta)\, ds + \int_0^t b_s(\zeta)\, d\tilde{W}_s \\
&= \eta_0 + \int_0^t a_s(\zeta)\, ds + \int_0^t b_s(\zeta)\alpha_s(\omega)\, ds + \int_0^t b_s(\zeta)\, dW_s \\
&= \eta_0 + \int_0^t a_s(\zeta)\, ds + \int_0^t b_s(\zeta) b_s^*(\zeta)[A_s(\omega) - a_s(\zeta)]\, ds + \int_0^t b_s(\zeta)\, dW_s \\
&= \eta_0 + \int_0^t A_s(\omega)\, ds + \int_0^t b_s(\zeta)\, dW_s = \zeta_t.
\end{aligned} \tag{1.2.62}$$

Hence $\zeta = \{(\zeta_t, \mathcal{F}_t)\}$ on $(\Omega, \mathcal{F}, \tilde{P})$ satisfies the same stochastic differential equation that $\eta = \{(\eta_t, \mathcal{F}_t)\}$ on $(\Omega, \mathcal{F}, P)$ does. Therefore

$$\tilde{P}(\zeta \in A) = P(\eta \in A) \tag{1.2.63}$$

and

$$\begin{aligned}
\mu_\eta(A) = P(\eta \in A) = \tilde{P}(\zeta \in A) &= \int_{[\zeta \in A]} Z_T(\omega)\, dP(\omega) \\
&= \int_A E[Z_T | \mathcal{F}_T^\zeta]_{\zeta = x}\, d\mu_\zeta(x).
\end{aligned} \tag{1.2.64}$$

Hence $\mu_\eta \ll \mu_\zeta$ and the relation (1.2.57) holds. One can check that $\mu_\zeta \ll \mu_\eta$ using the fact that $d\tilde{P}/dP = Z_T$ and $P(Z_T = 0) = 0$.

A detailed proof is given in Liptser and Shiryayev (1977).

2

Parametric Inference for Diffusion Type Processes from Continuous Paths

2.1 Introduction

We have seen how diffusion processes can be used for stochastic modeling with wide applications in physical sciences, biological sciences, medical sciences and more recently in economic and social sciences, for instance in studies on mathematical finance and in studies on community relationships. However, all these models involve unknown parameters or unknown functions which need to be estimated from observations on the processes. The parameters in the model may be finite-dimensional or infinite-dimensional. Then the parameters have to be estimated either from one or many continuous realizations of the process if a continuous realization is possible; or from discrete sampled data for the process if the process is not observable continuously due to limitations on the precision of measuring instruments or unavailability of observations at all the time points, as in medical studies, or for other reasons. In this chapter, we shall discuss different methods of estimation of the drift parameter for a diffusion type process when a continuous sample path is available.

Let us consider a stochastic differential equation of the type

$$dX_t = a(t, X_t, \theta)\, dt + \sigma(t, X_t)\, dW_t, \qquad X_0 = X_0, t \geq 0, \tag{2.1.1}$$

where $\{W_t, t \geq 0\}$ is the standard Wiener process. Here $a(\cdot, \cdot, \cdot)$ is the drift coefficient and $\sigma(\cdot, \cdot)$ is the diffusion coefficient of the process. We note that the function $\sigma(\cdot, \cdot)$ does not depend on the parameter θ. We assume that $\sigma(\cdot, \cdot)$ is either completely known or, if it is unknown, a constant (independent of the parameter θ). If σ is a constant, then it can be estimated from a sample path of the process using the fact that

$$\sum_{i=1}^{n}[X_{iT2^{-n}} - X_{(i-1)T2^{-n}}]^2 \rightarrow \sigma^2 T \text{ a.s.} \tag{2.1.2}$$

as $n \rightarrow \infty$ (cf. Basawa and Prakasa Rao 1980, p. 202). We assume that σ is known and set it equal to one if the process can be observed continuously. The problem of estimation of σ when only a discrete realization of the process is available will be discussed later in this book.

In the following discussion, we consider $\{X_t\}$ to be the solution of the stochastic differential equation

$$dX_t = a(t, X_t, \theta)\, dt + dW_t, \qquad X_0 = 0,\ t \geq 0. \tag{2.1.3}$$

We assume that sufficient conditions hold on the coefficient $a(\cdot, \cdot, \theta)$ so that there exists a unique solution to (2.1.3) for any fixed $\theta \in \Theta$.

2.2 Maximum Likelihood Method for the Estimation of the Drift Parameter

2.2.1 Diffusion Processes

Let $(\Omega, \mathcal{F}, P)$ be a probability space and $\{\mathcal{F}_t, t \geq 0\}$ be a filtration on $(\Omega, \mathcal{F})$. Let $\{X_t\}$ be adapted to $\{\mathcal{F}_t\}$ satisfying the stochastic differential equation (2.1.3). Suppose $\theta \in \Theta$ open in R^k. Let P_θ^T be the probability measure generated by the process $\{X_t, 0 \leq t \leq T\}$ on the space $(C[0,T], \mathcal{B}_T)$, where $C[0,T]$ denotes the space of continuous functions endowed with the supremum norm and $\mathcal{B}_T$ corresponding Borel σ-algebra. For convenience, let us write $X^T = \{X_t, 0 \leq t \leq T\}$. Let E_θ denote expectation with respect to P_θ^T and P_W^T be the probability measure induced by the standard Wiener process. We assume that the following conditions hold:

(A0) $P_{\theta_1} \neq P_{\theta_2}$ if $\theta_1 \neq \theta_2$ in Θ.

(A1) The process $\{X_t\}$ is the unique solution of the equation (2.1.3) and

$$P_\theta\left(\int_0^T a^2(t, X_t, \theta)\, dt < \infty\right) = 1. \tag{2.2.1}$$

Condition (2.2.1) ensures that $P_\theta^T \ll P_W^T$ for all $\theta \in \Theta$ and

$$\frac{dP_\theta^T}{dP_W^T} = \exp\left\{\int_0^T a(t, X_t, \theta)\, dX_t - \frac{1}{2}\int_0^T a^2(t, X_t, \theta)\, dt\right\}, \tag{2.2.2}$$

from results in Basawa and Prakasa Rao (1980) or Liptser and Shiryayev (1977).

Definition. A *maximum likelihood estimator* (MLE) $\hat{\theta}_T(X^T)$ of θ corresponding to the realization X^T is defined to be a measurable map $\hat{\theta}_T : (C_T, \mathcal{B}_T) \to (\Theta, \tau)$ such that

$$\frac{dP_{\hat{\theta}_T}^T}{dP_W^T} = \sup_{\theta \in \Theta} \frac{dP_\theta^T}{dP_W^T}, \tag{2.2.3}$$

where τ is the σ-algebra of Borel subsets of Θ.

If the parameter space Θ is compact and $\frac{dP_\theta^T}{dP_W^T}$ is continuous in θ, then it can be shown that there exists a measurable MLE – cf. Lemma 3.3 of Schmetterer (1974). Hereafter we assume the existence of such a measurable MLE.

Scalar parameter case

For any function $g(t, x, \theta)$, let prime denote differentiation with respect to θ and g_x, g_t denote the partial derivatives of g with respect to x and t, respectively, whenever they exist. We assume that the following conditions hold.

(A2) (i) Suppose that $a(t, x, \theta)$ is continuous in x. Let

$$F(t, x, \theta) = \int_0^x a(t, y, \theta)\, dy. \tag{2.2.4}$$

Suppose F is jointly continuous in $(t, x) \in [0, \infty) \times R$ with the partial derivatives F_x, F_{xx} and F_t. It is easy to see that $F_x = a$ and $F_{xx} = a_x$.

By Itô's formula (see Appendix B; cf. Kallianpur 1980, Stroock and Varadhan 1979), it follows that

$$\int_0^T a(t, X_t, \theta)\, dX_t = F(T, X_T, \theta) - \int_0^T f(t, X_t, \theta)\, dt, \tag{2.2.5}$$

where

$$f(t, x, \theta) = F_t(t, x, \theta) + \tfrac{1}{2} a_x(t, x, \theta). \tag{2.2.6}$$

The log-likelihood function $\ell_T(\theta)$ can be written in the form

$$\begin{aligned} \ell_T(\theta) &= \log \frac{dP_\theta^T}{dP_W^T} \\ &= F(T, X_T, \theta) - \int_0^T [f(t, X_t, \theta) + \tfrac{1}{2} a^2(t, X_t, \theta)]\, dt \end{aligned} \tag{2.2.7}$$

from (2.2.2), (2.2.5) and (2.2.6). An advantage of this form of the log-likelihood function is that it does not involve a stochastic integral and is amenable to computation.

(A2) (ii) Suppose that the integral defined by (2.2.4) and the integral on the right-hand side of (2.2.7) can be differentiated under the integral sign with respect to θ. Further assume that F' and F'' satisfy conditions similar to those of F stated in (A2)(i) (here prime denotes the derivative with respect to θ).

Under assumption (A2)(ii), $\ell_T(\theta)$ is twice differentiable with respect to θ and a further application of Itô's formula shows that

$$\ell_T'(\theta) = \int_0^T a'(t, X_t, \theta)\, dW_t^\theta, \tag{2.2.8}$$

and

$$\ell_T''(\theta) = \int_0^T a''(t, X_t, \theta)\, dW_t^\theta - \int_0^T [a'(t, X_t, \theta)]^2\, dt, \tag{2.2.9}$$

where

$$W_t^\theta \equiv X_t - \int_0^t a(s, X_s, \theta)\, ds, \qquad t \geq 0,$$

is a Wiener process under P_θ.

(A2) (iii) Suppose that $\ell_T''(\theta)$ is continuous in a neighborhood V_θ of θ for every $\theta \in \Theta$, and that

$$E_\theta \left\{ \int_0^T (a'(t, X_t, \theta))^2\, dt \right\} < \infty, \tag{2.2.10a}$$

and

$$E_\theta \left\{ \int_0^T (a''(t, X_t, \theta))^2\, dt \right\} < \infty. \tag{2.2.10b}$$

(A3) Suppose that for any $\theta \in \Theta$, there exists a neighborhood V_θ of θ in Θ such that

$$P_\theta \left(\int_0^\infty (a(t, X_t, \theta') - a(t, X_t, \theta))^2\, dt = \infty \right) = 1$$

for all $\theta' \in V_\theta - \{\theta\}$.

Let

$$I_T(\theta) = \int_0^T (a'(t, X_t, \theta))^2 \, dt \tag{2.2.11a}$$

and

$$Y_T(\theta) = \int_0^T (a''(t, X_t, \theta))^2 \, dt. \tag{2.2.11b}$$

(A4) Suppose there exists $m_T \uparrow \infty$ as $T \to \infty$ such that

(i) $\frac{I_T(\theta)}{m_T} \xrightarrow{p} \eta^2(\theta)$ in P_θ-probability as $T \to \infty$, where $P_\theta(\eta^2(\theta) > 0) > 0$, and

(ii) $\frac{Y_T(\theta)}{m_T} \xrightarrow{p} \zeta^2(\theta)$ in P_θ-probability as $T \to \infty$.

In general it is difficult to give sufficient conditions to check (A4). However, this can be done in some examples. We now state the main result concerning the consistency and asymptotic normality of the MLE.

Theorem 2.2.1 *Let the conditions (A0)–(A2) hold. Then there exists a solution of the likelihood equation $\ell'_T(\theta) = 0$ which is strongly consistent for θ as $T \to \infty$.*

Proof. It is easy to check that, for any $\delta > 0$ such that $\theta \pm \delta \in \Theta$,

$$\begin{aligned} \ell_T(\theta \pm \delta) - \ell_T(\theta) &= \frac{dP^T_{\theta\pm\delta}}{dP^T_\theta} \\ &= \int_0^T A_t^\theta \, dW_t^\theta - \tfrac{1}{2} \int_0^T (A_t^\theta)^2 \, dt, \end{aligned} \tag{2.2.12}$$

where

$$A_t^\theta = a(t, X_t, \theta \pm \delta) - a(t, X_t, \theta). \tag{2.2.13}$$

Let

$$K_T = \int_0^T (A_t^\theta)^2 \, dt.$$

Then

$$\frac{\ell_T(\theta \pm \delta) - \ell_T(\theta)}{K_T} = \frac{\int_0^T A_t^\theta \, dW_t^\theta}{\int_0^T (A_t^\theta)^2 \, dt} - \frac{1}{2}. \tag{2.2.14}$$

Under condition (A3), it follows that

$$\frac{\int_0^T A_t^\theta \, dW_t^\theta}{\int_0^T (A_t^\theta)^2 \, dt} \to 0 \text{ a.s. } [P_\theta] \qquad \text{as } T \to \infty \tag{2.2.15}$$

by Lepingle's law of large numbers (see also Lemma 17.4 of Liptser and Shiryayev 1978, p. 201). Hence

$$\frac{\ell_T(\theta \pm \delta) - \ell_T(\theta)}{K_T} \to -\frac{1}{2} \text{ a.s. } [P_\theta] \qquad \text{as } T \to \infty.$$

Note that $K_T > 0$ a.s. $[P_\theta]$ for large T again by (A3). Hence, for almost every $\omega \in \Omega$, δ and θ, there exists T_0 such that for $T \geq T_0$,

$$\ell_T(\theta \pm \delta) < \ell_T(\theta). \tag{2.2.16}$$

Since $\ell_T(\theta)$ is continuous on the compact set $[\theta-\delta, \theta+\delta]$, it has a local maximum and the maximum is attained at an element $\hat{\theta}_T \in (\theta-\delta, \theta+\delta)$ in view of (2.2.16). Since $\ell_T(\theta)$ is differentiable with respect to θ, it follows that $\ell'_T(\hat{\theta}_T) = 0$ and $|\hat{\theta}_T - \theta| < \delta$. This proves that $\hat{\theta}_T \to \theta$ a.s. $[P_\theta]$ as $T \to \infty$.

The next theorem deals with the asymptotic normality of the MLE. We first state a lemma.

Lemma 2.2.2 *Suppose that conditions (A1), (A2) and (A4) hold. Then:*

(i) $\frac{\ell'_T(\theta)}{\sqrt{I_T(\theta)}} \xrightarrow{\mathcal{L}} N(0,1)$ *as* $T \to \infty$, *where the convergence is with respect to any probability measure* $\mu \ll P^\theta_A$ *in which* $A = [\eta^2(\theta) > 0]$ *and* P^θ_A *denotes the conditional probability measure under* θ, *given A;*

(ii) $P^\theta_A - \lim_{T\to\infty} I_T(\theta) = \infty$;

(iii) $P^\theta_A - \lim_{T\to\infty} \frac{1}{I_T(\theta)} \int_0^T a''(t, X_t, \theta)\, dW^\theta_t = 0$.

This lemma is a consequence of the central limit theorem and the law of large numbers for square-integrable martingales. We omit the proof. It follows also as a consequence of Feigin (1976).

Theorem 2.2.3 *Suppose conditions (A1)–(A4) hold. Then*

$$\sqrt{I_T(\theta)}(\hat{\theta}_T - \theta) \xrightarrow{\mathcal{L}} N(0,1), \qquad \text{as } T \to \infty,$$

conditionally with respect to any probability measure $\mu \ll P^\theta_A$.

Proof. In view of assumption (A2), applying Taylor's expansion to $\ell'_T(\theta)$ around $\hat{\theta}_T$, it follows that

$$\begin{aligned}\ell'_T(\theta) &= \ell'_T(\hat{\theta}_T) + (\theta - \hat{\theta}_T)\ell''_T(\hat{\theta}_T + \beta_T(\theta - \hat{\theta}_T)) \\ &= (\theta - \hat{\theta}_T)\ell''_T(\hat{\theta}_T + \beta_T(\theta - \hat{\theta}_T)),\end{aligned} \tag{2.2.17}$$

where $|\beta_T| \le 1$ for T sufficiently large. Note that $I_T(\theta) > 0$ for T large with probability approaching one by (A4) on the set $A = [\eta^2(\theta) > 0]$, under P^θ_A. Hence

$$\frac{\ell'_T(\theta)}{\sqrt{I_T(\theta)}} = \frac{(\theta - \hat{\theta}_T)\ell''_T(\hat{\theta}_T + \beta_T(\theta - \hat{\theta}_T))}{\sqrt{I_T(\theta)}}. \tag{2.2.18}$$

Since $\hat{\theta}_T \to \theta$ a.s. as $T \to \infty$ by Theorem 2.2.1 and since $\ell''_T(\theta)$ is continuous in θ by (A2), it follows that

$$\ell''_T(\hat{\theta}_T + \beta_T(\theta - \hat{\theta}_T)) - \ell''_T(\theta) \to 0 \text{ in } P_\theta\text{-probability} \tag{2.2.19}$$

as $T \to \infty$. Furthermore $I_T(\theta) \to \infty$ in P^θ_A-measure as $T \to \infty$ by Lemma 2.2.2(ii). Hence

$$\frac{\ell'_T(\theta)}{\sqrt{I_T(\theta)}} - \frac{(\theta - \hat{\theta}_T)\ell''_T(\theta)}{\sqrt{I_T(\theta)}} \to 0 \text{ in } P^\theta_A\text{-probability} \tag{2.2.20}$$

as $T \to \infty$. This relation, together with Lemma 2.2.1(i), leads to the relation

$$\frac{(\theta - \hat{\theta}_T)\ell''_T(\theta)}{\sqrt{I_T(\theta)}} \xrightarrow{\mathcal{L}} N(0,1) \qquad \text{as } T \to \infty, \tag{2.2.21}$$

conditionally with respect to any probability measure $\mu \ll P_A^\theta$. But

$$\frac{\ell_T''(\theta)}{I_T(\theta)} \to -1 \text{ in } P_A^\theta\text{-measure} \qquad \text{as } T \to \infty$$

by Lemma 2.2.2(iii). Hence

$$(\theta - \hat{\theta}_T)\sqrt{I_T(\theta)} \xrightarrow{\mathcal{L}} N(0,1) \qquad \text{as } T \to \infty$$

conditionally with respect to any probability measure $\mu << P_A^\theta$ or, equivalently,

$$\sqrt{I_T(\theta)}(\theta - \hat{\theta}_T) \xrightarrow{\mathcal{L}} N(0,1) \qquad \text{as } T \to \infty$$

conditionally with respect to any probability measure $\mu \ll P_A^\theta$.

Example 2.2.1
Consider the stochastic differential equation

$$dX_t = \theta t X_t\, dt + dW_t, \qquad X(0) = 0,\ t \geq 0,\ \theta > 0.$$

It is easy to check that

$$X_T = e^{\theta T^2/2} \int_0^T e^{-\theta s^2/2}\, dW_s.$$

We leave it to the reader to check that conditions (A0)–(A4) hold in this example and the MLE is strongly consistent asymptotically normal after random normalization as in Theorem 2.2.3.

Remarks. It is easy to extend the results given above to the case of a stochasatic differential equation of the form

$$dX(t) = a(t, X(t), \theta)\, dt + b(t, X(t))\, dW(t), \qquad X(0) = X_0,\ t \geq 0,$$

by imposing suitable regularity conditions on $(a(t,x,\theta)/b(t,x))$. The linear case when $a(t,x,\theta) = \theta g(t,x)$ has been discussed in detail in Basawa and Prakasa Rao (1980) and Prakasa Rao (1985). Kutoyants (1984a) gives a discussion for the case of nonlinear stochastic differential equations of the type

$$dX(t) = a(\theta, X(t))\, dt + b(X(t))\, dW(t), \qquad X(0) = X_0,\ t \geq 0.$$

We return to the discussion of this problem later in this chapter.

The techniques used above, especially formula (2.2.7), leading to conversion of a stochastic integral into a Lebesgue integral, cannot be extended to the problem of estimation of a multidimensional parameter. We now describe an alternate approach to the problem.

Vector parameter case

Let $\{X_t, t \geq 0\}$ be a real-valued stationary ergodic process satisfying the stochastic differential equation

$$dX_t = a(\theta, X_t)\, dt + dW_t, \qquad X_0 = X_0,\ t \geq 0,$$

where $\{W_t, t \geq 0\}$ is a standard Wiener process, $E(X_0^2) < \infty$ and $a(\theta, x)$ is a real-valued function defined on $\Theta \times R$, with $\Theta = \{\theta \in R^d, |\theta| \leq 1\}$ and $d \geq 1$. Here $|\cdot|$ denotes the ordinary Euclidean norm on R^d. Suppose $\theta_0 \in \Theta^0$ (the interior of the set Θ). Suppose the following conditions hold on $a(\cdot,\cdot)$:

(B1) $a(\theta, x)$ is continuous on $\Theta \times R$.

(B2) (i) $|a(\theta, x)| \leq L(\theta)(1 + |x|), \theta \in \Theta, x \in R$ where $\sup\{L(\theta) : \theta \in \Theta\} < \infty$;

(ii) $|a(\theta, x) - a(\theta, y)| \leq L(\theta)|x - y|, \theta \in \Theta, x, y \in R$;

(iii) $|a(\theta, x) - a(\phi, x)| \leq J(x)|\theta - \phi|, \theta, \phi \in \Theta, x \in R$, where $J(\cdot)$ is continuous and

$$E|J(X_0)|^{d+\alpha_0} < \infty \text{ for some } d + \alpha_0 \geq 2$$

(here E denotes the expectation with respect to the true probability measure).

(B3) $I(\theta) = E(a(\theta, X_0) - a(\theta_0, X_0))^2 > 0$ for $\theta \neq \theta_0$.

(B4) the partial derivatives $a_\theta^{(i)}$ of a with respect θ_i exist for $i = 1, 2, \ldots, d$, where $\theta = (\theta_1, \theta_2, \ldots, \theta_d)$ (let $a_\theta^{(i)}(\theta^*, x)$ be the partial derivative of $a(\cdot, \cdot)$ with respect to θ_i evaluated at θ^*).

(B5) $|a_\theta^{(i)}(\theta, x) - a_\theta^{(i)}(\phi, x)| \leq C(x)|\phi - \theta|^\alpha, \theta, \phi \in \Theta, x \in R$ for some $\alpha > 0$, and

$$E[|C(X_0)|^{d+\alpha_1}] < \infty$$

for some $d + \alpha_1 \geq 2, i = 1, \ldots, d$.

(B6) $E[a_\theta^{(i)}(\theta, X_0)]^2 < \infty, 1 \leq i \leq d$.

(B7) $\sum_{i=1}^d |a_\theta^{(i)}(\theta, x)| \leq M(\theta)(1 + |x|)$ for all θ in a neighborhood V_{θ_0} of θ_0 and

$$\sup_{\theta \in V_{\theta_0}} M(\theta) < \infty.$$

Lemma 2.2.4 *Let the processes* $\{X_t, t \geq 0\}$ *and* $\{W_t, t \geq 0\}$ *be as defined above and* g *be a function on* $\Theta \times R$ *such that*

$$|g(\theta, x) - g(\phi, x)| \leq J(x)|\theta - \phi|, \qquad \theta, \phi \in \Theta, x \in R. \tag{2.2.22}$$

Define

$$Y_T(\theta) = \int_0^T g(\theta, X_t)\, dW_t, \qquad T \geq 0.$$

Suppose $Y_T(\theta)$ *exists as a stochastic integral. If* $k \geq 2$, *then there exists a constant* $C_k > 0$ *such that*

$$E(\sup_{0 \leq t \leq T} |Y_T(\theta) - Y_T(\phi)|^k) \leq C_k T^{k/2} |\theta - \phi|^k E|J(X_0)|^k.$$

Proof. Since $\{Y_t(\theta) - Y_t(\phi), 0 \leq t \leq T\}$ is a martingale for each θ and ϕ, it follows from Burkholder's inequality (see Stroock and Varadhan 1979; or Hall and Heyde 1980) that there exists a constant $C_k > 0$ such that

$$\begin{aligned}
&E(\sup_{0 \leq t \leq T} |Y_t(\theta) - Y_t(\phi)|^k) \\
&\leq C_k\, E\left\{\int_0^T [g(\theta, X_t) - g(\phi, X_t)]^2\, dt\right\}^{k/2} \\
&\leq C_k T^{\frac{k}{2}-1}\, E\left[\int_0^T |g(\theta, X_t) - g(\phi, X_t)|^k\, dt\right] \text{ (by Hölder's inequality)} \\
&\leq C_k T^{k/2} |\theta - \phi|^k E|J(X_0)|^k
\end{aligned}$$

by (2.2.22) and the stationarity of the process $\{X_t\}$.

We now state a result for the oscillation of functionals defined on R^d and taking values in a Banach space. For a proof of the following lemma, see Strook (1982, p. 7) or Strook and Varadhan (1979, p. 60).

Lemma 2.2.5 *Suppose p and ψ are strictly increasing continuous functions on $[0, \infty)$, with $p(0) = \psi(0) = 0$ and $\lim_{t\to\infty}\psi(t) = \infty$. Let $f : R^d \to L$, where L is a normed linear space, be a function strongly continuous on $\overline{B(a, 2\rho)}$, $\rho > 0$, where $B(a, \rho)$ denotes the sphere of radius ρ with center $a \in R^d$ and $\overline{B(a, \rho)}$ denotes its closure. Further suppose that*

$$\int_{B(a,\rho)}\int_{B(a,\rho)} \psi\left(\frac{\|f(\theta) - f(\phi)\|}{p(|\theta - \phi|)}\right) d\theta\, d\phi \le \lambda < \infty.$$

Then, for all $\theta, \phi \in B(a, \rho)$,

$$\|f(\theta) - f(\phi)\| \le 8 \int_0^{|\theta-\phi|} \psi^{-1}\left(\frac{4^{d+2}\lambda}{\gamma^2 u^{2d}}\right) p(du),$$

where

$$\gamma = \inf_{\theta\in B(0,1)} \inf_{1<\rho\le 2} \frac{|B(\theta, \psi) \cap B(0, 1)|}{\rho^d}.$$

Here $|A|$ denotes the Lebesgue measure of the set A and $\|\cdot\|$ denotes the norm of the Banach space L.

As a corollary to Lemma 2.2.5, we have the following result.

Lemma 2.2.6 *Let $\psi(x) = x^r$ and $p(x) = x^{\gamma/r}$, where $r > 0$ and $\gamma > 2d$ in Lemma 2.2.5. Then*

$$\|f(\theta) - f(\phi)\| \le c(r, \gamma, d)|\theta - \phi|^{\frac{\gamma-2d}{r}} \lambda^{1/r}$$

for some constant $c(r, \gamma, d) > 0$.

The next lemma gives a bound on the probability of oscillation for a random function taking values in a Banach space.

Lemma 2.2.7 *Suppose $\{Y(\theta), \theta \in R^d\}$ is a family of random variables taking values in a normed linear space. Suppose that, for every ω, $Y(\theta, \omega)$ is continuous on $\overline{B(a, \rho)}$ and there exist $r > 0$, $C > 0$ and $\alpha > 0$ such that*

$$E\|Y(\theta) - Y(\phi)\|^r \le C|\theta - \phi|^{d+\alpha}, \theta, \phi \in B(a, \rho). \tag{2.2.23}$$

Then, for all $\gamma \in (2d, 2d + \alpha)$ and $\lambda > 0$,

$$P\left[\sup_{\theta,\phi\in B(a,\rho)} \frac{\|Y(\theta) - Y(\phi)\|}{|\theta - \phi|^\beta} \ge c(r, \gamma, d)\lambda^{1/r}\right] \le \frac{CA}{\lambda} \tag{2.2.24}$$

for

$$\beta = \frac{\gamma - 2d}{r}, A = \int_{B(a,\rho)}\int_{B(a,\rho)} |\theta - \phi|^{d+\alpha-\gamma}\, d\theta\, d\phi. \tag{2.2.25}$$

Proof. In view of (2.2.23),

$$E\left[\int_{B(a,\rho)}\int_{B(a,\rho)}\left(\frac{\|Y(\theta)-Y(\phi)\|}{|\theta-\phi|^{\gamma/r}}\right)^r d\theta\, d\phi\right] \leq CA.$$

Hence, for any $\lambda > 0$,

$$P\left[\int_{B(a,\rho)}\int_{B(a,\rho)}\left(\frac{\|Y(\theta)-Y(\phi)\|}{|\theta-\phi|^{\gamma/r}}\right)^r d\theta\, d\phi \geq \lambda\right] \leq \frac{CA}{\lambda}$$

by Chebyshev's inequality. If

$$\int_{B(a,\rho)}\int_{B(a,\rho)}\left(\frac{\|Y(\theta)-Y(\phi)\|}{|\theta-\phi|^{\gamma/r}}\right)^r d\theta\, d\phi \geq \lambda,$$

then, by Lemma 2.2.6, it follows that

$$\|Y(\theta)-Y(\phi)\| \leq c(r,\gamma,d)|\theta-\phi|^{\frac{\gamma-2d}{r}}\lambda^{1/r}$$

for all $\theta, \phi \in B(a, \rho)$. Hence

$$P\left[\sup_{\theta,\phi\in B(a,\rho)}\frac{\|Y(\theta)-Y(\phi)\|}{|\theta-\phi|^{\beta}} \geq c(r,\gamma,d)\lambda^{1/r}\right] \leq \frac{CA}{\lambda}$$

where

$$\beta = \frac{\gamma-2d}{r}, \qquad A = \int_{B(a,\rho)}\int_{B(a,\rho)} |\theta-\phi|^{d+\alpha-\gamma}\, d\theta\, d\phi.$$

As a corollary to Lemma 2.2.7, we have the following result.

Lemma 2.2.8 *Suppose the conditions stated in Lemma 2.2.7 hold. Further suppose that there exists some $\theta_0 \in B(a, \rho)$ such that $Y(\theta_0) = 0$. Then*

$$P\left[\sup_{\theta\in B(a,\rho)}\|Y(\theta)\| \geq (2\rho)^{\beta}c(r,\gamma,d)\lambda^{1/r}\right] \leq \frac{CA}{\lambda}.$$

Let P_θ^T be the measure generated by the process $X_T = \{X(t), 0 \leq t \leq T\}$ on $C[0, T]$ when θ is the parameter. We have seen earlier that, under the conditions stated earlier, $P_\theta^T \ll P_{\theta_0}^T$ and

$$\ell_T(\theta) \equiv \log\frac{dP_\theta^T}{dP_{\theta_0}^T} = \int_0^T [a(\theta,X_t)-a(\theta_0,X_t)]\,dX_t - \frac{1}{2}\int_0^T [a(\theta,X_t)-a(\theta_0,X_t)]^2\,dt. \tag{2.2.26}$$

Let

$$Z_T^*(\theta) = \int_0^T v(\theta, X_t)\,dX_t, \tag{2.2.27}$$

$$Z_T(\theta) = \frac{1}{\sqrt{T}}Z_T^*(\theta), \tag{2.2.28}$$

where

$$v(\theta,x) = a(\theta,x) - a(\theta_0,x). \tag{2.2.29}$$

Let

$$I_T(\theta) = \int_0^T v^2(\theta, X_t)\, dX_t. \tag{2.2.30}$$

Then

$$\ell_T(\theta) = Z_T^*(\theta) - \tfrac{1}{2} I_T(\theta). \tag{2.2.31}$$

We now study the properties of the process $\{Z_t^*(\theta), t \geq 0\}$.

Lemma 2.2.9 *Suppose that condition (B2) holds. Then there exist positive constants c_1 and c_2 such that, for every $\lambda > 0$,*

$$P\{\sup_{\theta} \sup_{0 \leq t \leq T} |Z_t^*(\theta)| \geq c_1 \lambda^{1/(d+\alpha_0)}\} \leq c_2 \frac{T^{\frac{d+\alpha_0}{2}}}{\lambda}. \tag{2.2.32}$$

Proof. Note that

$$|v(\theta, x) - v(\phi, x)| \leq J(x)|\theta - \phi|.$$

Applying Lemma 2.2.4 with $g(\theta, x) = v(\theta, x)$ and Lemma 2.2.7 with $r = d + \alpha_0$ and then Lemma 2.2.8, we have the result.

Lemma 2.2.10 *Under condition (B2), for any $\gamma > \frac{1}{d+\alpha_0}$, there exists $H > 0$ such that*

$$\limsup_{T \to \infty} \sup_{\theta} \frac{|Z_T^*(\theta)|}{T^{1/2}(\log T)^\gamma} \leq H \ a.s. \tag{2.2.33}$$

Proof. Let $H' > 0$. Define

$$A_n = \{\sup_{2^{n-1} < t \leq 2^n} \sup_{\theta} |Z_t^*(\theta)| \geq H' 2^{n/2} n^\gamma\}, \qquad n \geq 1.$$

Then, by the stationarity of the process $\{X_t, t \geq 0\}$,

$$\begin{aligned} P(A_n) &= P[\sup_{0 < t < 2^{n-1}} \sup_{\theta} |Z_t^*(\theta)| \geq H' 2^{n/2} n^\gamma] \\ &\leq \frac{c_2'(2^{n-1})^{\frac{d+\alpha_0}{2}}}{(H' 2^{n/2} n^\gamma)^{d+\alpha_0}}, \end{aligned}$$

for some constant $c_2' > 0$, by Lemma 2.2.9. Note that $\sum_{n=1}^\infty P(A_n) < \infty$ by the choice of γ. This implies that $P(A_n$ occurs infinitely often$) = 0$ by the Borel–Cantelli lemma. Hence

$$\sup_{\theta} |Z_t^*(\theta)| \leq H' 2^{n/2} n^\gamma, \qquad \text{for } 2^{n-1} \leq t \leq 2^n,$$

except for finitely many n with probability one. This implies that (2.2.33) holds for some $H > 0$ depending on γ.

Lemma 2.2.11 *Under conditions (B2) and (B3), for every $\delta > 0$,*

$$\inf_{|\theta - \theta_0| \geq \delta} \frac{I_T(\theta)}{T} \xrightarrow{a.s.} \lambda, \qquad \text{as } T \to \infty, \tag{2.2.34}$$

under P_{θ_0} for some $\lambda > 0$ depending on δ.

Proof. By the ergodic theorem,

$$\frac{1}{T} I_T(\theta) \to I(\theta) \text{ a.s.} \qquad \text{as} T \to \infty \tag{2.2.35}$$

under P_θ. Furthermore, it is easy to check that

$$|I_T(\theta) - I_T(\phi)| \le C^* T |\theta - \phi| \text{ a.s.},$$

as $T \to \infty$, for some constant $C^* > 0$, under the conditions assumed and $\{P_\theta, \theta \in \Theta\}$ are absolutely continuous with respect to each other. Hence

$$\frac{I_T(\theta)}{T} \to I(\theta) \text{ a.s.}$$

under P_{θ_0} uniformly in $\theta \in \bar{\Theta}$ as $T \to \infty$. But

$$I(\theta) = E[a(\theta, X_0) - a(\theta_0, X_0)]^2$$

which is zero for $\theta = \theta_0$ and $I(\theta) > 0$ for $\theta \neq \theta_0$ by (B3). Hence, for any $\delta > 0$,

$$\inf_{|\theta - \theta_0| \ge \delta} \frac{I_T(\theta)}{T} \overset{a.s.}{\to} \lambda \qquad \text{as } T \to \infty$$

under P_{θ_0} for some $\lambda > 0$ depending on δ.

Strong consistency of the MLE

We now prove the strong consistency of the MLE $\hat{\theta}_T$.

Theorem 2.2.12 *Under assumptions (B2) and (B3),*

$$\hat{\theta}_T \to \theta_0 \text{ a.s. } [P_{\theta_0}] \qquad \text{as } T \to \infty,$$

Proof. Let $R_T(\theta) = I_T(\theta) - 2\, Z_T^*(\theta)$. Note that $R_T(\theta_0) = 0$ and $\hat{\theta}_T$ minimizes $R_T(\theta)$. Lemma 2.2.11 implies that, for any $\delta > 0$, there exists $\lambda > 0$ depending on δ such that

$$\inf_{|\theta - \theta_0| > \delta} I_T(\theta) \ge T\lambda \text{ a.s. } [P_{\theta_0}] \qquad \text{as } T \to \infty;$$

and with P_{θ_0}-probability one, for any $\gamma > \frac{1}{d + \alpha_0}$, there exists $H > 0$ depending on γ such that

$$\sup_\theta |Z_T^*(\theta) \le H T^{1/2} (\log T)^\gamma$$

for large T by the Lemma 2.2.10. Hence

$$\inf_{|\theta - \theta_0| \ge \delta} R_T(\theta) \ge \lambda^* T > 0 \text{ a.s. } [P_{\theta_0}] \qquad \text{as } T \to \infty,$$

for some $\lambda^* > 0$ depending on δ and γ. Since $\hat{\theta}_T$ minimizes $R_T(\theta)$ and $R_T(\theta_0) = 0$, it follows that $|\hat{\theta}_T - \theta_0| < \delta$ a.s. $[P_{\theta_0}]$ as $T \to \infty$. Hence $\hat{\theta}_T \to \theta_0$ a.s. $[P_{\theta_0}]$ as $T \to \infty$.

Asymptotic normality of the MLE

We assume that the regularity conditions (B1)–(B7) hold. Since $\hat{\theta}_T$ is strongly consistent by Theorem 2.2.12, $\hat{\theta}_T \in V_{\theta_0}$ for large T. Expanding $a(\theta, x)$ around θ_0, we have

$$a(\theta, x) = a(\theta_0, x) + (\theta - \theta_0)' \nabla a_\theta(\theta^*, x),$$

where $|\theta^* - \theta_0| \leq |\theta - \theta_0|$,

$$\nabla a_\theta(\theta^*, x) = \begin{pmatrix} a_\theta^{(1)}(\theta^*, x) \\ \vdots \\ a_\theta^{(d)}(\theta^*, x) \end{pmatrix}$$

and α' denotes the transpose of α for $\alpha \in R^d$.

Lemma 2.2.13 *For any $A_T > 0$,*

$$\sup_{|\psi| \leq A_T} \left| I_T(\theta) - \frac{1}{T} \int_0^T [\psi' \nabla a_\theta(\theta_0, X_t)]^2 \, dt \right| \leq M A_T^{2+\alpha} T^{-1-\alpha} \text{ a.s. } [P_{\theta_0}], \qquad (2.2.36)$$

for some constant $M > 0$, where $\theta = \theta_0 + \psi T^{-1/2}$.

Proof. Note that

$$\begin{aligned} I_T(\theta) &= \int_0^T [a(\theta, X_t) - a(\theta_0, X_t)]^2 \, dt \\ &= \int_0^T [(\theta - \theta_0)' \nabla a_\theta(\theta_0, X_t)]^2 \, dt \\ &\quad + \int_0^T \{[(\theta - \theta_0)' \nabla a_\theta(\theta^*, X_t)]^2 - [(\theta - \theta_0)' \nabla a_\theta(\theta_0, X_t)]^2\} \, dt. \end{aligned}$$

The integrand in the second expression of the above relation can be written in the form

$$(\theta - \theta_0)'(\nabla a_\theta(\theta^*, X_t) + \nabla a_\theta(\theta_0, X_t))(\theta - \theta_0)'(\nabla a_\theta(\theta^*, X_t) - \nabla a_\theta(\theta_0, X_t)).$$

Applying conditions (B5) and (B7), we obtain that

$$\left| I_T(\theta) - \int_0^T [(\theta - \theta_0)' \nabla a_\theta(\theta_0, X_t)]^2 \, dt \right| \leq 2M |\theta - \theta_0|^{2+\alpha} \int_0^T C(X_t)(1 + |X_t|) \, dt.$$

Noting that $\psi = T^{1/2}(\theta - \theta_0)$ and that $|\psi| \leq A_T$ and applying the ergodic theorem for the integral on the right-hand side of the above inequality (observing that $E[C(X_0)(1 + |X_0|)] < \infty$), we obtain inequality (2.2.36).

Lemma 2.2.14 *Let*

$$\nu_T(\psi, x) = a(\theta_0 + \psi T^{-1/2}, x) - a(\theta_0, x) - \psi T^{-1/2} \nabla a_\theta(\theta_0, x).$$

Then there exist constants $c_1 > 0$ and $c_2 > 0$ such that

$$P_{\theta_0}\left(\sup_{|\psi|\le A_T}\left|\int_0^T \nu_T(\psi, X_t)\,dW_t\right| \ge c_1\lambda^{1/(d+\alpha_0)}\right) \le \frac{c_2(A_T T^{-1/2})^{\alpha(d+\alpha_0)}}{\lambda} \tag{2.2.37}$$

for every $\lambda > 0$.

Proof. Note that

$$\nu_T(\psi, x) - \nu_T(\psi_1, x) = (\psi - \psi_1)'\nabla\nu_T(\zeta, x),$$

where $|\zeta - \psi| \le |\zeta - \psi_1|$. Hence, if $|\psi| \le A_t$ and $|\psi_1| \le A_T$, then, by using (B5), it follows that

$$|\nu_T(\psi, x) - \nu_T(\psi_1, x)| \le |\psi - \psi_1|(A_T T^{-1/2})^{\alpha}C(x).$$

The result now follows by the arguments similar to those given for proving Lemma 2.2.9.

We now prove the asymptotic normality of the MLE under conditions (B1)–(B7).

Theorem 2.2.15 *In addition to conditions (B1)–(B7), suppose that the matrix*

$$J(\theta_0) \equiv ((E[a_\theta^{(i)}(\theta_0, X_0)a_\theta^{(i)}(\theta_0, X_0)]))_{d\times d} \tag{2.2.38}$$

in nonsingular. Then

$$T^{1/2}(\hat\theta_T - \theta_0) \xrightarrow{\mathcal{L}} N_d(0, J^{-1}(\theta_0)), \tag{2.2.39}$$

as $T \to \infty$ under P_{θ_0}.

Proof. By the ergodic theorem,

$$\frac{1}{T}\int_0^T \nabla a_\theta(\theta_0, X_t)\nabla a_\theta(\theta_0, X_t)' \to J(\theta_0) \text{ a.s. } [P_{\theta_0}] \qquad \text{as } T \to \infty,$$

and, by the central limit theorem for stochastic integrals,

$$\frac{1}{\sqrt{T}}\int_0^T \nabla a_\theta(\theta_0, X_t)\,dW_t \xrightarrow{\mathcal{L}} N_d(0, J(\theta_0)) \qquad \text{as } T \to \infty.$$

Choose $A_T = \log T$ in Lemmas 2.2.13 and 2.2.14 and let $T \to \infty$. We observe that the asymptotic distribution of $\hat\theta_T$ which minimizes $R_T(\theta)$ is the same as the distribution of $\hat\psi$, where $\hat\psi$ is the minimizer of $\psi' J(\theta_0)\psi - 2\psi' Z$ in which Z is multivariate normal with mean 0 and covariance matrix $J(\theta_0)$. But $\hat\psi = J^{-1}(\theta_0)Z$, and hence

$$T^{1/2}(\hat\theta_T - \theta_0) \xrightarrow{\mathcal{L}} N_d(0, J^{-1}(\theta_0)), \qquad \text{as } T \to \infty,$$

under the probability measure P_{θ_0}.

2.2.2 Diffusion Type Processes

A stochastic process satisfying the stochastic differential equation

$$dX(t) = a(\theta, t, X)\,dt + \sigma(t, X)\,dW(t), \qquad X(0) = \eta,\ 0 \le t \le T, \tag{2.2.40}$$

where $\{W(t), t \ge 0\}$ is the standard Wiener process is called a *diffusion type process* with drift coefficient $a(\theta, t, x)$ and diffusion coefficient $\sigma(t, x)$. The drift and diffussion coefficients at time t depend on the entire path up to t. The problem of interest is the estimation of the parameter θ based on an observation of the process X. We assume that either $\sigma(\cdot, \cdot)$ is known or, if unknown, then it is a constant. For reasons mentioned earlier, we can reduce the problem to the study of the special case, namely,

$$dX(t) = a(\theta, t, X)\,dt + dW(t), \qquad X(0) = X_0,\ 0 \le t \le T. \tag{2.2.41}$$

Here X_0 is $\mathcal{F}_0$-measurable and $P(|X_0| < \infty) = 1$. Let P_θ^T be the probability measure induced on the measurable space $(C[0, T], \mathcal{B}[0, T])$ of continuous functions on $[0, T]$ by the process X when θ is the parameter. Let P_0^T be the probability measure induced by Y defined by $Y(t) = X_0 + W(t)$. Here $\mathcal{B}[0, T]$ is the σ-algebra generated by the supremum norm on $C[0, T]$. If

$$P\left[\int_0^T a(\theta, t, X)^2\,dt < \infty\right] = 1$$

and

$$P\left[\int_0^T a(\theta, t, Y)^2\,dt < \infty\right] = 1,$$

then the probability measures P_0^T and P_θ^T are equivalent and

$$\frac{dP_\theta^T}{dP_0^T}(X) = \exp\left\{\int_0^T a(\theta, t, X)\,dX(t) - \tfrac{1}{2}\int_0^T a(\theta, t, X)^2\,dt\right\}.$$

The following result is an analogue of the classical Cramér–Rao inequality.

2.2.3 Cramér–Rao Lower Bound

Suppose that the following conditions hold:

(i) $P_\theta\{\int_0^T a(\theta, t, X)^2\,dt < \infty\} = 1, \theta \in \Theta = (\alpha, \beta) \subset R$.

(ii) $a(\theta, t, x)$ is absolutely continuous with respect to θ for almost all $t \in [0, T]$ and

$$E_\theta\left\{\int_0^T a_\theta(\theta, t, X)^2\,dt\right\} < \infty,$$

where $a_\theta(\theta, t, x)$ denotes the partial derivative of $a(\theta, t, x)$ with respect to θ.

(iii) $I(\theta_1, \theta_2) = E_{\theta_1}\{\int_0^T a_\theta(\theta_2, t, X)^2\,dt\} > 0, \theta_1, \theta_2 \in \Theta$, and $I(\theta, \theta)$ is continuous in θ.

Theorem 2.2.16 *Let $\theta^* = \theta^*(X_T)$ be any estimator of θ such that $Q(\theta) = E_\theta(\theta^* - \theta)^2$ is bounded over a compact set in Θ. Let $b(\theta) = E_\theta(\theta^*) - \theta$. Then, under conditions (i)–(iii), for almost all $\theta \in \Theta$, $b(\theta)$ is differentiable and*

$$E_\theta(\theta^* - \theta)^2 \ge \frac{\left(1 + \frac{db(\theta)}{d\theta}\right)^2}{I(\theta, \theta)} + b^2(\theta). \tag{2.2.42}$$

Proof. Let θ be the true parameter. Define $\rho_i = dP^T_{\theta_i}/dP^T_\theta$, $i = 1, 2$.

Note that $E_\theta(\rho_i \zeta) = E_{\theta_i}(\zeta)$ for any random variable ζ.

Now

$$\begin{aligned} E_{\theta_2}(\theta^*) - E_{\theta_1}(\theta^*) &= E_\theta[\theta^*(\rho_2 - \rho_1)] \\ &= E_\theta\{[\theta^* - \tfrac{1}{2}(E_{\theta_2}(\theta^*) + E_{\theta_1}(\theta^*))](\rho_2 - \rho_1)\} \end{aligned}$$

since $E_\theta(\rho_2 - \rho_1) = 0$. Hence

$$\begin{aligned} (E_{\theta_2}(\theta^*) - E_{\theta_1}(\theta^*))^2 &= \{E_\theta([\theta^* - \tfrac{1}{2}(E_{\theta_2}(\theta^*) + E_{\theta_1}(\theta^*))](\rho_2 - \rho_1))\}^2 \\ &= E_\theta\{[\theta^* - \tfrac{1}{2}(E_{\theta_2}(\theta^*) + E_{\theta_1}(\theta^*))](\sqrt{\rho_2} + \sqrt{\rho_1})\}^2 \\ &\quad \times E_\theta(\sqrt{\rho_2} - \sqrt{\rho_1})^2 \\ &\le [2E_\theta\{[\theta^* - \tfrac{1}{2}(E_{\theta_2}(\theta^*) + E_{\theta_1}(\theta^*))]^2 \rho_1\} \\ &\quad + 2E_\theta\{[\theta^* - \tfrac{1}{2}(E_{\theta_2}(\theta^*) + E_{\theta_1}(\theta^*))]^2 \rho_2\}]2(1 - E_\theta\sqrt{\rho_1\rho_2}) \\ &\quad \text{(by the inequality } (X+Y)^2 \le 2(X^2 + Y^2)) \\ &= \{2E_{\theta_1}[\theta^* - \tfrac{1}{2}(E_{\theta_2}(\theta^*) + E_{\theta_1}(\theta^*))]^2 \\ &\quad + 2E_{\theta_2}[\theta^* - \tfrac{1}{2}(E_{\theta_2}(\theta^*) + E_{\theta_1}(\theta^*))]^2\}2(1 - E_\theta\sqrt{\rho_1\rho_2}). \end{aligned}$$

Note that

$$\begin{aligned} &E_{\theta_1}[\theta^* - \tfrac{1}{2}(E_{\theta_2}(\theta^*) + E_{\theta_1}(\theta^*))]^2 \\ &= E_{\theta_1}[\theta^* - E_{\theta_1}(\theta^*) - \tfrac{1}{2}(E_{\theta_2}(\theta^*) - E_{\theta_1}(\theta^*))]^2 \\ &= Q(\theta_1) - b^2(\theta_1) + \tfrac{1}{4}(E_{\theta_2}(\theta^*) - E_{\theta_1}(\theta^*))^2 \end{aligned}$$

and

$$E_{\theta_2}[\theta^* - \tfrac{1}{2}(E_{\theta_2}(\theta^*) + E_{\theta_1}(\theta^*))]^2 = Q(\theta_2) - b^2(\theta_2) + \tfrac{1}{4}(E_{\theta_2}(\theta^*) - E_{\theta_1}(\theta^*))^2.$$

Therefore

$$\begin{aligned} &(E_{\theta_2}(\theta^*) - E_{\theta_1}(\theta^*))^2 \\ &\le \{2Q(\theta_1) + 2Q(\theta_2) - 2b^2(\theta_1) - 2b^2(\theta_2) + (E_{\theta_2}(\theta^*) - E_{\theta_1}(\theta^*))^2\} \\ &\quad \times 2(1 - E_\theta\sqrt{\rho_1\rho_2}). \end{aligned}$$

It is known that

$$\frac{\rho_2}{\rho_1} = \frac{dP^T_{\theta_2}}{dP^T_{\theta_1}} = \exp\left\{\int_0^T (\theta_2 - \theta_1)(dX(\tau) - a_1\, d\tau) - \frac{1}{2}\int_0^T (a_2 - a_1)^2\, d\tau\right\}$$

since $dX(t) = a(\theta, t, X)\, dt + dW(t)$, $t \ge 0$. Here $a_i \equiv a(\theta_i, t, X)$.

Now

$$\begin{aligned} 1 - E_\theta\sqrt{\rho_1\rho_2} &= 1 - E_\theta\left\{\frac{dP^T_{\theta_1}}{dP^T_\theta} \cdot \frac{dP^T_{\theta_2}}{dP^T_\theta}\right\}^{1/2} \\ &= 1 - E_{\theta_1}\left\{\frac{dP^T_{\theta_2}}{dP\theta^T_1}\right\}^{1/2} \\ &= 1 - E_{\theta_1}(V(T)), \end{aligned}$$

where

$$V(T) = \left(\frac{dP_{\theta_2}^T}{dP_{\theta_1}^T}\right)^{1/2} = \exp\left\{\frac{1}{2}\int_0^T (a_2 - a_1)(dX(\tau) - a_1\, d\tau) - \frac{1}{4}\int_0^T (a_2 - a_1)^2\, d\tau\right\}.$$

Observe that

$$V(T) = \exp(Y(T)),$$

where

$$dY(t) = \tfrac{1}{2}(a_2 - a_1)(dX(t) - a_1\, dt) - \tfrac{1}{4}(a_2 - a_1)^2\, dt$$

and hence, by the Itô formula,

$$\begin{aligned}
dV(t) &= \exp(Y(t))\, dY(t) + \tfrac{1}{2}\exp(Y(t))\tfrac{1}{4}(a_2 - a_1)^2\, dt \\
&= V(t)\, dY(t) + \tfrac{1}{8}V(t)(a_2 - a_1)^2\, dt \\
&= V(t)[\tfrac{1}{2}(a_2 - a_1)(dX(t) - a_1\, dt) - \frac{1}{4}(a_2 - a_1)^2\, dt] \\
&\quad + \tfrac{1}{8}(a_2 - a_1)^2\, dt \\
&= \tfrac{1}{2}(a_2 - a_1)V(t)(dX(t) - a_1\, dt) - \tfrac{1}{8}(a_2 - a_1)^2 V(t)\, dt.
\end{aligned}$$

But $V(0) = 1$. Hence

$$V(T) = 1 - \frac{1}{8}\int_0^T (a_2 - a_1)^2 V(t)\, dt + \frac{1}{2}\int_0^T (a_2 - a_1)V(t)(dX(t) - a_1\, dt).$$

Hence

$$E_{\theta_1} V(T) = 1 - \frac{1}{8}E_{\theta_1}\int_0^T (a_2 - a_1)^2 V(t)\, dt,$$

since $E_{\theta_1}\int_0^T (a_2 - a_1)V(t)\, dW(t) = 0$.

Therefore

$$\begin{aligned}
1 - E_\theta\sqrt{\rho_1\rho_2} &= \frac{1}{8}E_{\theta_1}\left\{\int_0^T (a_2 - a_1)^2 V(t)\, dt\right\} \\
&= \frac{1}{8}E_{\theta_1}\left\{\int_0^T (a_2 - a_1)^2\left(\frac{dP_{\theta_2}^T}{dP_{\theta_1}^T}\right)^{1/2} dt\right\} \\
&\le \frac{1}{16}E_{\theta_1}\left[\int_0^T (a_2 - a_1)^2\left(\frac{dP_{\theta_2}^T}{dP_{\theta_1}^T} + 1\right) dt\right] \\
&\le \frac{1}{16}E_{\theta_2}\left[\int_0^T (a_2 - a_1)^2\, dt\right] + \frac{1}{16}E_{\theta_1}\left[\int_0^T (a_2 - a_1)^2\, dt\right].
\end{aligned}$$

But

$$\begin{aligned}
E_{\theta_1}\left[\int_0^T (a_2 - a_1)^2\, dt\right] &= E_{\theta_1}\left\{\int_0^T\left[\int_{\theta_1}^{\theta_2} a_\theta(\phi)\, d\phi\right]^2 dt\right\} \\
&\le E_{\theta_1}\left\{\int_0^T\left[\left(\int_{\theta_1}^{\theta_2} a_\theta^2(\phi)\, d\phi\right)|\theta_2 - \theta_1|\right] dt\right\}
\end{aligned}$$

$$= |\theta_2 - \theta_1| \int_{\theta_1}^{\theta_2} \left\{ E_{\theta_1} \int_0^T a_\theta^2(\phi)\, dt \right\} d\phi$$

$$= |\theta_2 - \theta_1| \int_{\theta_1}^{\theta_2} I(\theta_1, \phi)\, d\phi.$$

Similarly,

$$E_{\theta_2}\left[\int_0^T (a_2 - a_1)^2\, dt \right] \le |\theta_2 - \theta_1| \int_{\theta_1}^{\theta_2} I(\theta_2, \phi)\, d\phi.$$

Therefore

$$2\{1 - E_\theta \sqrt{\rho_1 \rho_2}\} \le \tfrac{1}{8} |\theta_2 - \theta_1| \int_{\theta_1}^{\theta_2} \{I(\theta_1, \phi) + I(\theta_2, \phi)\}\, d\phi$$

and

$$\begin{aligned}(E_{\theta_2}(\theta^*) - E_{\theta_1}(\theta^*))^2 \le \{&2Q(\theta_1) + 2Q(\theta_2) - 2b^2(\theta_1)\\ &-2b^2(\theta_2) + (E_{\theta_2}(\theta^*) - E_{\theta_1}(\theta^*))^2\}\\ &\times \tfrac{1}{8}|\theta_2 - \theta_1| \int_{\theta_1}^{\theta_2} \{I(\theta_1, y) + I(\theta_2, y)\}\, dy.\end{aligned}$$

Let $\theta_2 = \theta$. Divide the inequality by $(\theta_1 - \theta)^2$ on both sides and let $\theta_1 \to \theta$. Then it follows that

$$Q(\theta) \ge \frac{[1 + \dot{b}(\theta)]^2}{I(\theta, \theta)} + b^2(\theta),$$

proving the Cramér–Rao inequality where $\dot{b}(\theta)$ denotes the derivative of $b(\theta)$ with respect to θ.

Remark. For a multiparameter extension of this inequality, see Kutoyants (1984a), to whom the above result is due.

Local asymptotic normality for the scheme of series

Suppose $(\Omega, \mathcal{F}, P)$ is a probability space and, for every $\varepsilon \in (0, 1]$, let $\mathcal{F}^{(\varepsilon)} = \{\mathcal{F}_t^{(\varepsilon)}, 0 \le t \le T_\varepsilon\}$ be a filtration on $(\Omega, \mathcal{F})$. Let $X_\varepsilon = \{X_\varepsilon(t), 0 \le t \le T_\varepsilon\}$ be a random process of diffusion type satisfying the stochastic differential equation

$$dX_\varepsilon(t) = a_\varepsilon(\theta, t, X_\varepsilon)\, dt + dW_\varepsilon(t), \qquad X_\varepsilon(0) = \eta_\varepsilon,\ 0 \le t \le T_\varepsilon, \tag{2.2.43}$$

where η_ε is an $\mathcal{F}_0^{(\varepsilon)}$-measurable random variable, $P\{|\eta_\varepsilon| < \infty\} = 1$ and $\theta \in \Theta = (\alpha, \beta)$. Here $W_\varepsilon(t), 0 \le t \le T_\varepsilon$, is the standard Wiener process. Let $P_\theta^{(\varepsilon)}$ be the probability measure generated by the process X_ε when θ is the true parameter. Let $\mathcal{P}_\theta$ be the family of nonanticipating functions $f(t, X_\varepsilon), 0 \le t \le T_\varepsilon$, for which $P_\theta^{(\varepsilon)}\{\|f(X_\varepsilon)\| < \infty\} = 1$, where $\|f(X_\varepsilon)\|^2 = \int_0^{T_\varepsilon} f^2(t, X_\varepsilon)\, dt$.

Recall that the family of probability measures $\{P_\theta^{(\varepsilon)}, \theta \in \Theta\}$ is locally asymptotically normal (LAN) at θ_0 as $\varepsilon \to 0$ if there exists a function $\phi_\varepsilon(\theta_0)$ such that, for every $u \in R$,

$$Z_\varepsilon(u) = \exp\{u\Delta_\varepsilon(\theta_0, X_\varepsilon) - \tfrac{1}{2}u^2 + \psi_\varepsilon(\theta_0, u, X_\varepsilon)\}, \tag{2.2.44}$$

where

$$Z_\varepsilon(u) = \frac{dP^{(\varepsilon)}_{\theta_0+\phi_\varepsilon(\theta_0)u}}{dP^{(\varepsilon)}_{\theta_0}}(X_\varepsilon), \tag{2.2.45a}$$

$$\mathcal{L}\{\Delta_\varepsilon(\theta, X_\varepsilon)|P^{(\varepsilon)}_{\theta_0}\} \to N(0,1) \text{ as } \varepsilon \to 0 \tag{2.2.45b}$$

and

$$P^{(\varepsilon)}_{\theta_0} - \lim_{\varepsilon\to 0} \psi_\varepsilon(\theta_0, u, X_\varepsilon) = 0. \tag{2.2.45c}$$

Remarks. Here $\phi_\varepsilon(\theta_0)$ is the normalizing function. If the family $\{P^{(\varepsilon)}_\theta, \theta \in \Theta\}$ is LAN at every $\theta_0 \in \Theta$, then it is LAN in Θ. If (2.2.45*a*) and (2.2.45*b*) hold uniformly for $\theta \in K$, K compact in Θ, then $\{P^{(\varepsilon)}_\theta, \theta \in \Theta\}$ is uniformly LAN in Θ. Loosely speaking, this means that, for values θ asymptotically close to θ_0 (for ε small), the likelihood ratio $\frac{dP^{(\varepsilon)}_\theta}{dP^{(\varepsilon)}_{\theta_0}}(X_\varepsilon)$ has the properties of the process

$$Z(u) = \exp\{u\zeta - \tfrac{1}{2}u^2\},$$

where ζ is the standard normal random variable. In problems of estimation $\phi_\varepsilon(\theta)=(I_\varepsilon(\theta))^{-1/2}$, where $I_\varepsilon(\theta)$ is the Fisher information. The following theorem is a consequence of results in Chapter 6 of Prakasa Rao (1999). We omit the proof.

Theorem 2.2.17 (*Local asymptotic normality*) *Suppose that*

$$a_\varepsilon(\theta_1, \cdot, X_\varepsilon) \in \mathcal{P}_{\theta_2}, \quad \textit{for all } \theta_1, \theta_2 \in \Theta,$$

and there exists $q_\varepsilon(\theta, \cdot, X_\varepsilon) \in \mathcal{P}_\theta$ *and* $\phi_\varepsilon(\theta) \to 0$ *such that*

(C1) $P_\theta - \lim_{\varepsilon\to 0}\phi_\varepsilon(\theta)\|q_\varepsilon(\theta, X_\varepsilon)\| = 1,$

(C2) $P_\theta - \lim_{\varepsilon\to 0}\|a_\varepsilon(\theta + \phi_\varepsilon(\theta)u, X_\varepsilon) - a_\varepsilon(\theta, X_\varepsilon) - u\phi_\varepsilon(\theta)q_\varepsilon(\theta, X_\varepsilon)\| = 0,$ *where*

$$\|f\|^2 = \int_0^{T_\varepsilon} f^2(t)\,dt.$$

Then the family $\{P^{(\varepsilon)}_\theta, \theta \in \Theta\}$ *is LAN in* Θ *with the normalizing function* $\phi_\varepsilon(\theta)$ *and*

$$\Delta_\varepsilon(\theta, X_\varepsilon) = \phi_\varepsilon(\theta)\int_0^{T_\varepsilon} q_\varepsilon(\theta, t, X_\varepsilon)\,dW_\varepsilon(t). \tag{2.2.46}$$

Note. A basic result in proving the above theorem is the central limit theorem for Itô stochastic integrals.

As a consequence of local asymptotic normality, one can obtain the Hájek–Leĉam inequality giving the *asymptotic minimax lower bound* for the risk of an estimator. For a proof, see Kutoyants (1984a).

Theorem 2.2.18 (*Hájek–Lecâm inequality*) *Suppose the conditions stated in Theorem 2.2.17 for LAN hold. Let $\ell(\cdot)$ be a symmetric function, continuous at zero, such that $\{x : \ell(x) < c\}$ is convex for all $c > 0$, and, for any $h > 0$, $\ell(x)$ does not increase faster than $\exp(h|x|^2)$ as $|x| \to \infty$. Then, for every $\gamma \in (0, 1)$,*

$$\liminf_{\varepsilon\to 0} \inf_{\theta_\varepsilon^*} \sup_{|\theta-y|<\phi_\varepsilon^\gamma(\theta)} E_y \ell((\theta_\varepsilon^* - y)\phi_\varepsilon(\theta)^{-1}) \geq E\ell(\zeta), \tag{2.2.47}$$

where ζ is a Gaussian random variable with zero mean and unit variance.

Remark. If $\ell(x) = x^2$, the inequality reduces to

$$\liminf_{\varepsilon\to 0} \inf_{\theta_\varepsilon^*} \sup_{|\theta-y|<\phi_\varepsilon^\gamma(\theta)} E_y \left\{\frac{\theta_\varepsilon^* - y}{\phi_\varepsilon(\theta)}\right\}^2 \geq 1. \tag{2.2.48}$$

Remark. An estimator θ_ε^* is said to be *asymptotically efficient* if for every $\theta \in \Theta$ and some $\gamma \in (0, 1)$,

$$\lim_{\varepsilon\to 0} \sup_{|\theta-y|<\phi_\varepsilon^\gamma(\theta)} E_y \left\{\frac{\theta_\varepsilon^* - y}{\phi_\varepsilon(\theta)}\right\}^2 = 1. \tag{2.2.49}$$

Example 2.2.2
Consider the stochastic differential equation

$$dX(t) = -\theta X(t)\, dt + dW(t), \qquad X(0) = 0,\ 0 \leq t \leq T,\ \theta \in (\alpha, \beta),\ \alpha > 0. \tag{2.2.50}$$

Here the family of measures $P_\theta^{(T)}, \theta \in (\alpha, \beta), \alpha > 0$ is LAN as $T \to \infty$ with the normalizing function $\phi_T(\theta) = \sqrt{2\theta}T^{-1/2}$.

Furthermore, the solution $X(\cdot)$ of (2.2.50) is then ergodic in the sense that there exists a stationary measure $\mu_\theta(\cdot)$ such that

$$\lim_{T\to\infty} \frac{1}{T} \int_0^T f(X(t))\, dt = \int_{-\infty}^{\infty} f(x)\, \mu_\theta(x) \text{ a.s. } [P_\theta].$$

If $\theta \in (\alpha, \beta)$, $\beta < 0$, then the family $P_\theta^{(T)}, \theta \in (\alpha, \beta)$, is LAN as $T \to \infty$ with the normalizing function $\phi_T(\theta) = 2\theta e^{\theta T}$.

Maximum likelihood estimation (scalar parameter case)

Consider the stochastic differential equation

$$dX_\varepsilon(t) = a_\varepsilon(\theta, t, X_\varepsilon)\, dt + dW_\varepsilon(t), \qquad X_\varepsilon(0) = 0,\ 0 \leq t \leq T_\varepsilon,\ \theta \in \Theta = (\alpha, \beta). \tag{2.2.51}$$

It is required to estimate θ from a sample path of X_ε on $[0, T_\varepsilon]$. The likelihood function is given by

$$L(\theta, X_\varepsilon) = \exp\left\{\int_0^{T_\varepsilon} a_\varepsilon(\theta, t, X_\varepsilon)\, dX_\varepsilon - \tfrac{1}{2}\|a_\varepsilon(\theta, X_\varepsilon)\|^2\right\} \tag{2.2.52}$$

and the MLE $\hat{\theta}_\varepsilon$ is defined as a solution of

$$L(\hat{\theta}_\varepsilon, X_\varepsilon) = \sup_{\theta\in\Theta} L(\theta, X_\varepsilon). \tag{2.2.53}$$

Introduce the notation

$$\dot{a}_\varepsilon(\theta, t, X_\varepsilon) = \frac{\partial}{\partial\theta} a_\varepsilon(\theta, t, X_\varepsilon), \tag{2.2.54}$$

$$\phi_\varepsilon(\theta) = \{E_\theta \|\dot{a}_\varepsilon(\theta, X_\varepsilon)\|^2\}^{-1/2} \tag{2.2.55}$$

$$\Delta a_\varepsilon(u) = a_\varepsilon(\theta + \phi_\varepsilon(\theta)u, t, X_\varepsilon) - a_\varepsilon(\theta, t, X_\varepsilon) \tag{2.2.56}$$

and

$$h_\theta(u) = -\log E_\theta[\exp\{-\lambda\|\Delta_\varepsilon(u)\|^2\}]. \tag{2.2.57}$$

Suppose the following conditions hold:

(D1) $E_\theta \|\dot{a}_\varepsilon(\theta, X_\varepsilon)\|^2 < \infty, \quad \theta \in \Theta.$

(D2) $\lim_{\varepsilon\to 0}\phi_\varepsilon(\theta) = 0.$

(D3) There exists $c_0 > 0$ such that for every $\theta, \theta_1, \theta_2 \in \Theta$, $\varepsilon \in (0, 1]$,

$$\phi_\varepsilon^2(\theta_1) E_\theta \|\dot{a}_\varepsilon(\theta_2, t, X_\varepsilon)\|^2 < c_0.$$

(D4) $P_\theta - \lim_{\varepsilon\to 0} \phi_\varepsilon(\theta)\|\dot{a}_\varepsilon(\theta, X_\varepsilon)\| = 1.$

(D5) There exists $\delta \in (0, 1)$ such that

$$\lim_{\varepsilon\to 0} \sup_{|\theta - y| < \phi_\varepsilon^\delta(\theta)} \phi_\varepsilon(\theta)^2 E_\theta \|\dot{a}_\varepsilon(y, X_\varepsilon) - \dot{a}_\varepsilon(\theta, X_\varepsilon)\|^2 = 0.$$

(D6) There exist $\gamma > 0, \lambda > 0, c > 0$ such that

$$h_\theta(u) \geq c|u|^\gamma.$$

Theorem 2.2.19 *Under the above conditions,*

$$P_\theta - \lim_{\varepsilon\to 0} \hat{\theta}_\varepsilon = \theta, \tag{2.2.58}$$

$$\mathcal{L}\{(\hat{\theta}_\varepsilon - \theta)\phi_\varepsilon(\theta)^{-1} | P_\theta^{(\varepsilon)}\} \to N(0, 1), \quad \text{as } \varepsilon \to 0 \tag{2.2.59}$$

and

$$\lim_{\varepsilon\to 0} E_\theta |(\hat{\theta}_\varepsilon - \theta)\phi_\varepsilon(\theta)^{-1}|^p = \frac{2^{p/2}}{\sqrt{\pi}} \Gamma\left(\frac{p+1}{2}\right), \quad p > 0. \tag{2.2.60}$$

A detailed proof is given by Kutoyants (1984a, p. 85).

Remark. Suppose that, in addition to (E1)–(E6):

(D7) $\lim_{\varepsilon\to 0} \sup_{|\theta-\theta_0| < \phi_\varepsilon(\theta_0)^\delta} |\phi_\varepsilon(\theta)\phi_\varepsilon(\theta_0)^{-1} - 1| = 0$ holds for all θ and θ_0.

Then $\hat{\theta}_\varepsilon$ is asymptotically efficient.

We briefly discuss an approach to asymptotic theory of the MLE through the weak convergence of the log-likelihood ratio process.

Weak convergence of the likelihood ratio process

Let $(\Omega, \mathcal{F}, P)$ be a probability space and $\Theta = (\alpha, \beta)$, $-\infty < \alpha < \beta < \infty$. For every $\theta \in \Theta$ and $0 < \varepsilon \le 1$, let $X_\varepsilon = \{X_\varepsilon(t), 0 \le t \le T_\varepsilon\}$ be a process inducing a measure $P_\theta^{(\varepsilon)}$. Suppose the measures $\{P_\theta^{(\varepsilon)}, \theta \in \Theta\}$ are equivalent. Then the MLE $\hat{\theta}_\varepsilon$ is defined by

$$\frac{dP_{\hat{\theta}_\varepsilon}^{(\varepsilon)}}{dP_{\theta_0}^{(\varepsilon)}}(X_\varepsilon) = \sup_{\theta \in \Theta} \frac{dP_\theta^{(\varepsilon)}}{dP_{\theta_0}^{(\varepsilon)}}(X_\varepsilon) \tag{2.2.61}$$

for some arbitrary but fixed $\theta_0 \in \Theta$.

Let θ be the true parameter and $\phi_\varepsilon(\theta)$ be a normalizing function such that the normalized likelihood ratio

$$Z_\varepsilon(u) = \frac{dP_{\theta+\phi_\varepsilon(\theta)u}^{(\varepsilon)}}{dP_\theta^{(\varepsilon)}}(X_\varepsilon), \qquad u \in U_{\theta,\varepsilon} = \{u : \theta + \phi_\varepsilon(\theta)u \in \Theta\}, \tag{2.2.62}$$

has a distribution converging weakly to a limiting distribution as $\varepsilon \to 0$. In other words,

$$\mathcal{L}(Z_\varepsilon(u)|P_\theta^{(\varepsilon)}) \xrightarrow{w} \mathcal{L}\{Z(u)\} \qquad \text{as } \varepsilon \to 0, \tag{2.2.63}$$

where $Z(u)$ is some random process. If

$$Z_\varepsilon(\hat{u}_\varepsilon) = \sup_{u \in U_{\theta,\varepsilon}} Z_\varepsilon(u) \tag{2.2.64}$$

then $\hat{\theta}_\varepsilon = \theta + \phi_\varepsilon(\theta)\hat{u}_\varepsilon$. Suppose the limiting process $Z(u)$ has a unique maximum a.s., that is,

$$Z(\hat{u}) = \sup_{u \in R} Z(u). \tag{2.2.65}$$

The process $Z_\varepsilon(u)$ is defined on $U_{\theta,\varepsilon}$. We extend the process $Z_\varepsilon(u)$ to $u \in R$ in a smooth fashion. Let $\mathcal{X}$ be the space of realizations of $Z_\varepsilon(u)$ and $Z(\cdot)$. Suppose $\mathcal{X}$ is a complete separable metric space and $\mathcal{F}$ the Borel σ-algebra associated with it. Let $\nu_\theta^{(\varepsilon)}$ and ν_θ be the measures generated by $Z_\varepsilon(\cdot)$ and $Z(\cdot)$ on $(\mathcal{X}, \mathcal{F})$ when θ is the true parameter. Suppose that

$$\nu_\theta^{(\varepsilon)} \xrightarrow{w} \nu_\theta \qquad \text{as } \varepsilon \to 0. \tag{2.2.66}$$

Then, it is well known that the distribution of any continuous functional on $\mathcal{X}$ of $Z_\varepsilon(\cdot)$ converges weakly to the distribution of this functional of $Z(\cdot)$. Hence

$$\begin{aligned} P_\theta^{(\varepsilon)}\left\{\frac{\hat{\theta}_\varepsilon - \theta}{\phi_\varepsilon(\theta)} \le x\right\} &= P_\theta^{(\varepsilon)}\{\hat{u}_\varepsilon \le x\} \\ &= P_\theta^{(\varepsilon)}\{\sup_{u \le x} Z_\varepsilon(u) > \sup_{u > x} Z_\varepsilon(u)\} \\ &\to P_\theta\{\sup_{u \le x} Z(u) > \sup_{u > x} Z(u)\} \\ &= P_\theta(\hat{u} \le x). \end{aligned} \tag{2.2.67}$$

In the smooth case, it can be shown that

$$Z(u) = \exp\{u\zeta - \tfrac{1}{2}u^2\}, \qquad \text{where } \zeta \text{ is } N(0, 1) \tag{2.2.68}$$

and hence

$$P_\theta\{\hat{u} \le x\} = P\{\zeta \le x\} = \frac{1}{\sqrt{2\pi}} \int_{-\infty}^{x} e^{-y^2/2}\, dy,$$

that is,

$$[\phi_\varepsilon(\theta)]^{-1}(\hat{\theta}_\varepsilon - \theta) \xrightarrow{\mathcal{L}} N(0, 1) \qquad \text{as } \varepsilon \to 0. \tag{2.2.69}$$

The following theorem gives sufficient conditions for (2.2.66) to hold.

Theorem 2.2.20 *Suppose the following conditions hold:*

(i) The family of measures $\{P_\theta^{(\varepsilon)}, \theta \in \theta\}$ *is LAN as* $\varepsilon \to 0$.
(ii) There exists $c > 0$ *such that*

$$E_\theta |Z_\varepsilon^{1/2}(u_2) - Z_\varepsilon^{1/2}(u_1)|^2 \le c\, |u_2 - u_1|^2, \qquad u_1, u_2, \in R. \tag{2.2.70}$$

(iii) There exist $c > 0$ *and* $\gamma > 0$ *such that, for some* $\varepsilon_0 > 0$ *and* $0 < \varepsilon < \varepsilon_0$,

$$P_\theta^{(\varepsilon)}\{Z_\varepsilon(u) > e^{-cg(u)}\} \le e^{-cg(u)}, \tag{2.2.71}$$

where $g(u) = |u|^\gamma$.

Then the weak convergence in (2.2.66) holds for the process $\{Z_\varepsilon(u)\}$.

For a proof, see Kutoyants (1984a, p. 183).

Remarks. Condition (i) ensures the convergence of the finite dimensional distributions of the process $Z_\varepsilon(u)$. Condition (ii) gives the tightness of the processes when $Z_\varepsilon(u)$ is restricted to a compact set in R. Condition (iii) implies that $\phi_\varepsilon(\theta)^{-1}(\hat{\theta}_\varepsilon - \theta) = O_p(1)$. In fact, from condition (iii) we obtain the existence of $c > 0$ and $d > 0$ such that, for every $\delta > 0$,

$$P_\theta^{(\varepsilon)}\{|\hat{\theta}_\varepsilon - \theta| > \delta\} \le c \exp\{-dg(\delta\phi_\varepsilon(\theta)^{-1})\}$$

(cf. Kutoyants 1984, p. 188).

Remarks. The stochastic model described by (2.2.43) is known as the scheme of series: for every $\varepsilon \in (0, 1]$, a diffusion type process is observed given by the stochastic differential equation

$$dX_\varepsilon(t) = a_\varepsilon(\theta, t, X_\varepsilon)\, dt + \sigma_\varepsilon(t, X_\varepsilon)\, dW_\varepsilon(t), \qquad X_\varepsilon(0) = \eta_\varepsilon,\ 0 \le t \le T_\varepsilon \tag{2.2.72}$$

where $a_\varepsilon(\cdot, t, \cdot)$, $\sigma_\varepsilon(t, \cdot)$, η_ε and T_ε arbitrarily vary with ε and growth and Lipschitzian type conditions are imposed on the coefficients $a_\varepsilon(\theta, t, X_\varepsilon)$ and $\sigma_\varepsilon(t, X_\varepsilon)$. The problem of estimation of the parameter θ can be studied through different limit processes, for example, either by letting the time of observation $T \to \infty$, or the diffusion coefficient $\sigma_\varepsilon(t, X_\varepsilon) \to 0$ or the trend coefficient $a_\varepsilon(\theta, t, X_\varepsilon) \to \infty$ or a combination of these. We have considered the passage to the limit as $\varepsilon \to 0$ in this Section in connection with the problem of estimation of θ. If $T_\varepsilon = \frac{1}{\varepsilon}$, then we obtain the traditional approach we have discussed earlier, namely, $T \to \infty$. If $\sigma_\varepsilon(t, X_\varepsilon) = \varepsilon$, then the limit properties are studied as the diffusion coefficient approaches zero.

Example 2.2.3 (*Diffusion process with small diffusion coefficient*)
Let X_ε be the solution of the stochastic differential equation

$$dX_\varepsilon(t) = a(t, \theta, X_\varepsilon(t))\, dt + \varepsilon\, dW(t), \qquad X_\varepsilon(0) = x,\ 0 \le t \le T. \tag{2.2.73}$$

It is required to estimate $\theta \in \Theta \subset R$ based on the sample path of the process $X_\varepsilon(t)$ over $[0, T]$ and study its properties as $\varepsilon \to 0$. This model corresponds to small disturbances of the dynamical system

$$\frac{dX_0(t)}{dt} = a(t, \theta, X_0(t)), \qquad X_0(0) = x,\ 0 \le t \le T.$$

Suppose the following conditions hold:

(E1) The function $a(t, \theta, y)$ is Lipschitz in y, continuously differentiable in θ and the derivative $a_\theta(t, \theta, y)$ is continuous in y.

(E2) For all $\theta' \neq \theta$ in Θ,

$$\int_0^T [a(t, \theta', X_0(t)) - a(t, \theta, X_0(t)]^2\, dt > 0.$$

(E3) $I(\theta) = \int_0^T a_\theta^2(t, \theta, X_0(t))\, dt > 0.$

(E4) $\sup_{\theta, \theta' \in \Theta} E_\theta \|a_\theta(t, \theta', X_\varepsilon(t))\|^2 < c < \infty.$

Under the above conditions, it can be checked that the MLE $\hat\theta_\varepsilon$ is consistent, asymptotically normal uniformly over $\theta \in$ compact set $K \subset \Theta$, that is,

$$\frac{1}{\varepsilon}(\hat\theta_\varepsilon - \theta) \xrightarrow{\mathcal{L}} N(0, I(\theta)^{-1}) \qquad \text{as } \varepsilon \to 0, \tag{2.2.74}$$

and the moments of $|\hat\theta_\varepsilon - \theta|\varepsilon^{-1}$ converge to the corresponding absolute moments of a random variable Z having the normal distribution with mean 0 and variance $I(\theta)^{-1}$. Details are given in Kutoyants (1984b). For instance, if

$$dX_\varepsilon(t) = \theta X_\varepsilon(t)\, dt + \varepsilon\, dW(t), \qquad X_\varepsilon(0) = x,\ 0 \le t \le T,$$

and $\theta > 0$, $x \neq 0$, then the MLE has the form

$$\hat\theta_\varepsilon = \left\{\int_0^T X_\varepsilon(t)^2\, dt\right\}^{-1} \int_0^T X_\varepsilon(t)\, dX_\varepsilon(t)$$

and conditions (E1)–(E3) hold with $I(\theta) = \frac{x^2}{2\theta}[e^{2\theta T} - 1] > 0$.

Example 2.2.4 (*Kalman–Bucy filter*)
Let $\{(\zeta_t, \eta_t), 0 \le t \le T\}$ be a two-dimensional diffusion process

$$\begin{aligned} d\zeta_t &= \theta \eta_t\, dt + \sigma\, dW_t^{(1)}, \qquad \zeta_0 = 0, \\ d\eta_t &= -a\eta_t\, dt + b\, dW_t^{(2)}, \qquad \eta_0 = 0, \end{aligned} \tag{2.2.75}$$

where $\{W_t^{(1)}\}$ and $\{W_t^{(2)}\}$ are two independent standard Wiener processes with known parameters $\sigma \neq 0$, $a > 0$ and $b \in R$, with unknown $\theta \in (\alpha, \beta)$, $\alpha > 0$. Suppose the component ζ_t

is observable and the problem is to estimate θ from the observation $\zeta^{(T)} = \{\zeta_t, 0 \le t \le T\}$. It is known that

$$d\zeta_t = a(t, \theta, \zeta)\, dt + \sigma\, dW_t, \qquad \zeta_0 = 0,\ 0 \le t \le T, \tag{2.2.76}$$

where $a(t, \theta, \zeta) = E_\theta(\theta\eta_t | \zeta_s, 0 \le s \le t)$ and $\{W_t\}$ is a standard Wiener process by the innovation theorem (cf. Liptser and Shiryayev 1977, Theorem 7.17). Hence the initial problem of estimation of θ reduces to the problem of estimation by the diffusion type process solving (2.2.76). For details, see Kutoyants (1984b).

Example 2.2.5 (*Diffusion process with ergodic properties*)
Let the observed process $\zeta^{(T)} = \{\zeta_t, 0 \le t \le T\}$ be the solution of the differential equation

$$d\zeta_t = a(\theta, \zeta_t)\, dt + \sigma(\zeta_t)\, dW_t, \qquad \zeta_0 = x,\ 0 \le t \le T,\ \theta \in \Theta, \tag{2.2.77}$$

where $\{W_t, 0 \le t \le T\}$ is the standard Wiener process. Suppose the process $\{\zeta_t\}$ is ergodic. Let

$$B(\theta, x) = 2\int_0^x \frac{a(\theta, y)}{\sigma^2(y)}\, dy, \tag{2.2.78a}$$

$$G(\theta) = \int_{-\infty}^{\infty} e^{B(\theta,x)}\, dx, \text{ and} \tag{2.2.78b}$$

$$\mu_\theta(x) = \frac{1}{G(\theta)}\int_{-\infty}^{x} e^{B(\theta,y)}\, dy. \tag{2.2.78c}$$

Let ζ be a random variable such that $P_\theta[\zeta \le x] = \mu_\theta(x)$. Let $a_\theta(\theta, x)$ be the derivative of $a(\theta, x)$ with respect to θ and $I(\theta) = E_\theta\left[\frac{a_\theta^2(\theta,\zeta)}{\sigma^2(\zeta)}\right]$. Define

$$\phi_T(\theta) = T^{-1/2} I(\theta)^{-1/2}.$$

Let $L_2(\mu_\theta)$ be the space of square-integrable functions with respect to μ_θ and $\|f\|_2^2 = \int_{-\infty}^{\infty} f^2(x)\, \mu_\theta(dx)$. Suppose the following conditions hold:

(F1) $\sup_\theta G(\theta) < \infty$.

(F2) $\frac{a(\theta,\cdot)}{\sigma(\cdot)} \in L_2(\mu_\theta)$, differentiable in θ, and the derivative $\frac{a_\theta(\theta,y)}{\sigma(y)} \in L_2(\mu_\theta)$, continuous in y in $L_2(\mu_\theta)$ at $y = \theta$.

(F3) For all $\theta, \theta_1 \in \Theta$,

$$\|a_\theta(\theta_1, \cdot)/\sigma(\cdot)\|_2 < c < \infty, \qquad I(\theta) > 0.$$

(F4) The convergence

$$P_\theta - \lim_{T\to\infty} \frac{1}{T}\int_0^T \left[\frac{a_\theta(\theta, \zeta_t)}{\sigma(\zeta_t)}\right]^2 dt = I(\theta)$$

is uniform on $\theta \in K$, K compact in Θ.

(F5) There exist constants $\gamma > 0$, $\lambda > 0$ and $C > 0$ such that, for all $T > 0$ and $\theta \in \Theta$,

$$h_\theta^{(T)}(u) \ge C|u|^2,$$

where

$$h_\theta^{(T)}(u) = -\log E_\theta[\exp\{-\lambda\|\Delta f_t^{(T)}(u)\|_2^2\}],$$

with

$$f_t(\cdot,\cdot) = \frac{a(\cdot,\cdot)}{\sigma(\cdot)},$$

$$\Delta f_t^{(T)}(u) = f_t(\theta + \phi_T(\theta)u, X_t) - f_t(\theta, X_t).$$

Under conditions (F1)–(F5), it can be shown that the MLE $\hat{\theta}_T$, as $T \to \infty$, uniformly on $\theta \in K$, is consistent and asymptotically normal, that is,

$$T^{1/2}(\hat{\theta}_T - \theta) \xrightarrow{\mathcal{L}} N(0, I(\theta)^{-1}) \text{ as } T \to \infty;$$

is asymptotically efficient. Furthermore, the absolute moments of $T^{1/2}(\hat{\theta}_T - \theta)$ converge to the corresponding absolute moments of a Gaussian random variable with mean θ and variance $I(\theta)^{-1}$. This result follows from Theorem 2.2.19. An alternate set of conditions are given in Theorems 2.2.1 and 2.2.3 for strong consistency and asymptotic normality.

Local asymptotic mixed normality for the scheme of series

We have already discussed the LAN approximation of experiments for diffusion type models. We now discuss diffusion type processes and their approximations by local asymptotic mixed normal (LAMN) experiments.

Let us consider a statistical experiment $\mathcal{H} = (E, \mathcal{E}, P_\theta; \theta \in \Theta)$ where $(E, \mathcal{E})$ is a measurable space equipped with a family $\{P_\theta, \theta \in \Theta\}$ of probability measures and Θ denotes the parameter space. We assume that Θ is open and contained in R^d. Let us again consider a scheme of series, that is, a family $\mathcal{D}$ of experiments $\mathcal{D} = \{\mathcal{H}_\varepsilon, \varepsilon \in (0, 1]\}$ parameterized by the same parameter space $\Theta : \mathcal{H}_\varepsilon = (E_\varepsilon, \mathcal{E}_\varepsilon, P_\theta^{(\varepsilon)}, \theta \in \Theta)$ as $\varepsilon \to 0$.

A *norming* $\boldsymbol{Q}$ for $\mathcal{D}$ is a family $\{\boldsymbol{Q}_\varepsilon(\theta), \varepsilon \in (0, 1]\}$ of nonsingular $d \times d$ matrices satisfying

$$\|\boldsymbol{Q}_\varepsilon(\theta)\| \to 0 \qquad \text{as } \varepsilon \to 0. \tag{2.2.79}$$

(Here $\|\boldsymbol{A}\| = \sup\{\|\boldsymbol{A}\boldsymbol{x}\| : \|\boldsymbol{x}\| = 1\}$ for any square matrix $\boldsymbol{A}$.)

Given a norming for $\mathcal{D}$, the *localized likelihood field* $Z_{\theta,\varepsilon}$ is defined by

$$Z_{\theta,\varepsilon}(\boldsymbol{u}) = \frac{dP^{(\varepsilon)}_{\theta+\boldsymbol{Q}_\varepsilon(\theta)\boldsymbol{u}}}{dP^{(\varepsilon)}_\theta}, \qquad \boldsymbol{u} \in U_\varepsilon(\theta), \tag{2.2.80}$$

where $U_\varepsilon(\theta) = \{\boldsymbol{u} \in R^d : \theta + \boldsymbol{Q}_\varepsilon(\theta)\boldsymbol{u} \in \Theta\}$.

Definition. A family of experiments $\mathcal{D}$ is called *$\boldsymbol{Q}$-locally approximable at* $\theta \in \Theta$ if there is a random field $Z_\theta = \{Z_\theta(\boldsymbol{u}), \boldsymbol{u} \in R^d\}$ defined on some probability space $(\Omega, \mathcal{F}, P)$ such that $Z_\theta(\boldsymbol{u}) \geq 0$ a.e. and $EZ_\theta(\boldsymbol{u}) = 1$ for each $\boldsymbol{u} \in R^d$ such that

$$Z_{\theta,\varepsilon} \underset{P_\theta^{(\varepsilon)}}{\overset{\text{f.d.}}{\to}} Z_\theta \qquad \text{as } \varepsilon \to 0.$$

$\mathcal{D}$ is said to be *$\boldsymbol{Q}$-locally approximable* if the above holds for any $\theta \in \Theta$ and $\mathcal{D}$ is said to be *locally approximable (LA)* if there exists a suitable norming $\boldsymbol{Q}$ for which the above is true. If $\mathcal{D}$ is $\boldsymbol{Q}$-locally approximable, then $(E, \boldsymbol{Q}, Z)$ is called a *localized model.*

Here $Z_\varepsilon \underset{P^{(\varepsilon)}}{\overset{\text{f.d.}}{\rightarrow}} Z_0$ denotes the convergence in law (weak convergence) of the finite-dimensional distributions of Z_ε under $P^{(\varepsilon)}$ to the corresponding finite-dimensional distribution of Z_0 as $\varepsilon \to 0$ and as the domain of definition U_ε of Z_ε increases to the domain of definition U_0 of Z_0. If Z_ε are $C_0(R^d)$-valued random variables, then $Z_\varepsilon \overset{P^{(\varepsilon)}}{\rightarrow} Z_0$ denotes the weak convergence of Z_ε on $C_0(R^d)$ to Z_0.

In the following discussion, we assume that the localized likelihood field is almost surely continuous.

Definition. A family of experiments $\mathcal{D}$ is said to be *local asymptotic mixed normal* at $\theta \in \Theta$ if there is a norming $\boldsymbol{Q}$ such that, for each $\boldsymbol{u} \in R^d$ and sufficiently small ε,

$$
\begin{aligned}
Z_{\theta,\varepsilon}(\boldsymbol{u}) &= \frac{dP^{(\varepsilon)}_{\theta+\boldsymbol{Q}_\varepsilon(\theta)\boldsymbol{u}}}{dP^{(\varepsilon)}_\theta} \\
&= \exp\{\boldsymbol{\Delta}_\varepsilon(\theta)^T\boldsymbol{u} - \tfrac{1}{2}\boldsymbol{u}^T\boldsymbol{G}_\varepsilon(\theta)\boldsymbol{u} + \psi_\varepsilon(\theta,\boldsymbol{u})\}, \qquad (2.2.81)
\end{aligned}
$$

where the families $\{\boldsymbol{\Delta}_\varepsilon(\theta)\}$, $\{\boldsymbol{G}_\varepsilon(\theta)\}$ and $\{\psi_\varepsilon(\theta,\boldsymbol{u})\}$ are R^d, $R^{d\times d}$ and R-valued random variables respectively defined on the spaces $(E_\varepsilon, \mathcal{E}_\varepsilon, P^{(\varepsilon)}_\theta)$ for $\varepsilon \in (0,1]$ such that the following conditions hold with R^d and $R^{d\times d}$ random vectors $\boldsymbol{\Delta}(\theta)$ and $\boldsymbol{G}(\theta)$, respectively defined on a probability space $(\Omega, \mathcal{F}, P)$:

(i) $\boldsymbol{G}(\theta)$ and $\boldsymbol{G}_\varepsilon(\theta)$ are symmetric positive definite matrices a.s. with respect to P and $P^{(\varepsilon)}_\theta$ respectively for $\varepsilon \in (0,1]$;
(ii) conditionally on $\boldsymbol{G}(\theta)$, the random vector $\boldsymbol{\Delta}(\theta)$ has the distribution $N_d(0, \boldsymbol{G}(\theta))$;
(iii) $\psi_\varepsilon(\theta,\boldsymbol{u}) \to 0$ in $P^{(\varepsilon)}_\theta$-probability as $\varepsilon \to 0$ for each $\boldsymbol{u} \in R^d$; and
(iv) $\mathcal{L}((\boldsymbol{\Delta}_\varepsilon(\theta), \boldsymbol{G}_\varepsilon(\theta))|P^{(\varepsilon)}_\theta) \to \mathcal{L}((\boldsymbol{\Delta}(\theta), \boldsymbol{G}(\theta)))$ as $\varepsilon \to 0$.

The family $\mathcal{D}$ is said to be LAMN if it is LAMN at each $\theta \in \Theta$ and $\mathcal{D}$ is said to be LAMN in the extended sense if (i)–(iv) hold but with 'positive definite' in (i) replaced by 'nonnegative definite' at each $\theta \in \Theta$.

Local asymptotic mixed normality in diffusion type models

Let $(\Omega^{(\varepsilon)}, \mathcal{F}^{(\varepsilon)}, P^{(\varepsilon)})$, $\varepsilon \in (0,1]$, be a family of complete probability spaces endowed with the filtrations $\{\mathcal{F}^{(\varepsilon)}_t, t \geq 0\}$, $\varepsilon \in (0,1]$, and independent standard Wiener processes $\{W^{(\varepsilon)}_t, \varepsilon \in [0,1]\}$. Let $\{a^{(\varepsilon)}(\theta, x), \theta \in \Theta\}$, $\varepsilon \in (0,1]$, denote a family of nonanticipative functionals on $[0, T_\varepsilon] \times C[0, T_\varepsilon]$. Consider the family of stochastic differential equations

$$
dX^{(\theta,\varepsilon)}_t = a^{(\varepsilon)}(\theta, X^{(\theta,\varepsilon)})_t\, dt + \sigma_\varepsilon\, dW^{(\varepsilon)}_t, \qquad X^{(\theta,\varepsilon)}_0 = X^{(\theta,\varepsilon)}, \ \sigma_\varepsilon > 0. \qquad (2.2.82)
$$

Let $P^{(\varepsilon)}_\theta$ denote the corresponding probability measure on $C[0, T_\varepsilon]$. We assume that the measures $\{P^{(\varepsilon)}_\theta, \theta \in \Theta\}$ are equivalent for any fixed $\varepsilon \in (0,1]$. In fact, for $\theta \neq \theta' \in \Theta$,

$$
\begin{aligned}
L_{T_\varepsilon}(\theta', \theta) &\equiv \frac{dP^{(\varepsilon)}_{\theta'}}{dP^{(\varepsilon)}_\theta}(X^{(\theta,\varepsilon)}) \\
&= \exp\left\{-\frac{1}{2}\int_0^{T_\varepsilon} \Delta a^{(\varepsilon)}[\theta', \theta]^2 \sigma_\varepsilon^{-2}\, d\lambda + \int_0^{T_\varepsilon} \Delta a^{(\varepsilon)}[\theta', \theta]\sigma_\varepsilon^{-1}, dW^{(\varepsilon)}\right\}
\end{aligned}
\qquad (2.2.83)
$$

where

$$\Delta a^{(\varepsilon)}[\theta', \theta] = a^{(\varepsilon)}(\theta', X^{(\theta,\varepsilon)}) - a^{(\varepsilon)}(\theta, X^{(\theta,\varepsilon)})$$

and λ is the Lebesgue measure.

Theorem 2.2.21 *Let $(\Omega^{(\varepsilon)}, \mathcal{F}^{(\varepsilon)}, P^{(\varepsilon)})$ be the same probability space for every $\varepsilon \in (0, 1]$, and denote it by $(\Omega, \mathcal{F}, P)$. Suppose there exist a norming $\{\boldsymbol{Q}_\varepsilon(\theta), \varepsilon \in (0, 1]\}$ and a family $\{\boldsymbol{d}^{(\varepsilon)}(\theta, x)\}$ of nonanticipative functionals on $[0, T_\varepsilon] \times C[0, T_\varepsilon], \varepsilon \in (0, 1]$, satisfying the following conditions.*

(G1) $P\{\int_0^{T_\varepsilon} \|\boldsymbol{d}^{(\varepsilon)}(\theta, X)\|^2 \, d\lambda < \infty\} = 1, \varepsilon \in (0, 1].$

(G2) For each $\boldsymbol{u} \in R^d$,

$$\int_0^{T_\varepsilon} (a^{(\varepsilon)}(\theta + \boldsymbol{Q}_\varepsilon(\theta)\boldsymbol{u}) - a^{(\varepsilon)}(\theta) - \boldsymbol{d}^{(\varepsilon)}(\theta, X)^T \boldsymbol{Q}_\varepsilon(\theta)\boldsymbol{u})^2 \sigma_\varepsilon^{-2} \, d\lambda$$

$$\xrightarrow{P^{(\varepsilon)}} 0, \qquad \text{as } \varepsilon \to 0$$

where $a^{(\varepsilon)}(\theta') = a^{(\varepsilon)}(\theta', X^{(\theta,\varepsilon)})$.

(G3) There is a family $\{\psi_\varepsilon\}$ of mappings $[0, 1] \to [0, T_\varepsilon]$ satisfying $\psi_\varepsilon(0) = 0, \psi_\varepsilon(1) = T_\varepsilon$, such that:

(i) for some $\zeta \subset \mathcal{F}$, there is a continuous $d \times d$ matrix-valued increasing process $\boldsymbol{G}(\theta) = (\boldsymbol{G}(\theta)_t, 0 \leq 1)$ on the space $(\Omega, \mathcal{F}, P)$ for which

$$\boldsymbol{Q}_\varepsilon(\theta)^T \left\{ \int_0^{\psi_\varepsilon} \boldsymbol{d}^{(\varepsilon)}(\theta, X)^T \sigma_\varepsilon^{-2} \boldsymbol{d}^{(\varepsilon)}(\theta, X) d\lambda \right\} \boldsymbol{Q}_\varepsilon(\theta) \to \boldsymbol{G}(\theta)_t$$

in P-probability as $\varepsilon \to 0$ for every $t \in [0, 1]$ (here $\boldsymbol{A}^T$ denotes the transpose of $\boldsymbol{A}$); and

(ii) there is a mapping $\tau : (0, 1] \to (0, 1]$ with $\tau(\varepsilon) \downarrow 0$ as $\varepsilon \downarrow 0$ such that the filtration $F^{(\varepsilon)} = \{\mathcal{F}_t^{(\varepsilon)}\}$ satisfies

$$\varepsilon \geq \varepsilon' \Rightarrow \mathcal{F}^{(\varepsilon)}_{\psi_\varepsilon(\tau_\varepsilon)} \subset \mathcal{F}^{(\varepsilon')}_{\psi_{\varepsilon'}(\tau_{\varepsilon'})}$$

and $\zeta = \vee_{\varepsilon \in (0,1]} \mathcal{F}^{(\varepsilon)}_{\psi_\varepsilon(\tau_\varepsilon)}$.

Then the family of experiments $\mathcal{D}$ is LAMN at θ with asymptotic variance $\boldsymbol{G}(\theta)_1$ if, in addition, $\int_0^{T_\varepsilon} \boldsymbol{d}^{(\varepsilon)}(\theta, X)^T \sigma_\varepsilon^{-2} \boldsymbol{d}^{(\varepsilon)}(\theta, X) \, d\lambda$ and $\boldsymbol{G}(\theta)_1$ are positive definite almost surely.

Remarks. For a proof, see Dietz (1989), who gives further verifiable sufficient conditions for the locally asymptotic mixed normality and studies the asymptotic properties of the MLE. We omit the details. An alternate set of sufficient conditions for locally asymptotic mixed normality for diffusion type processes is given in Linkov (1989).

Linkov (1989) proved the following result concerning local asymptotic mixed normality for processes of diffusion type.

Let $\{X_t, t \geq 0\}$ be a process of diffusion type, that is, a solution of the stochastic differential equation

$$dX_t = a_t(\theta) \, dt + b_t \, dW_t, \qquad X_0 = \zeta, \; t \geq 0, \tag{2.2.84}$$

where ζ is an $\mathcal{F}_0$-measurable random variable independent of $\theta \in \Theta \subset R^k$ and $W(\cdot)$; $a_t(\theta) = a(t, X, \theta)$ and $b_t = b(t, X)$ are measurable functions of X; and θ is the unknown parameter. Suppose that the coefficients $a_t(\theta)$ and b_t satisfy the existence and uniqueness conditions for a strong solution of the equation. Further suppose that conditions hold so that $\{P_\theta^t, \theta \in \Theta\}$ are mutually equivalent for $t > 0$, where P_θ^t is the measure generated by the process X on $C[0, t]$. We know that

$$\begin{aligned}\ell_t(\theta', \theta) &= \log \frac{dP_{\theta'}^t}{dP_\theta^t}\\ &= \int^t \lambda_s(\theta', \theta)\, dW_s - \tfrac{1}{2}\int_0^t \lambda_s^2(\theta', \theta)\, ds, \end{aligned}\tag{2.2.85}$$

where $\lambda_s(\theta', \theta) = b_s^+(a_s(\theta') - a_s(\theta))$ and $b_s^+ = b_s^{-1}$ if $b_s \neq 0$ and $b_s^+ = 0$ if $b_s = 0$.

Theorem 2.2.22 *Suppose the following conditions hold:*

(i) $a_s(\theta) \in C_0^1(\Theta)$ *and for all y in a neighborhood of* θ,

$$P_\theta\left\{\int_0^T \|g_s(y)\|^2\, ds < \infty\right\} = 1, \qquad g_s(y) = \frac{\partial}{\partial y}(a_s(y) b_s^T),$$

with $\frac{\partial}{\partial \boldsymbol{y}} = \left(\frac{\partial}{\partial y_1}, \ldots, \frac{\partial}{\partial y_k}\right)^T$, $\boldsymbol{y} = (y_1, \ldots, y_k)^T$.

(ii) As $t \to \infty$,

$$\mathcal{L}\left(Q_t^2 \int_0^t g_s(\theta) g_s(\theta)^T\, ds | P_\theta\right) \xrightarrow{w} \mathcal{L}(\boldsymbol{K}^2 | P_\theta), \tag{2.2.86}$$

where $Q_t \to 0$ *as* $t \to \infty$ *and* $\boldsymbol{K}$ *is a symmetric positive definite random matrix such that*

$$P_\theta(\boldsymbol{\lambda}^T \boldsymbol{K} \boldsymbol{\lambda} > 0) = 1 \qquad \text{for all } \boldsymbol{\lambda} \in R^k,\ \|\boldsymbol{\lambda}\| \neq 0.$$

(iii) For all $\boldsymbol{u} \in R^k$,

$$P_\theta - \lim_{t\to\infty} Q_t^2 \sup_{v\in[0,1]} \int_0^t \|g_s(\theta + vQ_t\boldsymbol{u}) - g_s(\theta)\|^2\, ds = 0, \tag{2.2.87}$$

where $P_\theta - \lim_{t\to\infty}$ *denotes limit in probability under* P_θ *as* $t \to \infty$.

Then, for all $\boldsymbol{u} \in R^k$,

$$\ell_t(\theta + Q_t\boldsymbol{u}, \theta) = \boldsymbol{u}^T \boldsymbol{\Delta}_t - \tfrac{1}{2}\boldsymbol{u}^T \boldsymbol{K}_t^2 \boldsymbol{u} + r_t(\boldsymbol{u}), \tag{2.2.88}$$

where

$$\mathcal{L}((\boldsymbol{\Delta}_t, \boldsymbol{K}_t^2) | P_\theta) \xrightarrow{w} \mathcal{L}((\boldsymbol{K}\boldsymbol{\eta}, \boldsymbol{K}^2) | P_\theta) \qquad \text{as } t \to \infty, \tag{2.2.89a}$$

$$P_\theta - \lim_{t\to\infty} r_t(\boldsymbol{u}) = 0, \tag{2.2.89b}$$

and $\boldsymbol{\eta} = (\eta_1, \ldots, \eta_k)^T$ *is a Gaussian random vector with independent standard normal components under* P_θ *with* $\boldsymbol{\eta}$ *independent of the random matrix* $\boldsymbol{K}$.

Example 2.2.6 (*Generalized Ornstein–Uhlenbeck Process (Dietz 1992)*)
Consider the stochastic differential equation of the form

$$dX_t = \rho\alpha\left\{\int_0^t e^{-\alpha s}X_{t-s}\,ds\right\}dt + \sigma\,dW_t, t \in [0,T], X_0 = X_0, \tag{2.2.90}$$

where $\sigma > 0$ is a known constant and let $\theta = (\alpha, \rho) \in R_+ \times R$. The process $\{X_t, t \geq 0\}$ is a diffusion type process solving an equation of the type

$$dX_t = a(t, X)\,dt + b(t, X)\,dW_t$$

with X_0 given, and the nonanticipating coefficients a and b here depend on the whole 'past' of the process. Note that the process $\{X_t\}$ is not a Markov process. The linear stochastic differential equation

$$dX_t = \rho X_t\,dt + \sigma\,dW_t, \qquad t \in [0,T],\ X_0 = X_0, \tag{2.2.91}$$

gives rise to the classical Ornstein–Uhlenbeck process with parameters ρ and $\sigma > 0$ and with Markov property which can be considered as the limit of the processes defined by (2.2.90) as $\alpha \to \infty$. Observe that the drift term in (2.2.90) depends on the complete observation of the process over the interval $[0, t]$, whereas the drift term in (2.2.91) depends on the observation of the process at time t. This suggests the interpretation that there is a 'loss of memory' property in the limiting case of (2.2.90). In fact the solution $X^{(\rho,\alpha)}$ of (2.2.90) converges to X^ρ of (2.2.91) as $\alpha \to \infty$ in a strong sense (Dietz 1987). It is known that the family of measures $\{P_\rho\}$ generated by (2.2.91) is LAN for $\rho < 0$ ('ergodic' case) and LAMN if $\rho > 0$ ('non ergodic' case), while at $\rho = 0$ there is a singularity (see Feigin 1979; Basawa and Prakasa Rao 1980; Basawa and Scott 1983; Kutoyants 1984a).

Using Itô's formula, it can be shown (Dietz 1987) that the solution of (2.2.90) can be explicitly given as

$$X_t = Y_t(\theta)X_0 + \int_0^t Y_{t-s}(\theta)\sigma\,dW_s, \tag{2.2.92}$$

where

$$y_t(\theta) = \begin{cases} \dfrac{\lambda_2 e^{\lambda_1 t} - \lambda_1 e^{\lambda_2 t}}{\lambda_2 - \lambda_1} & \text{if } \alpha \neq 0 \text{ and } \rho \neq -\frac{\alpha}{4} \\ e^{-(\alpha/2)t}\left(1 + \frac{\alpha}{2}t\right) & \text{if } \alpha = 0 \text{ or } \rho = -\frac{\alpha}{4}, \end{cases} \tag{2.2.93}$$

and $\lambda_1 = -\frac{\alpha}{2} - \Delta, \lambda_2 = -\frac{\alpha}{2} + \Delta$ with $\Delta = \sqrt{\frac{\alpha^2}{4} + \rho\alpha}$. Note that λ_1 and λ_2 are the roots of the equation

$$\lambda^2 + \alpha\lambda - \rho\alpha = 0.$$

Furthermore, the process $\{X_t\}$ is almost surely continuous and admits moments up to the same order as X_0 does. In fact $\{X_t\}$ is a Gaussian process if X_0 is a normal random variable. In addition,

$$E(X_t)^2 = y_t^2(\theta)EX_0^2 + \sigma^2\int_0^t y_s^2(\theta)\,ds, \qquad t \geq 0,$$

and

$$\operatorname{cov}(X_t, X_s) = y_s(\theta)y_t(\theta)\operatorname{var}(X_0) + \sigma^2\int_0^{t\wedge s} y_{s-r}(\theta)y_{t-r}(\theta)\,dr, \qquad s,t \geq 0$$

(cf. Dietz 1987). Let

$$a_t(\theta, x) = \rho c_t(\alpha, x),$$

with

$$c_t(\alpha, x) = \alpha \int_0^t e^{-\alpha(t-s)} x_s \, ds, \qquad x \in C[0, T].$$

Equation (2.2.90) can be written as

$$dX_t = a_t(\theta, X)\, dt + \sigma \, dW_t, \qquad X_0 = X_0, \; t \in [0, T].$$

Let P_θ^T be the measure induced by the process $\{X_t\}$ on $C[0, T]$ when θ is the parameter. The process $a_t(\theta, X)$ is almost surely continuous and

$$\int_0^T a_t(\theta', X)^2 \, dt < \infty$$

a.s. $[P_\theta]$ for all θ' and θ. Hence the measures $\{P_\theta^T, \theta \in \Theta\}$ are equivalent with

$$\begin{aligned} L_T(\theta', \theta) &\equiv \frac{dP_{\theta'}^T}{dP_\theta^T}(X) \\ &= \exp\left\{ -\frac{1}{2} \int_0^T \left\{ \frac{a^2(\theta', X) - a^2(\theta, X)}{\sigma^2} \right\} dt \right. \\ &\quad \left. + \int_0^T \frac{(a(\theta', X) - a(\theta, X))}{\sigma^2} \, dX_t \right\}. \end{aligned}$$

It can be checked that the MLE $\hat{\theta}_T$ is given by

$$\hat{\theta}_T = \underset{\theta' \in \Theta}{\arg\max} \exp\left\{ -\tfrac{1}{2} \int_0^T a(\theta', X)^2 \, d\lambda + \int_0^T a(\theta', X) \, dX_t \right\}, \tag{2.2.94}$$

where X denotes the process corresponding to the true parameter θ. The following result is due to Dietz (1992). We omit the details. Consider the family of experiments $\mathcal{D} = (C[0, T], \mathcal{C}_T, P_\theta^T, \theta \in \Theta)$, $T > 0$, where $\mathcal{C}_T$ is the σ-algebra generated by the supremum norm on $C[0, T]$:

(i) In the ergodic case, that is, for $\theta \in (0, \infty) \times (-\infty, 0)$ the family of experiments $\mathcal{D}$ is LAN, with norming

$$Q_T(\theta) = T^{-1/2} \boldsymbol{\Phi}(\theta)^{-1/2}, \qquad \text{where } \boldsymbol{\Phi}(\theta) = \boldsymbol{B}(\theta) \boldsymbol{D}(\theta) \boldsymbol{B}(\theta)^T,$$

with

$$\boldsymbol{B}(\theta) = \begin{cases} \dfrac{1}{2\Delta} \begin{pmatrix} \lambda_1 - \rho & \rho - \lambda_2 & 2\Delta \\ -\alpha & \alpha & 0 \end{pmatrix} & \text{if } \rho \neq -\frac{\alpha}{4}, \\ \begin{pmatrix} -1 & -1 & \alpha/4 \\ 0 & 0 & \alpha \end{pmatrix} & \text{if } \rho = -\frac{\alpha}{4}, \end{cases}$$

$$\boldsymbol{D}(\theta) = \begin{cases} \begin{pmatrix} -\frac{1}{2\lambda_1} & \frac{1}{\alpha} & \frac{1}{\alpha - \lambda_1} \\ \frac{1}{\alpha} & -\frac{1}{2\lambda_2} & \frac{1}{\alpha - \lambda_2} \\ \frac{1}{\alpha - \lambda_1} & \frac{1}{\alpha - \lambda_2} & \frac{1}{2\alpha} \end{pmatrix} & \text{if } \rho \neq -\frac{\alpha}{4}, \\ \frac{1}{\alpha} \begin{pmatrix} 1 & \frac{2}{3} & \frac{1}{\alpha} \\ \frac{2}{3} & \frac{1}{2} & \frac{4}{9\alpha} \\ \frac{1}{\alpha} & \frac{4}{9\alpha} & \frac{2}{\alpha^2} \end{pmatrix} & \text{if } \rho = -\frac{\alpha}{4}. \end{cases}$$

(ii) In the nonergodic case, that is, for $\theta \in (0, \infty) \times (0, \infty)$, $\mathcal{D}$ is LAMN in the extended sense with norming

$$\boldsymbol{Q}_T(\theta) = \sigma\sqrt{2\lambda_2}e^{-\lambda_2 T}\boldsymbol{I},$$

where $\boldsymbol{I}$ is the identity matrix and the corresponding asymptotic variance is

$$\boldsymbol{G}(\theta) = Z^2(\theta)\tilde{\boldsymbol{G}}(\theta),$$

where

$$Z(\theta) = X_0 + \int_0^\infty e^{-\lambda_2 s}\sigma\, dW_s \tag{2.2.95}$$

and

$$\tilde{\boldsymbol{G}}(\theta) = \frac{1}{4\Delta^2}\begin{pmatrix} (\rho-\lambda_2)^2 & \alpha(\rho-\lambda_2) \\ \alpha(\rho-\lambda_2) & \alpha^2 \end{pmatrix}. \tag{2.2.96}$$

Note that λ_1, λ_2 and Δ depend on θ. The matrix $\tilde{\boldsymbol{G}}(\theta)$ has rank equal to 1 for all θ and the family of experiments $\mathcal{D}$ is LAMN only in the extended sense.

Let

$$Z_{\theta,T}(\boldsymbol{u}) = \frac{dP^T_{\theta+\boldsymbol{Q}_T(\theta)\boldsymbol{u}}}{dP^T_\theta}.$$

It follows from the earlier discussion that, if $\rho \neq 0$, then

$$Z_{\theta,T}(\boldsymbol{u}) \xrightarrow{\mathcal{D}} \exp\{\boldsymbol{\Delta}(\theta)\boldsymbol{u} - \tfrac{1}{2}\boldsymbol{u}^T\boldsymbol{G}(\theta)\boldsymbol{u}\} \tag{2.2.97}$$

and the MLE $\hat{\boldsymbol{u}}$ for $\boldsymbol{u}$ in the limiting experiment is given by the equation

$$\boldsymbol{\Delta}(\theta)^T = \boldsymbol{G}(\theta)\hat{\boldsymbol{u}}.$$

In the ergodic case ($\rho < 0$), $\boldsymbol{G}(\theta)$ is nonsingular and the equation has a unique solution

$$\hat{\boldsymbol{u}} = \boldsymbol{G}(\theta)^{-1}\boldsymbol{\Delta}(\theta)^T \tag{2.2.98}$$

and, under some additional conditions,

$$\boldsymbol{Q}_T^{-1}(\hat{\theta}_T - \theta) \xrightarrow{\mathcal{L}} N(0, \boldsymbol{G}(\theta)^{-1}) \quad \text{as } T \to \infty. \tag{2.2.99}$$

Remarks. (i) Consider the Ornstein–Uhlenbeck process satisfying the stochastic differential equation

$$dX_t = -\theta X_t\, dt + dW_t, \qquad X_0 = 0,\ t \geq 0, \tag{2.2.100}$$

where $\theta \in (\alpha, \infty)$, $\alpha > 0$. We have seen earlier that the MLE $\hat{\theta}_T$ of θ based on $X^T = \{X_t, 0 \leq t \leq T\}$ has an asymptotically normal distribution as $T \to \infty$. Mishra and Prakasa Rao (1985b) used the Skorokhod embedding technique and obtained the rate of convergence for the normal approximation. They proved that

$$P_\theta(|\hat{\theta}_T - \theta) > \varepsilon) = O(T^{-1/5})$$

and

$$\sup_x \left| P_\theta\left\{\left(\frac{T}{2\theta}\right)^{1/2}(\hat{\theta}_T - \theta) \leq x\right\} - \Phi(x)\right| = O(T^{-1/5}).$$

Recently Bishwal and Bose (1995) proved that

$$\sup_{x\in R}\left|P_\theta\left(\left(\frac{T}{2\theta}\right)^{1/2}(\hat{\theta}_T-\theta)\le x\right)-\Phi(x)\right| = O(T^{-1/2}(\log T)^{1/2}),$$

and that there exists a constant $C > 0$, not depending on $\varepsilon > 0$, such that

$$P_\theta(|\hat{\theta}_T-\theta|>\varepsilon) = O(e^{-CT}).$$

(ii) Consider the stochastic differential equation

$$dX_t = a(\theta, X_t)\,dt + \varepsilon dW_t, \qquad X(0) = x_0,\ 0 \le t \le T. \tag{2.2.101}$$

Let $\hat{\theta}_\varepsilon$ denote an MLE of θ based on $X^T = \{X_t, 0 \le t \le T\}$. Note that $\hat{\theta}_\varepsilon$ is a solution of the equation

$$L(\hat{\theta}_\varepsilon, \theta_0; X^T) = \sup_{\theta\in\Theta} L(\theta, \theta_0; X^T), \tag{2.2.102}$$

where θ_0 is an arbitrary but fixed value of θ, and that $\hat{\theta}_\varepsilon$ satisfies

$$\frac{\partial}{\partial\theta}\log L(\theta,\theta_0; X^T)|_{\theta=\hat{\theta}_\varepsilon} = 0 \tag{2.2.103}$$

under some regularity conditions. Here $L(\theta, \theta_0; X^T)$ is the likelihood function. Let A be a measurable set such that the MLE $\hat{\theta}_\varepsilon$ is one of the solutions of (2.2.102) for $\omega \in A_1$, and A_2 be a measurable set such that (2.2.103) has a unique solution a.s. Then, on $A_1 \cap A_2$, the solutions of (2.2.102) and (2.2.103) coincide with probability one and give the MLE on this set. Equation (2.2.103) defines the MLE implicitly by the relation $\hat{\theta}_\varepsilon = f(\varepsilon)$ for suitable f. Applying Taylor's formula, we obtain

$$\hat{\theta}_\varepsilon = \sum_{j=0}^{k}\psi_j(0)\frac{\varepsilon^j}{j!} + \frac{\varepsilon^{k+1}}{(k+1)!}\psi_{k+1}(\varepsilon_0) \tag{2.2.104}$$

for $\omega \in A_1 \cap A_2$, where $\psi_j(\varepsilon) = \frac{\partial^j f(\varepsilon)}{\partial\varepsilon^j}$. Kutoyants (1982) studies sufficient conditions for the validity of the expansion of the MLE in the form (2.2.104) under which there exists a measurable set A_3 such that

$$P[\cup_{j=1}^{3}\bar{A}_j] \le C\exp\{-c|\varepsilon|^{-1}\}, \tag{2.2.105}$$

where C, c and γ are positive constants and, on $\omega \in A_1 \cap A_2 \cap A_3$, the last term in (2.2.104) is small compared to the remaining terms. Here $\bar{A}$ denotes the complement of the set A. Earlier work in this area is due to Burnashev (1977).

(iii) Consider the stochastic differential equation

$$dX_t = a(\theta, X_t, t)\,dt + \sigma(X_t, t)\,dW_t, \qquad X_0 = 0. \tag{2.2.106}$$

Suppose that θ_0 is the true parameter. Assume that the usual conditions hold, so that the log-likelihood function is given by

$$\ell_t(\theta, X) = \int_0^t g(\theta, X_s, s)\,dW_s - \tfrac{1}{2}\int_0^t g^2(\theta, X_s, s)\,ds, \tag{2.2.107}$$

where

$$g(\theta, x, s) = \frac{a(\theta, x, s) - a(\theta_0, x, s)}{\sigma(x, s)}. \tag{2.2.108}$$

Let $U_t = \nabla \ell_t$. Then $\{U_t(\theta_0)\}$ is a continuous martingale with respect to the filtration $\{\mathcal{F}_t\}$ generated by $\{W_t, t \geq 0\}$. Let $I_t = \langle U \rangle_t$. Then I_t is an increasing process. Suppose that $I_t \uparrow \infty$ a.s. Let $\tilde{\theta}_t$ be an estimator of θ_0. Then $\tilde{\theta}_t$ is called *first-order efficient* if there exists a nonrandom constant $\gamma = \gamma(\theta_0)$ such that

$$I_t^{1/2} |\tilde{\theta}_t - \theta_0 - \gamma I_t^{-1} U_t| \xrightarrow{p} 0 \qquad \text{as } t \to \infty. \tag{2.2.109}$$

It is known that an MLE $\hat{\theta}_t$ is first-order efficient with $\gamma = 1$ (cf. Hall and Heyde 1980; Basawa and Prakasa Rao 1980). If $\tilde{\theta}_t$ is first-order efficient, then

$$I_t^{1/2} (\tilde{\theta}_t - \theta_0) \xrightarrow{\mathcal{L}} N(0, \gamma^2) \qquad \text{as } t \to \infty. \tag{2.2.110}$$

Let h be a nondecreasing continuous function on $[0, \infty)$ such that:

(H1)

$$\lim_{t \to \infty} \frac{\sqrt{t \log \log t}}{h(t)} = 0.$$

Further suppose that:

(H2) there exists $b = b(h) \in (0, \infty)$, such that

$$\limsup_{t \to \infty} \frac{t}{h^2(t)} \inf_{s \geq 0} \frac{h^2(t+s)}{s} = b.$$

Conditional law The following result is due to Levanony (1994), giving a large-deviation law for the tail probabilities of $\tilde{\theta}_t - \theta_0$.

Theorem 2.2.23 *Suppose an estimator $\tilde{\theta}_t$ of θ_0 is first-order efficient with a constant γ and let $h \in C[0, \infty)$, $h(t) \uparrow \infty$, satisfy conditions (H1) and (H2). Assume further that:*

(H3) $\lim_{t \to \infty} a_t \log P(\sup_{r \geq t} I_r^{1/2} |\tilde{\theta}_r - \theta_0 - \gamma I_r^{-1} U_r| > \varepsilon / \sqrt{a_t} | \mathcal{F}_t) = -\infty$ *a.s.*

for any $\mathcal{F}_t$-adapted process $\{a_t\}$, $0 < a_t \to 0$ and for all $\varepsilon > 0$. Let

$$\phi_t = \sup_{s \geq 0} \frac{s}{h^2(I_t + s)}.$$

Then

$$\lim_{t \to \infty} \phi_t \log P\left(\sup_{r \geq t} \frac{I_r}{h(I_r)} |\tilde{\theta}_r - \theta_0| > \lambda | \mathcal{F}_t \right) = -\frac{1}{2} \frac{\lambda^2}{\gamma^2} \text{ a.s.}$$

for all $\lambda > 0$.

Proof of this result is based on a large-deviation result for continuous local martingales (cf. Prakasa Rao 1999). We omit the details. We assume that condition (H3) holds in the following examples and the estimator $\tilde{\theta}_t$ is the MLE $\hat{\theta}_t$.

Example 2.2.7
Let $h(s) = s^{(1+\nu)}/2$. Then

$$\phi_t = \sup_{s \geq o} \frac{s}{h^2(I_t + s)} = \frac{1}{C(\nu) I_t^{\nu}},$$

where $C(\nu) = \frac{(1+\nu)^{1+\nu}}{\nu^{\nu}}$, and

$$P\Big(\sup_{r\geq t} I_r^{(1-\nu)/2} |\hat{\theta}_r - \theta_0| > \lambda | \mathcal{F}_t\Big) \simeq \exp\left(-\frac{C(\nu)}{2} I_t^{\nu} \lambda^2\right).$$

If $\nu = 1$, then $C(\nu) = 4$ and

$$P\Big(\sup_{r\geq t} |\hat{\theta}_r - \theta_0| > \lambda | \mathcal{F}_t\Big) \simeq \exp(-2 I_t \lambda^2).$$

Example 2.2.8
Let $h(s) = \sqrt{s \log s}$, where $s > 1$. Then condition (H2) holds with

$$b = \lim_{t\to\infty} \inf_{s\geq 0} \frac{(t+s)\log(t+s)}{s \log t} = 1,$$

which implies that

$$P\left(\sup_{s\geq t} \left(\frac{I_r}{\log I_r}\right)^{1/2} |\hat{\theta}_r - \theta_0| > \lambda | \mathcal{F}_t\right) \simeq \exp\{-\tfrac{1}{2}\lambda^2 \log I_t\} = I_t^{-\lambda^2/2}.$$

Example 2.2.9 (*Ergodic case*)
Suppose equation (2.2.106) is time homogeneous, that is, the drift and the diffusion functions $a(\cdot,\cdot,\cdot)$ and $b(\cdot,\cdot)$ do not depend on t. Let μ_θ be the invariant measure for each θ, assuming that it exists, such that

$$\lim_{T\to\infty} \frac{1}{T} \int_0^T f(\theta', X_s^{\theta})\, ds = \int_{-\infty}^{\infty} f(\theta', x)\, \mu_\theta(dx) \text{ a.s.},$$

for every $f : R^2 \to R$.

Let $h(t) = t$. Then

$$\phi_t = \frac{1}{4I_t} = \frac{1}{4\int_0^T a_\theta^2(\theta_0, X_s)\, ds}$$

and it follows from Theorem 2.2.23 that

$$\lim_{t\to\infty} \frac{1}{t} \log P\Big(\sup_{r\geq t} |\hat{\theta}_r - \theta_0| > \lambda | \mathcal{F}_t\Big) = -\frac{2\lambda^2}{\gamma^2} E_{\theta_0}(a_\theta^2(\theta_0, Z)),$$

where Z is a random variable with measure μ_{θ_0}.

Unconditional law Let θ be fixed and $\tilde{\theta}_t(X^\theta)$ be an estimator of θ based on the observation of the process $\{X_s^\theta\}$ over $[0, t]$ when θ is the true parameter. Let $\theta_0 \in \Theta$,

$$g(\theta, \cdot, \cdot) = \frac{a(\theta, \cdot, \cdot) - a(\theta_0, \cdot, \cdot)}{\sigma(\cdot, \cdot)}$$

and

$$G_t(\theta, X^\theta) = \int_0^t g(\theta, X_s^\theta, s)\, dW_s.$$

Then $G(\theta, X^\theta)$ is a martingale with its increasing process

$$\langle G(\theta, X^\theta)\rangle_t = \int_0^t g^2(\theta, X_s^\theta, s)\, ds,$$

and

$$\ell_t(\theta, X^\theta) = \log \frac{dP_t^\theta}{dP_t^{\theta_0}}(X^\theta) = G_t(\theta, X^\theta) + \tfrac{1}{2}\langle G(\theta, X^\theta)\rangle_t, \tag{2.2.111}$$

where P_t^θ is the measure induced by X^θ on $C[0, t]$.

Suppose the following condition holds:

(H4) There exists non random $\{\alpha_t\}$, $0 < \alpha_t \to 0$ such that

(i) $\liminf_{t\to\infty} \alpha_t \langle G(\theta, X^\theta)\rangle_t > 0$ a.s. for all $\theta \neq \theta_0$, and
(ii) $\limsup_{t\to\infty} \alpha_t \langle G(\theta, X^\theta)\rangle_{t+T} < \infty$ a.s. for all θ and $T < \infty$.

Define

$$K(\theta, T) = \inf\{a \mid \lim_{t\to\infty} P(\alpha_t \ell_{t+T}(\theta, X^\theta) \leq a) = 1\}.$$

Condition (H4), relation (2.2.111) and the martingale law of large numbers leading to

$$\frac{G_t(\theta, X^\theta)}{\langle G(\theta, X^\theta)\rangle_t} \to 0 \text{ a.s.} \qquad \text{as } t \to \infty$$

imply that

$$\liminf_{t\to\infty} \alpha_t \ell_{t+T}(\theta, X^\theta) = \tfrac{1}{2}\liminf_{t\to\infty} \alpha_t \langle G(\theta, X^\theta)\rangle_{t+T} > 0 \text{ a.s.}$$

for all $T < \infty$ and $\theta \neq \theta_0$. We can redefine $K(\theta, T)$ as

$$K(\theta, T) = \inf\{a > 0 \mid \lim_{t\to\infty} P(\tfrac{1}{2}\alpha_t \langle G(\theta, X^\theta)\rangle_{t+T} \leq a) = 1\}.$$

The following lemma follows from Theorem 2.1 of Bahadur *et al.* (1980).

Lemma 2.2.24 *Suppose that condition (H4) holds. Let* $\{A_t^T(X^\theta), t \geq 0\}$ *be a sequence of* $\mathcal{F}_{t+T}$*-measurable sets such that* $\liminf_{t\to\infty} P(A_t^T(X^\theta)) > 0$ *for all* $T \in [0, \infty)$. *Then* $\liminf_{t\to\infty} \alpha_t \log P(A_t^T(X)) \geq -K(\theta, T)$ *for all* $T \in [0, \infty)$.

As an application of the above lemma, the following result is proved by Levanony (1994) giving a large-deviation type lower bound. We omit the proof.

Theorem 2.2.25 *Suppose that condition (H4) holds and $\tilde{\theta}_t$ is a uniformly strongly consistent estimator. Then*

$$\liminf \alpha_t \log P(\sup_{r\geq t} |\tilde{\theta}_r - \theta_0| > \lambda) \geq - \inf_{|\theta-\theta_0|>\lambda} K(\theta, 0)$$
$$\geq - \inf_{|\theta-\theta_0|>\lambda} \bar{K}(\theta),$$

where $\bar{K}(\theta) = \frac{1}{2}$ ess. sup lim sup$_{t\to\infty}$ $\alpha_t \langle G(\theta, X^\theta)\rangle_t$.

As an application of Theorem 2.2.25, we have the following:

Example 2.2.10 (*Example 2.2.9 continued*)
Suppose that ergodic case holds. Let $\alpha_t = 1/t$. Then

$$\lim_{t\to\infty} \frac{1}{t}\int_0^t g^2(\theta, X_s^\theta)\, ds = E_\theta(g^2(\theta, Z)) = K(\theta, 0) = \bar{K}(\theta),$$

where Z has the invariant measure $\mu(\theta)$. An application of Theorem 2.2.25 gives

$$\liminf_{t\to\infty} \frac{1}{t} \log P\Big(\sup_{r\geq t} |\tilde{\theta}_r - \theta_0| > \lambda\Big) \geq -\tfrac{1}{2} \inf_{|\theta-\theta_0|>\lambda} E_\theta(g^2(\theta, Z))$$

for any uniformly consistent estimator $\tilde{\theta}$. For sufficiently small λ,

$$E_\theta(g^2(\theta, Z)) = |\theta - \theta_0|^2 E_\theta a_\theta^2(\theta_0, Z) + o(|\theta - \theta_0|^2).$$

Since $E_\theta a_\theta^2(\theta_0, Z) \to E_{\theta_0} a_\theta^2(\theta_0, Z)$ $as \theta \to \theta_0$, it follows that

$$\lim_{\lambda\downarrow 0} \liminf_{t\to\infty} \frac{1}{\lambda^2 t} \log P\Big(\sup_{r\geq t} |\tilde{\theta}_r - \theta_0| > \lambda\Big) \geq -\tfrac{1}{2} E_{\theta_0} a_\theta^2(\theta_0, Z).$$

2.3 Minimum Contrast Method for the Estimation of the Drift Parameter

Let us first consider some properties of the diffusion process $\{X_t\}$ satisfying the stochastic differential equation

$$dX_t = a(X_t)\, dt + \sigma(X_t)\, dW_t, \qquad t \geq 0. \tag{2.3.1}$$

Let

$$B(x) = 2\int_0^x \frac{a(y)}{\sigma^2(y)}\, dy, \qquad p(x) = \int_0^x \exp(-B(y))\, dy \tag{2.3.2}$$

and

$$m(x) = 2\int_0^x \frac{1}{\sigma^2(y)} \exp(B(y))\, dy. \tag{2.3.3}$$

Lemma 2.3.1 *The following results hold:*

(i) *If $m(+\infty) - m(-\infty) = M < \infty$, then the stationary distribution of the process $\{X_t, t \geq 0\}$ exists and is equal to $\frac{1}{M}[m(x) - m(-\infty)]$.*

(ii) Let $M < \infty$ and f be a Borel measurable function. If

$$\frac{1}{M}\int_{-\infty}^{\infty} f(y)\,dm(y) = A < \infty,$$

then

$$\lim_{t\to\infty}\frac{1}{t}\int_0^t f(X_s)\,ds = A \ a.s.$$

(iii) Suppose that (ii) holds with $A = 0$ and

$$0 < \frac{2}{M}\int_{-\infty}^{\infty} f(y)\int_y^0\int_{-\infty}^{s} f(z)\,dm(z)\,dp(s)\,dm(y) = D < \infty.$$

Then

$$\lim_{t\to\infty}((Dt)^{-1/2}\int_0^t f(X_s)\,ds \le y) = \frac{1}{\sqrt{2\pi}}\int_{-\infty}^{y} e^{-s^2/2}\,ds,\ y\in R.$$

(iv) Let $M < \infty$, $p(\infty) = p(-\infty) = \infty$. Let ψ be a Borel measurable function, nonincreasing in $(-\infty, 0)$ and nondecreasing in $(0,\infty)$. If

$$\int_1^{\infty}\frac{1}{p(y)}\,d\psi(y) < \infty, \qquad \int_{-\infty}^{-1}\frac{1}{p(y)}\,d\psi(y) < \infty,$$

then

$$\lim_{t\to\infty}\frac{\psi(X_t)}{t} = 0 \ a.s.$$

(v) Let

$$\lim_{t\to\infty}\frac{1}{t}\int_0^t \phi_s^2\,ds = \sigma \qquad a.s.,$$

where $\{\phi_s, s \ge 0\}$ is a progressively measurable process. Then

$$\lim_{t\to\infty}\frac{1}{t}\int_0^t \phi_s\,dW_s = 0 \ a.s.$$

and

$$P\left((t\sigma)^{-1/2}\int_0^t \phi_s\,dW_s \le x\right) \to \frac{1}{\sqrt{2\pi}}\int_{-\infty}^{x} e^{-y^2/2}\,dy \qquad as\ t\to\infty,\ x\in R.$$

For proofs of (i)–(iii), see Mandl (1968, p. 93). Part (iv) is proved in Lanska (1979) and (v) is proved in Dufkova (1977).

Consider the stochastic differential equation

$$dX_t = a(X_t,\theta)\,dt + \sigma(X_t)\,dW_t, \qquad X_0 = X_0,\ t \ge 0 \tag{2.3.4}$$

where $\theta \in \Theta$ open in R. Suppose $a(x,\theta)$ and $\sigma(x)$ are continuous in (x,θ) and x, respectively. Further assume that:

(J1) (i) $\sigma(x) > 0$, $\frac{d\sigma(x)}{dx}$, $\frac{\partial a(x,\theta)}{\partial x}$ are continuous in x and (x,θ), respectively, and

(ii) $a'(x,\theta)$, $\frac{\partial a'(x,\theta)}{\partial x}$ are continuous in (x,θ), where $a'(x,\theta)$ denotes the derivative of $a(x,\theta)$ with respect to θ.

We saw earlier that the log-likelihood function is given by

$$\ell_t(\theta) = \int_0^t \frac{a(X_s,\theta)}{\sigma^2(X_s)}\,dX_s - \frac{1}{2}\int_0^t \frac{a^2(X_s,\theta)}{\sigma^2(X_s)}\,ds. \tag{2.3.5}$$

This can also be written in the form

$$\ell_t(\theta) = F(X_t,\theta) - F(X_0,\theta) + \int_0^t f(X_s,\theta)\,ds \tag{2.3.6}$$

by Itô's formula, where

$$F(x,\theta) = \int_0^x \frac{a(y,\alpha)}{\sigma^2(y)}\,dy \tag{2.3.7}$$

and

$$f(x,\theta) = -\frac{1}{2}\left(\frac{a^2(x,\theta)}{\sigma^2(x)} + \frac{\partial a(x,\theta)}{\partial x}\right) + \frac{a(x,\theta)\frac{\partial\sigma(x)}{\partial x}}{\sigma(x)}. \tag{2.3.8}$$

Differentiating (2.3.6) with respect to θ, we have

$$\begin{aligned}\ell_t'(\theta) &= F'(X_t,\theta) - F'(X_0,\theta) + \int_0^t f'(X_s,\theta)\,ds \\ &= \int_0^t \frac{a'(x_s,\theta)}{\sigma^2(X_s)}\,dX_s - \int_0^t \frac{a'(X_s,\theta)a(X_s,\theta)}{\sigma^2(X_s)}\,ds. \end{aligned}\tag{2.3.9}$$

The following results can be proved as an application of Lemma 2.3.1 (cf. Lanska 1979).

Theorem 2.3.2 *Suppose that conditions (J1)(i)–(ii) hold. Further suppose that $a(x,\theta) \neq a(x,\theta_0)$ for $\theta \neq \theta_0$ in Θ. In addition, assume that:*

(J1) (iii)

$$\int_{-\infty}^{\infty} \frac{a^2(x,\theta)}{\sigma^2(x)}\,m_0(dx) < \infty, \qquad \theta \in \Theta,$$

where $m_0(\cdot)$ is the stationary measure corresponding to θ_0.

Then there exists an estimator $\hat\theta_t$ such that

$$\hat\theta_t \to \theta_0 \text{ a.s.} \qquad \text{as } t \to \infty$$

and, for t sufficiently large, $\ell_t'(\hat\theta_t) = 0$.

Theorem 2.3.3 *Suppose that conditions (J1)(i)–(iii) hold. In addition, suppose that:*

(J1) (iv)

$$a''(x,\theta),\quad \frac{\partial a''(x,\theta)}{\partial x} \text{ exist and are continuous in } (x,\theta);$$

(v)

$$\frac{1}{M_0}\int_{-\infty}^{\infty}\left[\frac{a'(x,\theta_0)}{\sigma(x)}\right]^2 m_0(dx) = \sigma^2 > 0; \tag{2.3.10}$$

(vi)

$$\frac{1}{M_0}\int_{-\infty}^{\infty}\left[\frac{a''(x,\theta_0)}{\sigma(x)}\right]^2 m_0(dx) < \infty;$$

(vii) there exists a neighborhood V_{θ_0} of θ_0 such that for all $\theta \in V_{\theta_0}$,

$$|F''(x,\theta) - F''(x,\theta_0)| \le \psi_1(x),$$
$$|f''(x,\theta) - f''(x,\theta_0)| \le \psi_2(x)\, Q(|\theta - \theta_0|).$$

where $\psi_1(x)$ satisfies the hypothesis of Lemma 2.3.1(iv) with

$$\int_{-\infty}^{\infty} \psi_2(x)\, m_0(dx) < \infty \quad and \quad \lim_{y\to 0} Q(y) = Q(0) = 0.$$

Then

$$\sqrt{t}(\hat{\theta}_t - \theta_0) \xrightarrow{\mathcal{L}} N\left(0, \frac{1}{\sigma^2}\right) \qquad as\ t \to \infty.$$

We shall now introduce the minimum contrast method for estimation of the parameter θ. We first prove a lemma.

Lemma 2.3.4 *Let us define f and F as above. Suppose that:*

(J2) *(i)* $\lim_{x\to\pm\infty} \frac{a(x,\theta)}{\sigma^2(x)} \exp\left\{2\int_0^x \frac{a(y,\theta_0)}{\sigma^2(y)}\,dy\right\} = 0,\ \theta \in \Theta;$

(ii) $\int_{-\infty}^{\infty} \frac{a^2(x,\theta)}{\sigma^2(x)}\, m_0(dx) < \infty,\ \theta \in \Theta;$

(iii) $\int_{-\infty}^{\infty} f(x,\theta)\, m_0(dx) < \infty,\ \theta \in \Theta.$

Then

$$\frac{1}{t} F(X_t,\theta) \to 0 \text{ a.s.} \qquad as\ t \to \infty \tag{2.3.11}$$

and $\int_{-\infty}^{\infty} f(x,\theta)\, dm_0(x)$ attains its minimum at θ_0.

Proof. In view of (J2)(ii), it follows by Lemma 2.3.1 that

$$\frac{1}{t}\ell_t(\theta) \to \frac{1}{M_0}\int_{-\infty}^{\infty} \frac{a(x,\theta)}{\sigma^2(x)}\left(\frac{a(x,\theta_0) - a(x,\theta)}{2}\right) m_0(dx) \text{ a.s.} \tag{2.3.12}$$

as $t \to \infty$. Furthermore,

$$\frac{1}{t}\int_0^t f(X_s,\theta)\, ds \to \frac{1}{M_0}\int_{-\infty}^{\infty} f(x,\theta)\, m_0(dx) \text{ a.s.} \tag{2.3.13}$$

as $t \to \infty$. Integrating by parts, it follows that

$$\int_{-\infty}^{\infty} f(x,\theta)\, m_0(dx) = \int_{-\infty}^{\infty} \frac{a(x,\theta)}{\sigma^2(x)}\left(\frac{a(x,\theta_0) - a(x,\theta)}{2}\right) m_0(dx). \tag{2.3.14}$$

From these observations and from the form of $\ell_t(\theta)$ given by (2.3.6), it follows that

$$\frac{1}{t}F(X_t,\theta) \to 0 \text{ a.s.} \qquad \text{as } t \to \infty.$$

It is clear from (2.3.14) that $\int_{-\infty}^{\infty} f(x,\theta)\, m_0(dx)$ attains its maximum at $\theta = \theta_0$.

In view of Lemma 2.3.2, one can define a contrast function generalizing the log-likelihood function in the following way.

Definition. A *contrast function* $g(x,\theta)$ is real-valued function defined on $R \times \Theta$ such that

$$-\infty < \int_{-\infty}^{\infty} g(x,\theta_0)\, m_0(dx) < \int_{-\infty}^{\infty} g(x,\theta)\, m_0(dx) < \infty, \qquad \theta,\theta_0 \in \Theta. \tag{2.3.15}$$

Define a contrast

$$M_t(\theta) = G(X_t,\theta) - G(X_0,\theta) + \int_0^t g(X_s,\theta)\, ds, \tag{2.3.16}$$

where $G(\cdot,\cdot)$ is a real-valued function defined on $R \times \Theta$ such that

$$\frac{1}{t} G(X_t,\theta) \to 0 \text{ a.s.} \qquad \text{as } t \to \infty. \tag{2.3.17}$$

Note that the choice of $G = F$ and $g = f$ reduces (2.3.16) to the log-likelihood function, and (2.3.17) holds by Lemma 2.3.4.

Definition. An estimator $\tilde{\theta}_t$ is said to be a *minimum contrast estimator* (MCE) if it minimizes the contrast $M_t(\theta)$ over $\theta \in \Theta$.

The following results can be proved regarding consistency and asymptotic normality of an MCE.

Theorem 2.3.5 *Let $G'(x,\theta)$ and $g'(x,\theta)$ be continuous in (x,θ), where prime denotes differentiation with respect to θ. Then there exists a solution $\tilde{\theta}_t$ of $M_t'(\theta) = 0$ for T sufficiently large and $\tilde{\theta}_t \to \infty$ a.s. as $t \to \infty$. Furthermore, suppose that g'' and G'' exist and are continuous in (x,θ) and that g, G satisfy the following conditions:*

(i) $\lim_{t\to\infty} t^{-1/2} G'(X_t,\theta_0) = 0$ *in probability;*
(ii) $\lim_{t\to\infty} t^{-1} G''(X_t,\theta_0) = 0$ *a.s.;*
(iii) $\int_{-\infty}^{\infty} g'(x,\theta_0)\, m_0(dx) = 0$;
(iv) $0 < \frac{2}{M_0} \int_{-\infty}^{\infty} g'(y,\theta_0) \int_y^0 \int_{-\infty}^{s} g'(z,\theta_0)\, m_0(dz)\, p_0(ds)\, m_0(dy) = D_0 < \infty$;
(v) $|\int_{-\infty}^{\infty} g''(x,\theta_0)\, m_0(dx)| < \infty$;
(vi) there exists a neighborhood V_{θ_0} of θ_0 such that for all $\theta \in V_{\theta_0}$,

$$|G''(x,\theta) - G''(x,\theta_0)| \le \psi_1(x),$$
$$|g''(x,\theta) - g''(x,\theta_0)| \le \psi_2(x) Q(|\theta - \theta_0|)$$

where $\psi(x)$ satisfies the hypotheses of the Lemma 2.3.1 (iv) with

$$\int_{-\infty}^{\infty} \psi_2(x)\, m_0(dx) < \infty \quad \textit{and} \quad \lim_{y\to 0} Q(y) = Q(0) = 0.$$

Then

$$\sqrt{t}(\tilde{\theta}_t - \theta_0) \xrightarrow{\mathcal{L}} N\left(0, \frac{1}{D_0}\right) \qquad \textit{as } t \to \infty.$$

For proofs of Theorems 2.3.2–2.3.5, we refer the reader to Lanska (1979). Note that the MCE reduces to the MLE if $G = F$ and $g = f$, where F and f are as given by (2.3.7) and (2.3.8), respectively.

Example 2.3.1
Consider the stochastic differential equation

$$dX_t = -\theta X_t\, dt + \sqrt{2}\, dW_t, \qquad \theta > 0.$$

Then

$$m_0(x) = \int_0^x \exp(-\theta_0 y^2/2)\, dy, \qquad p_0(x) = \int_0^x \exp(\theta_0 y^2/2)\, dy$$

and the MLE $\hat{\theta}_T$ is given by

$$\hat{\theta}_t = \frac{\int_0^t X_s\, dX_s}{\int_0^t X_s^2\, ds} = \theta_0 + \sqrt{2}\frac{\int_0^t X_s\, dW_s}{\int_0^t X_s^2\, ds}.$$

It can be checked that the conditions of Theorem 2.3.2 and 2.3.3 hold or it can be shown by other methods directly (cf. Basawa and Prakasa Rao 1980) that

$$\sqrt{t}(\hat{\theta}_t - \theta_0) \xrightarrow{\mathcal{L}} N(0, 2\theta_0) \qquad \text{as } t \to \infty.$$

Let us now consider the contrast function $g_1(x, \theta) = (\theta x)^2/2 - \theta$, with $G_1(x, \theta) \equiv 0$. Then

$$M_t(\theta) = \frac{\theta^2}{2}\int_0^t X_s^2\, ds - \theta t \quad \text{and} \quad \tilde{\theta}_t = t\left(\int_0^t X_s^2\, ds\right)^{-1}.$$

If we choose the contrast function $g_2(x, \theta) = (x^2 - 1/\theta)^2$, with $G_2(x, \theta) \equiv 0$, then

$$M_t(\theta) = \int_0^t \left(X_s^2 - \frac{1}{\theta}\right)^2 ds \quad \text{and} \quad \tilde{\theta}_t = t\left(\int_0^t X_s^2\, ds\right)^{-1}$$

again.

2.4 Maximum Probability Estimation for the Drift Parameter

Consider the stochastic differential equation

$$dX_t = a(X_t, \theta)\, dt + \sigma(X_t)\, dW_t, \qquad X_0 = X_0,\ t \geq 0. \tag{2.4.1}$$

Suppose the equation has a unique stationary ergodic solution on $[0, T]$ for every $T > 0$ and for every $\theta \in \Theta$ open contained in R. Assume that the functional form of $a(\cdot, \cdot)$ and $\sigma(\cdot)$ is known. Let us again consider the problem of estimation of the parameter θ from a continuous path of the process $\{X_t\}$ observed over $[0, T]$. We indicated earlier that $\sigma(\cdot)$ can be estimated if it is a constant but unknown from the quadratic variation of the process. We will assume in the following that $\sigma(\cdot) \equiv 1$. Following the notation given in Section 2.2, it is known that $P_\theta^T \ll P_W^T$ under a condition analogous to condition (2.2.1) and

$$\begin{aligned}\ell_T(\theta) = \log L_T(\theta) &= \log \frac{dP_\theta^T}{dP_W^T} \\ &= \int_0^T a(X_t, \theta)\, dX_t - \tfrac{1}{2}\int_0^T a^2(X_t, \theta)\, dt. \end{aligned}\tag{2.4.2}$$

Let

$$Z_T(\theta) = \int_{\theta - T^{1/2}}^{\theta + T^{1/2}} \exp(\ell_T(\theta))\, dt. \tag{2.4.3}$$

Then $Z_T(\theta)$ is the integrated likelihood.

Definition. Any measurable function $\tilde{\theta}_T$ for which the integrated likelihood $Z_T(\theta)$ is maximized with respect to θ is called a *maximum probability estimator* (MPE) of θ based on the realization of the process $\{X_t\}$ over $[0, T]$.

We assume the following:

(K1) Suppose $\ell_T(\theta)$ is differentiable twice with respect to θ and that differentiation under the integral sign is valid in (2.4.2).

It is easy to check that

$$\phi_T(\theta) \equiv \frac{d \log L_T(\theta)}{d\theta} = \int_0^T a'(X_t, \theta)\, dX_t - \int_0^T a(X_t, \theta) a'(X_t, \theta)\, dt \tag{2.4.4}$$

and

$$\begin{aligned}\psi_T(\theta) &\equiv \frac{d^2 \log L_T(\theta)}{d\theta^2} \\ &= \int_0^T a''(X_t, \theta)\, dX_t - \int_0^T \{a(X_t, \theta) a''(X_t, \theta) + a'^2(X_t, \theta)\}\, dt,\end{aligned} \tag{2.4.5}$$

where prime denotes differentiation of the function with respect to θ. In particular,

$$E_\theta \left[\frac{d \log L_T(\theta)}{d\theta} \right] = 0 \tag{2.4.6}$$

and

$$\begin{aligned} E_\theta \left[\frac{d \log L_T(\theta)}{d\theta} \right]^2 &= -E_\theta \left[\frac{d^2 \log L_T(\theta)}{d\theta^2} \right] \\ &= \int_0^T E_\theta [a'(X_t, \theta)]^2\, dt \\ &= T\sigma^2(\theta), \end{aligned} \tag{2.4.7}$$

where $\sigma^2(\theta) \equiv E_\theta[a'(X_0, \theta)]^2$. This follows by the stationarity of the process $\{X_t\}$. We assume that $\sigma^2(\theta) > 0$. Let

$$\Delta_T(\theta) = T^{-1/2} \phi_T(\theta), \qquad \theta_T = \theta + hT^{-1/2} \text{ for } h \in R. \tag{2.4.8}$$

Since Θ is open, $\theta_T \in \Theta$ for large T. Define

$$\begin{aligned}\Lambda_T(\theta) &\equiv \ell_T(\theta_T) - \ell_T(\theta) \\ &= \int_0^T [a(X_t, \theta_T) - a(X_t, \theta)]\, dX_t - \tfrac{1}{2} \int_0^T [a^2(X_t, \theta_T) - a^2(X_t, \theta)]\, dt.\end{aligned} \tag{2.4.9}$$

Hence

$$\begin{aligned}&\Lambda_T(\theta) - h\Delta_T(\theta) \\ &= \int_0^T [a(X_t, \theta_T) - a(X_t, \theta)]\, dX_t - \tfrac{1}{2} \int_0^T [a^2(X_t, \theta_T) - a^2(X_t, \theta)]\, dt \\ &\quad -hT^{-1/2} \left[\int_0^T a'(X_t, \theta)\, dX_t - \int_0^T a(X_t, \theta) a'(X_t, \theta)\, dt \right].\end{aligned} \tag{2.4.10}$$

If θ is the true parameter, then

$$dX_t = a(X_t, \theta)\,dt + dW_t,$$

and (2.4.10) can be written in the form

$$\begin{aligned}
&\Lambda_T(\theta) - h\Delta_T(\theta)\\
&= \int_0^T [a(X_t, \theta_T) - a(X_t, \theta)]\,dW_t + \int_0^T [a(X_t, \theta_T) - a(X_t, \theta)]a(X_t, \theta)\,dt\\
&\quad -\tfrac{1}{2}\int_0^T [a^2(X_t, \theta_T) - a^2(X_t, \theta)]\,dt - hT^{-1/2}\int_0^T a'(X_t, \theta)\,dW_t\\
&= \int_0^T [a(X_t, \theta_T) - a(X_t, \theta) - hT^{-1/2}a'(X_t, \theta)]\,dW_t\\
&\quad -\tfrac{1}{2}\int_0^T [a(X_t, \theta_T) - a(X_t, \theta)]^2\,dt\\
&= \tfrac{1}{2}h^2T^{-1}\int_0^T a''(X_t, \theta_T^*)\,dW_t - \tfrac{1}{2}h^2T^{-1}\int_0^T [a'(X_t, \tilde{\theta}_T)]^2\,dt,
\end{aligned}$$

for some $\theta^*{}_T$ and $\tilde{\theta}_T$ such that $|\theta_T^* - \theta| \le |\theta_T - \theta| = hT^{-1/2}$ and $|\tilde{\theta}_T - \theta| \le |\theta_T - \theta| = hT^{-1/2}$. Note that

$$\begin{aligned}
\psi_T(s) &= \frac{d^2 \log L_T(\theta)}{d\theta^2}\,|_{\theta=s}\\
&= \int_0^T a''(X_t, s)\,dX_t - \int_0^T [a(X_t, s)a''(X_t, s) + a'^2(X_t, s)]\,dt
\end{aligned}$$

and if θ is the true parameter, then

$$\begin{aligned}
\psi_T(s) = &\int_0^T a''(X_t, s)\,dW_t\\
&- \int_0^T [\{a(X_t, s) - a(X_t, \theta)\}a''(X_t, s) + a'^2(X_t, s)]\,dt.
\end{aligned}$$

Let

$$\eta_T(s) = \frac{1}{T}(\psi_T(s) - E_\theta(\psi_T(s))).$$

Lemma 2.4.1 *Suppose $a''(x, \theta)$ is differentiable with respect to θ and $a'''(x, \theta)$ is Lipschitzian in θ in a closed interval $I(\theta)$ containing θ, with Lipschitzian coefficient $c(x)$ such that*

$$E_\theta[a'''(X_0, s)]^2 < \infty, \qquad s \in I(\theta)$$

and

$$E_\theta[c(X_0)]^2 < \infty.$$

Then

$$\frac{1}{T}\sup_{s\in I(\theta)}\left|\int_0^T a''(X_t, s)\,dW_t\right| \to 0 \qquad \text{as } T \to \infty$$

a.s. $[P_\theta]$.

This lemma follows from the uniform ergodic theorem for stochastic integrals proved in the Appendix B.

Lemma 2.4.2 *Suppose that the functions $a(x, s_1)a''(x, s_2)$ and $a'(x, s_1)a'(x, s_2)$ are uniformly bounded for s_1, s_2 in a closed interval $I(\theta)$ of θ by an integrable function of x. Then*

$$\sup_{(s_1,s_2)\in I(\theta)\times I(\theta)} \left| \frac{1}{T}\int_0^T \{a'(X_t, s_1)a''(X_t, s_2) - E_\theta[a'(X_t, s_1)a'(X_t, s_2)]\}\, dt \right| \to 0$$

as $T \to \infty$ a.s. $[P_\theta]$. Furthermore, $E_\theta[a(X_0, s_1)a''(X_0, s_2)]$ and $E_\theta[a'(X_0, s_1)\, a'(X_0, s_2)]$ are continuous in (s_1, s_2).

This lemma is a consequence of the uniform ergodic theorem proved in Appendix A and the dominated convergence theorem.

As a consequence of Lemmas 2.4.1 and 2.4.2, the following result can be proved.

Lemma 2.4.3 *Suppose the conditions stated in the Lemmas 2.4.1 and 2.4.2 hold. Then*

(i) $\sup_{s\in I(\theta)}|\eta_T(s)| \to 0$ *as* $T \to \infty$ *a.s.* $[P_\theta]$;

(ii) $\int_0^1 |\eta_T(\theta + \lambda\tau_T)|\, d\lambda \to 0$ *as* $T \to \infty$ *a.s.* $[P_\theta]$ *whenever* $\tau_T \to 0$ *as* $T \to \infty$; *and*

(iii) $\frac{1}{T}\int_0^1 \psi_T(\theta+\lambda\tau_T)\, d\lambda \to -\sigma^2(\theta)$ *as* $T \to \infty$ *a.s.* $[P_\theta]$ *whenever* $\tau_T \to 0$ *as* $T \to \infty$.

Proof. Part (i) of this lemma is a consequence of the Lemmas 2.4.1 and 2.4.2. Part (ii) follows from (i). In view of (ii), it is sufficient to prove that

$$\frac{1}{T}\int_0^1 E_\theta[\psi_T(\theta + \lambda\tau_T)]\, d\lambda \to -\sigma^2(\theta)$$

as $T \to \infty$ to prove (iii). By the stationarity of the process X_t,

$$\frac{1}{T}E_\theta[\psi_T(s)] = -E_\theta\{[a(X_0, s) - a(X_0, \theta)]a''(X_0, s) + a'^2(X_0, s)\}.$$

It is clear that

$$\frac{1}{T}E_\theta(\psi_T(\theta + \lambda\tau_T)) \to -\sigma^2(\theta) \qquad \text{as } T \to \infty \tag{2.4.11}$$

by the continuity proved in Lemma 2.4.2. Furthermore, the integral is dominated in s by an integrable function. Hence the result stated in part (iii) is proved by the dominated convergence theorem.

2.4.1 Existence of a $\sqrt{t}$-Consistent Estimator for θ

If $\bar{\theta}_T$ is an MPE, then

$$\left.\frac{dZ_T(\theta)}{d\theta}\right|_{\theta=\bar{\theta}_T} = 0. \tag{2.4.12}$$

Let us consider the equation

$$\frac{dZ_T(\theta)}{d\theta} = L_T(\theta + T^{-1/2}) - L_T(\theta - T^{-1/2}) = 0, \tag{2.4.13}$$

where $Z_T(\theta)$ is as defined by (2.4.3). For any h in the neighborhood of θ,

$$\begin{aligned}\log L_T(h \pm T^{-1/2}) &= \log L_T(\theta) + (h-\theta \pm T^{-1/2})\frac{d\log L_T(\theta)}{d\theta}\\ &\quad + \frac{1}{2}(h-\theta\pm T^{-1/2})^2 \int_0^1 \left.\frac{d^2 \log L_T(\phi)}{d\phi^2}\right|_{\phi=\theta+\lambda(h-\theta\pm T^{-1/2})} d\lambda\\ &= \log L_T(\theta) + (h-\theta\pm T^{-1/2})\phi_T(\theta)\\ &\quad + \frac{1}{2}(h-\theta\pm T^{-1/2})^2 \int_0^1 \psi_T(\theta+\lambda(h-\theta\pm T^{-1/2}))\, d\lambda.\end{aligned} \tag{2.4.14}$$

Let

$$\begin{aligned} I_T^{\pm} &\equiv \frac{1}{T}\int_0^1 \psi_T(\theta+\lambda(h-\theta\pm T^{-1/2}))\, d\lambda\\ &\equiv I_T((h-\theta\pm T^{-1/2})+\theta). \end{aligned} \tag{2.4.15}$$

Replacing θ by h in equation (2.4.13) and using (2.4.14), we have

$$\begin{aligned}&(h-\theta+T^{-1/2})\phi_T(\theta)+\tfrac{1}{2}(h-\theta+T^{-1/2})^2 T I_T^{+}\\ &= (h-\theta-T^{-1/2})\phi_T(\theta)+\tfrac{1}{2}(h-\theta-T^{-1/2})^2 T I_T^{-},\end{aligned}$$

and hence

$$\begin{aligned}&4T^{-1/2}\phi_T(\theta)+\{T^{1/2}(h-\theta)\}^2(I_T^{+}-I_T^{-})\\ &+2T^{1/2}(h-\theta)(I_T^{+}+I_T^{-})+(I_T^{+}-I_T^{-})=0.\end{aligned} \tag{2.4.16}$$

Let $\gamma_T(h)$ denote the expression on the left-hand side of equation (2.4.16) for any fixed θ. Then, for any constant $M > 0$,

$$\begin{aligned}&\gamma_T(\theta+MT^{-1/2})\\ &= 4T^{-1/2}\phi_T(\theta)+M^2\{I_T(\theta+(M+1)T^{-1/2})-I_T(\theta+(M-1)T^{-1/2})\}\\ &\quad +2M\{I_T(\theta+(M+1)T^{-1/2})+I_T(\theta+(M-1)T^{-1/2})\}\\ &\quad +\{I_T(\theta+(M+1)T^{-1/2})-I_T(\theta+(M-1)T^{-1/2})\}\end{aligned} \tag{2.4.17}$$

and

$$\begin{aligned}&\gamma_T(\theta-MT^{-1/2})\\ &= 4T^{-1/2}\phi_T(\theta)+M^2\{I_T(\theta-(M-1)T^{-1/2})-I_T(\theta-(M+1)T^{-1/2})\}\\ &\quad -2M\{I_T(\theta-(M-1)T^{-1/2})+I_T(\theta-(M+1)T^{-1/2})\}\\ &\quad +\{I_T(\theta-(M-1)T^{-1/2})-I_T(\theta-(M+1)T^{-1/2})\}.\end{aligned} \tag{2.4.18}$$

Applying the central limit theorem for stochastic integrals (cf. Basawa and Prakasa Rao 1980), it can be checked that

$$T^{-1/2}\phi_T(\theta) \xrightarrow{\mathcal{L}} N(0,\sigma^2(\theta)) \qquad \text{as } T\to\infty, \tag{2.4.19}$$

when θ is the true parameter. On the other hand, all the functions $I_T(\cdot)$ appearing in (2.4.17) and (2.4.18) converge to $-\sigma^2(\theta)$ a.s. $[P_\theta]$ by part (iii) of Lemma 2.4.3. Hence, for any fixed $M > 0$,

$$\gamma_T(\theta + MT^{-1/2}) \xrightarrow{\mathcal{L}} N(-4M\sigma^2(\theta), 16\,\sigma^2(\theta)) \tag{2.4.20}$$

and

$$\gamma_T(\theta - MT^{-1/2}) \xrightarrow{\mathcal{L}} N(4M\sigma^2(\theta), 16\,\sigma^2(\theta)) \tag{2.4.21}$$

as $T \to \infty$, when θ is the true parameter. Since $\sigma^2(\theta) > 0$, it follows that, for $\varepsilon > 0$ and for large T,

$$P_\theta[\gamma_T(\theta + MT^{-1/2}) < 0] > 1 - \tfrac{1}{2}\varepsilon \tag{2.4.22}$$

and

$$P_\theta[\gamma_T(\theta - MT^{-1/2}) > 0] > 1 - \tfrac{1}{2}\varepsilon \tag{2.4.23}$$

for large M. Since $\gamma_T(h)$ is continuous in h, it follows that there exists $\bar{\theta}_T$ such that $\gamma_T(\bar{\theta}_T) = 0$ and $\bar{\theta}_T \in [\theta - M_T^{-1/2}, \theta + MT^{-1/2}]$ with probability greater than $1 - \varepsilon$, and we have the following result.

Theorem 2.4.4 *Under the assumptions stated above, for all sufficiently large T with P_θ-probability approaching one, there exists a solution $\bar{\theta}_T$ for the equation (2.4.12) such that*

$$\sqrt{T}(\bar{\theta}_T - \theta) = O_p(1). \tag{2.4.24}$$

2.4.2 Existence of a Maximum Probability Estimator

Let $\bar{\theta}_T$ be chosen as given by the Theorem 2.4.4. We will now prove that

$$\left.\frac{d^2 Z_T(\theta)}{d\theta^2}\right|_{\theta=\bar{\theta}_T} < 0 \tag{2.4.25}$$

with probability under P_θ approaching one as $T \to \infty$. It is sufficient to show that

$$\left.\frac{dL_T(\theta + T^{-1/2})}{d\theta}\right|_{\theta=\bar{\theta}_T} < \left.\frac{dL_T(\theta - T^{-1/2})}{d\theta}\right|_{\theta=\bar{\theta}_T} \tag{2.4.26}$$

with probability approaching one under P_θ as $T \to \infty$ in view of (2.4.13).

It is easy to see that

$$\begin{aligned}\frac{dL_T(\theta \pm T^{-1/2})}{d\theta} &= L_T(\theta \pm T^{-1/2})\frac{d\log L_T(\theta \pm T^{-1/2})}{d\theta}\\ &= L_T(\theta \pm T^{-1/2}) \times \left[\int_0^T a'(X_t, \theta \pm T^{-1/2})\,dX_t\right.\\ &\quad \left. - \int_0^T a(X_t, \theta \pm T^{-1/2})a'(X_t, \theta \pm T^{-1/2})\,dt\right].\end{aligned} \tag{2.4.27}$$

Since $L_T(\bar{\theta}_T - T^{-1/2}) = L_T(\bar{\theta}_T + T^{-1/2})$ with probability approaching one under P_θ as $T \to \infty$, it is sufficient to prove that

$$\begin{aligned}
&\int_0^T a'(X_t, \bar{\theta}_T + T^{-1/2})\, dX_t - \int_0^T a(X_t, \bar{\theta}_T + T^{-1/2}) a'(X_t, \bar{\theta}_T + T^{-1/2})\, dt \\
&< \int_0^T a'(X_t, \bar{\theta}_T - T^{-1/2})\, dX_t - \int_0^T a(X_t, \bar{\theta}_T - T^{-1/2}) a'(X_t, \bar{\theta}_T - T^{-1/2})\, dt
\end{aligned} \tag{2.4.28}$$

with probability approaching one as $T \to \infty$. Since

$$dX_t = a(X_t, \theta)\, dt + dW_t,$$

under the probability measure P_θ, it follows that (2.4.28) holds provided

$$\begin{aligned}
&\int_0^T a'(X_t, \bar{\theta}_T + T^{-1/2})\, dW_t \\
&\quad - \int_0^T [a(X_t, \bar{\theta}_T + T^{-1/2}) - a(X_t, \theta)] a'(X_t, \bar{\theta}_T + T^{-1/2})\, dt \\
&< \int_0^T a'(X_t, \bar{\theta}_T - T^{-1/2})\, dW_t \\
&\quad - \int_0^T [a(X_t, \bar{\theta}_T - T^{-1/2}) - a(X_t, \theta)] a'(X_t, \bar{\theta}_T - T^{-1/2})\, dt
\end{aligned} \tag{2.4.29}$$

with probability under P_θ approaching one as $T \to \infty$. Note that $\bar{\theta}_T \xrightarrow{p} \theta$ from Theorem 2.4.4. Applying Taylor's expansion, the inequality (2.4.29) holds provided

$$\begin{aligned}
&\int_0^T a'(X_t, \bar{\theta}_T + T^{-1/2})\, dW_t \\
&\quad -(\bar{\theta}_T + T^{-1/2} - \theta) \int_0^T a'(X_t, \theta_T^* + T^{-1/2}) a'(X_t, \bar{\theta}_T + T^{-1/2})\, dt \\
&< \int_0^T a'(X_t, \bar{\theta}_T - T^{-1/2})\, dW_t \\
&\quad -(\bar{\theta}_T - T^{-1/2} - \theta) \int_0^T a'(X_t, \theta_T^{**} - T^{-1/2}) a'(X_t, \bar{\theta}_T - T^{-1/2})\, dt,
\end{aligned} \tag{2.4.30}$$

where $|\theta_T^* + T^{-1/2} - \theta| < |\bar{\theta}_T + T^{-1/2} - \theta|$ and $|\theta_T^{**} - T^{-1/2} - \theta| < |\bar{\theta}_T - T^{-1/2} - \theta|$ with probability under P_θ approaching one as $T \to \infty$. Note that $\theta_T^* \xrightarrow{p} \theta$ as $T \to \infty$. Dividing by $T^{1/2}$, inequality (2.4.30) can be written in the form

$$\begin{aligned}
&\frac{2}{T} \int_0^T a''(X_t, \hat{\theta}_t)\, dW_t \\
&\quad -[(\bar{\theta}_T - \theta)\sqrt{T}] \frac{1}{T} \int_0^T a'(X_t, \theta_T^* + T^{-1/2}) a'(X_t, \bar{\theta}_T + T^{-1/2})\, dt
\end{aligned}$$

$$-\frac{1}{T}\int_0^T a'(X_t, \theta_T^* + T^{-1/2})a'(X_t, \bar{\theta}_T + T^{-1/2})\,dt$$

$$< -[(\bar{\theta}_T - \theta)\sqrt{T}]\frac{1}{T}\int_0^T a'(X_t, \theta_T^{**} + T^{-1/2})a'(X_t, \bar{\theta}_T + T^{-1/2})\,dt$$

$$+\frac{1}{T}\int_0^T a'(X_t, \theta_T^{**} - T^{-1/2})a'(X_t, \bar{\theta}_T - T^{-1/2})\,dt, \tag{2.4.31}$$

where $\bar{\theta}_T - T^{-1/2} < \hat{\theta}_T < \bar{\theta}_T + T^{-1/2}$. Since

$$\left|\frac{1}{T}\int_0^T a''(X_t, \hat{\theta}_T)\,dW_t\right| \le \frac{1}{T}\sup_{\phi\in I(\theta)}\left|\int_0^T a''(X_t, \phi)\,dW_t\right|,$$

and the last term tends to zero a.s. $[P_\theta]$ as $T \to \infty$ by the Lemma 2.4.1, it follows that $\frac{1}{T}\int_0^T a''(X_t, \hat{\theta}_T\,dW_t \to 0$ a.s. $[P_\theta]$ as $T \to \infty$.

Furthermore, Lemma 2.4.2 and the dominated convergence theorem imply that

$$\left|\frac{1}{T}\int_0^T a'(X_t, \theta_T^* + T^{-1/2})a'(X_t, \bar{\theta}_T + T^{-1/2})\,dt - \sigma^2(\theta)\right| \to 0 \text{ a.s. } [P_\theta]$$

and

$$\left|\frac{1}{T}\int_0^T a'(X_t, \theta_T^{**} - T^{-1/2})a'(X_t, \bar{\theta}_T - T^{-1/2})\,dt - \sigma^2(\theta)\right| \to 0 \text{ a.s. } [P_\theta]$$

as $T \to \infty$. From these facts, together with the observation that $\sqrt{T}(\bar{\theta}_T - \theta) = O_p(1)$ and $\sigma^2(\theta) > 0$, it follows that inequality (2.4.31) holds with P_θ-probability aproaching one as $T \to \infty$. Hence we have the following theorem.

Theorem 2.4.5 *Under the assumption stated in Theorem 2.4.4, there exists a maximum probability estimator $\bar{\theta}_T$ of θ which is $\sqrt{T}$-consistent with probability approaching one as $T \to \infty$.*

2.4.3 Asymptotic Normality of the Maximum Probability Estimator

Let $\bar{\theta}_T$ be an MPE as obtained in Theorem 2.4.5. Since

$$\left.\frac{dZ_T(\theta)}{d\theta}\right|_{\theta=\bar{\theta}_T} = 0, \tag{2.4.32}$$

it follows from (2.4.16) that

$$4T^{-1/2}\phi_T(\theta) + \{\sqrt{T}(\bar{\theta}_T - \theta)\}^2(\hat{I}_T^+ - \hat{I}_T^-)$$
$$+2T^{1/2}(\bar{\theta}_T - \theta)(\hat{I}_T^+ + \hat{I}_T^-) + (\hat{I}_T^+ - \hat{I}_T^-) = 0, \tag{2.4.33}$$

where

$$\hat{I}_T^{\pm} = \frac{1}{T}\int_0^1 \psi_T(\theta + \lambda(\bar{\theta}_T - \theta \pm T^{-1/2}))\,d\lambda. \tag{2.4.34}$$

We observed earlier that

$$\hat{I}_T^{\pm} \xrightarrow{P} -\sigma^2(\theta) < 0 \text{ under } P_\theta \qquad \text{as } T \to \infty \tag{2.4.35}$$

since $\bar{\theta}_T \xrightarrow{p} \theta$ as $T \to \infty$. Relation (2.4.33) implies that

$$T^{1/2}(\bar{\theta}_T - \theta) = -\frac{2T^{-1/2}\phi_T(\theta)}{\hat{I}_T^+ + \hat{I}_T^-} - \frac{1}{2}\{\sqrt{T}(\bar{\theta}_T - \theta)\}^2 \left(\frac{\hat{I}_T^+ - \hat{I}_T^-}{\hat{I}_T^+ + \hat{I}_T^-}\right)$$
$$-\frac{1}{2}\left(\frac{\hat{I}_T^+ - \hat{I}_T^-}{\hat{I}_T^+ + \hat{I}_T^-}\right). \tag{2.4.36}$$

Since $T^{-1/2}\phi_T(\theta) \xrightarrow{\mathcal{L}} N(0, \sigma^2(\theta))$, $\sqrt{T}(\bar{\theta}_T - \theta) = O_p(1)$ and (2.4.35) holds, it follows that

$$T^{1/2}(\bar{\theta}_T - \theta) \xrightarrow{\mathcal{L}} N(0, \sigma^{-2}(\theta)) \qquad \text{as } T \to \infty.$$

Theorem 2.4.6 *Under the assumptions stated above, the MPE $\bar{\theta}_T$ given in Theorem 2.4.4 is asymptotically normal. In fact*

$$T^{1/2}(\bar{\theta}_T - \theta) \xrightarrow{\mathcal{L}} N\left(0, \frac{1}{\sigma^2(\theta)}\right) \qquad \text{as } T \to \infty. \tag{2.4.37}$$

Remarks. It is clear the $\bar{\theta}_T$ is asymptotically efficient in the classical sense that its asymptotic variance is the Cramér–Rao lower bound. It can also be shown that it is also asymptotically efficient in the sense of Weiss and Wolfowitz (1974) who studied the MPE in the classical case. For details, see Prakasa Rao (1982).

Example 2.4.1
Consider the class of diffusion processes

$$dX(t) = [a(t, X(t)) + \theta b(t, X(t))]dt + \sigma(t, X(t))\, dW(t), \qquad t \geq 0$$

where $a(\cdot,\cdot)$, $b(\cdot,\cdot)$ and $\sigma(\cdot,\cdot)$ are known functions. Let θ_0 be the true parameter. It is easy to check that

$$\log L_T(\theta) = (\theta - \theta_0)\int_0^T \frac{b(t, X(t))}{\sigma(t, X(t))}\, dW(t)$$
$$-\frac{1}{2}(\theta - \theta_0)^2 \int_0^T \left\{\frac{b(t, X(t))}{\sigma(t, X(t))}\right\}^2 dt$$
$$= (\theta - \theta_0)\alpha_T - \frac{1}{2}(\theta - \theta_0)^2\beta_T \text{ (say)}$$

and

$$Z_T(\theta) = \int_{\theta - T^{-1/2}}^{\theta + T^{-1/2}} L_T(t)\, dt.$$

It can be checked that $\frac{dZ_T(\theta)}{d\theta} = 0$ if and only if $\alpha_T = \beta_T(\theta - \theta_0)$. Let $\bar{\theta}_T$ be the solution of the equation $\alpha_T = \beta_T(\theta - \theta_0)$. It can again be seen that

$$\frac{d^2 Z_T(\theta)}{d\theta^2}\Big|_{\theta = \bar{\theta}_T} < 0,$$

showing that $\bar{\theta}_T$ is the MPE for θ. In this example, the MPE and the MLE coincide.

2.5 Bayes Method for the Estimation of the Drift Parameter

Consider the stochastic differential equation

$$dX_t = a(\theta, X_t)\,dt + dW_t, \qquad X_0 = X_0,\ E(X_0^2) < \infty,\ t \geq 0, \tag{2.5.1}$$

where $\{W(t), t \geq 0\}$ is the standard Wiener process and $\theta \in \Theta$ open in R. We now study the Bayes estimation of the parameter θ under a suitable class of loss functions. Assume that conditions (B1)–(B7) of Section 2.2 hold. Let Λ be a prior probability measure on $(\Theta, \mathcal{B})$ where $\mathcal{B}$ is the σ-algebra of Borel subsets of Θ. Suppose that Λ has a density $\lambda(\cdot)$ with respect to the Lebesgue measure and the density $\lambda(\cdot)$ is continuous and positive in a neighborhood of the true parameter θ_0. Following the notation used in Section 2.2, let

$$\begin{aligned}\ell_T(\theta) &= \log \frac{dP_\theta^T}{dP_{\theta_0}^T}\\ &= \int_0^T [a(\theta, X_t) - a(\theta_0, X_t)]\,dX_t - \tfrac{1}{2}\int_0^T [a(\theta, X_t) - a(\theta_0, X_t)]^2\,dt.\end{aligned} \tag{2.5.2}$$

The posterior density of θ given $X^T \equiv \{X(t), 0 \leq t \leq T\}$ is

$$p(\theta | X(t), 0 \leq t \leq T) = \frac{\frac{dP_\theta^T}{dP_{\theta_0}^T}(X^T)\lambda(\theta)}{\int_\Theta \frac{dP_\theta^T}{dP_\theta^T}(X^T)\lambda(\theta)\,d\theta}. \tag{2.5.3}$$

Let θ_T be an MLE satisfying the properties given in Theorems 2.2.12 and 2.2.15.

Let $t = \sqrt{T}(\theta - \theta_T)$. Then the posterior density of $\sqrt{T}(\theta - \theta_T)$ is given by

$$p^*(t|X^T) = T^{-1/2} p(\theta_T + tT^{-1/2}|X^T). \tag{2.5.4}$$

Under conditions (B) of Section 2.2, all the stochastic integrals occuring in the present discussion can be defined pathwise, and differentiation of the stochastic integral with respect to θ can be justified following the results in Karandikar (1983).

Since

$$\frac{d}{d\theta}\left[\log \frac{dP_\theta^T}{d\theta_0^T}\right]\Bigg|_{\theta=\theta_T} = 0, \tag{2.5.5}$$

it follows that

$$\int_0^T a'(\theta_T, X_s)\,dW(s) = \int_0^T a'(\theta_T, X_s)[a(\theta_T, X_s) - a(\theta_0, X_s)]\,ds, \tag{2.5.6}$$

where prime denotes differentiation with respect to θ. Let

$$\gamma_T(t) = \frac{dP_{\theta_T+tT^{-1/2}}^T / dP_{\theta_0}^T}{dP_{\theta_T}^T / dP_{\theta_0}^T}(X^T), \tag{2.5.7}$$

$$C_T = \int_{-\infty}^{\infty} \gamma_T(t)\lambda(\theta_T + tT^{-1/2})\,dt, \tag{2.5.8}$$

$$I_T(\theta) = \int_0^T [a(\theta, X_s) - a(\theta_0, X_s)]^2\,ds \tag{2.5.9}$$

and

$$\beta = E_{\theta_0}([a'(\theta_0, X_0)]^2) \equiv I(\theta_0). \tag{2.5.10}$$

Suppose $\beta > 0$. It is easy to see that

$$p^*(t|X^T) = C_T^{-1}\gamma_T(t)\lambda(\theta_T + tT^{-1/2}).$$

Let $K(\cdot)$ be a nonnegative measurable function such that

(L1) There exists a number ε, $0 < \varepsilon < \beta$, such that

$$\int_{-\infty}^{\infty} K(t)\exp\left(-(\beta-\varepsilon)\frac{t^2}{2}\right) dt < \infty,$$

and

(L2) For every $h > 0$ and every $\delta > 0$,

$$e^{-T\delta}\int_{|t|>h} K(T^{1/2}t)\lambda(\theta_T + t)\,dt \to 0 \text{ a.s. } [P_{\theta_0}] \qquad \text{as } T \to \infty.$$

We now state and sketch a proof of the theorem concerning the asymptotic behavior of the posterior density $p^*(t|X^T)$.

Theorem 2.5.1 *Given conditions (B1)–(B7), (L1) and (L2) and the additional condition that*

(B8) the second derivative $\frac{\partial^2 a(\theta,x)}{\partial\theta^2}$ *exists, is continuous and satisfies conditions analogous to (B5) and (B7),*

then

$$\lim_{T\to\infty}\int_{-\infty}^{\infty} K(t)\left|p^*(t|X^T) - \left(\frac{\beta}{2\pi}\right)^{1/2} e^{-\frac{1}{2}\beta t^2}\right| dt = 0 \text{ a.s. } [P_{\theta_0}]$$

for any nonnegative measurable $K(\cdot)$ *satisfying (L1) and (L2).*

We first prove a few lemmas which will be used later.

Lemma 2.5.2 *Under conditions (B2) and (B3), for every* $\delta > 0$,

$$\lim_{T\to\infty}\inf_{|\theta-\theta_0|\geq\delta}\frac{I_T(\theta 0)}{T} = \lambda(\delta) > 0 \text{ a.s.}$$

For a proof of this lemma, see Lemma 2.2.11.

Lemma 2.5.3 *Under condition (B8),*

$$\lim_{T\to\infty}\frac{1}{T}\int_0^T a''(\theta_0, X_s)\,dW(s) = 0 \text{ a.s. } [P_{\theta_0}],$$

where $a''(\theta, x)$ *denotes the second partial derivative of* $a(\theta, x)$ *with respect to* θ.

Proof. Note that

$$g(t) = \int_0^t a''(\theta_0, X_s)\, dW(s), \qquad t \geq 0,$$

is a continuous martingale. Hence by the martingale inequality and the stationarity of the process $\{X(t), t \geq 0\}$,

$$\begin{aligned} P[\sup_{0 \leq t \leq T} |g(t)| \geq \lambda] &\leq \frac{T}{\lambda^2} E[a''(\theta_0, X_0)]^2 \\ &= \frac{T}{\lambda^2} \beta_1 \text{ (say)}. \end{aligned}$$

Let

$$A_n = \left\{ \sup_{2^{n-1} \leq t < 2^n} \left| \frac{1}{t} g(t) \right| \geq 2^{-n/4} \right\}.$$

Then

$$\begin{aligned} P(A_n) &\leq P\Big[\sup_{2^{n-1} \leq t < 2^n} |g(t)| \geq 2^{n-1} 2^{-n/4} \Big] \\ &\leq P\Big[\sup_{0 \leq t < 2^n} |g(t)| \geq 2^{n-1} 2^{-n/4} \Big] \\ &\leq \frac{2^n}{(2^{n-1} 2^{-n/4})^2} \beta_1. \end{aligned}$$

Hence $\sum_n P(A_n) < \infty$, and the lemma follows from the Borel–Cantelli lemma.

Let

$$V(\theta, x) = a''(\theta, x) - a''(\theta_0, x)$$

and

$$Z(t, \theta) = \int_0^t V(\theta, X_s)\, dW(s).$$

It can be seen that $\mathbf{Z}^T(\theta) = \{Z(t, \theta) \leq t \leq T\}$ is a.s. continuous for all $T > 0$ as a function of θ from Θ into $C[0, T]$. This follows from the arguments given in Section 2.2.

Lemma 2.5.4 *Under the conditions stated above,*

(i) *there exist constants $C_1 > 0$ and $C_2 > 0$ independent of T such that $P\{\sup_\theta \sup_{0 \leq t \leq T} |Z(t, \theta)| \geq C_1 \lambda^{1/(d+\alpha_0)}\} \leq C_2 \frac{T^{(d+\alpha_0)/2}}{\lambda}$;*

(ii) *for every $\gamma > 1/(d + \alpha_0)$, there exists H such that*

$$\lim_{T \to \infty} \sup_{\theta \in \Theta} \frac{|Z(T, \theta)|}{T^{1/2} (\log T)^\gamma} \leq H \text{ a.s.};$$

(iii) *$\lim_{T \to \infty} \sup_{\theta \in \Theta} \frac{|Z(T,\theta)|}{T} = 0$ a.s.;*

(iv) *$\lim_{T \to \infty} \sup_{\theta \in \Theta} \frac{1}{T} \int_0^T a''(\theta, X_s) dW_s = 0$ a.s.*

Proof. Parts (i) and (ii) are consequences of Lemmas 2.2.9 and 2.2.10. Part (iii) follows from (ii), and part (iv) follows from (iii) and Lemma 2.5.3.

Lemma 2.5.5 *Under the conditions (B1)–(B8),*

(a) *for each fixed t,*

$$\lim_{T\to\infty} \log \gamma_T(t) = -\tfrac{1}{2}\beta t^2 \text{ a.s.};$$

(b) *for every $\varepsilon, 0 < \varepsilon < \beta$, there exist δ_0 and T_0 such that*

$$\gamma_T(t) \le \exp\left(-\tfrac{1}{2}t^2(\beta-\varepsilon)\right)$$

for $|t| \le \delta_0 T^{1/2}$ and $T \ge T_0$ a.s.;

(c) *for every $\delta > 0$, there exist a positive ε and T_0 such that*

$$\sup_{|t|\ge\delta T^{1/2}} \gamma_T(t) \le \exp\left(-\tfrac{1}{4}T\varepsilon\right) \qquad \text{for } T \ge T_0 \text{ a.s.}$$

Proof. Note that

$$\begin{aligned}\log\gamma_T(t) &= \int_0^T [a(\theta_T + tT^{-1/2}, X_s) - a(\theta_T, X_s)]\,dW_s \\ &\quad -\tfrac{1}{2}\int_0^T [a(\theta_T + tT^{-1/2} X_s) - a(\theta_0, X_s)]^2\,ds \\ &\quad +\tfrac{1}{2}\int_0^T [a(\theta_T, X_s) - a(\theta_0, X_s)]^2\,ds. \end{aligned} \tag{2.5.11}$$

Applying the mean-value theorem and the relation (2.5.6), it follows that

$$\log\gamma_T(t) = J_1 + J_2 + J_3 + J_4, \tag{2.5.12}$$

where

$$\begin{aligned} J_1 &= -\frac{t^2}{2T}\int_0^T [a'(\theta_0, X_s)]^2\,ds, \\ J_2 &= \frac{t^2}{2T}\int_0^T \{[a'(\theta_0, X_s)]^2 - [a'(\theta_T^{**}, X_s)]^2\}\,ds, \\ J_3 &= \frac{t^2}{2T}\int_0^T [a''(\theta_T^{*}, X_s)]\,dW_s, \\ J_4 &= -\int_0^T [a(\theta_T, X_s) - a(\theta_0, X_s)] \\ &\quad \times [a(\theta_T + tT^{-1/2}, X_s) - a(\theta_T, X_s) - tT^{-1/2}a'(\theta_T, X_s)]\,ds, \end{aligned}$$

and with $\max(|\theta_T^* - \theta_T|, |\theta_T^{**} - \theta_T|) \le |t|/\sqrt{T}$.

By the ergodic theorem, it follows that

$$J_1 \to \frac{-t^2}{2}\beta \text{ a.s.} \qquad \text{as } T\to\infty.$$

Furthermore, by the ergodic theorem, there exist constants M and T_0 such that

$$\frac{1}{T}\int_0^T c(X(s))(1+|X(s)|)\,ds \le M \text{ a.s.} \qquad \text{for } T \ge T_0. \tag{2.5.13}$$

Assumptions (B5) and (B7) and the consistency of the estimator θ_T, together with inequality (2.5.13), prove that

$$J_2 \to 0 \text{ a.s.} \qquad \text{as } T \to \infty.$$

Again, an application of the mean-value theorem proves that

$$|J_4| \le c_0|\theta_T - \theta_0|\frac{t^2}{T}\int_0^T c(X(s))(1+|X(s)|)\,ds$$

and hence $J_4 \to 0$ a.s. as $T \to \infty$. Lemma 2.5.3 implies that $J_3 \to 0$ a.s. as $T \to \infty$. Combining all the above facts, we obtain that (a) holds.

Fix $\varepsilon_1 > 0$. There exists a T_1 such that, for all $T \ge T_1$,

$$-\frac{1}{2}\frac{t^2}{T}\int_0^T [a'(\theta_0, X_s)]^2\,ds \le -\frac{1}{2}t^2(\beta - \varepsilon_1) \text{ a.s.} \tag{2.5.14}$$

By Lemma 2.5.4, there exists a T_2 such that, for all $T \ge T_2$,

$$\sup_\theta \frac{1}{T}\int_0^T a''(\theta, X_s)\,dW_s \le \frac{\varepsilon_1}{2} \text{ a.s.} \tag{2.5.15}$$

Now

$$\begin{aligned}
|J_2| &\le \frac{t^2}{2T}|\theta_T^{**} - \theta_0|\int_0^T c(X(s))(1+|X(s)|)\,ds \\
&\le \frac{t^2}{2T}\left(\frac{|t|}{\sqrt{T}} + |\theta_T - \theta_0|\right)\int_0^T c(X(s))(1+|X(s)|)\,ds \\
&\le \frac{t^2}{2T}(\delta_0 + |\theta_T - \theta_0|)\int_0^T c(X(s))(1+|X(s)|)\,ds,
\end{aligned}$$

for $|t|/\sqrt{T} \le \delta_0$. Using (2.5.13) and choosing δ_0 suitably with the additional property that θ_T is strongly consistent, it follows that there exist a δ_0 and T_3 such that

$$\frac{|t|}{\sqrt{T}} \le \delta_0 \quad \text{and} \quad T \ge T_3 \to |J_2| \le \frac{t^2}{2T}\varepsilon_1 \text{ a.s.} \tag{2.5.16}$$

A similar analysis using the mean-value theorem shows that there exist a T_4 and a δ_1 such that

$$\frac{|t|}{\sqrt{T}} \le \delta_1 \quad \text{and} \quad T \ge T_4 \to |J_4| \le \frac{t^2}{2T}\varepsilon_1 \text{ a.s.} \tag{2.5.17}$$

Combining the estimates (2.5.14)–(2.5.17), we obtain (b).

Note that

$$\begin{aligned}
\frac{\log \gamma_T(t)}{T} &= \frac{1}{T}\int_0^T [a(\theta_T + tT^{-1/2}, X_s) - a(\theta_T, X_s)]\,dW_s \\
&\quad -\frac{1}{2}\frac{1}{T}\int_0^T [a(\theta_T + tT^{-1/2}, X_s) - a(\theta_0, X_s)]^2\,ds \\
&\quad +\frac{1}{2}\frac{1}{T}\int_0^T [a(\theta_T, X_s) - a(\theta_0, X_s)]^2\,ds \\
&= A_1(t,T) + A_2(t,T) + A_3(T) \qquad \text{(say).}
\end{aligned}$$

Then $A_3(T) \to 0$ a.s. as $T \to \infty$ by the arguments given earlier. Furthermore,

$$\sup_t |A_1(t, T)| \le 2 \sup_\theta \frac{1}{T} | \int_0^T a(\theta, X_s)\, dW_s| \to 0 \text{ a.s.}$$

Finally, the strong consistency of θ_T implies that there exists a T_0 such that for all $T \ge T_0$, $|\theta - \theta_T| \le \delta$ a.s. Hence, if $|t|/\sqrt{T} \ge \delta$ and $T \ge T_0$, then $|\theta_T + tT^{-1/2} - \theta_0| > \delta/2$ and

$$A_2 \le -\frac{1}{2} \inf_{|\theta-\theta_0|\ge\delta/2} \frac{I_T(\theta)}{T} \to -\frac{1}{2}\lambda\left(\frac{\delta}{2}\right) \text{ a.s.}$$

Combining these estimates for A_1, A_2 and A_3, we obtain (c). This completes the proof.

Lemma 2.5.6 *Under the conditions (B1)–(B8) and (L1) and (L2),*

(a) there exists a $\delta_0 > 0$ *such that*

$$\lim_{T\to\infty} \int_{|t|\le\delta_0 T^{1/2}} K(t) \left| \gamma_T(t)\lambda(\theta_T + tT^{-1/2}) - \lambda(\theta_0) e^{-\frac{1}{2}\beta t^2} \right| dt = 0 \ a.s.\ [P_{\theta_0}];$$

(b) for every $\delta > 0$,

$$\lim_{T\to\infty} \int_{|t|>\delta T^{1/2}} K(t) \left| \gamma_T(t)\lambda(\theta_T + tT^{-1/2}) - \lambda(\theta_0) e^{-\frac{1}{2}\beta t^2} \right| dt = 0 \ a.s.\ [P_{\theta_0}].$$

The proof of Lemma 2.5.6 is analogous to the proofs of Lemmas 3.2 and 3.3 in Chapter 10 of Basawa and Prakasa Rao (1980) (cf. Borwanker *et al.* 1971). We omit the details.

The main Theorem 2.5.1 is now a consequence of the above lemmas, following arguments given in Basawa and Prakasa Rao (1980, Chapter 10). A special case of this theorem is known as the Bernstein–von Mises theorem.

Corollary 2.5.7 *If conditions (B1)–(B8) hold and* $\int_{-\infty}^{\infty} |\theta|^m \lambda(\theta) d\theta < \infty$ *for some* $m \ge 0$, *then*

$$\lim_{T\to\infty} \int_{-\infty}^{\infty} |t|^m \left| p^*(t|X^T) - \left(\frac{\beta}{2\pi}\right)^{1/2} e^{-\frac{1}{2}\beta t^2} \right| dt = 0 \ a.s.$$

Remark. The case $m = 0$ gives the Bernstein–von Mises theorem showing that the posterior density converges to the normal density in L_1-mean under some regularity conditions.

Let $\ell(\theta, \phi)$ be a loss function defined on $\Theta \times \Theta$. Assume that $\ell(\theta, \phi) = \ell(|\theta - \phi|) \ge 0$ and $\ell(\cdot)$ is nondecreasing. Suppose $R(\cdot)$ is a nonnegative function, and K and G are functions such that

(M1) $R(T)\ell\left(\frac{t}{\sqrt{T}}\right) \le G(t)$ for all t and $T \ge 0$;

(M2) $R(T)\ell\left(\frac{t}{\sqrt{T}}\right) \to K(t)$ uniformly on bounded intervals of T as $T \to \infty$;

(M3) $\int_{-\infty}^{\infty} K(t + m)e^{-\frac{1}{2}\beta t^2}\, dt$ has a strict minimum at $m = 0$;

(M4) $G(\cdot)$ satisfies conditions (L1) and (L2).

Definition. An estimator $\hat{\theta}_T$ is said to be a *regular Bayes estimator* based on $X^T = \{X(t), 0 \le t \le T\}$ if it minimizes

$$B_T(\phi) = \int_\Theta \ell(\theta, \phi) p(\theta|X^T)\, d\theta.$$

Theorem 2.5.8 *Under the conditions (B1)–(B8) and (M1)–(M4):*

(i) $T^{1/2}(\theta_T - \hat{\theta}_T) \to 0$ *a.s.* $[P_{\theta_0}]$ *as* $T \to \infty$;

(ii) $\lim_{T\to\infty} R(T)B_T(\theta_T) = \lim_{T\to\infty} R(T)B_T(\hat{\theta}_T) = \left(\frac{\beta}{2\pi}\right)^{1/2} \int_{-\infty}^{\infty} K(t)e^{-\frac{1}{2}\beta t^2}\, dt$, $\beta = I(\theta_0)$.

The proof of this theorem, as a consequence of Theorem 2.5.7, can be found in Basawa and Prakasa Rao (1980) or Borwanker *et al.* (1971). We omit the details. As a corollary, we obtain the following major result giving the asymptotic properties of the Bayes estimators.

Corollary 2.5.9 *Under the conditions (B1)–(B8) and (M1)–(M4):*

(i) $\hat{\theta}_T \to \theta_0$ *a.s* $[P_{\theta_0}]$ *as* $T \to \infty$;

(ii) $\sqrt{T}(\hat{\theta}_T - \theta_0) \xrightarrow{\mathcal{L}} N(0, I(\theta_0)^{-1})$ *as* $T \to \infty$ *under* P_{θ_0}.

This corollary follows from the properties of the MLE θ_T discussed in Section 2.2 and Theorem 2.5.8. In other words, the Bayes estimator is asymptotically normal and asymptotically efficient for a suitable class of loss functions and for smooth priors satisfying the main condition that the prior density is continuous and positive in a neighborhood of the true parameter θ_0.

Remarks. The Bernstein–von Mises theorem for a class of diffusion processes governed by a stochastic linear differential equation was first proved in Prakasa Rao (1981). The rate of convergence in the Bernstein–von Mises theorem for such processes was investigated in Mishra and Prakasa Rao (1987; 1991). Mishra (1989) extended the result for nonhomogeneous diffusion processes satisfying a stochastic differential equation of the type

$$dX_t = a(\theta, t, X_t)\, dt + dW_t, \qquad X_0 = 0,\ t \geq 0,$$

following the techniques in Prakasa Rao (1981) and Mishra and Prakasa Rao (1985a). Prakasa Rao (1988a) discussed the law of iterated logarithem for fluctuations of posterior distributions for a class of diffusion processes and developed a sequential test of power one for testing the drift parameter θ for processes governed by a stochastic differential equation of the form

$$dX_t = [a(t, X) + \theta b(t, X)]\, dt + \sigma(t, X)\, dW_t$$

where a, b and σ are known. The Bernstein–von Mises theorem for diffusion fields

$$dX_\zeta = a_\zeta(X)\, d\zeta + \sigma_\zeta(X)\, dW_\zeta, \qquad \zeta \in R_+^2,$$

where W_ζ is a two-dimensional Wiener field, was proved in Prakasa Rao (1984). An alternate approach for the study of the Bayes estimation for drift parameter θ for diffusion type processes is discussed in Kutoyants (1984a; 1984b) using the weak convergence of the log-likelihood ratio process as presented in Section 2.2.

2.6 Minimum Distance Method for the Estimation of the Drift Parameter

Consider the stochastic differential equation

$$dX_t = a_t(\theta, X)\, dt + \varepsilon\, dW_t, \qquad X_0 = x_0,\ 0 \leq t \leq T, \tag{2.6.1}$$

where $a_t(\cdot, X)$ is a known measurable nonanticipative functional, $\theta \in \Theta$, an open bounded set in R^d, $d \geq 1$, W_t is a standard Wiener process and $\varepsilon \in (0, 1]$. The solution $\{X_t\}$ of (2.6.1) is a diffusion type process. The problem is again to estimate the parameter θ based on the observation of the process X_t over $[0, T]$ and study the asymptotic properties of the estimator as $\varepsilon \to 0$.

In earlier sections, we studied the properties of the maximum likelihood and the Bayes estimators for the parameter θ. We now consider a different approach to the problem by minimizing a suitable norm; for instance, we may estimate θ by $\hat{\theta}_\varepsilon$, where

$$\hat{\theta}_\varepsilon = \arg\min_{\theta\in\Theta} \|X - x(\theta)\|, \tag{2.6.2}$$

in which

$$\|f\|^2 = \int_0^T |f(t)|^2 \mu(dt) \tag{2.6.3}$$

for some finite measure $\mu(\cdot)$ and $x_t(\theta)$ is a solution of the differential equation

$$\frac{dx_t}{dt} = a_t(\theta, x), x_0, \qquad 0 \leq t \leq T. \tag{2.6.4}$$

Another approach is to estimate θ by θ_ε^* under the uniform metric:

$$\theta_\varepsilon^* = \arg\min_{\theta\in\Theta} \sup_{0\leq t\leq T} |X_t - x_t(\theta)|. \tag{2.6.5}$$

If the equation

$$\sup_{0\leq t\leq T} |X_t - x_t(\theta_\varepsilon^*)| = \inf_{\theta\in\Theta} \sup_{0\leq t\leq T} |X_t - x_t(\theta)|$$

has many solutions, then we call the solution with the minimal norm the *minimum distance estimator* (MDE). We now study the properties of the estimator θ_ε^*.

2.6.1 L_∞-norm (MDE)

Suppose the following conditions hold:

(N1) the functional $a_t(\theta, \cdot)$ satisfies the following inequalities for all $t \in [0, T]$, and $x, y \in C[0, T]$,

(i) $|a_t(\theta, x) - a_t(\theta, y)| \leq L_1 \int_0^t |x_s - y_s|\, ds + L_2|x_t - y_t|$,

(ii) $|a_t(\theta, x)| \leq L_1 \int_0^t (1 + |x_s|)\, ds + L_2(1 + |x_t|)$,

where L_1 and L_2 are positive constants; and

(N2) for any $\nu > 0$,

$$g(\nu) = \inf_{|\theta-\theta_0|\geq\nu} \sup_{0\leq t\leq T} |x_t(\theta) - x_t(\theta_0)| > 0, \tag{2.6.6}$$

where $x_t(\theta)$ is the solution of equation (2.6.4).

Let $P_\theta^{(\varepsilon)}$ be the probability measure induced by the process $\{X_t, 0 \leq t \leq T\}$ on $(C[0, T], \mathcal{B}[0, T])$ when θ is the parameter. Here $\mathcal{B}([0, T])$ is the σ-algebra of Borel subsets under the supremum norm.

Theorem 2.6.1 *Under the conditions (N1) and (N2),*

$$P_{\theta_0}^{(\varepsilon)}\{|\theta_\varepsilon^* - \theta_0| \geq \nu\} \leq 2\exp\left\{-\gamma\frac{g^2(\nu)}{\varepsilon^2}\right\} \tag{2.6.7}$$

for every $\nu > 0$*, for some positive constant* γ.

Proof. Under the condition (N1)(ii), it can be checked that

$$\sup_{0\leq t\leq T} |X_t - x_t(\theta_0)| \leq C\varepsilon \sup_{0\leq t\leq T} |W_t|, \tag{2.6.8}$$

for some constant $C > 0$, using the Gronwall–Bellman lemma (Appendix D) (Kutoyants 1984b, Lemma 3.4.4). Let $\|\cdot\|$ be the uniform norm and

$$H_0 = H_0(\nu) = \{\omega : \inf_{|\theta-\theta_0|<\nu} \|X - x(\theta)\| < \inf_{|\theta-\theta_0|\geq\nu} \|X - x(\theta)\|\}. \tag{2.6.9}$$

Then, for all $\omega \in H_0$, the MDE $\theta_\varepsilon^* \in \{\theta : |\theta - \theta_0| < \nu\}$.

Furthermore, we have that

$$\begin{aligned}
P_{\theta_0}^{(\varepsilon)}\{H_0^C\} &\leq P_{\theta_0}^{(\varepsilon)}\Big\{\inf_{|\theta-\theta_0|<\nu} (\|X - x(\theta_0)\| + \|x(\theta) - x(\theta_0)\|) \\
&\geq \inf_{|\theta-\theta_0|\geq\nu} (\|x(\theta) - x(\theta_0)\| - \|X - x(\theta_0)\|)\Big\} \\
&\leq P_{\theta_0}^{(\varepsilon)}\Big\{\|X - x(\theta_0)\| \geq g(\nu) - C\varepsilon \sup_{0\leq t\leq T} |W_t|\Big\} \\
&\leq P\Big\{2C\varepsilon \sup_{0\leq t\leq T} |W_t| \geq g(\nu)\Big\} \\
&= P\left\{\sup_{0\leq t\leq T} |W_t| \geq \frac{g(\nu)}{2C\varepsilon}\right\} \\
&\leq 4P\left\{W_T > \frac{g(\nu)}{2C\varepsilon}\right\} \\
&\leq 2\exp\left\{\frac{-g(\nu)^2}{8TC^2\varepsilon^2}\right\},
\end{aligned} \tag{2.6.10}$$

using the properties of the norms, the relation

$$\inf_{|\theta-\theta_0|<\nu} \|x(\theta) - x(\theta_0)\| = 0,$$

inequality (2.6.8) and the following property of the Wiener process:

$$P\Big\{\sup_{0\leq t\leq T} W_t > a\Big\} = 2P\{W_T > a\} \leq \exp\left\{\frac{-a^2}{2T}\right\}.$$

Hence

$$P_{\theta_0}^{(\varepsilon)}\{|\theta_\varepsilon^* - \theta_0| \geq \nu\} \leq 2\exp\left\{-\gamma\frac{g(\nu)^2}{\varepsilon^2}\right\},$$

for some positive constant γ. This completes the proof.

In order to study the limit behavior of the normed difference $u_\varepsilon = \varepsilon^{-1}(\theta_\varepsilon^* - \theta_0)$, we consider a special case of the model. Suppose

$$a_t(\theta, X) = V(\theta, t, X_t) + \int_0^t K(\theta, t, s, X_s)\,ds, \tag{2.6.11}$$

where V and K satisfy the following condition.

(N3) The functions $V(\theta, t, x)$ and $K(\theta, t, s, x)$ have two continuous bounded derivatives with respect to θ and x.

Let $\dot{x}_t(\theta)$ denotes the vector of derivatives of $x_t(\theta)$ with respect to θ. It can be checked that the derivative exists under condition (N3). Let

$$I_t(\theta) = \dot{x}_t(\theta)\dot{x}_t(\theta)^T,$$

where T denotes the transposition. Suppose that

(N4)

$$\inf_{\theta\in\Theta}\inf_{|e|=1}\sup_{0\le t\le T}(e, I_t(\theta)e) > 0,$$

where e is a unit vector in R^d and $(\cdot,\cdot)$ denotes the inner product.

Let us introduce a Gaussian process $X_t^{(1)} = X_t^{(1)}(\theta)$ satisfying the stochastic differential equation

$$dX_t^{(1)} = \left[V_x'(\theta, t, X_t)X_t^{(1)} + \int_0^t K_x'(\theta, t, s, X_s)X_s^{(1)}\,ds\right]dt + dW_t, X_0^{(1)} = 0,$$
$$0 \le t \le T, \tag{2.6.12}$$

which is in fact a derivative, with probability one, of X_t with respect to ε at $\varepsilon = 0$ (cf. Kutoyants and Pilibossian 1994b). Here V_x' and K_x' are the derivatives of $V(\theta, t, x)$ and $K(\theta, t, s, x)$ with respect to x.

Define the random variable $\zeta = \zeta(\theta_0)$ by the relation

$$\|X^{(1)} - (\zeta, \dot{x}(\theta_0))\| = \inf_{u\in R^d}\|X^{(1)} - (u, \dot{x}(\theta_0))\|. \tag{2.6.13}$$

(N5) Assume that the equation (2.6.13) has a *unique* solution ζ with probability one.

Theorem 2.6.2 *Under the conditions (N1)–(N5),*

$$\varepsilon^{-1}(\theta_\varepsilon^* - \theta_0) \xrightarrow{\mathcal{L}} \zeta \quad \text{as } \varepsilon \to 0 \tag{2.6.14}$$

under P_{θ_0}.

Proof. Let us first localize the problem. Let $\nu = \nu_\varepsilon = \varepsilon\lambda_\varepsilon \to 0$, where $\lambda_\varepsilon \to \infty$ as $\varepsilon \to 0$. Define H_0 as in (2.6.9). Note that $|\theta_\varepsilon^* - \theta_0| < \nu_\varepsilon$ whenever $\omega \in H_0$. Let

$$F(u) = \sup_{0\le t\le T}|x_t(\theta_0 + u) - x_t(\theta_0)|^2.$$

It follows from condition (N3) that

$$\sup_{0\le t\le T} |x_t(\theta_0+u) - x_t(\theta_0) - (u, \dot{x}_t(\theta_0))| = O(|u|^2).$$

Define

$$k_0 = k(\theta_0) = \inf_{|e|=1} \sup_{0\le t\le T} (e, I_t(\theta_0)e).$$

Note that $k_0 > 0$ by condition (N4). Hence we can find a neighborhood V of zero such that

$$\inf_{u\in V} \frac{F(u)}{|u|^2} \ge \frac{1}{2}k_0,$$

and, for $u \in V$,

$$F(u) \ge \frac{1}{2}k_0|u|^2.$$

Condition (N2) implies that $F(u) > 0$ for $u \notin V$. Hence there exists $k > 0$ such that

$$F(u) \ge k|u|^2 \qquad \text{for all } u \in \Theta - \{\theta_0\}.$$

Thus

$$\inf_{|u|>\nu_\varepsilon} \sup_{0\le t\le T} |x_t(\theta_0+u) - x_t(\theta_0)|^2 \ge k\nu_\varepsilon^2.$$

Hence

$$g(\nu) \ge \sqrt{k}\nu_\varepsilon,$$

and Theorem 2.6.1 implies that

$$\begin{aligned} P_{\theta_0}^{(\varepsilon)}(|\theta_\varepsilon^* - \theta_0| \ge \nu_\varepsilon) &\le 2\exp\left\{-\gamma k\frac{\nu_\varepsilon^2}{\varepsilon^2}\right\} \\ &\le 2\exp\{-\gamma k\lambda_\varepsilon^2\}, \end{aligned}$$

and the last term tends to zero as $\varepsilon \to 0$. Let us now consider the behavior of the norm $\|X - x(\theta)\|$ for $\theta \in \{\theta : |\theta - \theta_0| < \nu_\varepsilon\}$. Let $\theta = \theta_0 + \varepsilon u$. Then

$$\begin{aligned} \varepsilon^{-1}\|X - x(\theta)\| &= \left\| \frac{X - x(\theta_0)}{\varepsilon} - \frac{x(\theta) - x(\theta_0)}{\varepsilon} \right\| \\ &= \|x^{(1)} - (u, \dot{x}(\theta_0)) - r + q\|, \end{aligned}$$

where

$$q_t = \frac{X_t - x_t(\theta_0)}{\varepsilon} - x_t^{(1)}$$

and

$$r_t = \frac{x_t(\theta_0+\varepsilon u) - x_t(\theta_0)}{\varepsilon} - (u, \dot{x}_t(\theta_0)).$$

In view of (N3), one can apply Taylor's formula to obtain

$$\begin{aligned} \sup_{0\le t\le T} \left| \frac{x_t(\theta_0+\varepsilon u) - x_t(\theta_0)}{\varepsilon} - (u, \dot{x}_t(\theta_0)) \right| &= \sup_{0\le t\le T} |(u, (\dot{x}_t(\theta_0^*) - \dot{x}_t(\theta_0)))| \\ &\le |u| \sup_{0\le t\le T} |\dot{x}_t(\theta_0^*) - \dot{x}_t(\theta_0)| \\ &\le C\varepsilon|u|^2, \end{aligned}$$

and hence

$$\sup_{|u|\le\lambda_\varepsilon}\sup_{0\le t\le T}|r_t|\le C\varepsilon\lambda_\varepsilon^2.$$

Following the relations (2.6.1), (2.6.8), (2.6.11) and (2.6.12), it follows that

$$\begin{aligned}
|q_t| &= \left|\frac{X_t - x_t(\theta_0)}{\varepsilon} - x_t^{(1)}\right| \\
&= \left|\int_0^t \left[\frac{a_v(\theta_0, X) - a_v(\theta_0, x)}{\varepsilon} - V_x'(\theta_0, v, x_v)x_v^{(1)} \right.\right.\\
&\quad \left.\left. - \int_0^v K_x'(\theta_0, v, h, x_h)x_h^{(1)}\,dh\right] dv\right| \\
&\le \int_0^t \left|\frac{V(\theta_0, s, X_s) - V(\theta_0, s, x_s)}{\varepsilon} - V_x'(\theta_0, s, x_s)x_s^{(1)}\right| ds \\
&\quad + \int_0^t\int_0^s \left|\frac{K(\theta_0, s, v, X_s) - K(\theta_0, s, v, x_v)}{\varepsilon} - K_x'(\theta_0, s, v, x_v)x_v^{(1)}\right| dv\,ds \\
&\le \int_0^t \left|V_x'(\theta_0, s, \tilde{X}_s)\left(\frac{X_s - x_s}{\varepsilon}\right) - V_x'(\theta_0, s, x_s)x_s^{(1)}\right| ds \\
&\quad + \int_0^t\int_0^s \left|K_x'(\theta_0, s, v, \tilde{\tilde{X}}_v)\frac{X_v - x_v}{\varepsilon} - K_x'(\theta_0, s, v, x_v)x_v^{(1)}\right| dv\,ds \\
&\le \int_0^t |V_x'(\theta_0, s, \tilde{X}_s)|\left|\frac{X_s - x_s}{\varepsilon} - x_s^{(1)}\right| ds \\
&\quad + \int_0^t |V_x'(\theta_0, s, \tilde{X}_s) - V_x'(\theta_0, s, x_s)|\,|x_s^{(1)}|\,ds \\
&\quad + \int_0^t\int_0^s |K_x'(\theta_0, s, v, \tilde{\tilde{X}}_v)|\left|\frac{X_v - x_v}{\varepsilon} - x_v^{(1)}\right| dv\,ds \\
&\quad + \int_0^t\int_0^s |K_x'(\theta_0, s, v, \tilde{\tilde{X}}_v) - K_x'(\theta_0, s, v, x_v)|\,|x_v^{(1)}|\,dv\,ds \\
&\le C_1\int_0^t |q_s|\,ds + C_2\int_0^t\int_0^s |q_v|\,dv\,ds + C_3\varepsilon\sup_{0\le t\le T}|W_t|\sup_{0\le t\le T}|x_t^{(1)}|, \qquad (2.6.15)
\end{aligned}$$

for some constants $C_i > 0$, $i = 1, 2, 3$.

In view of (2.6.12), condition (N3) and the Gronwall–Bellman lemma in Appendix D, we obtain that

$$\sup_{0\le t\le T}|\varepsilon^{-1}(X_t - x_t(\theta_0)) - x_t^{(1)}| \le C\varepsilon\sup_{0\le t\le T}|W_t|^2. \qquad (2.6.16)$$

Let us now consider

$$\begin{aligned}
&\sup_{|u|<\lambda_\varepsilon}\sup_{0\le t\le T}\left|\frac{X_t - x_t(\theta_0+\varepsilon u)}{\varepsilon} - (x_t^{(1)} - (u, \dot{x}_t(\theta_0)))\right| \\
&\le \sup_{|u|<\lambda_\varepsilon}\sup_{0\le t\le T}\left\{\left|\frac{X_t - x_t(\theta_0)}{\varepsilon} - x_t^{(1)}\right| + \left|\frac{x_t(\theta_0+\varepsilon u) - x_t(\theta_0)}{\varepsilon} - (u, \dot{x}_t(\theta_0))\right|\right\} \\
&\le \sup_{0\le t\le T}|q_t| + \sup_{|u|<\lambda_\varepsilon}\sup_{0\le t\le T}|r_t| \\
&\le C\varepsilon\sup_{0\le t\le T}|W_t|^2 + C\varepsilon\lambda^2\varepsilon,
\end{aligned}$$

from (2.6.15) and (2.6.16). Hence, if we choose λ_ε such that $\varepsilon\lambda_\varepsilon^2 \to 0$, then

$$\sup_{|u|\le\lambda_\varepsilon}\left|\frac{\|X - x(\theta_0+\varepsilon_u)\|}{\varepsilon} - \|x^{(1)} - (u, \dot{x}(\theta_0))\|\right| \to 0$$

with probability one. Hence, for $t \in [0, T]$, we have the uniform convergence of continuous functions of u towards a continuous function, and the minimizer, as a continuous functional of the trajectory, converges to the minimizer of the limit process in view of (N5). Therefore

$$\begin{aligned}\arg\inf_{|u|<\lambda_\varepsilon}\|X - x(\theta_0+\varepsilon_u)\| &\sim \arg\inf_{|u|<\lambda_\varepsilon}\| x^{(1)} - (u, \dot{x}(\theta_0))\| \\ &\xrightarrow{\mathcal{L}} \zeta \qquad \text{as } \varepsilon \to 0,\end{aligned}$$

where ζ is as defined by equation (2.6.13).

One can consider other norms instead of the uniform norm for the problem of estimation of θ. We make a few remarks about such results.

2.6.2 L_2-norm (MDE)

Let us consider the diffusion type process

$$dX_t = a_t(\theta, X)\,dt + \varepsilon\,dW_t, \qquad X_0 = x_0,\ 0 \le t \le T, \tag{2.6.17}$$

where $\theta \in \Theta \subset R^d$ and the problem is to estimate the value of θ from a realization of the process $X^T = \{X_t, 0 \le t \le T\}$. Let H be a Hilbert space with $|\cdot|_H$ as the norm and $(\cdot,\cdot)$ its inner product. For every $\varepsilon \in [0, 1]$ and $\theta \in \Theta$, consider an H-valued random element $Z_\varepsilon(\theta)$. Define θ_ε^* to be the MDE of θ if

$$\inf_{\theta\in\Theta}|Z_\varepsilon(\theta)|_H = |Z_\varepsilon(\theta_\varepsilon^*)|_H,$$

and suppose this equation has a unique solution for sufficiently small ε.

Example 2.6.1 (*Diffusion process with ergodic property*)
Suppose the process $\{X_t\}$ satisfying the stochastic differential equation

$$dX_t = a(\theta, X_t)\,dt + \sigma(X_t)\,dW_t, \qquad 0 \le t \le T, \tag{2.6.18}$$

is stationary and ergodic with stationary distribution $F_\theta(x)$. Define the empirical distribution

$$F_T^*(x) = \frac{1}{T}\int_0^T I[X_t \le x]\,dt,$$

where $I(A)$ denotes the indicator function of the set A. Let $H = L_2(\mu)$, where μ is a finite measure and

$$Z_\varepsilon(\theta) = F_T^* - F_\theta, \qquad \varepsilon = \frac{1}{T}.$$

Then

$$\hat{\theta}_T = \arg\min_{\theta\in\Theta}|F_T^* - F_\theta|_{L_2(\mu)} \tag{2.6.19}$$

gives an MDE $\hat{\theta}_T$ for θ.

Dietz and Kutoyants (1997) considered a stochastic differential equation of the form

$$dX_t = a(\theta, X_t)\,dt + dW_t, \qquad t \geq 0,\ \theta \in \Theta \subset R^d, \tag{2.6.20}$$

with a given initial condition X_0, known function $a(\cdot,\cdot)$ and $\theta \in \Theta$, a nonempty bounded open subset of R^d. Let

$$X(\theta)_t = X_0 + \int_0^t a(\theta, X_u)\,du. \tag{2.6.21}$$

They study the performance of the MDEs $\hat{\theta}_T$ defined by

$$\hat{\theta}_T = \arg\inf_{\theta\in\Theta} \int_0^T (X_t - X(\theta)_t)^2\,\eta^T(dt), \qquad T > 0 \tag{2.6.22}$$

as $T \to \infty$, where $\{\eta^T, T \geq 0\}$ is a family of measures for which a generalized version of the Toeplitz lemma holds. They prove that the estimator $\hat{\theta}_T$ is strongly consistent under some regularity conditions and is asymptotically normal for $d = 1$. If $a(\theta, x)$ is of the form $a_1(\theta) + a_2(x)$ – the 'signal plus noise' model – one can obtain an efficient estimator for θ by choosing $\eta^T, T \geq 0$, appropriately.

Consider a family of measures $\{\eta^T, T \geq 0\}$ on $([0,\infty), \mathcal{B}_{[0,\infty]})$ satisfying the following conditions: (i) for all $T > 0$, η^T is a probability measure with $\eta^T((0,T]) = 1$; (ii) for all $K > 0$, $\eta^T((0,K]) \to 0$ as $T \to \infty$. We say that $\{\eta_T\}$ satisfies the TOE property. It is easy to see that under these conditions, the following version of the Toeplitz lemma holds:

$$\int_{[0,\infty]} f_t\,\eta^T(dt) \to f_\infty \qquad \text{as } T \to \infty, \tag{2.6.23}$$

for every bounded and measurable function $f : [0,\infty) \to R$ for which $f_\infty \triangleq \lim_{t\to\infty} f_t$ exists. Examples of such families include:

$$\eta^T(dt) = \frac{1}{T} I_{[0,T]}(t)\,dt, \qquad T > 0; \tag{2.6.24a}$$

$$\eta^T(dt) = T^{1-T} t^{T-1} I_{[0,T]}(t)\,dt, \qquad T > 0; \tag{2.6.24b}$$

$$\eta^T(dt) = I_{[C(T),C(T)+1]}(t)\,dt, \qquad T > 1. \tag{2.6.24c}$$

Here $C : (1,\infty) \to [0,\infty)$ with the properties

$$C(T) \leq T - 1 \text{ for } T > 1, \text{ and } C(T) \to \infty \text{ as } T \to \infty;$$

and I_A denotes the indicator function of the set A.

Example 2.6.2 (*Shifted coefficient model*)
Consider the stochastic differential equation

$$dX_t = c(X_t - \theta)\,dt + dW_t, \qquad X_0 = X_0,\ t \geq 0, \tag{2.6.25}$$

where $\theta \in \Theta \subset R$, Θ bounded and nonempty. Here $a(\theta, x) = c(x - \theta)$. Suppose the following conditions hold:

(P1) There exists $L > 0$ such that

$$|c(x) - c(y)| \leq L|x - y|,$$
$$|c(x)| \leq L(1 + |x|)$$

for all x, y.

(P2) $m(A) = \int_A \exp\{2\int_0^x c(y)\,dy\}\,dx$, $A \in \mathcal{B}(R)$ defines a finite measure on $(R, \mathcal{B}(R))$.

With $\mu_0(A) = m(A)/m(R)$, $A \in \mathcal{B}(R)$, consider the function h defined by

$$h(y) = \int_R (c(x) - c(x-y))\,\mu_0(dx), \qquad y \in R.$$

(P3) The function $h(\cdot)$ is strictly monotone in a neighborhood U of zero and satisfies the condition $|h(y)| \geq \bar{C} > 0$ for some constant $\bar{C}$ and for every $y \notin U$.

(P4) The function $c(\cdot)$ is twice continuously differentiable with derivatives c', c'' satisfying

$$|c'(x) - c'(y)| \leq M|x-y|,$$
$$|c''(x) - c''(y)| \leq M|x-y|^k$$

for all $x, y \in R$, for some constants $M > 0$ and $k > 0$.

(P5) $E_{\mu_0}(c') \neq 0$.

Let

$$\psi_T = \int_{(0,T]} t^2 \eta^T(dt), \tag{2.6.26}$$

and

$$\zeta^T(dt) = \frac{1}{\psi_T} t^2 \eta^T(dt). \tag{2.6.27}$$

It can be checked that (ζ^T) has the TOE property if $\{\eta^T\}$ does. Define

$$\zeta_t^T = \int_{(0,t]} s\,\eta^T(ds), \qquad \sigma_T^2 = \int_{(0,T]} (\zeta_T^T - \zeta_s^T)^2\,ds \tag{2.6.28}$$

and

$$\phi_T = \psi_T \sigma_T^{-1}. \tag{2.6.29}$$

Suppose the following condition holds.

(Q) There is some constant $C > 0$ such that

$$\lim_{T\to\infty} \frac{\sigma_T^2}{T(\zeta_T^T)^2} \geq C. \tag{2.6.30}$$

The TOE condition implies that the mass of the η^Ts is shifted out of every bounded part of the nonnegative real axis. The additional condition (R) requires this shifting at a certain rate. This condition holds for the families $\{\eta^T\}$ given above in (2.6.24a) and (2.6.24b). It will hold for the family defined by (2.6.24c) if $\liminf_{T\to\infty} T^{-1}c(T) > 0$.

Dietz and Kutoyants (1997) proved that under conditions (P1)–(P5), if $\{\eta^\tau\}$ satisfies the TOE property and condition (Q), then the MDE $\hat{\theta}^T$ satisfies the property

$$\phi_T(\hat{\theta}^T - \theta_0) \stackrel{\mathcal{L}}{\to} N(0, [E_{\mu_0}(c')]^{-2}) \tag{2.6.31}$$

as $T \to \infty$. We omit the details.

In general the MDE is not asymptotically efficient and the MLE has a lower asymptotic variance. However, if $a(\theta, x)$ can be represented in the form $a(\theta, x) = a_1(\theta) + a_2(x)$, then the MDE corresponding to the family of measures $\eta^T(dt) = T^{1-T} t^{T-1} I_{[0,T]}(t)\,dt$, $T > 0$, is asymptotically efficient. In particular, for the shifted Ornstein–Uhlenbeck model

$$dX_t = \rho(X_t - \theta)\,dt + dW_t, \qquad t \geq 0,\ \rho < 0,\ X_0 = x_0, \tag{2.6.32}$$

the MDE corresponding to the above family of measure is asymptotically efficient in the sense of minimizing the asymptotic variance in the class of asymptotically normal estimators (cf. Dietz and Kutoyants 1997).

Example 2.6.3 (*Small diffusions*)
Let the diffusion type process

$$dX_t = a_t(\theta, X)\,dt + \varepsilon\,dW_t \qquad X_0 = x_0,\ 0 \le t \le T \tag{2.6.33}$$

be observed and $\varepsilon \to 0$. Denote by $x_t(\theta)$, the solution of the equation

$$\frac{dx_t}{dt} = a_t(\theta, x), \qquad x_t = x_0 \text{ at } t = 0,\ 0 \le t \le T. \tag{2.6.34}$$

Let μ be a measure on $[0, T]$ with $\mu([0, T]) < \infty$, and introduce the MDE θ_ε^* by the relation

$$\theta_\varepsilon^* = \arg\min_{\theta\in\Theta} \|X - x(\theta)\|_{L_2(\mu)}. \tag{2.6.35}$$

For this scheme of observations, we can introduce another MDE in the following way. Let

$$\tilde{X}_t(\theta) = x_0 + \int_0^t a_s(\theta, X)\,ds, \tag{2.6.36}$$

and define θ_ε^{**} by

$$\theta_\varepsilon^{**} = \arg\min_{\theta\in\Theta} \|X - \tilde{X}(\theta)\|_{L_2(\mu)}. \tag{2.6.37}$$

Let

$$\tilde{a}_\varepsilon = \frac{1}{Q_\varepsilon}\int_0^T G\left(\frac{t-\tau}{Q_\varepsilon}\right) dX_\tau \tag{2.6.38}$$

be a kernel type nonparametric estimator of $a_t(\theta, X)$ (cf. Prakasa Rao 1983b, Kutoyants 1991). Define

$$\theta_\varepsilon^{***} = \arg\min_{\theta\in\Theta} \|a(\theta, X) - \tilde{a}\|_{L_2(\mu)}. \tag{2.6.39}$$

This gives a third example of an MDE.

Example 2.6.4 (*Partially observed linear system*)
Consider the system

$$\begin{aligned} dX_t &= a_t(\theta)Y_t\,dt + \varepsilon\,dW_t, & X_0 &= 0,\ 0 \le t \le T,\\ dY_t &= b_t(\theta)Y_t\,dt + \varepsilon\,dV_t, & Y_0 &= y_0 \ne 0,\ 0 \le t \le T, \end{aligned} \tag{2.6.40}$$

where W and V are independent standard Wiener processes. Define

$$x_t(\theta) = y_0 \int_0^t a_s(\theta) \exp\left(\int_0^s b_r(\theta)\,dr\right) ds \tag{2.6.41}$$

and

$$\theta_\varepsilon^* = \arg\min_{\theta\in\Theta} \|X - x(\theta)\|_{L_2(\mu)} \tag{2.6.42}$$

or

$$\theta_\varepsilon^{**} = \arg\min_{\theta\in\Theta} \|X - \hat{X}(\theta)\|_{L_2(\mu)}, \tag{2.6.43}$$

where

$$\hat{X}_t(\theta) = x_0 + \int_0^t a_s(\theta)m_s(\theta)\,ds, \tag{2.6.44}$$

with

$$m_t(\theta) = E_\theta(Y_t|X_s, 0 \le s \le t), \tag{2.6.45}$$

and μ is a finite measure on $[0, T]$. The estimators θ_ε^* and θ_ε^{**} are MDEs for θ.

We now describe briefly the properties of MDE θ_ε^* given in Example 2.6.3. Suppose the function $a(\cdot,\cdot)$ satisfies the Lipschitz condition

$$|a_t(\theta, x) - a_t(\theta, y)| \le L \int_0^t |x_s - y_s|\,dK_s + L|x_t - y_t| \tag{2.6.46}$$

for all $\theta \in \Theta$ and $x(\cdot), y(\cdot) \in C[0, T]$, where $K(\cdot)$ is a nondecreasing right-continuous bounded function. Define

$$g(\delta) = \inf_{\theta_0 \in \Theta} \inf_{|\theta - \theta_0| > \delta} \|x(\theta) - x(\theta_0)\|^2. \tag{2.6.47}$$

The following result follows from arguments similar to those in Theorem 2.6.1. We omit the proof.

Theorem 2.6.3 *If, for every $\delta > 0$, the function $g(\delta) > 0$, then the MDE θ_ε^* is uniformly consistent as $\varepsilon \to 0$, and*

$$\sup_{\theta_0 \in \Theta} P_{\theta_0}^{(\varepsilon)}\{|\theta_\varepsilon^* - \theta_0| > \delta\} \le C \exp\left\{-K \frac{g(\delta)^2}{\varepsilon^2}\right\} \tag{2.6.48}$$

for some constants $C > 0$ and $K > 0$.

Suppose that the coefficient $a_t(\theta, x)$ in (2.6.33) is of the form

$$a_t(\theta, x) = V(\theta, t, x_t) + \int_0^t G(\theta, t, s, x_s)\,dK_s. \tag{2.6.49}$$

Let us denote the derivatives of the process X_t and x_t by $X_t^{(1)}$ and $\dot{x}_t$ with respect to ε and θ, respectively at the point $\varepsilon = 0$. Then $X_t^{(1)}$ is a Gaussian process and $\dot{x}_t$ is a vector of nonrandom functions. The following result proves the asymptotic normality of the MDE θ_ε^*. Its proof is analogous of that of Theorem 2.6.2. We omit the details.

Theorem 2.6.4 *Let the functions $V(\cdot)$ and $G(\cdot)$ have uniformly continuous derivatives with respect to θ and x, $g(\delta) > 0$ for $\delta > 0$ and the matrix $J(\theta) = (\dot{x}, \dot{x}^T)$ be positive definite. Here α^T denotes the transpose of α. Then*

$$\varepsilon^{-1}(\theta_\varepsilon^* - \theta) \xrightarrow{\mathcal{L}} N(0, \Gamma(\theta)) \text{ as } \varepsilon \to 0 \tag{2.6.50}$$

where $\Gamma(\theta) = J^{-1}(\theta)B(\theta)J(\theta)$ and

$$B(\theta) = \int_0^T \int_0^T \dot{x}_t \dot{x}_s^T E_\theta(X_t^{(1)} X_s^{(1)})\,d\mu_t\,d\mu_s.$$

Asymptotic expansions

Kutoyants (1991) obtained the following theorem giving a stochastic expansion for the MDE θ_ε^*.

Theorem 2.6.5 *(Asymptotic expansion; $d = 1$) Let the functions $V(\cdot)$ and $G(\cdot)$ have $k+2$, $k \geq 1$, continuous bounded derivatives with respect to θ and x, $g(\delta) > 0$ for $\delta > 0$ and $\inf_\theta J(\theta) > 0$. Then there exist a set A, and random variables $\psi_1, \ldots, \psi_k, \eta, \zeta$, such that the MDE admits the representation*

$$\theta_\varepsilon^* = \theta + \left\{ \sum_{j=1}^{k} \psi_j \varepsilon^j + \eta \varepsilon^{k+(1/2)} \right\} I(A) + \zeta I(A^c), \tag{2.6.51}$$

where $|\eta| < 1$ and

$$\sup_{\theta \in \Theta} P_\theta^{(\varepsilon)}\{A^c\} \leq C_1 \exp\{-k_1 \varepsilon^{-\gamma_1}\}, \tag{2.6.52}$$

$$\sup_{\theta \in \Theta} P_\theta^{(\varepsilon)}\{|\zeta| > \varepsilon^\delta\} < C_2 \exp\{-k_2 \varepsilon^{-\gamma_2}\} \tag{2.6.53}$$

for some positive constants C_i, k_i, γ_i, $i = 1, 2$ and for some $0 < \delta < \frac{1}{2}$. In fact

$$\psi_1 = J^{-1}(\dot{x}, X^{(1)}), \qquad J = J(\theta), \tag{2.6.54}$$

$$\psi_2 = J^{-3}\{2J(\dot{x}, X^{(1)})(\ddot{x}, X^{(1)}) - 3(\dot{x}, \ddot{x})(\dot{x}, X^{(1)})^2 + J^2(\dot{x}, X^{(2)})\}, \tag{2.6.55}$$

(where $\ddot{x}_t$ and $X_t^{(2)}$ are the second derivatives of x_t and X_t, respectively).

Robustness

Consider the stochastic differential eqution

$$dX_t = a(\theta, X_t)\,dt + \varepsilon\,dW_t, \qquad X_0 = x_0,\ 0 \leq t \leq T, \tag{2.6.56}$$

where $\theta \in \Theta \subset R^d$ and $a(\theta, x) > 0$ for all x. Let $x_t = x_t(\theta)$ be a solution of the equation

$$\frac{dx_t}{dt} = a(\theta, x_t), \qquad \text{with } x_t = x_0 \text{ for } t = 0,\ 0 \leq t \leq T. \tag{2.6.57}$$

Define an MDE by

$$\theta_\varepsilon^* = \arg\min_{\theta \in \Theta} \|X - x(\theta)\|. \tag{2.6.58}$$

Suppose $a(\cdot, \cdot)$ satisfies the Lipschitz condition

$$|a(\theta, x) - a(\theta, y) \leq L|x - y| \tag{2.6.59}$$

and $x_t(\theta)$ is differentiable with respect to θ in $L_2(\mu)$ in the Fréchet sense. Then

$$\|x(\theta) - x(\theta_0) - (\theta - \theta_0, \dot{x}(\theta_0))\| = o(|\theta - \theta_0|). \tag{2.6.60}$$

Suppose the matrix $J = ((\dot{x}(\theta_0), \dot{x}(\theta_0)^T))$ is positive definite. Fix $\theta_0 \in \Theta$ to be the true parameter. Let us find the lower bound for any arbitrary estimator when the real observation differs slightly from the proposed model. Consider a neighborhood of the measure $P_{\theta_0}^{(\varepsilon)}$ through a perturbation of $a(\cdot, \cdot)$ defined by

$$a(h, \theta_0, x) = a(\theta_0, x) + \varepsilon h(x) a(\theta_0, x)^{-1}. \tag{2.6.61}$$

Consider the differential equation

$$\frac{dx_t^{(h)}}{dt} = a(h, \theta_0, x_t^{(h)})\, dt, \qquad x_0^{(h)} = x_0 \tag{2.6.62}$$

and the process

$$dX_t^{(h)} = a(h, \theta_0, X_t^{(h)}) + \varepsilon\, dW(t), \qquad X_0^{(h)} = x_0,\ 0 \le t \le T. \tag{2.6.63}$$

Let

$$H_m = \{h : \sup_x |h(x) a(\theta_0, x)^{-1}| < m\}, \qquad m = 1, 2, \ldots, \tag{2.6.64}$$

and define θ_{h_ε} by the relation

$$\inf_{\theta \in \Theta} \|x^{(h)} - x(\theta)\| = \|x^{(h)} - x(\theta_{h_\varepsilon})\|. \tag{2.6.65}$$

Then $x(\theta_{h_\varepsilon})$ is the closest element to $x^{(h)}$ of the family $x(\theta), \theta \in \Theta$. Consider the problem of the estimation of θ_{h_ε} from the observation on (2.6.63).

Let us introduce the loss function $\ell(x) = \ell_0(|x|^2)$, where ℓ_0 is a nondecreasing nonnegative function on $[0, \infty)$.

The following result, due to Kutoyants (1991), gives a locally asymptotic minimax lower bound.

Theorem 2.6.6 *Under the conditions stated above,*

$$\lim_{n \to \infty} \liminf_{\varepsilon \to 0} \inf_{\tilde{\theta}_\varepsilon} \sup_{h \in H_m} \int \ell(\varepsilon^{-1}(\tilde{\theta}_\varepsilon - \theta_{h_\varepsilon}))\, dP_h^{(\varepsilon)} \ge E_{\theta_0}(\ell(\zeta)), \tag{2.6.66}$$

where ζ is $N(0, \Gamma(\theta_0))$, $\Gamma(\theta_0) = J^{-1} B J$, with

$$B = \int_0^T \int_0^T \dot{x}_t(\theta_0) \dot{x}_s(\theta_0)^T E_{\theta_0}(X_t^{(1)} X_s^{(1)})\, d\mu_t\, d\mu_s. \tag{2.6.67}$$

Furthermore, ζ admits the representation

$$\zeta = J^{-1}(\theta_0) \int_0^T \dot{x}_t(\theta_0) X_t^{(1)}\, d\mu_t. \tag{2.6.68}$$

Definition. The estimator $\tilde{\theta}_\varepsilon$ for which the equality holds in (2.6.66) is said to be *locally asymptotically minimax* at θ.

Remarks. Suppose the function $\ell(x) \le \exp(\alpha x^2)$ for any $\alpha > 0$. Then it can be shown that the MDE θ_ε^* defined by (2.6.58) is locally asymptotically minimax.

2.6.3 L_1-norm (MDE)

Consider the problem of estimating the parameter θ of the Ornstein–Uhlenbeck process given by

$$dX_t = \theta X_t\, dt + \varepsilon\, dW_t, \qquad X_0 = x_0,\ 0 \le t \le T \tag{2.6.69}$$

where $\theta \in \Theta = (\alpha, \beta)$, $\varepsilon \in (0, 1]$, $x_0 > 0$ and W_t is a standard Wiener process. We are interested in the problem of estimation of θ as $\varepsilon \to 0$. Let θ_0 denote the true parameter. Let us consider the minimum distance estimate θ^*_ε defined by the L_1 norm, namely

$$\tilde{\theta}_\varepsilon = \arg \inf_{\theta \in \Theta} \int_0^T |X_t - x_t(\theta)|\, dt, \tag{2.6.70}$$

where $x_t(\theta)$ is a solution of the differential equation

$$\frac{dx_t}{dt} = \theta x_t \tag{2.6.71}$$

with $x_t = x_0$ at $t = 0$. Note that $x_t(\theta) = x_0 \exp(\theta t)$. Let

$$g(\nu) = \inf_{|\theta - \theta_0| > \nu} \int_0^T |x_t(\theta) - x_t(\theta_0)|\, dt. \tag{2.6.72}$$

Note that $g(\nu) > 0$ if $\nu > 0$. The following results hold.

Theorem 2.6.7 *Under the conditions stated above, for any $\nu > 0$,*

$$P^{(\varepsilon)}_{\theta_0}\{|\tilde{\theta}_\varepsilon - \theta_0| \ge \nu\} \le 2 \exp\left\{-K \frac{g(\nu)^2}{\varepsilon^2}\right\} \tag{2.6.73}$$

where $K = \exp\{-2|\theta_0|T\}/(2T)^3$.

Theorem 2.6.8 *Under the conditions stated above,*

$$\varepsilon^{-1}(\tilde{\theta}_\varepsilon - \theta_0) \xrightarrow{\mathcal{L}} \zeta_T \text{ under } P_{\theta_0} \quad \text{as } \varepsilon \to 0, \tag{2.6.74}$$

where

$$\zeta_T = \arg \inf_{-\infty < u < \infty} \int_0^T |Y_t - u t x_0 e^{\theta_0 t}|\, dt \tag{2.6.75}$$

and

$$Y_t = e^{\theta_0 t} \int_0^t e^{-\theta_0 s}\, dW_s.$$

Remarks. Note that

$$X_t - x_t(\theta_0) = \theta_0 \int_0^t (X_s - x_s(\theta_0))\, ds + \varepsilon W_t, \tag{2.6.76}$$

and the solution of the equation

$$u_t = \theta_0 \int_0^t u_s\, ds + \varepsilon W_t \tag{2.6.77}$$

is

$$X_t - x_t(\theta_0) = \varepsilon e^{\theta_0 t} \int_0^t e^{-\theta_0 s}\, dW_s. \tag{2.6.78}$$

Hence the 'derivative' of the process X_t with respect to ε is the Gaussian process

$$Y_t = e^{\theta_0 t} \int_0^t e^{-\theta_0 s}\, dW_s. \tag{2.6.79}$$

The distribution of ζ_T for fixed T is unknown. However, the following result holds as $T \to \infty$.

Theorem 2.6.9 *If $\alpha > 0$, then as $T \to \infty$,*

$$\zeta_T T x_0 \sqrt{2\theta_0} \xrightarrow{\mathcal{L}} N(0, 1). \tag{2.6.80}$$

Remark. The above results are due to Kutoyants and Pilibossian (1994a; 1994b). We omit the proofs.

2.7 *M*-estimation Method for the Estimation of Drift Parameter

We have seen several methods of estimation of drift parameter in the earlier sections. These include the method of maximum likelihood which is the 'best' in the sense that the estimators so obtained are asymptotically efficient by virtue of achieving either the Cramér–Rao lower bound or the Hájek–Le Cam lower bound under some regularity conditions. However, one can hardly obtain clean data generated by the model in a strict sense. The data might be either contaminated by some noise or there might be misspecification of the model itself. We have discussed M-estimation or minimum contrast estimation for the one-dimensional case in Section 2.3 (cf. Lanska 1979). We now consider the asymptotic behavior of the estimators for multidimensional diffusion models. Note that the technique used in Section 2.3 of representing the estimating equation pathwise via Lebesgue integrals in the one-dimensional case cannot be used in the multidimensional case. In the following discussion, we will supress the summation sign for simplicity.

Let $\boldsymbol{X}_t = (X_t^1, \ldots, X_t^d)$ be a diffusion process defined on a probability space $(\Omega, \mathcal{F}, P)$. Suppose a parameter θ is estimated based on an observation of the process $\boldsymbol{X}_t$ over $[0, T]$ using the functional Q_T defined from $C[0, T] \times \bar{\Theta}$ to R^d by

$$Q_T(\boldsymbol{X}, \theta) \equiv \exp\left\{ \int_0^T a_i(\boldsymbol{X}_t, \theta)\, dX_t^i - \tfrac{1}{2} \int_0^T R(\boldsymbol{X}_t, \theta)\, dt \right\}, \tag{2.7.1}$$

where a_i and R are known functions defined on $R^d \times \bar{\Theta}$, and Θ is a bounded convex domain in R^k. An estimator $\hat{\theta}_T$ is called an *M-estimator* of the parameter θ if the functional $Q_T(\boldsymbol{X}, \theta)$ attains its maximum at $\hat{\theta}_T$ in $\bar{\Theta}$.

Suppose that the process $\{\boldsymbol{X}_t\}$ satisfies the stochastic differential equation

$$dX_t^i = \sum_{\alpha=1}^{r} V_\alpha^i(\boldsymbol{X}_t)\, dW_t^\alpha + V_0^i(\boldsymbol{X}_t)\, dt, \qquad 1 \le i \le d,\ \boldsymbol{X}_0 = \eta \tag{2.7.2}$$

where $\{W_t^\alpha\}$ are independent standard Wiener processes. Notice that the true model *need not* correspond to any member of the parametric family of probability measures defined by $\theta \in \Theta$. We will now study the behavior of the M-estimator $\hat{\theta}_T$ defined above when the observed process is generated by (2.7.2).

Let θ_0 be a fixed point in Θ and consider the ratio

$$Z_T(\boldsymbol{u}, \theta_0) = Q_T(\boldsymbol{X}, \theta_0 + b^{-1/2}\boldsymbol{u}) / Q_T(\boldsymbol{X}, \theta_0), \qquad b \in R,\ \boldsymbol{u} \in R^k, \tag{2.7.3}$$

where $\boldsymbol{u} \in R^k$ such that $\theta_0 + b^{-1/2}\boldsymbol{u} \in \Theta$. Note that

$$\begin{aligned} Z_T(\boldsymbol{u}) &\equiv Z_T(\boldsymbol{u}, \theta_0) \\ &= \exp\left\{ \int_0^T \Delta a_i \, dX_t^i - \frac{1}{2} \int_0^T \Delta R \, dt \right\}, \end{aligned} \tag{2.7.4}$$

where $\Delta a_i = a_i(\boldsymbol{X}, \theta_0 + b^{-1/2}\boldsymbol{u}) - a_i(\boldsymbol{X}, \theta_0)$, $\Delta R = R(\boldsymbol{X}, \theta_0 + b^{-1/2}\boldsymbol{u}) - R(\boldsymbol{X}, \theta_0)$. We assume that $b = b_T$ is a positive function such that $b_T \to \infty$ as $T \to \infty$. We write $\theta = (\theta^1, \ldots, \theta^k)$.

We will use the following notation:

1. $\partial_i = \frac{\partial}{\partial x^i}$, $\delta_i = \frac{\partial}{\partial \theta^i}$; $\partial = (\partial_1, \ldots, \partial_d)$, $\delta = (\delta_1, \ldots, \delta_k)$;
2. $V_\alpha = V_\alpha^i \partial_i$, $v^{ij}(\boldsymbol{x}) = V_\alpha^i(\boldsymbol{x}) V_\alpha^j(\boldsymbol{x})$;
3. $L^i = \frac{1}{2} v^{ij} \partial_j + V_0^i$;
4. for any function $f(\boldsymbol{x}, \theta)$ and $\boldsymbol{u} \in R^k$,

$$\Delta f(\boldsymbol{x}, \theta_1, \theta_2) = \Delta f(\theta_1, \theta_2) = f(\boldsymbol{x}, \theta_2) - f(\boldsymbol{x}, \theta_1),$$
$$D^{(1)} f(\theta, \boldsymbol{u}) = f(\boldsymbol{x}, \theta + b^{-1/2}\boldsymbol{u}) - f(\boldsymbol{x}, \theta) - b^{-1/2} u^m \delta_m f(\boldsymbol{x}, \theta)$$

   and

$$D^{(2)} f(\theta, \boldsymbol{u}) = D^{(1)} f(\theta, \boldsymbol{u}) - (2b)^{-1} u^m u^n \delta_m \delta_n f(\boldsymbol{x}, \theta);$$

5. $L = L^i \partial_i$; and
6. $H = H(\theta) = H(\boldsymbol{x}, \theta) = \frac{1}{2} R(\boldsymbol{x}, \theta) - a_i(\boldsymbol{x}, \theta) V_0^i(\boldsymbol{x})$.

We assume that the following condition holds:

(R) The functions a_i and R are defined on $R^d \times \bar{\Theta}$, twice differentiable in θ and these derivatives are continuous.

In addition, we assume that, for the true model, there exists $\theta_0 \in \Theta$ satisfying the following conditions:

(S0) $\sup_{\theta \in \bar{\Theta}} \left| \frac{1}{b_T} \int_0^T [a_i(\boldsymbol{X}_t, \theta) - a_i(\boldsymbol{X}_t, \theta_0)] [dX_t^i - V_0^i(\boldsymbol{X}_t)\, dt] \right| \xrightarrow{p} 0$ as $T \to \infty$;

(S1) Let

$$C_T = \sup_{\theta \in \bar{\Theta}} \left| \frac{1}{b_T} \int_0^T \delta H(\boldsymbol{X}_t, \theta)\, dt \right|.$$

Suppose that $\lim_{\ell \to \infty} \sup_{T \geq 0} P(|C_T| > \ell) = 0$.

(T0) For each $\theta \in \bar{\Theta}$, there exists a random variable $\tilde{\Gamma}(\theta)$ such that

$$\frac{1}{b_T} \int_0^T [H(\boldsymbol{X}_t, \theta) - H(\boldsymbol{X}_t, \theta_0)]\, dt \xrightarrow{p} \tilde{\Gamma}(\theta), \qquad \text{as } T \to \infty,$$

and $\tilde{\Gamma}$ attains its minimum value zero in $\bar{\Theta}$ *only* at θ_0.

(T1) There exists a random matrix $\Gamma_T = ((\Gamma_{mnT}))$ such that

$$\Gamma_{mnT} \equiv \frac{1}{b_T} \int_0^T \delta_m \delta_n H(\boldsymbol{X}_t, \theta_0)\, dt \xrightarrow{p} \Gamma_{mn}, \qquad \text{as } T \to \infty,$$

and $\Gamma = ((\Gamma_{mn}))$ is positive definite a.s.

(T2) There exists a positive definite random matrix $\Phi_T = ((\phi_{mnT}))$ such that, at θ_0,

$$\phi_{mnT} \equiv \frac{1}{b_T}\int_0^T (\delta_m a_i + \partial_i G_m)(\delta_n a_j + \partial_j G_n) v^{ij}\, dt$$

$$\xrightarrow{p} \phi_{mn}, \qquad \text{as } T \to \infty,$$

where $G_m(x)$ are solutions of the partial differential equation

$$LG_m(x) = \delta_m H(x, \theta_0).$$

Suppose that $G_m(x)$ have continuous partial differentials $\partial_i \partial_j G_m$, and

$$b_T^{-1/2} G_m(X_T) \xrightarrow{p} 0 \qquad \text{as } T \to \infty,\ 1 \le m \le k.$$

(U0) For each $\varepsilon > 0$, there exists a random variable $\bar{\delta} = \bar{\delta}(\varepsilon) > 0$ a.s., such that

$$\lim_{T\to\infty} P\left[\sup_{\|u\| \le \delta b_T^{1/2}} \frac{1}{1 + \|u\|^2}\left|\int_0^T D^{(1)} a_i(X_t, \theta_0, u)[dX_t^i - V_0^i(X_t)]\, dt\right| > \varepsilon\right] = 0.$$

(U1) For each $\varepsilon > 0$, there exists a random variable $\bar{\delta} = \bar{\delta}(\varepsilon) > 0$ a.s. such that

$$\lim_{T\to\infty} P\left[\sup_{\|u\| \le \delta b_T^{1/2}} \frac{1}{1 + \|u\|^2}\left|\int_0^T D^{(2)} H(X_t, \theta_0, u)\, dt\right| > \varepsilon\right] = 0.$$

2.7.1 Asymptotic Behavior of *M*-estimators

Theorem 2.7.1 *Suppose that conditions (S0), (S1) and (T0) hold. Then $\hat{\theta}_T \xrightarrow{p} \theta_0$ as $T \to \infty$ (here the convergence in probability is under the true probability measure P).*

Proof. Let

$$Y_t(\theta) = \frac{1}{b_T}\log Q_T(X, \theta) - \frac{1}{b_T}\log Q_T(X, \theta_0),$$

$$M_T(\theta) = \int_0^T \Delta a_i(\theta_0, \theta)[dX_t^i - V_0^i\, dt],$$

$$N_T(\theta) = \int_0^T \Delta H_i(\theta_0, \theta)\, dt.$$

Then

$$Y_T(\theta) = \frac{1}{b_T} M_T(\theta) - \frac{1}{b_T} N_T(\theta) \tag{2.7.5}$$

and, for any $\theta_1, \theta_2 \in \bar{\Theta}$,

$$\left|\frac{1}{b_T} N_T(\theta_2) - \frac{1}{b_T} N_T(\theta_1)\right| \le \left|\frac{1}{b_T}\int_0^T \delta H_i(\theta_3)\, dt\right| \|\theta_2 - \theta_1\|$$
$$\le C_T \|\theta_2 - \theta_1\|, \tag{2.7.6}$$

for some θ_3 such that $\|\theta_3 - \theta_1\| < \|\theta_2 - \theta_1\|$, where $\{C_T\}$ is a sequence of stochastically bounded nonnegative random variables independent of θ_1 and θ_2 by conditions (S1). The inequality (2.7.6), together with the fact that $N_T(\theta_0) = 0$, implies that the family of probability measures of $\{N_T(\theta)/b_T, \theta \in \bar{\Theta}\}$ on $C(\bar{\Theta})$ with the supremum norm is tight. Hence the limit $\tilde{\Gamma}(\theta)$ defined by the condition (T0) are continuous functions on $\bar{\Theta}$ with probability one.

Let $\varepsilon > 0$ and $\eta > 0$. For any $\tau > 0$, there exists a finite subset $\{\theta_n\}$ in Θ such that the spheres $\{\theta \in \bar{\Theta} : \|\theta - \theta_n\| < \tau\}$ cover $\bar{\Theta}$. For each $\theta \in \bar{\Theta}$, let us choose $\theta_{n(\theta)}$ which is one of the closest points to θ in the subset $\{\theta_n\}$. For sufficiently small τ,

$$P\Big(|C_T| \sup_{\theta\in\bar{\Theta}} \|\theta - \theta_{n(\theta)}\| > \eta/3\Big) < \frac{\varepsilon}{3}$$

and

$$P\Big(\sup_{\theta\in\bar{\Theta}} |\tilde{\Gamma}(\theta_{n(\theta)}) - \tilde{\Gamma}(\theta)| > \eta/3\Big) < \frac{\varepsilon}{3}$$

by the continuity. Since

$$P\left(\sum_n \left|\frac{1}{b_T} N_T(\theta_n) - \tilde{\Gamma}(\theta_n)\right| > \eta/3\right) < \frac{\varepsilon}{3}$$

for large T, from (T0) and the finiteness of the subset $\{\theta_n\}$, we have

$$\begin{aligned}
P\left(\sup_{\theta\in\bar{\Theta}} \left|\frac{1}{b_T} N_T(\theta) - \tilde{\Gamma}(\theta)\right| > \eta\right) &\le P\left(\sup_{\theta\in\bar{\Theta}} \left|\frac{1}{b_T} N_T(\theta) - \frac{1}{b_T} N_T(\theta_{n(\theta)})\right| > \eta/3\right) \\
&\quad + P\left(\sup_{\theta\in\bar{\Theta}} \left|\frac{1}{b_T} N_T(\theta_{n(\theta)}) - \tilde{\Gamma}(\theta_{n(\theta)})\right| > \eta/3\right) \\
&\quad + P\left(\sup_{\theta\in\bar{\Theta}} \left|\tilde{\Gamma}(\theta_{n(\theta)}) - \tilde{\Gamma}(\theta)\right| > \eta/3\right) \\
&< \varepsilon.
\end{aligned} \tag{2.7.7}$$

Hence

$$\sup_{\theta\in\bar{\Theta}} \left|\frac{1}{b_T} N_T(\theta) - \tilde{\Gamma}(\theta)\right| \xrightarrow{p} 0 \text{ as } T \to \infty, \tag{2.7.8}$$

under the probability measure P.

In order to prove the theorem, it is sufficient to prove that for any neighborhood U of θ_0,

$$\lim_{T\to\infty} P\Big(\sup_{\theta\in U} Y_T(\theta) - \sup_{\theta\in U^c} Y_T(\theta) > 0\Big) = 1,$$

where U^c denotes the complement of the set U.

Let $\varepsilon > 0$. For any $\omega \in \Omega$ and $T > 0$, there exists $\theta_1(\omega, T) \in U^c$ such that

$$\sup_{\theta\in U^c} Y_T(\theta) < Y_T(\theta_1(\omega, T)) + \varepsilon.$$

Hence

$$P(\sup_{\theta\in U} Y_T(\theta) - \sup_{\theta\in U^c} Y_T(\theta) > 0)$$

$$
\begin{aligned}
&= P\Big(\sup_{\theta\in U} Y_T(\theta) > \sup_{\theta\in U^c} Y_T(\theta)\Big)\\
&\geq P\Big(\sup_{\theta\in U} Y_T(\theta) > Y_T(\theta_1(\omega,T)) + \varepsilon\Big)\\
&\geq P(0 > Y_T(\theta_1(\omega,T)) + \varepsilon)\ (\text{since } Y_T(\theta_0) = 0 \text{ and } \theta_0 \in U)\\
&= P(-Y_T(\theta_1(\omega,T)) > \varepsilon)\\
&\geq P(\tilde{\Gamma}(\theta_1(\omega,T)) > 2\varepsilon) - P(Y_T(\theta_1(\omega,T)) + \tilde{\Gamma}(\theta_1(\omega,T)) > \varepsilon)\\
&\geq P\Big(\inf_{\theta\in U^c} \tilde{\Gamma}(\theta) > 2\varepsilon\Big) - P\Big(\sup_{\theta\in\Theta} |Y_T(\theta) + \tilde{\Gamma}(\theta)| > \varepsilon\Big). \qquad (2.7.9)
\end{aligned}
$$

In view of (S0) and (2.7.8), it follows that

$$
\sup_{\theta\in\Theta} |Y_T(\theta) + \tilde{\Gamma}(\theta)| \leq \sup_{\theta\in\Theta}\left|\frac{1}{b_T} M_T(\theta)\right| + \sup_{\theta\in\Theta}\left|\frac{1}{b_T} N_T(\theta) - \tilde{\Gamma}(\theta)\right|,
$$

and the last two terms tend to zero in probability as $T \to \infty$ under P. Hence, from (2.7.9), it follows that

$$
\liminf_{T\to\infty}\Big(\sup_{\theta\in U} Y_T(\theta) - \sup_{\theta\in U^c} Y_T(\theta) > 0\Big) \geq P\Big(\inf_{\theta\in U^c} \tilde{\Gamma}(\theta) > 2\varepsilon\Big).
$$

Let $\varepsilon \downarrow 0$ on the right-hand side of the above inequality. Then the term on the right-hand side tends to 1 as θ_0 is the unique point at which $\tilde{\Gamma}(\theta)$ is minimum from (E0). This completes the proof of the theorem.

2.7.2 Local Asymptotic Mixed Normality

The asymptotic behavior of the ratio $Z_T(\boldsymbol{u})$ defined by (2.7.4) is given by the following result.

Theorem 2.7.2 *Suppose that conditions (T1), (T2) and (U0), (U1) hold. Then*

$$
\log Z_T(\boldsymbol{u}) = u^m \Delta_{mT} - \tfrac{1}{2} u^m u^n \Gamma_{mnT} + \rho_T(u), \qquad (2.7.10)
$$

where $\rho_T(u) \xrightarrow{P} 0$, $(\Phi_T, \Gamma_T) \xrightarrow{P} (\phi, \Gamma)$ *under* P, *and*

$$
(\Delta_T, \Phi_T, \Gamma_T) \xrightarrow{\mathcal{L}} (\Phi^{1/2} N, \Phi, \Gamma) \qquad (2.7.11)
$$

as $T \to \infty$ *under* P, *where* $N \simeq N_k(0, I_k)$ *independent of* (Φ, Γ). *Here* I_k *is the identity matrix of order* $k \times k$.

Proof. Note that

$$
\begin{aligned}
Z_T(\boldsymbol{u}) &= Q_T(\boldsymbol{X}, \theta_0 + b_T^{-1/2}\boldsymbol{u}) / Q_T(\boldsymbol{X}, \theta_0)\\
&= \exp\left\{\int_0^T \Delta a_i(\theta_0, \theta_0 + b_T^{-1/2}\boldsymbol{u})[dX_t^i - V_0^i\,dt] - \int_0^T \Delta H_i(\theta_0, \theta_0 + b_T^{-1/2}\boldsymbol{u})\,dt\right\},
\end{aligned}
$$

and hence, by Itô's lemma,

$$
\log Z_T(\boldsymbol{u}) = b_T^{-1/2} u^m \int_0^T \delta_m a_i(\theta_0)[dX_t^i - V_0^i\,dt]
$$

$$
\begin{aligned}
&+\int_0^T D^{(1)} a_i(\theta_0, \boldsymbol{u})[dX_t^i - V_0^i\,dt] \\
&-b_T^{-1/2} u^m \int_0^T \delta_m H(\theta_0)\,dt - (2b_T)^{-1} u^m u^n \int_0^T \delta_m \delta_n H(\theta_0)\,dt \\
&-\int_0^T D^{(2)} H(\theta_0, \boldsymbol{u})\,dt \\
&= b_T^{-1/2} u^m \int_0^T \delta_m a_i(\theta_0)[dX_t^i - V_0^i\,dt] \\
&-b_T^{-1/2} u^m G_m(X_T) \\
&+b_T^{-1/2} u^m G_m(X_0) + b_T^{-1/2} u^m \int_0^T \partial_i G_m(X_t)[dX_t^i - V_0^i\,dt] \\
&-(2b_T)^{-1} u^m u^n \int_0^T \delta_m \delta_n H(\theta_0)\,dt \\
&+\int_0^T D^{(1)} a_i(\theta_0, \boldsymbol{u})[dX_t^i - V_0^i\,dt] \\
&-\int_0^T D^{(2)} H(\theta_0, \boldsymbol{u})\,dt \,(\text{by (E2)}) \\
&= u^m \Delta_{mT} - \tfrac{1}{2} u^m u^n \Gamma_{mnT} + \rho_T(\boldsymbol{u}),
\end{aligned}
$$

where

$$
\Delta_{mT} = b_T^{-1/2} \int_0^T [\delta_m a_i(\theta_0) + \partial_i G_m(X_t)][dX_t^i - V_0^i\,dt],
$$

$$
\Gamma_{mnT} = b_T^{-1} \int_0^T \delta_m \delta_n H(\theta_0)\,dt
$$

and

$$
\begin{aligned}
\rho_T(\boldsymbol{u}) = &-b_T^{-1/2} u^m G_m(X_T) + b_T^{-1/2} u^m G_m(X_0) \\
&+\int_0^T D^{(1)} a_i(\theta_0, \boldsymbol{u})[dX_t^i - V_0^i\,dt] \\
&-\int_0^T D^{(2)} H(\theta_0, \boldsymbol{u})\,dt.
\end{aligned}
$$

For any $\varepsilon > 0$, let $\bar{\delta} = \bar{\delta}(\varepsilon/(1 + \|\boldsymbol{u}\|^2))$ as given by (U0). Then

$$
\begin{aligned}
&P\left(\left|\int_0^T D^{(2)} H(\theta_0, \boldsymbol{u})\,dt\right| > \varepsilon\right) \\
&\le P(\|\boldsymbol{u}\| > \bar{\delta} b_T^{1/2}) + P\left(\sup_{\|\boldsymbol{\zeta}\| \le \bar{\delta} b_T^{1/2}} \frac{1}{1 + \|\boldsymbol{\zeta}\|^2}\left|\int_0^T D^2 H(\theta_0, \boldsymbol{\zeta})\,dt\right| > \frac{\varepsilon}{1 + \|\boldsymbol{u}\|^2}\right),
\end{aligned}
$$

and hence

$$
\int_0^T D^{(2)} H(\theta_0, \boldsymbol{u})\,dt \xrightarrow{p} 0 \qquad \text{as } T \to \infty
$$

under P by (T0). Similarly, it follows that

$$\int_0^T D^{(1)} a_i(\theta_0, \boldsymbol{u})[dX_t^i - V_0^i\, dt] \xrightarrow{P} 0 \qquad \text{as } T \to \infty$$

under P from (U1). Therefore, conditions (U0), (U1) and (T2) imply that for each $\boldsymbol{u}$,

$$\rho_T(\boldsymbol{u}) \xrightarrow{P} 0 \qquad \text{as } T \to \infty$$

under P. On the other hand, (T1) and (T2) prove that

$$(\Phi_T, \Gamma_T) \xrightarrow{P} (\Phi, \Gamma) \qquad \text{as } T \to \infty$$

under P. By the martingale central limit theorems and the stability of the weak convergence, it follows that

$$(\Delta_T, \Phi_T, \Gamma_T) \xrightarrow{\mathcal{L}} (\Phi^{1/2} N, \Phi, \Gamma) \qquad \text{as } T \to \infty$$

under P, where $N \sim N_k(0, I_k)$ independent of (Φ, Γ) (cf. Prakasa Rao 1999). See also Aldous and Eagleson (1978) or Feigin (1985). This proves Theorem 2.7.2.

Remark. It is easy to check the convergence of the finite-dimensional distributions of the process $\{Z_T(\boldsymbol{u})\}$ following Theorem 2.7.2.

Let us now consider the asymptotic behavior of the M-estimators discussed above. Define

$$B_c = \{\boldsymbol{u} \in R^k : \|\boldsymbol{u}\| \le c\},$$
$$B_{cT} = \{\boldsymbol{u} \in R^k : \|\boldsymbol{u}\| \le c, \theta_0 + b_T^{-1/2}\boldsymbol{u} \in \bar{\Theta}\}$$

and

$$W_T(\delta, c) = \sup_{\substack{\|u_1 - u_2\| \le \delta \\ u_1, u_2 \in B_{cT}}} |\log Z_T(\boldsymbol{u}_2) - \log Z_T(\boldsymbol{u}_1)|.$$

Note that, for large T, $B_{cT} = B_c$.

Lemma 2.7.3 *Suppose that conditions (T1), (T2), (U0) and (U1) hold. Then, for each $\varepsilon > 0$ and $c > 0$,*

$$\overline{\lim_{T\to\infty}}\ P(W_T(\delta, c) > \varepsilon) \to 0 \qquad \text{as } \delta \to 0. \tag{2.7.12}$$

Proof. Theorem 2.7.2 implies that

$$|\log Z_T(\boldsymbol{u}_2) - \log Z_T(\boldsymbol{u}_1)| \le \|\boldsymbol{u}_2 - \boldsymbol{u}_1\|\ |\Delta_T| + \beta\|\boldsymbol{u}_2 - \boldsymbol{u}_1\|\ |\Gamma_T| + \sum_{i=1}^{2} |\rho_T(\boldsymbol{u}_i)|,$$

where β is a constant. Hence

$$P(W_T(\delta, c) > \varepsilon) \le P\left(|\Delta_T| > \frac{\varepsilon}{4\delta}\right) + P\left(|\Gamma_T| > \frac{\varepsilon}{4\beta\delta}\right) + 2\,P\left(\sup_{u \in B_{cT}} |\rho_T(\boldsymbol{u})| > \frac{\varepsilon}{4}\right).$$

Let $\bar{\delta} = \bar{\delta}(\varepsilon/16(1 + c^2))$ as given by (U1). Then

$$P\left(\sup_{\|\boldsymbol{u}\| \le c} \left| \int_0^T D^{(2)} H(\theta_0, \boldsymbol{u})\, dt \right| > \frac{\varepsilon}{16}\right)$$
$$\le P(c > \bar{\delta} b_T^{1/2}) + P\left(\sup_{\|\boldsymbol{u}\| \le \bar{\delta} b_T^{1/2}} (1 + \|\boldsymbol{u}\|^2)^{-1} \left| \int_0^T D^{(2)} H(\theta_0, \boldsymbol{u})\, dt \right| > \frac{\varepsilon}{16(1 + c^2)}\right)$$

and the terms on the right-hand side of the above inequality tend to zero as $T \to \infty$ by (U1) and the fact that $b_T \to \infty$ as $T \to \infty$. Similarly, it can be shown that

$$P\left(\sup_{\|u\|\le c}\left|\int_0^T D^{(2)}a_i(\theta_0,u)[dX_t^i - V_0^i\,dt]\right| > \frac{\varepsilon}{16}\right) \xrightarrow{p} 0 \qquad \text{as } T \to \infty$$

by (U0). Since

$$\lim_{T\to\infty} P\left(\sup_{u\in B_{CT}}\left|\rho_T(u)\right| > \frac{\varepsilon}{4}\right) = 0$$

by the above results and condition (T2), we have

$$\overline{\lim_{T\to\infty}}\; P(W_T(\delta,c) > \varepsilon) \le \overline{\lim_{T\to\infty}}\; P\left(|\Delta_T| > \frac{\varepsilon}{4\delta}\right) + \overline{\lim_{T\to\infty}}\; P\left(|\Gamma_T| > \frac{\varepsilon}{4\beta\delta}\right).$$

Since the sequence of random variables Δ_T and Γ_T converge in distribution, it follows that

$$\lim_{\delta\to 0}\overline{\lim_{T\to\infty}}\; P(W_T(\delta,c) > \varepsilon) = 0.$$

Lemma 2.7.4 *Suppose the conditions (T1), (T2), (U0) and (U1) hold. Then, for* $c > 0$,

$$\lim_{N\to\infty}\overline{\lim_{T\to\infty}}\; P\left(\sup_{u\in B_{cT}} |\log Z_T(u)| > N\right) = 0. \tag{2.7.13}$$

Proof. Since Γ is positive definite a.s., for any $\varepsilon_1 > 0$, there exists an $\varepsilon_2 > 0$ such that

$$P(A(\varepsilon_2)) \ge 1 - \varepsilon_1,$$

where

$$A(\varepsilon_2) = \{\omega : \varepsilon_2\|u\|^2 \le \tfrac{1}{4}u^m u^n \Gamma_{mn}\}.$$

Let

$$r_T(u) = \tfrac{1}{2}u^m u^n(\Gamma_{mn} - \Gamma_{mnT}) + \rho_T(u).$$

Then

$$\begin{aligned}
&(1+\|u\|^2)^{-1}|r_T(u)| \\
&\le \tfrac{1}{2}\sum_{m,n}|\Gamma_{mn} - \Gamma_{mnT}| + \tfrac{1}{2}b_T^{-1/2}\sum_m |G_m(X_T)| + \tfrac{1}{2}b_T^{-1/2}\sum_m |G_m(X_0)| \\
&\quad + (1+\|u\|^2)^{-1}\left|\int_0^T D^{(1)}a_i(\theta_0,u)[dX_t^i - V_0^i\,dt]\right| \\
&\quad + (1+\|u\|^2)^{-1}\left|\int_0^T D^{(2)}H(\theta_0,u)\,dt\right|.
\end{aligned}$$

Conditions (T2), (U0) and (U1) show that, for any $\varepsilon_3 > 0$, there exists $\bar{\delta} = \bar{\delta}(\varepsilon_3/4)$ such that $\lim_{T\to\infty} P(S(T,\bar{\delta})) = 1$, where

$$S(T,\bar{\delta}) = \left\{\omega : \sup_{\|u\|\le\bar{\delta}b_T^{1/2}} (1+\|u\|^2)^{-1}|r_T(u)| < \varepsilon_3\right\}.$$

Let $\varepsilon_2 > \varepsilon_3$. For $\omega \in S(T, \bar{\delta}) \cap A(\varepsilon_2)$ and for $\|u\| \leq \bar{\delta} b_T^{1/2}$, it follows that

$$\begin{aligned}\log Z_T(u) &= u^m \Delta_{mT} - \tfrac{1}{2} u^m u^n \Gamma_{mn} + r_T(u)\\ &\leq u^m \Delta_{mT} - \tfrac{1}{2} u^m u^n \Gamma_{mn} + \varepsilon_3(1 + \|u\|^2)\\ &\leq \|u\| |\Delta_T| - \varepsilon_2 \|u\|^2 + \varepsilon_3.\end{aligned}$$

Then

$$\begin{aligned}P\left[\sup_{r \leq \|u\| \leq \bar{\delta} b_T^{1/2}} Z_T(u) \geq \exp\left(-\varepsilon_2 \frac{r^2}{2}\right)\right]\\ &\leq P\{S(T, \bar{\delta})^c\} + P\{A(\varepsilon_2)^c\}\\ &\quad + P\left\{\sup_{r \leq \|u\| \leq \bar{\delta} b_T^{1/2}} (\|u\|\, |\Delta_T| - \varepsilon_2 \|u\|^2) + \varepsilon_3 \geq -\varepsilon_2 \frac{r^2}{2}\right\}\\ &\leq P\{|\Delta_T| > 2\varepsilon_2 r\} + P\left\{r|\Delta_T| - \varepsilon_2 r^2 + \varepsilon_3 \geq -\varepsilon_2 \frac{r^2}{2}\right\}\\ &\quad + \varepsilon_1 + o(1)\\ &\leq 2P\left\{|\Delta_T| > \varepsilon_2 \frac{r}{2} - \frac{\varepsilon_3}{r}\right\} + \varepsilon_1 + o(1).\end{aligned}$$

Let $\varepsilon > 0$, $\gamma > 0$ and $\varepsilon_1 = \varepsilon_3 = \frac{\varepsilon}{3}$. Choose r large enough so that $\exp(-\varepsilon_2 r^2/2) < \gamma$ and

$$\overline{\lim_{T \to \infty}}\, P\left\{|\Delta_T| > \varepsilon_2 \frac{r}{2} - \frac{\varepsilon_1}{r}\right\} < \frac{\varepsilon}{3}.$$

Then

$$\overline{\lim_{T \to \infty}}\, P\left\{\sup_{r < \|u\| \leq \delta b_T^{1/2}} Z_T(u) \geq \gamma\right\} \leq \varepsilon.$$

Now

$$\begin{aligned}\overline{\lim_{T \to \infty}}\, P\left[\sup_{\delta b_T^{1/2} \leq \|u\|} Z_T(u) \geq \gamma\right] &= \overline{\lim_{T \to \infty}}\, P\left[\sup_{\delta \leq \|h\|} Y_T(\theta_0 + h) \geq \frac{1}{b_T} \log \gamma\right]\\ &= P\left[\sup_{\delta \leq \|h\|} (-\tilde{\Gamma}(\theta_0 + h)) \geq 0\right]\\ &= 0,\end{aligned}$$

from (T0) and the proof of Theorem 2.7.1. Hence

$$\overline{\lim_{T \to \infty}}\, P\left[\sup_{r \leq \|u\|} Z_T(u) \geq \gamma\right] \leq \varepsilon.$$

This completes the proof (cf. Strasser 1985, Section 2.4).

Let $C_0(R^k)$ be the Banach space of continuous functions on R^k vanishing at infinity with the supremum norm. Let us extend $Z_T(u)$ to a function of $C_0(R^k)$ with maximum inside $B_T = \{u : \theta_0 + b_T^{1/2} u \in \bar{\Theta}\}$. The following theorem follows from the weak convergence of the finite-dimensional distributions of the random field $\{Z_T(u), u \in R^k\}$ and Lemmas 2.7.3 and 2.7.4.

Theorem 2.7.5 *Under the conditions stated above,*

$$Z_T(\cdot) \xrightarrow{\mathcal{L}} Z(\cdot) \text{ in } C_0(R^k) \qquad \text{as } T \to \infty, \tag{2.7.14}$$

where

$$\log Z(\boldsymbol{u}) = u^m \Delta_m - \tfrac{1}{2} u^m u^n \Gamma_{mn} \tag{2.7.15}$$

in which $(\Delta_m) = \Phi^{1/2} N$ *and* N *are as given in Theorem 2.7.2.*

This theorem can be used for investigating the asymptotic properties of the M-estimators and the test statisties related to them.

Let $\hat{\boldsymbol{u}}_T$ be a point at which $Z_T(\boldsymbol{u})$ attains its maximum in $\{\boldsymbol{u} : \theta_0 + b_T^{-1/2} \boldsymbol{u} \in \bar{\Theta}\}$ and $\hat{\boldsymbol{u}}$ be a point at which $Z(\boldsymbol{u})$ is maximum in R^k. From the weak convergence of the random fields $\{Z_T(\cdot)\}$, it follows that

$$\hat{\boldsymbol{u}}_T \xrightarrow{\mathcal{L}} \hat{\boldsymbol{u}} = \Gamma^{-1} \Delta$$

and we have the following result giving the asymptotic behavior of an M-estimator.

Theorem 2.7.6 *Suppose the conditions (T0)–(T2), (U0) and (U1) hold. Then*

$$b_T^{1/2}(\hat{\theta}_T - \theta_0) \xrightarrow{\mathcal{L}} \Gamma^{-1} \Delta \qquad \text{as } T \to \infty \tag{2.7.16}$$

under P, where $\Delta = \Phi^{1/2} N$, $N \simeq N_k(0, I_k)$ *independent of* (Φ, Γ). *In particular, if* Φ *and* Γ *are nonrandom, then the limiting distribution is* $N_k(0, \Gamma^{-1} \Phi \Gamma^{-1})$.

2.7.3 Estimation by a Misspecified Model

It is known that if the parametric model chosen contains the true model, then the MLE is consistent, asymptotically normal and asymptotically efficient, for instance, in the sense that it is locally asymptotically minimax (cf. Kutoyants 1984a) under some regularity conditions. Since the MLE is sensitive to contamination of the data or misspecification of the model, it would be of interest to study how the difference between the true model and the assumed parametric model affects the asymptotic behavior of the MLE. Let us consider the ergodic case for simplicity.

Suppose the observer's parametric model is given by

$$dX_t^i = \sum_{\beta=1}^{r} A_\beta^i(X_i)\, dW^\beta(t) + A_0^i(X_i, \theta)\, dt, \qquad 1 \le i \le d,\ X_0 = \eta. \tag{2.7.17}$$

Then the likelihood function based on the observation $\{X_t : 0 \le t \le T\}$ is given by the expression

$$\exp\left[\int_0^T A_0^i \alpha_{ij}\, dX_t^j - \frac{1}{2}\int_0^T A_0^i \alpha_{ij} A_0^j\, dt\right]$$

where $\alpha_{ij} = (\alpha^{ij})^+$ is the symmetric generalized inverse and $\alpha^{ij} = \sum_{\beta=1}^{r} A_\beta^i A_\beta^j$ (cf. Basawa and Prakasa Rao 1980; Liptser and Shiryayev 1977). In our notation of (2.7.1),

$$a_i = A_0^j \alpha_{ij} \quad \text{and} \quad R = A_0^i \alpha_{ij} A_0^j.$$

We have the following result as a consequence of Theorem 2.7.6.

Theorem 2.7.7 *Let X be an ergodic and stationary process with the invariant distribution $\nu(dx)$. Suppose there exist a θ_0 at which the mapping*

$$\theta \to \int_{R^d} \left[\tfrac{1}{2} A_0^i(\theta)\alpha_{ij} A_0^j(\theta) - A_0^i(\theta)\alpha_{ij} V_0^j - \tfrac{1}{2} A_0^i(\theta_0)\alpha_{ij} A_0^j(\theta_0) + A_0^i(\theta_0)\alpha_{ij} V_0^j \right] \nu(dx)$$

has a unique minimum; and a matrix Γ with components

$$\Gamma_{mn} = \int_{R^d} \gamma_{mn}\, \nu(dx),$$

where

$$\gamma_{mn} = (\delta_m A_0^i(\theta_0))\alpha_{ij}(\delta_n A_0^j(\theta_0)) + (\delta_m \delta_n A_0^i(\theta_0))\alpha_{ij}[A_0^j(\theta_0) - V_0^j] \tag{2.7.18}$$

is positive definite. Furthermore, suppose the following conditions hold:

(i) *V_0^i, V_α^i, α_{ij} and the maxima of A_0^i and their derivatives up to the third order with respect to θ belong to $\cap_{p>1} L^p(R^d, \nu)$.*

(ii) *There exist functions G_m such that*

$$(\partial_i G_m)^2 V^{ij} \in L^1(R^d, \nu)$$

and

$$LG_m = (\delta A_0^i(\theta_0))\alpha_{ij}[A_0^j(\theta_0) - V_0^j].$$

Then the weak convergence of the distribution of the random fields $Z_T(\cdot)$ holds and

$$T^{1/2}(\hat{\theta}_T - \theta_0) \overset{\mathcal{L}}{\to} N_k(0, \Gamma^{-1}\Phi\Gamma^{-1}) \qquad \text{as } T \to \infty$$

under P, where $\Phi = ((\Phi_{mn}))$ with

$$\Phi_{mn} = \int (\delta_m A_0^p(\theta_0)\alpha_{ip} + \partial_i G_m)(\delta_n A_0^q(\theta_0)\alpha_{jq} + \partial_j G_n) V^{ij}\, d\nu(x).$$

Remarks. If the parametric model contains the true model, then condition (ii) in Theorem 2.7.7 is not necessary and the MLE is consistent with asymptotic covariance matrix Γ^{-1}, the reciprocal of the Fisher information. The integrability in condition (i) can also be weakened.

Example 2.7.1 *(Nonergodic model)*
Consider the one-dimensional diffusion process

$$dX_t = -\zeta\theta X_t\, dt + dW_t, \qquad t \geq 0,$$

where θ is a positive parameter and ζ is a positive random variable independent of the σ-algebra generated by $\{W_t, t \geq 0\}$. Suppose we observe $\{X(t), 0 \leq t \leq T; \zeta\}$ and the problem is to estimate the parameter θ. The MLE is obtained by maximizing

$$Q(T; X, \theta) = \exp\left\{ \int_0^T (-\zeta\theta X_t)\, dX_t - \tfrac{1}{2} \int_0^T \zeta^2\theta^2 X_t^2\, dt \right\}.$$

Conditional on ζ, the process X is ergodic with the invariant measure $\nu_\zeta(dx)$ with density

$$\left(\frac{\zeta\theta}{\pi}\right)^{1/2} \exp(-\zeta\theta x^2).$$

If θ_0 is the true value, then

$$\frac{1}{T}\int_0^T \Delta H(\theta_0, \theta)\, dt \to \tilde{\Gamma} = \frac{\zeta}{4\theta_0}(\theta - \theta_0)^2 \text{ a.s.},$$

and the consistency of the MLE holds. The asymptotic distribution of the MLE is the mixture of the normal distributions $N(0, \frac{2\theta_0}{\zeta})$ by ζ. These results can also be obtained by using the explicit representation of the MLE which, in this case, is given by

$$\hat{\theta}_T = -\int_0^T \zeta X_t\, dX_t \Big/ \int_0^T \zeta^2 X_t^2\, dt.$$

If ζ is not observed and ζ is replaced by $E(\zeta)$, the mean of the random variable ζ, then $\hat{\theta}_T$ is not consistent.

Example 2.7.2 (*Nonergodic model*)
Consider a nonlinear diffusion process defined by

$$dX_t = [g(X_t) + \theta f(X_t)]\, dt + dW_t, \qquad X_0 = x_0 > 0,\ t \geq 0,$$

where $\theta > 0$ is unknown and g and f are functions such that

(i) g and f are positive, (ii) g and f have derivatives up to second order which are bounded, (iii) $g' + \theta f' \geq 0$ for large x, and (iv) $\lim_{x\to\infty} \frac{g(x)}{x} = 1 = \lim_{x\to\infty} \frac{f(x)}{x^\alpha}$ for some $\alpha, 0 < \alpha < 1$.

It is known that $X_t \to \infty$ as $t \to \infty$ a.s. (cf. Gihman and Skorohod 1972, p. 117) and there exists a positive random variable K such that

$$\frac{X_t}{\eta_t} \to K \text{ a.s.}, \qquad \text{as } t \to \infty,$$

where η_t is the inverse of the map

$$x \to \int_{x_0}^x [g(u) + \theta f(u)]^{-1}\, du$$

(cf. Keller *et al.* 1984; Yoshida 1990).

With $b_T = \int_0^T \eta_t^{2\alpha} dt$, $\tilde{\Gamma} = \frac{1}{2}(\theta - \theta_0)^2 K^{2\alpha}$ and $\Phi = \Gamma = K^{2\alpha}$, it can be shown that the MLE is consistent and asymptotically mixed normal with random variable $\Gamma^{-1} = K^{-2\alpha}$.

Example 2.7.3 (*M-estimation*)
Consider a one-dimensional ergodic and stationary diffusion process given by

$$dX_t = f(X_t, \theta)\, dt + \sigma(X_t)\, dW_t, \qquad X_0 = x_0,\ t \geq 0,$$

where $\theta \in \Theta$, bounded interval in R. Let $\nu(dx, \theta)$ be the stationary distribution when θ is the parameter. An M-estimator can be constructed by using the equation

$$M(T, \theta) \equiv \int_0^T \psi(X_t, \theta)\, dt = 0,$$

where ψ satisfies

$$\int_{-\infty}^{\infty} \psi(x,\theta)\,\nu(dx,\theta) = 0, \qquad \theta \in \Theta.$$

This M-estimator corresponds to the choice of Q given by

$$\log Q_T(X,\theta) = -\int_0^T H(X_t,\theta)\,dt,$$

where $H(x,\theta) \equiv \int_{-\infty}^{\theta} \psi(x,\theta')\,d\theta'$. Supose the true realization of the process X is generated by the stochastic differential equation

$$dX_t = v_1(X_t)\,dW_t + v_0(X_t)\,dt, \qquad X_0 = x_0,\; t \geq 0.$$

Let ν be the stationary distribution of the true model. Its density is proportional to

$$2[v_1(u)]^{-2}\exp\{B(u)\},$$

where

$$B(x) = \int_0^x \frac{2\,v_0(u)}{[v_1(u)]^2}\,du.$$

Let θ_0 be the solution of the equation

$$\int_{-\infty}^{\infty} \psi(x,\theta)\,\nu(dx) = 0.$$

Define

$$G(x,\theta) = -\left\{\int_0^x \exp(-B(y))\,dy\right\}\left\{\int_y^{\infty} 2\frac{\psi(u,\theta)}{[v_1(u)]^2}\exp(B(u))\,du\right\}.$$

Then

$$G_1(x) = G(x,\theta_0)$$

satisfies the equation

$$(v_0(x)\partial + \tfrac{1}{2}v_1(x)\partial + \tfrac{1}{2}v_1(x)\partial^2)G_1 = \psi(x,\theta_0).$$

Here Γ and Φ are given by

$$\Gamma = \int_{-\infty}^{\infty} \delta\psi(x,\theta_0)\,\nu(dx)$$

and

$$\Phi = \int_{-\infty}^{\infty} [\partial G_1(x)]^2 v_1^2(x)\,\nu(dx).$$

The asymptotic variance of the M-estimator is Φ/Γ^2.

For more work on the robust estimation for diffusion processes, see Yoshida (1988). For additional examples, see Yoshida (1990).

2.8 Recursive Estimation

Recursive methods for parameter estimation are of great importance where the model of the system is used on-line. In earlier sections, we have considered off-line parametric identification with the MLE and its variants.

Consider a process $\{X_t, t \geq 0\}$ satisfying the stochastic differential equation

$$dX_t = a(\theta, X_t, t)\,dt + b(X_t, t)\,dW_t, \qquad X_0 = 0, t \geq 0. \tag{2.8.1}$$

Let μ_t^θ be probability measure induced by the process $\{X_s, 0 \leq s \leq t\}$ on $(C[0,t], \mathcal{B}_t)$ where $\mathcal{B}_t$ is the Borel σ-algebra of $C[0,t]$ under the supremum norm. We saw earlier that, under suitable conditions,

$$\begin{aligned}\ell_t(\theta) = \log \frac{d\mu_t^\theta}{d\mu_t^{\theta_0}} &= \int_0^t \frac{a(\theta, X_s, s) - a(\theta_0, X_s, s)}{b^2(X_s, s)}\,dX_s \\ &\quad - \frac{1}{2}\int_0^t \frac{a^2(\theta, X_s, s) - a^2(\theta_0, X_s, s)}{b^2(X_s, s)}\,ds \\ &= \int_0^t g(\theta, X_s, s)\,dW_s - \frac{1}{2}\int_0^t g^2(\theta, X_s, s)\,ds, \end{aligned} \tag{2.8.2}$$

where

$$g(\theta, x, s) = \frac{a(\theta, x, s) - a(\theta_0, x, s)}{b(x, s)}. \tag{2.8.3}$$

For instance, if

$$E\left[\exp \tfrac{1}{2}\int_0^t g^2(\theta, X_s, s)\,ds\right] < \infty, \qquad \theta \in \Theta,\ t \geq 0, \tag{2.8.4}$$

then $\mu_t^\theta \ll \mu_t^{\theta_0}$ and $L_t(\theta) = d\mu_t^\theta / d\mu_t^{\theta_0}$ is a martingale. Let

$$G_t(\theta) = \int_0^t g(\theta, X_s, s)\,dW_s. \tag{2.8.5}$$

Then

$$\ell_t(\theta) = G_t(\theta) - \tfrac{1}{2}\langle G(\theta)\rangle_t. \tag{2.8.6}$$

Assume that $\ell_t(\theta)$ is continuously differentiable with respect to θ up to fourth order and that differentiation under the integral sign in (2.2.2) is valid up to four times. Sufficient conditions for this property to hold are given in Theorem 2.1 of Levanony *et al.* 1994 (cf. Karandikar 1983; Hutton and Nelson 1986). Then

$$\nabla_\theta^k \ell_t(\theta) = \int_0^t \nabla_\theta^k g(\theta, X_s, s)\,dW_s - \tfrac{1}{2}\int_0^t \nabla_\theta^k (g^2(\theta, X_s, s))\,ds, \qquad 1 \leq k \leq 4. \tag{2.8.7}$$

Let

$$U_t(\theta) = \nabla_\theta \ell_t(\theta),\ H_t(\theta) = \nabla_\theta^2 \ell_t(\theta), \qquad Q_t(\theta) = \nabla_\theta^3 \ell_t(\theta). \tag{2.8.8}$$

The global MLE satisfies the relation

$$\hat{\theta}_t = \arg\sup_{\theta \in \Theta} L_t(\theta).$$

Let

$$I_t = \int_0^t g^2(\theta_0, X_s, s)\,ds. \tag{2.8.9}$$

It is known from results discussed in the earlier sections that

$$\hat{\theta}_t \to \theta_0, \text{ a.s.} \quad \text{as } t \to \infty, \tag{2.8.10}$$

and

$$I_t^{1/2}(\hat{\theta}_t - \theta_0) \xrightarrow{\mathcal{L}} N(0,1), \quad \text{as } t \to \infty, \tag{2.8.11}$$

under some regularity conditions.

2.8.1 MLE Evolution Equation

We now derive the evolution equation of the MLE's trajectories using the generalized Itô differentiation rule known as the Itô–Ventzell formula. Let us at first assume that θ is a scalar parameter.

Suppose the MLE is a continuous semimartingale of the form

$$\tilde{\theta}_t = \tilde{\theta}_{t_0} + \int_{t_0}^t A_s\,ds + \int_{t_0}^t B_s\,dX_s. \tag{2.8.12}$$

Note that $\{U_t(\theta), t \geq 0\}$ is a continuous semimartingale for each θ. Furthermore, $U_t(\cdot) \in C^2$ for $t \geq 0$ a.s. and, together with its derivatives, is jointly continuous in (θ, t) following Theorem 2.1 of Levanony *et al.* (1994). Then, by Kunita (1990), one can apply the Itô–Ventzell formula for the composition of $\{\tilde{\theta}_t\}$ with random field $U_t(\cdot)$ to obtain

$$\begin{aligned} dU_t(\tilde{\theta}_t) &= a_\theta(\tilde{\theta}_t, X_t, t)[dX_t - a(\tilde{\theta}_t, X_t, t)\,dt] + H_t(\tilde{\theta}_t)\,d\tilde{\theta}_t + \tfrac{1}{2}Q_t(\tilde{\theta}_t)B_t^2\,dt \\ &\quad + a_{\theta\theta}(\tilde{\theta}_t, X_t, t)B_t\,dt, \qquad t \geq t_0. \end{aligned} \tag{2.8.13}$$

Recall that a_θ and $a_{\theta\theta}$ denote the first partial derivative and the second partial derivative with respect to θ respectively. Here $H_t(\theta) = \nabla_\theta^2 \ell_t(\theta)$ and $Q_t(\theta) = \nabla_\theta^3 \ell_t(\theta)$. Assume that $H_t(\tilde{\theta}_t) < 0$ for $t \geq t_0$. We conclude that the MLE which solves $U_t(\theta) = 0$ for all $t > 0$ should be a solution of the equation

$$d\tilde{\theta}_t = -H_t^{-1}(\tilde{\theta}_t)\{a_\theta(\tilde{\theta}_t)[dX_t - a(\tilde{\theta}_t)\,dt] + [\tfrac{1}{2}Q_t(\tilde{\theta}_t)B_t^2 + a_{\theta\theta}(\tilde{\theta}_t)B_t]\,dt\}, \qquad t \geq t_0 \tag{2.8.14}$$

(omitting the dependence of $a(\cdot,\cdot,\cdot)$ on X_t and t for simplicity), which, after equating coefficients with (2.2.12), gives

$$\begin{aligned} d\tilde{\theta}_t &= -H_t^{-1}(\tilde{\theta}_t)\{a_\theta(\tilde{\theta}_t)[dX_t - a(\tilde{\theta}_t)\,dt] \\ &\quad + [\tfrac{1}{2}Q_t(\tilde{\theta}_t)H_t^{-2}(\tilde{\theta}_t)a_\theta^2(\tilde{\theta}_t) - H_t^{-1}(\tilde{\theta}_t)a_\theta(\tilde{\theta}_t)a_{\theta\theta}(\tilde{\theta}_t)]\,dt\}, \qquad t \geq t_0, \end{aligned} \tag{2.8.15}$$

with the initial conditions $|\tilde{\theta}_{t_0}| < \infty$, $U_{t_0}(\tilde{\theta}_{t_0}) = 0$, $H_{t_0}(\tilde{\theta}_{t_0}) < 0$.

The choice of the initial time $t_0 > 0$ is imposed by the fact that $H_0(\theta) = 0$ for all θ. Levanony *et al.* (1994) prove that if the MLE $\hat{\theta}_t$ is a.s. continuous, then it does satisfy (2.8.15) at least for all large enough t, that is, for all $\varepsilon > 0$, there exists $t_0 = t_0(\varepsilon)$ finite such that the stochastic differential equation (2.8.15) describes the MLE path on $[t_0, \infty)$ with probability greater than $1-\varepsilon$. If, in addition $P(H_t(\hat{\theta}_t) < 0 \text{ for all } t > 0) = 1$, that is, the log-likelihood $\ell_t(\cdot)$ is strictly concave in some small neighborhood of the MLE for all $t > 0$, then (2.8.15) is the MLE evolution equation on $[t_0, \infty)$ a.s. for all $t_0 > 0$.

The above results also hold when θ is a vector parameter in R^d. By applying the vector-valued Itô–Ventzell formula (Kunita 1990) (cf. Chapter 1, Section 6 of Prakasa Rao 1999), the vector-valued MLE evolution equation becomes

$$\begin{aligned} d\hat{\theta}_t = -H_t^{-1}(\hat{\theta}_t)\Bigg\{ & a_\theta(\hat{\theta}_t)(dX_t - a(\hat{\theta}_t)\,dt) \\ & + \frac{1}{2}\sum_{i=1}^{d} \frac{\partial H_t(\hat{\theta}_t)}{\partial \theta^{(i)}} H_t^{-1}(\hat{\theta}_t) a_\theta(\hat{\theta}_t)(H_t^{-1}(\hat{\theta}_t)a_\theta(\hat{\theta}_t))^{(i)}\,dt \\ & - a_{\theta\theta}(\hat{\theta}_t) H_t^{-1}(\hat{\theta}_t) a_\theta(\hat{\theta}_t)\,dt \Bigg\} \end{aligned} \tag{2.8.16}$$

where $\theta, a_\theta \in R^d$, $a_{\theta\theta} \in R^{d\times d}$ and $y^{(i)}$ denotes the ith component of y. Similar results are obtained in Ljung *et al.* (1988) and Gerencsér *et al.* (1984).

Remarks. Equation (2.8.15) is not suitable for recursive estimation since it is valid for large t and requires knowledge of the MLE at the initial time. Levanony *et al.* (1994) discuss a modified algorithm which is insensitive to the initial conditions and is implementable for all $t_0 > 0$. We omit the details. A suitable version of this algorithm which is applicable if the log-likelihood is strictly concave has been given Levanony *et al.* (1994) in their work on continuous-time recursive identification.

2.9 Sequential Estimation

In earlier sections, we discussed different types of method of estimation of a parameter involved in a stochastic differential equation based on a realization of the process over a fixed time interval and where the time of observation is nonrandom. However, there are situations where the time of observation turns out to be random and the decision to terminate or continue the observations of the process depends possibly on the results of the observations made earlier. We will now discuss a method of sequential estimation for the estimation of the drift parameter for diffusion type processes.

Let X and Y be processes of diffusion type satisfying the differential equations

$$dX_t = A_t(X)\,dt + b_t(X)\,dW_t, \qquad X_0 = 0,\ 0 \le t \le T, \tag{2.9.1}$$

and

$$dY_t = a_t(Y)\,dt + b_t(Y)\,dW_t, \qquad X_0 = 0,\ 0 \le t \le T, \tag{2.9.2}$$

where $a(\cdot)$, $A(\cdot)$ and $b(\cdot)$ are nonanticipating. Suppose that:

(V1) $a(\cdot)$ and $b(\cdot)$ satisfy the Lipschitz condition,

(i)

$$\begin{aligned} &|a_t(x) - a_t(y)|^2 + |b_t(x) - b_t(y)|^2 \\ &\quad \le L_1 \int_0^t |x(s) - y(s)|^2\,dK(s) + L_2|x(t) - y(t)|^2, \end{aligned}$$

(ii)

$$a_t^2(x) + b_t^2(x) \le L_1 \int_0^t (1 + x^2(s))dK(s) + L_2(1 + x^2(t)),$$

for each $t \in [0, T]$, $x, y \in C[0, T]$, where L_1 and L_2 are constants and $K(\cdot)$ is a nondecreasing right-continuous function with $0 \le K(s) \le 1$.

In the following, $[P]$ denotes that the statement holds except possibly on a set of P-measure zero. Suppose that, for any $t \in [0, T]$, the equation

(V2) $b_t(x)\beta_t(x) = A_t(x) - a_t(x)[P]$ has a bounded solution $\beta_t(x)$ satisfying the conditions

(V3) $P(\int_0^T \beta_t^2(X)\,dt < \infty) = 1$ and

(V4) $P(\int_0^T \beta_t^2(Y)\,dt < \infty) = 1$.

Let μ_X be the probability measure generated by the solution X of (2.9.1) on $C[0, T]$. Then it follows that $\mu_X \simeq \mu_Y$ and

$$\frac{d\mu_Y}{d\mu_X}(t, X) = \exp\left\{-\int_0^t b_s^-(X)^2(A_s(X) - a_s(X))\,dX_s + \tfrac{1}{2}\int_0^t b_s^-(X)^2(A_s^2(X) - a_s^2(X))\,ds\right\}, \tag{2.9.3}$$

where b^- is defined by

$$b_t^-(x) = \begin{cases} b_t(x)^{-1}, & \text{if } b_t(x) \ne 0, \\ 0, & \text{if } b_t(x) = 0. \end{cases} \tag{2.9.4}$$

Note that

$$\beta_t(x) = b_t^-(x)(A_t(x) - a_t(x)). \tag{2.9.5}$$

Equation (2.9.3) gives the Radon–Nikodym derivative of μ_Y with respect to μ_X when the measures are restricted to $\mathcal{A}_T$, where $\mathcal{A}_T$ is the σ-algebra generated by the sets $\{x : x \in C[0, T], x(s) \in B, 0 \le s \le T\}$ and B is a Borel set in R.

We will discuss a result on the absolute continuity and the Radon–Nikodym derivatives of the measures on $C[0, T]$ induced by processes of the diffusion type which are terminated at a stopping time τ relative to the filtration $\{\mathcal{A}_T\}$. Let $\mu_{\tau,X}$ denote the restriction μ_X to the σ-algebra $\mathcal{A}_{T\wedge\tau}$ (recall that the σ-algebra $\mathcal{F}_\tau$ corresponding to a stopping time adapted to the filtration $\{\mathcal{F}_t\}$ is defined by $A \in \mathcal{F}_\tau \Leftrightarrow$ for all $t \in [0, \infty) : A \cap \tau^{-1}([0, t]) \in \mathcal{F}_t$).

We denote $\min(\alpha, \beta)$ by $\alpha \wedge \beta$.

Theorem 2.9.1 *Let X and Y be diffusion type stochastic processes satisfying the system of stochastic differntial equations*

$$dX_t = A_t(X)\,dt + b_t(X)\,dW_t, \qquad 0 \le t \le T,\ X_0 = 0,$$

and

$$dY_t = a_t(Y)\,dt + b_t(Y)\,dW_t, \qquad 0 \le t \le T,\ Y_0 = 0.$$

Let $\tau : C[0, T] \to [0, \infty]$ be a stopping time with respect to the filtration $\{\mathcal{A}_t\}$. Furthermore, suppose that the following conditions hold:

(V1)′ $a(\cdot)$ and $b(\cdot)$ satisfy the condition L;

(V2)′ for any $t \le T \wedge \tau(X)$, the equation

$$b_t(X)\beta_t(X) = A_t(X) - a_t(X)[P]$$

has a bounded solution;

(V3)′ $P(\int_0^{T\wedge\tau(X)} \beta_t^2(X)\,dt < \infty) = 1$; *and*

(V4)′ $P(\int_0^{T\wedge\tau(Y)} \beta_t^2(Y)\,dt < \infty) = 1$.

Then

$$\mu_{\tau,X} \simeq \mu_{\tau,Y}$$

and

$$\frac{d\mu_{\tau,Y}}{d\mu_{\tau,X}}(X) = \exp\left\{-\int_0^{\tau(X)\wedge T} b_t^-(X)^2(A_t(X) - a_t(X))\,dX_t + \tfrac{1}{2}\int_0^{\tau(X)\wedge T} b_t^-(X)^2(A_t^2(X) - a_t^2(X))\,dt\right\}, \quad (2.9.6)$$

where b^- is as defined by (2.9.4).

Proof. Let X^* be the unique solution of the equation

$$X_t^* = X_{t\wedge\tau(X)} + \int_0^t I(\tau(X) < s)a_s(X^*)\,ds + \int_0^t I(\tau(X) < s)b_s(X^*)\,dW_s.$$

Then $X_t^* = X_t$ for $t \le \tau(X)$ and, by Itô's lemma,

$$dX_t^* = A_t^*(X^*)\,dt + b_t(X^*)\,dW_t,$$

where

$$A_t^*(x) = a_t(x) + I(t \le \tau(x))(A_t(x) - a_t(x)).$$

Observe that

$$(A_t^*(x) - a_t(x))b_t^-(x) = I(t \le \tau(x))\beta_t(x)$$

and

$$\int_0^t \beta_s(X^*)I(s \le \tau(X^*))\,ds = \int_0^{t\wedge\tau(X^*)} \beta_s(X^*)\,ds = \int_0^{t\wedge\tau(X)} \beta_s(X)\,ds,$$

$$\int_0^t \beta_s^2(X^*)I(s \le \tau(X^*))\,ds = \int_0^{t\wedge\tau(X^*)} \beta_s^2(X^*)\,ds = \int_0^{t\wedge\tau(X)} \beta_s^2(X)\,ds,$$

$$\int_0^t \beta_s(X^*)I(s \le \tau(X^*))\,dW_s = \int_0^{t\wedge\tau(X^*)} \beta_s(X^*)\,dW_s = \int_0^{t\wedge\tau(X)} \beta_s(X)\,dW(s).$$

Hence, the conditions (V1)′–(V4)′ imply that $\mu_Y \simeq \mu_{X^*}$ and

$$\frac{d\mu_Y}{d\mu_{X^*}}(X^*) = \exp\left(-\int_0^{T\wedge\tau(X^*)} \beta_s(X^*)\,dW_s - \tfrac{1}{2}\int_0^{T\wedge\tau(X^*)} \beta_s^2(X^*)\,ds\right)$$
$$= \exp\left\{-\int_0^{T\wedge\tau(X^*)} b_s^-(X^*)^2(A_s(X^*) - a_s(X^*))\,dX_s^* + \tfrac{1}{2}\int_0^{T\wedge\tau(X^*)} b_s^-(X^*)^2(A_s^2(X^*) - a_s^2(X^*))\,ds\right\}.$$

The process

$$-\int_0^t b_s^-(X^*)^2(A_s(X^*) - a_s(X^*))\,dX_s^* + \tfrac{1}{2}\int_0^t b_s^-(X^*)^2(A_s^2(X^*) - a_s^2(X^*))\,ds$$

is progressively measurable with respect to $\{\mathcal{F}_t^{X^*}\}$, the filtration generated by the process X^*. Hence the random variable

$$\frac{d\mu_Y}{d\mu_{X^*}}(X^*)$$

is $\mathcal{F}_{\tau\wedge\tau(X^*)}^{X^*}$-measurable. Hence

$$\frac{d\mu_{\tau,Y}}{d\mu_{\tau,X^*}} = E\left(\frac{d\mu_Y}{d\mu_{X^*}}|\mathcal{A}_{T\wedge\tau}\right) = \frac{d\mu_Y}{d\mu_{X^*}}.$$

But $\mu_{X^*}(A) = \mu_X(A)$ for all $A \in \mathcal{A}_{T\wedge\tau}$. Hence $\mu_{\tau,X^*} \simeq \mu_{\tau,X}$. Hence $\mu_{\tau,X} \simeq \mu_{\tau,Y}$ and

$$\frac{d\mu_{\tau,Y}}{d\mu_{\tau,X}} = \frac{d\mu_{\tau,Y}}{d\mu_{\tau,X^*}} = \frac{d\mu_Y}{d\mu_{X^*}}.$$

Remark. If $P(\tau < \infty) = 1$, then $\tau \wedge T$ may be replaced by τ in (2.9.6).

2.9.1 Maximum Likelihood Estimation

Suppose we observe a process X satisfying the stochastic differential equation

$$dX_t = a_t(X;\theta)\,dt + b_t(X)\,dW_t, \qquad \theta \in \Theta \subset R^n, \tag{2.9.7}$$

continuously in the time interval $[0, \tau]$, where τ is a stopping time relative to $\{\mathcal{A}_t\}$. Let $\mu_{\theta,\tau}$ be the probability measure induced on $\mathcal{A}_\tau$ by the process X corresponding to θ and stopped at time τ. Let us assume that the following conditions hold for any fixed $\theta_0 \in \Theta$:

(W1) $a(\cdot)$ and $b(\cdot)$ satisfy the condition (L) for all θ;

(W2) for any $t \le \tau$ and $\theta \in \Theta$, the equation

$$b_t(x)\beta_t(x,\theta) = a_t(x;\theta) - a_t(x,\theta_0)[\mu_{\theta_0,\tau}]$$

has a solution;

(W3) for all $\theta \in \Theta$,

$$\mu_{\theta_0,\tau}\left(\int_0^\tau \beta_t^2(\theta)\,dt < \infty\right) = 1;$$

(W4) for all $\theta \in \Theta$,

$$\mu_{\theta,\tau}\left(\int_0^\tau \beta_t^2(\theta)\,dt < \infty\right) = 1.$$

Then equation (2.9.7) has a unique solution for all $\theta \in \Theta$, and by Theorem 2.9.1 the log-likelihood function is given by

$$\ell_\tau(\theta) = \int_0^{\tau(X)} b_s^-(X)^2 a_s(X;\theta)\,dX_s - \tfrac{1}{2}\int_0^{\tau(X)} b_s^-(X)^2 a_s(X;\theta)^2\,ds, \tag{2.9.8}$$

where we have omitted an additive constant.

The score function is given by

$$\dot{\ell}_\tau(\theta)_i = \frac{\partial}{\partial\theta_i}\int_0^\tau b_s^-(X)^2 a_s(X;\theta)^2\,dX_s$$
$$-\tfrac{1}{2}\frac{\partial}{\partial\theta_i}\int_0^\tau b_s^-(X)^2 a_s(X;\theta)^2\,ds, \qquad i=1,\ldots,n,$$

where $\theta = (\theta_1,\ldots,\theta_n)^T$. Under the condition that the differentiation with respect to θ_i under the stochastic integration is valid (cf. Karandikar 1983), we obtain that

$$\dot{\ell}_\tau(\theta)_i = \int_0^\tau b_s^-(X)^2\frac{\partial}{\partial\theta_i}a_s(X;\theta)\,dX_s - \int_0^\tau b_s^-(X)^2 a_s(X;\theta)\frac{\partial a_s}{\partial\theta_i}(X;\theta)\,ds$$
$$= \int_0^\tau b_s^-(X)\frac{\partial a_s}{\partial\theta_i}(X;\theta)\,dW_s. \tag{2.9.9}$$

Hence the stochastic process $\dot{\ell}_t(\theta)$ is a local martingale relative to $\{\mathcal{F}_t\}$. If, for any t,

$$E\left(\int_0^t [b_s^-(X)\frac{\partial a_s}{\partial\theta_i}(X;\theta)]^2\,ds\right) < \infty, \tag{2.9.10}$$

then $\dot{\ell}_t(\theta)$ is a square-integrable martingale relative to $\{\mathcal{F}_t\}$ with the quadratic characteristic

$$I_t(\theta)_i = \int_0^t \left[b_s^-(X)\frac{\partial a_s}{\partial\theta_i}(X;\theta)\right]^2 ds. \tag{2.9.11}$$

The conditional Fisher information matrix is defined by

$$I_t(\theta)_{ij} = \int_0^t b_s^-(X)^2\frac{\partial a_s}{\partial\theta_i}(X;\theta)\frac{\partial a_s}{\partial\theta_j}(X;\theta)\,ds, \tag{2.9.12}$$

and the process given by (2.9.11) forms the diagonal elements of this matrix.

If σ_t is an increasing sequence of stopping times and if (2.9.10) holds for any $t \geq 0$, then $\dot{\ell}_{\sigma_t}(\theta)_i$ is a square-integrable martingale relative to the filtration $\{\mathcal{F}_{\sigma_t}\}$ with the quadratic characteristic $I_{\sigma_t}(\theta)_i$.

Let us now consider a special case of (2.9.7), namely, the models with a drift coefficient of the form

$$a_t(x;\theta) = \sum_{j=1}^n \theta_j\alpha_t^{(j)}(x), \qquad \theta\in\Theta\subset R^n, \tag{2.9.13}$$

where $\alpha^{(i)}$ are linearly independent nonanticipating functionals. Conditions (W1)–(W4) are implied by the following conditions:

(X1) $\alpha^{(i)}(\cdot)$ and $b(\cdot)$ satisfy the Lipschitz condition for $1 \leq j \leq n$;

(X2) for any $t \leq \tau$, the equation

$$b_t(x)\beta_t^{(j)}(x) = \alpha_t^{(j)}(x)[\mu_{\theta_0,x}]$$

has a solution for all $1 \leq j \leq n$;

(X3) $\mu_{\theta,\tau}(\int_0^\tau \beta_t^{(j)2}\,dt < \infty) = 1$, $1 \leq j \leq n$, $\theta\in\Theta$.

Note that

$$\beta_t^{(j)}(x) = \alpha_t^{(j)}(x) b_t^{-}(x),$$

where b_t^{-} is as defined by (2.9.4). Let

$$C_{\tau,i} = \int_0^\tau b_t^{-}(X)^2 \alpha_t^{(i)}(X)\, dX_t$$

and

$$I_{\tau,i,j} = \int_0^\tau b_t^{-}(X)^2 \alpha_t^{(i)}(X) \alpha_t^{(j)}(X)\, dt.$$

Let

$$\boldsymbol{C}_\tau = (C_{\tau,1} C_{\tau,2}, \ldots, C_{\tau,n})^T$$

and

$$\boldsymbol{I}_\tau = ((I_{\tau,i,j})).$$

It is easy to see that the log likelihood function $\ell_\tau(\theta)$ and its derivatives are given by

$$\ell_\tau(\theta) = \boldsymbol{C}_\tau^T \theta - \tfrac{1}{2} \theta^T \boldsymbol{I}_\tau \theta, \tag{2.9.14a}$$

$$\dot{\boldsymbol{\ell}}_\tau(\theta) = \boldsymbol{C}_\tau - \boldsymbol{I}_\tau \theta, \tag{2.9.14b}$$

$$\ddot{\boldsymbol{\ell}}_\tau(\theta) = -\boldsymbol{I}_\tau. \tag{2.9.15}$$

It is clear that $\boldsymbol{I}_\tau$ is a symmetric nonnegative definite matrix. We assume that $\boldsymbol{I}_\tau$ is nonsingular almost surely. The solution of the likelihood equation is

$$\hat{\theta}_\tau = \boldsymbol{I}_\tau^{-1} \boldsymbol{C}_\tau. \tag{2.9.16}$$

If $\hat{\boldsymbol{\theta}}_\tau \in \Theta$, then it is the unique maximum likelihood estimate. Note that the statistical model is a curved exponential family with minimal sufficient statistic $(\boldsymbol{C}_\tau, \boldsymbol{I}_\tau)$ from (2.9.14a). If $n = 1$ and if I_τ has a finite variance for all θ, then the following quantities exist for all $\theta \in \Theta$:

$$i(\theta) = E_\theta I_\tau, \qquad v(\theta) = \text{var}_\theta\, I_\tau, \qquad u(\theta) = E_\theta\left(I_\tau \int_0^\tau b_t^{-}(X) \alpha_t(X)\, dW_t \right).$$

Since

$$C_\tau = \theta I_\tau + \int_0^\tau b_t^{-}(X) \alpha_t(X)\, dW_s,$$

it follows that

$$E_\theta C_\tau = \theta i(\theta),$$
$$\text{var}_\theta\, C_\tau = \theta^2 v(\theta) + i(\theta) + 2\theta u(\theta)$$

and

$$\text{cov}_\theta(C_\tau, I_\tau) = u(\theta) + \theta v(\theta).$$

Let us consider the stopping time defined by

$$\tau_s = \inf\{t : \max_j I_{t,j,j} \geq s\}. \tag{2.9.17}$$

An important property of this stopping rule is that it reduces the model to a linear exponential family with the minimal sufficient statistic C_{τ_s}.

We now study the large-sample properties of the estimate $\hat{\theta}_\tau$ for the case $n = 1$. We first state a lemma.

Lemma 2.9.2 *Let W be a standard Wiener process relative to a filtration $\{\mathcal{F}_t\}$, and let f_t be an $\{\mathcal{F}_t\}$-adapted nonanticipating stochastic process such that*

$$P\left(\text{there exists } \varepsilon > 0 \text{ such that } \int_0^{\varepsilon} f_t^2\,dt < \infty\right) = 1 \text{ and } P\left(\int_0^{\infty} f_t^2\,dt = \infty\right) = 1.$$

If

$$\tau_s = \inf\left\{t : \int_0^t f_u^2\,du \geq s\right\},$$

then

$$Z_s = \int_0^{\tau_s} f_t\,dW_t, \qquad s \geq 0,$$

is a Wiener process relative to $\{\mathcal{F}_{\tau_s}\}$.

This lemma follows from the arguments in Kutoyants (1984a) (see also Basawa and Prakasa Rao 1980; or Liptser and Shiryayev 1977).

The process $\int_0^t f_s^2\,ds$ is the *intrinsic time* of the process $\int_0^t f_s\,dW_s$. In the present context, since the score function $\dot{\ell}_t(\theta)$ can be expressed as an integral with respect to the Wiener process W, that is,

$$\dot{\ell}_t(\theta) = \int_0^t b_s^-(X)\alpha_s(X)\,dW_s, \tag{2.9.18}$$

the information process I_t is the intrinsic time of the score function process.

Theorem 2.9.3 *Suppose the conditions (X2) and (X3) hold for $\tau = \infty$ and that*

$$\mu_{\theta_0}(\text{ there exists } \varepsilon > 0 \text{ such that } I_\varepsilon < \infty) = 1, \tag{2.9.19a}$$

$$\mu_{\theta_0}(I_\infty = \infty) = 1 \tag{2.9.19b}$$

when θ_0 is the true parameter. Let

$$\tau_s = \inf\{t : I_t \geq s\}, s > 0 \tag{2.9.20}$$

Then $\mu_{\theta_0}(\tau_s < \infty) = 1$ and $\hat{\theta}_{\tau_s}$ is normally distributed with mean θ_0 and variance s^{-1}. In particular,

$$\hat{\theta}_{\tau_s} \xrightarrow{p} \theta_0 \qquad as\ s \to \infty. \tag{2.9.21}$$

Furthermore, if $\{\sigma_s\}$ is any increasing sequence of stopping times satisfying the above conditions, then $\hat{\theta}_{\sigma_s}$ is consistent for θ_0 as $s \to \infty$.

Proof. Note that conditions (X2) and (X3) imply that the likelihood function exists from Theorem 2.9.1. Furthermore,

$$\mu_{\theta_0}(\tau_s = \infty) = \mu_{\theta_0}(I_\infty < s) = 0$$

from (2.9.19b). Hence $\mu_{\theta_0}(\tau_s < \infty) = 1$.

From earlier calculations, it is easy to see that

$$\hat{\theta}_{\tau_s} - \theta_0 = \frac{\dot{\ell}_{\tau_s}(\theta_0)}{I_{\tau_s}} = \frac{\dot{\ell}_{\tau_s}(\theta_0)}{s}$$

and hence $\hat{\theta}_{\tau_s}$ is $N(\theta_0, s^{-1})$ from Lemma 2.9.2. In particular, $\hat{\theta}_{\tau_s}$ is consistent for θ_0 as $s \to \infty$.

Let $\{\sigma_s, s \geq 0\}$ be any family of finite stopping times such that (X3) holds and $\sigma_s \uparrow \infty$ or $I_{\sigma_s} \uparrow \infty$ a.s. μ_{θ_0} as $s \to \infty$. Then,

$$\hat{\theta}_{\tau_{I_{\sigma(s)}}} \xrightarrow{p} \theta_0 \qquad \text{as } s \to \infty.$$

Remark. As a corollary, we obtain that if the nonrandom stopping time $\sigma_s = s$ satisfies conditions (X2) and (X3), then the condition $\mu_{\theta_0}(I_\infty = \infty) = 1$ is sufficient to ensure the consistency for any sequence of stopping times increasing to infinity.

The following theorem gives the asymptotic distribution of the estimator $\hat{\theta}_{\sigma(s)}$.

Theorem 2.9.4 *Suppose $\{\sigma_s\}$ is a nondecreasing family of stopping times such that conditions (X2) and (X3) hold and such that the $\{I_{\sigma_s}, s \geq 0\}$ satisfies the following conditions:*

(i) I_{σ_s} is nondecreasing in s almost surely.
(ii) There exists a nonrandom function $m(t) \uparrow \infty$ as $t \to \infty$ such that

$$\frac{I_{\sigma_s}}{m(s)} \to \eta^2 \qquad \textit{as } s \to \infty,$$

where $P_{\theta_0}(\eta^2 > 0) = 1$.

Then

$$I_{\sigma_s}^{1/2}(\hat{\theta}_{\sigma_s} - \theta_0) \xrightarrow{\mathcal{L}} N(0, 1) \qquad \textit{as } s \to \infty.$$

Proof. It can be seen that $\dot{\ell}_t(\theta_0) = Z_{I_t}$ where $Z_s = \dot{\ell}_{\tau_s}(\theta_0)$ is a Wiener process relative to the filtration $\{\mathcal{F}_{\tau_t}\}$ by Lemma 2.9.1. Here τ_s is as defined by (2.9.20). Note that $I_{\sigma_s}^{1/2}(\hat{\theta}_{\sigma_s} - \theta_0) = I_{\sigma_s}^{-1/2}\dot{\ell}_{\sigma_s}(\theta_0)$. The result now follows from the following lemma due to Feigin (1976).

Lemma 2.9.5 *Let $\{Y_t, t \geq 0\}$ be a stochastic process with stationary independent increments and with second moment $EY_t^2 = t$, $t \geq 0$. Let $\{\sigma_t\}$ be a process nondecreasing in t almost surely. Suppose there exists a nonrandom function $m(t) \uparrow \infty$ as $t \to \infty$ such that*

$$\frac{\sigma_t}{m(t)} \xrightarrow{p} \eta^2 \qquad \textit{as } t \to \infty,$$

where $P(\eta^2 > 0) = 1$. Let $\{U_t\}$ be a process such that

$$U_t = Y_{\sigma_t} \textit{ a.s.}$$

Then

$$\sigma_t^{-1/2} U_t \xrightarrow{\mathcal{L}} N(0, 1) \qquad \textit{as } t \to \infty.$$

Example 2.9.1 (*Geometrical Brownian motion*)
Consider the stochastic differential equation

$$dX_t = \theta X_t\, dt + X_t\, dW_t, \qquad X_0 = a > 0,\ t \geq 0. \tag{2.9.22}$$

An application of Itô's lemma shows that

$$X_t = a \exp(W_t + (\theta - \tfrac{1}{2})t). \tag{2.9.23}$$

This process is called geometric Brownian motion and can be used for modeling the behavior of prices of assets (cf. Karlin and Taylor 1975, p. 363). Note that X_t is log-normally distributed. It can be checked that the conditions (X2) and (X3) hold for any stopping time τ with $P(\tau < \infty) = 1$ and $\beta_t = 1$. Furthermore, the likelihood function is

$$\begin{aligned} L_\tau(\theta) &= \exp\left(\int_0^\tau \theta\, dW_t + \tfrac{1}{2}\int_0^\tau \theta^2\, ds\right) \\ &= \exp(\theta W_\tau + \tfrac{1}{2}\theta^2 \tau) \\ &= \left(\frac{X_\tau}{a}\right)^\theta e^{\theta(1-\theta)\tau/2}. \end{aligned}$$

Hence the MLE $\hat{\theta}_\tau$ is given by

$$\hat{\theta}_\tau = \frac{\log(X_\tau/a)}{\tau} + \frac{1}{2}$$

and the observed Fisher information $I_\tau = \tau$.

Hence, by the remarks made after Theorem 2.9.2, it follows that $\hat{\theta}_{\sigma_s}$ is consistent for any sequence of stopping times σ_s increasing to infinity.

For the process described by (2.9.22), the stopping time defined by (2.9.20) is $\tau_t = t$ and

$$\hat{\theta}_t = \frac{W_t}{t} + \theta_0 \simeq N(\theta_0, t^{-1}),$$

where θ_0 is the true parameter. Note that, for the stopping rule $\tau_t = t$, the stochastic model is a regular exponential family of order 1 and the statistic X_t is minimal sufficient.

Another example of a natural stopping rule for the process $\{X_t\}$ is to stop the observation when the process X_t hits a barrier. If $\sigma_s = \inf\{t \geq 0 : X_t = s\}$, then $P(\sigma_s < \infty) = 1$ if $s > a$ and $\theta > \frac{1}{2}$ or $0 < s < a$ and $\theta < \frac{1}{2}$ following similar results for the Wiener process in view of (2.9.23). Since $X_{\sigma_s} = s$ for all s, we have an exponential family of order 1 with the minimal sufficient statistic $\sigma_s(X)$, and the MLE in this case is given by

$$\hat{\theta}_{\sigma_s} = \frac{\log(s/a)}{\sigma_s} + \frac{1}{2}.$$

Again using the results for the Wiener process, it follows that

$$\sigma_s \simeq N^-\left(-\frac{1}{2}, \left(\log\frac{s}{a}\right)^2, \left(\theta_0 - \frac{1}{2}\right)^2\right),$$

where $N^-(\lambda, \chi, \psi)$ denotes the generalized inverse Gaussian distribution (cf. Jørgensen 1982). For $\lambda = -\frac{1}{2}$, this is the usual inverse Gaussian distribution. It can be checked that

$$\frac{I_{\sigma_s}}{|\log(s/a)|} = \frac{\sigma_s}{|\log(s/a)|} \simeq N^-\left(-\frac{1}{2}, \left|\log\frac{s}{a}\right|, \left|\log\frac{s}{a}\right|\left(\theta_0 - \frac{1}{2}\right)^2\right).$$

From the properties of a generalized inverse Gaussian distribution, it follows that

$$\frac{I_{\sigma_s}}{|\log(s/a)|} \xrightarrow{p} \left|\theta_0 - \frac{1}{2}\right| > 0,$$

as $s \to \infty$ or $s \to 0$. Theorem 2.9.3 implies that

$$\sigma_s^{1/2}(\hat{\theta}_{\sigma_s} - \theta_0) \xrightarrow{\mathcal{L}} N(0, 1),$$

as $s \to \infty$ or $s \to 0$. In fact, the exact distribution of $\hat{\theta}_{\sigma_s}$ is given by

$$\hat{\theta}_{\sigma_s} \simeq \begin{cases} N^{-}\left(\frac{1}{2}, \left(\theta_0 - \frac{1}{2}\right)^2 \log\left(\frac{s}{a}\right), \log\frac{s}{a}\right) + \frac{1}{2}, & s > a \\ -N^{-}\left(\frac{1}{2}, \left(\theta_0 - \frac{1}{2}\right)^2 \log\left(\frac{a}{s}\right), \log\frac{a}{s}\right) + \frac{1}{2}, & s < a. \end{cases}$$

In particular, for $s \neq a$,

$$E(\hat{\theta}_{\sigma_s}) = \theta_0 + \frac{1}{\log\frac{s}{a}}.$$

Example 2.9.2 (*Process with proportional drift and diffusion coefficient*)
Consider the stochastic differential equation

$$dX_t = \theta a_t^2(X)\,dt + a_t(X)\,dW_t, \tag{2.9.24}$$

where the drift and diffusion are proportional. If $a_t(x) = \sqrt{X_t}$, and $X_0 \equiv a > 0$, then the solution is the diffusion branching process (cf. Cox and Miller 1965, p. 225).

Suppose the Lipschitz condition is satisfied by the drift and diffusion coefficients in (2.9.24) and

$$P\left(\int_0^t a_s^2(X)\,ds < \infty\right) = 1, \qquad t \geq 0.$$

It can be checked that the likelihood function exists and the statistic

$$\left(X_\tau - X_0, \int_0^\tau a_t^2(X)\,dt\right)$$

is minimal sufficient for any stopping time τ. The MLE is

$$\hat{\theta}_\tau = \frac{X_\tau - X_0}{\int_0^\tau a_t^2(X)\,dt}$$

provided $\hat{\theta}_\tau \in \Theta$ and the observed Fisher information is

$$I_\tau = \int_0^\tau a_t^2(X)\,dt.$$

Let us consider the stopping time $\sigma_s = \inf\{t : X_t = s\}$ (cf. Brown and Hewitt 1975, Basawa and Prakasa Rao 1980). Suppose the function $a(\cdot)$ is such that the process X has no absorbing barrier. Let $X_0 \equiv a > 0$. Then $P(\sigma_s < \infty) = 1$ if $s > a$ and $\theta > 0$ or if $s < a$ and $\theta < 0$. Furthermore,

$$I_{\sigma_s} \simeq N^{-}(-\tfrac{1}{2}, (s-a)^2, \theta^2)$$

and, as $s \to \pm\infty$,

$$\frac{I_{\sigma_s}}{|s-a|} \simeq N^{-}(-\tfrac{1}{2}, |s-a|, \theta^2|s-a|) \to |\theta|.$$

Theorem 2.9.3 implies that

$$I_{\sigma_s}^{1/2}(\hat{\theta}_{\sigma_s} - \theta_0) \xrightarrow{\mathcal{L}} N(0, 1) \qquad \text{as } s \to \infty.$$

The exact distribution of the MLE is

$$\hat{\theta}_{\sigma_s} = \frac{s - a}{\int_0^{\sigma_s} a_t^2(X)\,dt} \simeq \pm N^-(\tfrac{1}{2}, \theta^2|s - a|, |s - a|),$$

where we take plus or minus according to the sign of $s - a$.

Example 2.9.3 (*Exponential families of processes of the diffusion type*)
Consider the stochastic differential equation

$$dX_t = (\theta\mu_t(X) + \nu_t(X))\,dt + \sigma_t(X)\,dW_t, \qquad t \geq 0, \tag{2.9.25}$$

where $\theta \in \Theta \subset R$. Suppose that μ_t, ν_t and σ_t satisfy the Lipschitz condition implying the existence and uniqueness of solution $\{X_t\}$ for (2.9.25) for all $\theta \in \Theta$. Examples of this type are the Wiener processes with drift, the geometric Brownian motion discussed earlier and the Ornstein–Uhlenbeck process. A special case of such a process is obtained when $\mu_t \equiv 1$, $\sigma_t \equiv 1$ and $\nu_t(X) = -\rho X_t$, where ρ is a known constant. For an application of such processes for stochastic modeling, see Erlandsen and Sørensen (1984). Let

$$A_t = \int_0^t \sigma_s^-(X)^2\mu_s(X)\,dX_s - \int_0^t \sigma_s^-(X)^2\mu_s(X)\nu_s(X)\,ds$$

and

$$B_t = \int_0^t \sigma_s^-(X)^2\mu_s(X)^2\,ds,$$

where

$$\sigma_t^-(X) = \begin{cases} \sigma_t(x)^{-1}, & \text{if } \sigma_t(x) \neq 0, \\ 0, & \text{otherwise.} \end{cases}$$

Suppose the process is observed up to the stopping time

$$\tau(u) = \inf\{t : \alpha A_t + \beta B_t \geq u\},$$

where u, α, β are prescribed constants such that $\alpha^2 + \beta^2 > 0$, and $u > 0$ and $P(\tau(u) < \infty) = 1$ (cf. Stefanov 1985). Let

$$D_t = \alpha A_t + \beta B_t - u.$$

Since the processes A_t and B_t are continuous, it follows that for all $u > 0$ and for all $\theta \in \Theta$,

$$P_\theta(D_{\tau(u)} = 0) = 1,$$

and we have a noncurved exponential family by observing the process on $[0, \tau(u)]$. It is easy to check that the MLE $\hat{\theta}_{\tau(u)}$ is given by

$$\hat{\theta}_{\tau(u)} = A_{\tau(u)}B_{\tau(u)}^{-1},$$

and the following result holds.

Let $\Theta \subset \{\theta : \alpha\theta + \beta > 0\}$, and suppose that $P_\theta(\tau(u) < \infty) = 1$ for all $\theta \in \Theta$. Then, it can be checked that

$$\hat{\theta}_{\tau(u)} + \alpha^{-1}\beta \simeq N^-(\tfrac{1}{2}, u\alpha^{-1}(\theta + \alpha^{-1}\beta)^2, u\alpha^{-1}), \qquad \alpha \neq 0,$$

and

$$\hat{\theta}_{\tau(u)} \simeq N(\theta, \beta u^{-1}), \qquad \alpha = 0, \beta > 0$$

(for details, see Sørensen 1983; 1986).

Remarks. For the special case of linear drift in diffusion processes, see Novikov (1972) and Basawa and Prakasa Rao (1980).

2.10 Estimation from Incomplete Observations

We have discussed several methods of estimation of the drift coefficient for a diffusion process continuously observed throughout a time interval $[0, T]$ as $T \to \infty$ or as the diffusion coefficient goes to zero (cf. Kutoyants 1984a). However, in practice, it may be difficult to observe the sample path of a diffusion process in every detail. Here we shall derive a general condition ensuring the asymptotic sufficiency (cf. Le Cam 1986; Le Cam and Yang 1990) of incomplete observations of the sample path of the process $\{X_t\}$ on the time interval $[0, T]$ as the diffusion coefficient goes to zero. An example of incomplete observation is the case when the process is observed at discrete time instants. We will discuss methods of estimation for diffusion processes observed at discrete times in the next chapter.

2.10.1 Asymptotic Sufficiency

Consider an m-dimensional diffusion process defined by the stochastic differential equation

$$dX_t = a(X_t, \theta)\, dt + \varepsilon\sigma(X_t)\, dW_t, \qquad X_0 = x,\ t \geq 0, \tag{2.10.1}$$

where $\{W_t\}$ is a standard m-dimensional Wiener process, $\theta \in \Theta \subset R^k$ is an unknown parameter in the drift function $a(\cdot, \theta) : R^m \to R^m$ and the diffusion matrix $\sigma : R^m \to R^m \times R^m$, $x \in R^m$ and ε are known. Let $a(\cdot, \theta) = (a_1(\cdot, \theta), \ldots, a_m(\cdot, \theta))$.

Suppose the sample path X_t is partially observed on $[0, T]$ and assume that it is possible to construct on the basis of these incomplete observations a process $\{Y_t\}$ defined on $[0, T]$ which is smooth enough and satisfies the conditions

$$\{Y_t, 0 \leq t \leq T\} = F(\{X_t, 0 \leq t \leq T\}), \tag{2.10.2}$$

where F is a map as defined below, and

(Y1) $$\sup\{\varepsilon^{-1}\|Y_t - X_t\|, 0 \leq t \leq T\} \xrightarrow{p} 0 \qquad \text{as } \varepsilon \to 0,$$

with $\xrightarrow{p}$ denoting convergence in probability under the probability measure $P_{\theta,\varepsilon}$ induced by the process $\{X_t\}$ on the space $(C, \mathcal{C})$. Here $C = C([0, \infty) \to R^n)$ is the space of all continuous functions defined on $[0, \infty)$ with values in R^m; and $\mathcal{C} = \cup_{t \geq 0} \mathcal{C}_t$, where $\mathcal{C}_t$ is the σ-algebra generated by $\{X_s, 0 \leq s \leq t\}$. Let $(D, \mathcal{D})$ denote the space of R^m-valued functions defined on $[0, \infty)$ which are left-continuous with right-hand limits endowed with the σ-algebra generated by the Skorokhod topology on each compact set (see Billlingsley 1968). We assume that $F : (C, \mathcal{C}) \to (D, \mathcal{D})$ is measurable. Let $x_\theta(t)$ be the solution of (2.10.1) with $\varepsilon = 0$, that is

$$dx_\theta(t) = a(x_\theta(t), \theta)\, dt, \qquad x_\theta(0) = x. \tag{2.10.3}$$

Suppose the following conditions hold.

(Y2) (i) The functions $a(u, \theta)$ and $\sigma(u)$ have continuous partial derivatives up to second order in $R^m \times \Theta^0$, where Θ^0 denotes the interior of Θ.

(ii) For every $\theta \in \Theta$, there exists a positive constant L_θ such that, for all $u \in R^m$,

$$\|a(u, \theta)\|^2 + \|\sigma(u)\|^2 \leq L_\theta(1 + \|u\|^2),$$

where $\|\cdot\|$ is the Euclidean norm on R^m.

(Y3) There exists $V : R^m \times \Theta \to R$ such that $V(x, \theta) = 0$ and

$$\frac{\partial V(u, \theta)}{\partial u} = [e(u)]^{-1} a(u, \theta),$$

where

$$e(u) = \sigma(u)\sigma(u)^t.$$

(Y4) Θ is a compact subset of R^k.

(Y5) For all $\theta \neq \theta'$ in Θ, $a(x_\theta(\cdot), \theta) \neq a(x_\theta(\cdot), \theta')$.

(Y6) The $k \times k$ matrix $I_T(\theta)$ defined by

$$I_T(\theta) = \int_0^T \left(\frac{\partial a(x_\theta(s), \theta)}{\partial \theta}\right)^t e(x_\theta(s))^{-1} \frac{\partial a(x_\theta(s), \theta)}{\partial \theta} \, ds \tag{2.10.4}$$

exists and is positive definite on Θ^0.

Here $\partial a(u, \theta)/\partial \theta$ is the matrix of order $m \times k$ of partial derivatives $(\partial a_i(u, \theta)/\partial \theta^j)$, $1 \leq i \leq m$, $1 \leq j \leq k$, $\theta = (\theta^1, \ldots, \theta^k)$. Suppose θ_0 is the true value of the parameter and $\theta_0 \in \Theta^0$.

The probability measures $P_{\theta,\varepsilon}$ and $P_{\theta_0,\varepsilon}$ are equivalent on $\mathcal{C}_T$ under assumption (Y2) and

$$L_\varepsilon(\theta) = \log\left(\frac{dP_{\theta,\varepsilon}}{dP_{\theta_0,\varepsilon}}\bigg|\mathcal{C}_T\right) = \lambda_\varepsilon(\theta) - \lambda_\varepsilon(\theta_0), \tag{2.10.5}$$

where

$$\lambda_\varepsilon(\theta) = \varepsilon^{-2}\left(\int_0^T (e(X_s)^{-1} a(X_s, \theta)) \cdot dX_s - \frac{1}{2}\int_0^T v(X_s, \theta)\, ds\right), \tag{2.10.6}$$

where $\cdot$ denotes the inner product on R^m and $v(u, \theta)$ is defined by

$$v(u, \theta) = a(u, \theta)^t e(u)^{-1} a(u, \theta). \tag{2.10.7}$$

Let A be a sphere $B(\theta_0, r) = \{z \in R^k : \|z - \theta_0\| \leq r\}$ such that $A \subset \Theta$. Under the assumption (Y6), the matrix $I_T(\theta_0)$ is symmetric and positive definite. Let $I_T(\theta_0)^{-1/2}$ be the symmetric square root of $I_T(\theta_0)^{-1}$. Define the net of experiments

$$\zeta_\varepsilon = \{Q_{z,\varepsilon}, z \in A\}_{\varepsilon > 0}, \tag{2.10.8}$$

with

$$Q_{z,\varepsilon} = P_{\theta_0 + \varepsilon I_T(\theta_0)^{-1/2}}|\mathcal{C}_T.$$

It is easy to see that

$$\log \frac{dQ_{z,\varepsilon}}{dQ_{0,\varepsilon}} = \Lambda_\varepsilon(z) = \lambda_\varepsilon(\theta_0 + \varepsilon I_T(\theta_0)^{-1/2} z) - \lambda_\varepsilon(\theta_0). \tag{2.10.9}$$

Let G_0 be $N_k(0, J_k)$, where J_k denotes the identity matrix of order k. In other words G_0 is the joint distribution of k i.i.d. $N(0, 1)$ components. Let G_z be G_0 translated by z. Then $\zeta = \{G_z, z \in R^k\}$ is the standard Gaussian shift experiment on R^k (cf. Le Cam and Yang 1990).

Let

$$\tilde{Z}_\varepsilon = \varepsilon^{-1} I_T(\theta_0)^{-1/2}\left(\frac{\partial V(Y_T,\theta_0)}{\partial\theta} - \frac{1}{2}\int_0^T \frac{\partial V(Y_s,\theta_0)}{\partial\theta}\,ds\right), \tag{2.10.10}$$

where $v(\cdot,\cdot)$ is as given by (2.10.7) and $\{Y_T\}$ is the observed process. Note that $\tilde{Z}_\varepsilon$ depends on $\{Y_s, 0 \le s \le T\}$ but not on $\{X_s, 0 \le s \le T\}$.

Theorem 2.10.1 *Suppose that conditions (Y2)–(Y6) hold. Then we have the following:*

(i) The net of experiments $\zeta_\varepsilon = \{Q_{z,\varepsilon}, z \in A\}$, $\varepsilon > 0$ converges uniformly on the precompact subsets of A to the restriction to A of the standard Gaussian shift experiment ζ as $\varepsilon \to 0$.

(ii) Furthermore, if $\{Y_s, 0 \le s \le T\}$ satisfies condition (Y1), then, for all precompact subsets S of A,

$$\Lambda_\varepsilon(z) = z^t\tilde{Z}_\varepsilon - \tfrac{1}{2}\|z\|^2 + \tilde{R}(\varepsilon, z, \theta_0) \quad \text{for all } z \in S, \tag{2.10.11}$$

with

$$\mathcal{L}(\tilde{Z}_\varepsilon | P_{\theta_0,\varepsilon}) \to N_k(0, J_k), \quad \text{as } \varepsilon \to 0, \tag{2.10.12}$$

and

$$\sup_{z\in S}|\tilde{R}(\varepsilon, z, \theta_0)| \to 0 \text{ in } P_{\theta_0,\varepsilon}\text{-probability}, \quad \text{as } \varepsilon \to 0. \tag{2.10.13}$$

Before we give a proof of this theorem, we shall state a result due to Azencott (1982) giving a stochastic Taylor expansion of X_t in powers of ε.

Lemma 2.10.2 *Suppose the assumption (Y2) holds. Then there exists a continuous centered Gaussian process $\{g_\theta(t), t \ge 0\}$ and a process $\{R_\theta(t), t \ge 0\}$ such that*

$$X_t = x_\theta(t) + \varepsilon g_\theta(t) + \varepsilon^2 R_\theta(t) \tag{2.10.14}$$

with

$$\lim_{\substack{\varepsilon\to 0\\ r\to\infty}} P_{\theta,\varepsilon}\{\sup_{0\le t}\|R_\theta(s)\| \ge r\} = 0. \tag{2.10.15}$$

In fact, the Gaussian process $g_\theta(t)$ is a solution of the differential equation,

$$dg_\theta(t) = \frac{\partial a(x_\theta(t),\theta)}{\partial u} g_\theta(t)\,dt + \sigma(x_\theta(t))\,dW_t, \quad g_\theta(0) = 0, \tag{2.10.16}$$

where $\partial a(u,\theta)/\partial u$ denotes the $m \times m$ matrix $((\partial a_i(u,\theta)/\partial u_j))$ of partial derivatives.

Proof of Theorem 2.10.1. For simplicity, we consider the case $m = 1$. Consider the random vectors

$$Z_\varepsilon = I_T(\theta_0)^{-1/2}\left\{\int_0^T \frac{\partial a(X_s,\theta_0)}{\partial\theta^j}\left(\frac{\partial X_s - a(X_s,\theta_0)\,ds}{\varepsilon\sigma(X_s)^2}\right), 1 \le j \le k\right\}. \tag{2.10.17}$$

It follows from the general theory in Le Cam (1986) that property (i) stated in the theorem holds if the following conditions are satisfied: for all precompact subsets S of A,

$$\Lambda_\varepsilon(z) = z^t Z_\varepsilon - \tfrac{1}{2}\|z\|^2 + R(\varepsilon, z, \theta_0), \tag{2.10.18}$$

with

$$\mathcal{L}(Z_\varepsilon | P_{\theta_0,\varepsilon}) \to N_k(0, J_k) \tag{2.10.19}$$

and

$$\sup_{z\in S} |R(\varepsilon, z, \theta_0)| \to 0 \text{ in } P_{\theta_0,\varepsilon}\text{-probability as } \varepsilon \to 0. \tag{2.10.20}$$

Since the diffusion process $\{X_t\}$ satisfies (2.10.1), it follows that

$$Z_\varepsilon = I_T(\theta_0)^{-1/2} \int_0^T f(X_s, \theta_0)\, dW_s, \tag{2.10.21}$$

where

$$f(u, \theta) = \sigma(u)^{-1} \frac{\partial a(u, \theta)}{\partial \theta}. \tag{2.10.22}$$

In view of Lemma 2.10.1, it follows that

$$\int_0^T \|f(X_s, \theta_0) - f(x_{\theta_0}(s), \theta_0)\|^2\, ds \to 0 \text{ in } P_{\theta_0,\varepsilon}\text{-probability}$$

as $\varepsilon \to 0$ and hence, for fixed T,

$$Z_\varepsilon \xrightarrow{\mathcal{L}} I_T(\theta_0)^{-1/2} \int_0^T f(x_{\theta_0}(s), \theta_0)\, dW_s \text{ under } P_{\theta_0,\varepsilon},$$

as $\varepsilon \to 0$. Note that the random vector $I_T(\theta_0)^{-1/2} \int_0^T f(x_{\theta_0}(s), \theta_0)\, dW_s$ has a multivariate normal distribution with mean 0 and covariance matrix J_k.

Expanding $a(X_s, \theta)$ as a Taylor series, the remainder term in (2.10.18) can be written in the form

$$R(\varepsilon, z, \theta_0) = \tfrac{1}{2} z^t (A_1 + A_2 + A_3) z,$$

where the matrices A_i are defined by

$$A_1 = \varepsilon \int_0^T \sigma(X_s)^{-1} \left(\int_0^1 \frac{\partial^2 a(X_s, \theta_0 + \varepsilon t z)}{\partial \theta^2}\, dt \right) dW_s,$$

$$A_2 = \int_0^T \frac{ds}{\sigma(X_s)^2} \int_0^1 (a(X_s, \theta_0) - a(X_s, \theta_0 + \varepsilon t z)) \frac{\partial^2 a(X_s, \theta_0 + \varepsilon t z)}{\partial \theta^2}\, dt,$$

$$A_3 = \int_0^T (H(x_{\theta_0}(s), \theta_0) - H(X_s, \theta_0 + \varepsilon t z))\, ds,$$

with

$$H(u, \theta) = \frac{1}{\partial(u)^2} \frac{\partial a(u, \theta)}{\partial \theta} \frac{\partial a(u, \theta)^t}{\partial \theta}. \tag{2.10.23}$$

Note that $A = A_1(t)$ is a martingale with respect to $\mathcal{C}_t$. Hence by the Lenglart domination property (cf. Jacod and Shiryayev 1987), it follows that

$$P_{\varepsilon,r}(\|A_1\| \geq h) \leq P_{\varepsilon,r}(\|B_1\| \geq \eta) + \frac{\eta^2}{h}$$

for all $h, \eta > 0$, where

$$B_1 = \varepsilon^2 \int_0^T \frac{ds}{\sigma(X_s)^2} \left(\int_0^1 \frac{\partial^2 a(X_s, \theta_0 + \varepsilon t z)}{\partial \theta^2}\, dt \right)^2.$$

Let

$$J_\delta = [|X_s - x_{\theta_0}(s)| \le \delta, 0 \le s \le T]. \tag{2.10.24}$$

Lemma 2.10.1 implies that $P_{\theta_0,\varepsilon}(J_\delta) \to 1$ as $\varepsilon \to 0$. On the set J_δ, the norm of $\boldsymbol{B}_1$ is bounded by $\varepsilon^2 T K_1$ where K_1 is a constant depending on T, θ_0, r and δ. Similarly $\|\boldsymbol{A}_2\|$ is bounded on J_δ by $\varepsilon T K_2(T, \theta_0, r, \delta)$ and $\|\boldsymbol{A}_3\|$ by

$$|X_s - x_{\theta_0}(s)| \sup_{(u,\theta)\in B} \left| \frac{\partial \boldsymbol{H}(u,\theta)}{\partial u} \right| + \varepsilon r \sup_{(u,\theta)\in B} \left\| \frac{\partial \boldsymbol{H}(u,\theta)}{\partial \theta} \right\|,$$

where B is the compact set $\cup_{0\le s\le T}[x_{\theta_0}(s) - \delta, x_{\theta_0}(s) + \delta] \times (\|\theta - \theta_0\| \le r)$.

Combining these relations, we obtain (2.10.20).

In order to prove (2.10.19), it suffices to show that the difference $\|\tilde{\boldsymbol{Z}}_\varepsilon - \boldsymbol{Z}_\varepsilon\| \to 0$ in $P_{\theta_0,\varepsilon}$-probability as $\varepsilon \to 0$ uniformly for $z \in S$. By the Itô formula, one has

$$\begin{aligned}
\boldsymbol{Z}_\varepsilon - \tilde{\boldsymbol{Z}}_\varepsilon &= \frac{1}{\varepsilon}\left[\frac{\partial V(X_T,\theta_0)}{\partial \theta} - \frac{\partial V(Y_T,\theta_0)}{\partial \theta}\right] \\
&\quad - \frac{1}{\varepsilon}\int_0^T \left(\frac{\partial a(X_s,\theta_0)}{\partial \theta}\frac{a(X_s,\theta_0)}{\sigma^2(X_s)} - \frac{\partial a(Y_s,\theta_0)}{\partial \theta}\frac{a(Y_s,\theta_0)}{\sigma^2(Y_s)}\right) ds \\
&\quad - \frac{\varepsilon}{2}\int_0^T \frac{\partial}{\partial u}\left(\frac{1}{\sigma^2(X_s)}\frac{\partial v(X_s,\theta)}{\partial \theta}\right)\sigma^2(X_s)\, ds.
\end{aligned}$$

Hence

$$\|\boldsymbol{Z}_\varepsilon - \tilde{\boldsymbol{Z}}_\varepsilon\| \le \left(\sup_{0\le t\le T} \frac{|Y_t - X_t|}{\varepsilon}\right)(W_1 + W_2) + \varepsilon T W_3,$$

where

$$W_1 = \sup_{\substack{0\le t\le T\\ \theta\in\Theta}} \left\| \frac{\partial^2 V(X_s,\theta_0)}{\partial u \partial \theta} \right\|,$$

$$W_2 = \sup_{\substack{0\le t\le T\\ \theta\in\Theta}} \left\| \frac{\partial}{\partial u}\left(\frac{1}{\sigma^2(X_s)}\frac{\partial a(X_s,\theta_0)}{\partial \theta}\right) \right\|,$$

$$W_3 = \sup_{\substack{0\le t\le T\\ \theta\in\Theta}} \left\| \frac{\partial}{\partial u}\left(\frac{1}{\sigma^2(X_s)}\frac{\partial a(X_s,\theta_0)}{\partial \theta}\right)\sigma^2(X_s) \right\|.$$

It follows from Lemma 2.10.1 that all the three random variables $W_i, 1 \le i \le 3$, are bounded on the set J_δ, defined by (2.10.24), by constants independent of z. Therefore $\|\boldsymbol{Z}_\varepsilon - \tilde{\boldsymbol{Z}}_\varepsilon\|$ converges to zero uniformly by condition (Y1) and the above relations. This completes the proof.

Let us again consider the case when $m = 1$ in (2.10.1) for simplicity. As a consequence of the Theorem 2.10.1, it follows that the observation $\{Y_t, 0 \le t \le T\}$ is asymptotically sufficient in the sense of Le Cam (1986). In particular, the tests or estimates based on $\{Y_t, 0 \le t \le T\}$ perform as well as those based on $\{X_t, 0 \le t \le T\}$. Let us now consider estimators based on $\{Y_t, 0 \le t \le T\}$. Consider the approximate log-likelihood

$$L_\varepsilon^{(a)}(\theta) = \ell_\varepsilon(\theta) - \ell_\varepsilon(\theta_0), \tag{2.10.25}$$

where

$$\ell_\varepsilon(\theta) = \frac{1}{\varepsilon^2}\left[V(Y_T,\theta) - \frac{1}{2}\int_0^T \frac{a(Y_s,\theta)^2}{\sigma^2(Y_s)}\,ds\right]. \tag{2.10.26}$$

Let $\hat{\theta}_\varepsilon$ be the MLE of θ based on $L_\varepsilon(\theta)$ given by (2.10.5) and $\tilde{\theta}_\varepsilon$ be the approximate MLE based on $L_\varepsilon^{(a)}(\theta)$ given by (2.10.25). In other words,

$$\sup_{\theta\in\Theta} L_\varepsilon(\theta) = L_\varepsilon(\hat{\theta}_\varepsilon) \text{ and } \sup_{\theta\in\Theta} L_\varepsilon^{(a)}(\theta) = L_\varepsilon^{(a)}(\tilde{\theta}_\varepsilon). \tag{2.10.27}$$

Theorem 2.10.3 *Under the assumptions (Y1)–(Y6), the following properties hold:*

(a) $P_{\theta_0,\varepsilon}(\|\tilde{\theta}_\varepsilon - \theta_0\| > h) \to 0$ *as* $\varepsilon \to 0$ *for all* $h > 0$*;*

(b) $\mathcal{L}(\varepsilon^{-1}(\tilde{\theta}_\varepsilon - \theta_0)|P_{\theta_0,\varepsilon}) \to N_k(0, I(\theta_0)^{-1})$ *as* $\varepsilon \to 0$*; and*

(c) $\varepsilon^{-1}(\hat{\theta}_\varepsilon - \tilde{\theta}_\varepsilon) \xrightarrow{p} 0$ *in* $P_{\theta_0,\varepsilon}$*-probability as* $\varepsilon \to 0$.

Proof. Let

$$D_\delta = \{|Y_s - x_{\theta_0}(s)| \le \delta, 0 \le s \le T\}, \quad \text{and} \quad U_\varepsilon(\theta) = -\varepsilon^2 L_\varepsilon^{(a)}(\theta). \tag{2.10.28}$$

Lemma 2.10.1 and condition (Y1) imply that $P_{\theta_0,\varepsilon}(D_\delta) \to 1$ as $\varepsilon \to 0$. Hence $U_\varepsilon(\theta) \xrightarrow{p} K(\theta_0,\theta)$ under $P_{\theta_0,\varepsilon}$-probability as $\varepsilon \to 0$ where

$$\begin{aligned} K(\theta_0,\theta) &= V(x_{\theta_0}(T),\theta_0) - V(x_{\theta_0}(T),\theta) \\ &\quad - \frac{1}{2}\int_0^T \frac{a^2(x_{\theta_0}(s),\theta_0) - a^2(x_{\theta_0}(s),\theta)}{\sigma^2(x_{\theta_0}(s))}\,ds \\ &= \frac{1}{2}\int_0^T \frac{(a(x_{\theta_0}(s),\theta_0) - a(x_{\theta_0}(s),\theta))^2}{\sigma^2(x_{\theta_0}(s))}\,ds. \end{aligned} \tag{2.10.29}$$

The identifiability assumption (Y5) ensures that the function $K(\theta_0,\cdot)$ is positive for all $\theta \ne \theta_0$ and is zero at $\theta = \theta_0$. Hence the random function $U_\varepsilon(\theta)$ is a contrast function in the sense of Dacunha-Castelle and Dufflo (1983). The function $\theta \to K(\theta_0,\theta)$ is continuous. Hence, by Ducunha-Castelle and Dufflo (1983), $\tilde{\theta}_\varepsilon$ is consistent provided that, for all α, there exists $\eta > 0$ such that

$$\lim_{\varepsilon\to 0} P_{\theta_0,\varepsilon}(\omega(U_\varepsilon,\eta) > \alpha) = 0, \tag{2.10.30}$$

where

$$\omega(U_\varepsilon,\eta) = \sup_{\|\theta-\theta'\|\le\eta} |U_\varepsilon(\theta) - U_\varepsilon(\theta')| \tag{2.10.31}$$

is the modulus of continuity of $U_\varepsilon(\theta)$. Note that

$$|U_\varepsilon(\theta) - U_\varepsilon(\theta')| \le |v(Y_T,\theta) - v(Y_T,\theta')| + \int_0^T \frac{a^2(Y_s,\theta) - a^2(Y_s,\theta')}{2\sigma^2(Y_s)}\,ds.$$

Let $K = \{x_\theta(s), 0 \le s \le T, \theta \in \Theta\}$ and $K_\delta = \{x \in R :$ there exists $y \in K$ such that $|y - x| \le \delta\}$. On the compact set $K_\delta \times \Theta$, $a(u,\theta)$ is uniformly continuous and hence the two functions

$$\beta(\eta) = \sup_{K_\delta} \sup_{\|\theta-\theta'\|\le\eta} |v(u,\theta) - v(u,\theta')|$$

and

$$\gamma(\eta) = \sup_{K_\delta} \sup_{\|\theta-\theta'\|\le\eta} |a(u,\theta)^2 - a(u,\theta')^2|$$

satisfy

$$\lim_{\eta\to 0}\beta(\eta) = \lim_{\eta\to 0}\gamma(\eta) = 0.$$

Therefore,

$$\omega(u,\eta) \le \beta(\eta) + \tfrac{1}{2}T\gamma(\eta)$$

on the set D_δ defined by (2.10.28) which in turn implies the relation (2.10.30), proving part (a) of the theorem.

Note that the function $U_r(\theta)$ can be expanded, leading to the equation

$$0 = \frac{1}{\varepsilon}\frac{\partial U_\varepsilon}{\partial\theta}\bigg|_{\theta=\theta_0} + \frac{1}{\varepsilon}\sum_{\ell=1}^{k}\left\{(\tilde{\theta}_\varepsilon^\ell - \tilde{\theta}_0^\ell)\frac{\partial^2 U_\varepsilon}{\partial\theta^\ell\,\partial\theta^j}\bigg|_{\theta=\theta_0} + R_{\ell,j}(\theta_0,\tilde{\theta}_\varepsilon-\theta_0)\right\}, \tag{2.10.32}$$

for $1 \le j \le k$, where the remainder term is given by

$$R_{\ell,j}(\theta,h) = \int_0^1 \frac{\partial^2 U_\varepsilon(\theta+sh)}{\partial\theta^\ell\,\partial\theta^j}\,ds.$$

Observe that

$$\frac{1}{\varepsilon}\frac{\partial U_\varepsilon}{\partial\theta}\bigg|_{\theta=\theta_0} = -\boldsymbol{I}_T(\theta_0)^{1/2}\tilde{\mathbf{Z}}_\varepsilon,$$

which converges to $N_k(0,\boldsymbol{I}_T(\theta_0))$ as $\varepsilon\to 0$ by Theorem 2.10.1. Applying Lemma 2.10.2 and condition (Y1), one can show that

$$\left(\left(\frac{\partial^2 U_\varepsilon}{\partial\theta^\ell\,\partial\theta^j}\bigg|_{\theta=\theta_0}\right)\right)_{k\times k} \xrightarrow{p} \boldsymbol{I}_T(\theta_0) \qquad \text{as } \varepsilon\to 0.$$

Hence

$$\frac{1}{\varepsilon}(\tilde{\theta}_\varepsilon - \theta_0) = \boldsymbol{I}_T(\theta_0)^{1/2}\tilde{\mathbf{Z}}_\varepsilon + o_{P_{\theta_0},\varepsilon}(1). \tag{2.10.33}$$

This proves part (b) of the theorem. Theorem 2.10.1 implies that

$$\frac{1}{\varepsilon}(\hat{\theta}_\varepsilon - \theta_0) = \boldsymbol{I}_T(\theta_0)^{1/2}\mathbf{Z}_\varepsilon + o_{P_{\theta_0},\varepsilon}(1).$$

Hence $\varepsilon^{-1}(\tilde{\theta}_\varepsilon - \hat{\theta}_\varepsilon) = \boldsymbol{I}_T(\theta_0)^{1/2}(\tilde{\mathbf{Z}}_\varepsilon - \mathbf{Z}_\varepsilon) + o_{P_{\theta_0},\varepsilon}(1)$. The proof is complete in view of Theorem 2.10.1.

2.10.2 Examples of Incomplete Data

We now describe various examples of incomplete observations of a diffusion process $\{X_t, 0\le t\le T\}$ satisfying condition (Y1).

Discrete observations of the sample path

Suppose the sampling interval is Δ and the observations consist of $\{X_{k\Delta}, 0 \leq k \leq N\}$ where $N = T/\Delta$. For any arbitrary function f defined at the points $k\Delta, k = 0, 1, \ldots, N$, with values in R^m, let us associate with f the interpolated function $f^\Delta : [0, T] \to R^m$ given by

$$f^\Delta(t) = f(k\Delta) + \frac{(t - k\Delta)}{\Delta}(f((k+1)\Delta) - f(k\Delta)) \qquad \text{for } k\Delta \leq t < (k+1)\Delta. \quad (2.10.34)$$

Define the process

$$Y_t = X_t^\Delta, \qquad 0 \leq t \leq T \qquad (2.10.35)$$

from the discrete observation $X_{k\Delta}, 0 \leq k \leq N$.

Theorem 2.10.4 *Suppose that assumptions (Y2) and (Y3) hold. Then, if $\Delta = \Delta(\varepsilon)$ satisfies $\varepsilon^{-1/2}\Delta(\varepsilon) \to 0$ as $\varepsilon \to 0$, then the interpolated process $\{Y_t, 0 \leq t \leq T\}$ defined by (2.10.35) satisfies condition (Y1).*

Proof. The stochastic Taylor expansion of $\{X_t\}$, given by Lemma 2.10.2, shows that

$$Y_t - X_t = x^\Delta(t) - x(t) + \varepsilon(g^\Delta(t) - g(t)) + \varepsilon^2(R_2^\Delta(t) - R_2(t)),$$

where x^Δ, g^Δ and R_2^Δ are the interpolated functions on $[0, T]$ of x, g and R_2, respectively. Let

$$\omega(g, \Delta) = \sup_{|t-s| \leq \Delta} \|g(t) - g(s)\|.$$

It can be checked that

$$\frac{\|Y_t - X_t\|}{\varepsilon} \leq \frac{\Delta^2}{\varepsilon} \sup_{[0,T]} \left\| \frac{d^2x}{dt^2}(s) \right\| + 2\omega(g, \Delta) + 4 \sup_{[0,T]} \|\varepsilon R_2(t)\|. \qquad (2.10.36)$$

Condition (Y2) implies that the norms of the $m \times m$ matrices $\frac{d^2x}{dt^2}(s)$ are uniformly bounded on $[0, T]$. The process $g(t)$ is continuous and $\sup(\|\varepsilon R_2(t)\|, 0 \leq t \leq T)$ converges to zero by Lemma 2.10.2. Hence, if the sampling interval Δ satisfies $\varepsilon^{-1/2}\Delta \to 0$, then an application of inequality (2.10.36) proves the result.

Remark. Genon-Catalot (1990) has obtained the optimality of estimators based on the discrete observation of the process $\{X_t, 0 \leq t \leq T\}$ for one-dimensional diffusions having ε for the diffusion coefficient when the sampling interval satisfies $\Delta = \varepsilon^\alpha, 0 < \alpha \leq 2$. For related work, see Prakasa Rao (1983). We will come back to a discussion of the result in the next chapter.

Smoothed or filtered diffusion

In practice, only a smoothed path $\{Y_t, 0 \leq t \leq T\}$ may be observable instead of $\{X_t, 0 \leq t \leq T\}$ due to the recording device used. Let a be a nonnegative function belonging to $C^\ell, \ell \geq 2$ with compact support contained in $[-1, 1]$ such that

$$\int_{-1}^{1} a(t)\, dt = 1.$$

Define

$$\phi_\eta(t) = \frac{1}{\eta}\phi\left(\frac{t}{\eta}\right).$$

Suppose that

$$Y_t = X_t^\eta = (\phi_\eta * X_t) = \int_0^T a_\eta(t-s)X_s\,ds. \tag{2.10.37}$$

This process $\{Y_t\}$ is called a filtered diffusion.

Theorem 2.10.5 *If ϕ is symmetric and $\varepsilon^{-1/2}\eta \to 0$ as $\varepsilon \to 0$ or $\int_{-1}^{1} v(\phi(v))\,dv \neq 0$ and $\varepsilon^{-1}\eta \to 0$, then the process Y_t defined by (2.10.37) satisfies condition (Y1).*

We omit the proof.

First hitting times and positions of concentric spheres

Suppose that $m \geq 2$ and that $X_0 = x$. Let $r > 0$. The hitting time of the sphere $S(x, r)$ by the m-dimensional diffusion $\{X_t\}$ is

$$T_r = \inf\{t \geq 0 : \|X_t - x\| = r\}. \tag{2.10.38}$$

Observe that T is almost surely finite under assumption (Y2). Let $R > 0$ and consider the incomplete set of observations (T_r, X_{T_r}) for $0 \leq r \leq R$ of $\{X_t\}$. We compare these data with the set of complete observations of $\{X_t\}$ for $0 \leq t \leq T_R$. Define

$$r_t = \sup_{0\leq s\leq t} \|X_t - x\| \tag{2.10.39}$$

and set

$$Y_t = X_{T_{r_t}} \qquad \text{for } 0 \leq t \leq T_R. \tag{2.10.40}$$

We assume that

(K7) $(x_\theta(t) - x).a(x_\theta(t), \theta) > 0$ for all $\theta \in \Theta$ and $t > 0$.

Theorem 2.10.6 *Under assumptions (Y2), (Y3) and (Y7), the process $\{Y_t, 0 \leq t \leq T_R\}$ defined by (2.10.40) satisfies condition (Y1) on the time interval* $[0, T_R]$.

A proof is given in Genon-Catalot (1989).

Record process for a one-dimensional diffusion

Consider the case $m = 1$. An incomplete observation of $\{X_t\}$ which occurs in practice is the record process

$$M_t = \sup_{0\leq s\leq t} X_s. \tag{2.10.41}$$

Let $A > x = X_0$ and define T_a, for $a > x$, by

$$T_a = \inf\{t > 0 : X_t = a\}. \tag{2.10.42}$$

The time T_a is the first hitting time of a level $a > x$. We compare the observation of $\{X_t\}$ on $[0, T_A]$ with the process $\{M_t\}$ on $[0, T_A]$. Suppose the following condition holds:

(Y8) $a(u, \theta) > 0$ for $u \in R$ and $\theta \in \Theta$.

Theorem 2.10.7 *Suppose that conditions (Y2) and (Y8) hold. Then the process $\{M_t\}$ define by (2.10.41) satisfies condition (Y1) on* $[0, T_A]$.

For a proof, see Genon-Catalot and Larédo (1987).

2.11 Estimation from First Hitting Times

We shall first briefly discuss some probabilistic properties of the first hitting times of a diffusion process with drift $\mu(\varepsilon, u)$ and diffusion $\varepsilon\sigma(u)$. Let $\{W_t, t \geq 0\}$ be a standard Wiener process defined on a probability space $(\Omega, \mathcal{A}, P)$ and consider the diffusion $X^{(\varepsilon)} = \{X_t^{(\varepsilon)}, t \geq 0\}$ which is the solution on Ω of the stochastic differential equation

$$dX_t^{(\varepsilon)} = \mu(\varepsilon, X_t^{(\varepsilon)})\,dt + \varepsilon\sigma(X_t^{(\varepsilon)})\,dW_t, \qquad X_0^{(\varepsilon)} = x,\ x \in (\ell, r) \tag{2.11.1}$$

where the functions $\mu(\varepsilon, u)$ and $\sigma(u)$ satisfy

(Z1) $$\begin{cases} \mu : [0, \infty] \times (\ell, r) \to R, & \mu \in C^2([0, \infty) \times (\ell, r)) \\ \sigma : (\ell, r) \to (0, \infty), & \sigma \in C^2((\ell, r)), -\infty \leq \ell < r \leq \infty. \end{cases}$$

Then $X^{(\varepsilon)}$ is a continuous Markov process uniquely determined on $[0, \zeta^{(\varepsilon)}]$ where $\zeta^{(\varepsilon)} = \inf\{t \geq 0 : X_t^{(\varepsilon)} \notin (\ell, r)\}$ is the explosion time of $X^{(\varepsilon)}$ (Ikeda and Watanabe 1981, pp. 361–367). For $x \leq a < r$, define

$$T_a^{(\varepsilon)} = \inf\{t \geq 0 : X_t^{(\varepsilon)} \geq a\} \tag{2.11.2}$$

and

$$T_r^{(\varepsilon)} = \lim_{a \uparrow r} T_a^{(\varepsilon)} \tag{2.11.3}$$

(with the convention that the infimum over an empty set is defined to be $+\infty$). Suppose that

(Z2) $\mu(0, u) > 0$ for all $u \in (\ell, r)$.

Let $x(t)$ be the solution of the ordinary differential equation corresponding to $\varepsilon = 0$ in (2.11.1), namely,

$$dx(t) = \mu(0, x(t))\,dt, \qquad x(0) = x. \tag{2.11.4}$$

Define

$$t(a) = \int_x^0 \frac{du}{\mu(0, u)} = x^{-1}(a) \tag{2.11.5}$$

and $\zeta^{(0)} = t(r-)$. Then $\zeta^{(0)}$ is the explosion time of $x(t)$.

Let D be the space of all real-valued functions on $[x, r)$ which are left-continuous with right limits. Let $\mathcal{D}$ be the σ-algebra generated by the Skorokhod topology on each compact set (Billingsley 1968). The process $\{T_a^{(\varepsilon)}, x \leq a < r\}$ is a nondecreasing process with independent increments and with the sample paths in D under condition (Z1).

Theorem 2.11.1 *Under conditions (Z1) and (Z2), the process $\{T_a^{(\varepsilon)}, x \leq a < r\}$ converges in probability in D to $t(a)$ as $\varepsilon \to 0$, that is, for all $A \in (x, r)$,*

$$\lim_{\varepsilon \to 0} P\Big(\sup_{x \leq a \leq A} |T_a^{(\varepsilon)} - t(a)| \geq h\Big) = 0 \qquad \text{for all } h > 0.$$

For a proof, see Genon-Catalot and Larédo (1987).

Under (Z1) and (Z2), it is known that (cf. Lemma 2.10.2)

$$X_t^{(\varepsilon)} = x(t) + \varepsilon g_1(t) + \varepsilon^2 R_2^{(\varepsilon)}(t), \tag{2.11.6}$$

where

$$\lim_{\varepsilon \to 0 r \to \infty} P(\sup_{0 \leq s \leq T} |R_2^{(\varepsilon)}(s)| \geq r) = 0$$

for all $T \in (0, \zeta^{(0)})$. Furthermore,

$$g_1(t) = \mu(0, x(t))\left[\int_0^t \frac{\mu'_\varepsilon(0, x(s))}{\mu(0, x(s))}\, ds + \int_0^t \frac{\sigma(x(s))}{\mu(0, x(s))}\, dW_s\right]. \tag{2.11.7}$$

(cf. Azencott 1982). Let

$$\begin{aligned} G(a) &= \frac{-g_1(t(a))}{\mu(0, a)} \\ &= -\int_0^{t(a)} \frac{\mu'_\varepsilon(0, x(s))}{\mu(0, x(s))}\, ds - \int_0^{t(a)} \frac{\sigma(x(s))}{\mu(0, x(s))}\, dW_s. \end{aligned} \tag{2.11.8}$$

Note that $G(u)$ is a Gaussian process with continuous sample paths such that

$$E(G(a)) = -\int_x^a \frac{\mu'_\varepsilon(0, u)}{\mu^2(0, u)}\, du \tag{2.11.9}$$

and

$$\operatorname{cov}(G(a), G(b)) = \int_x^{a \wedge b} \frac{\sigma^2(u)}{\mu^3(0, u)}\, du. \tag{2.11.10}$$

Theorem 2.11.2 *Under the conditions (Z1) and (Z2), for all $A \in [x, r)$, $h > 0$,*

$$\lim_{\varepsilon \to 0} P\left(\sup_{x \le a \le A}\left|\left(\frac{T_a^{(\varepsilon)} - t(a)}{\varepsilon}\right) - G(a)\right| > h\right) = 0, \tag{2.11.11}$$

that is, the process $\{(\frac{T_a^{(\varepsilon)} - t(a)}{\varepsilon}), x \le a \le r\}$ converges in distribution to the Gaussian process $\{G(a), x \le a \le r\}$ as defined by (2.11.8)–(2.11.10). Furthermore, for all $A \in [x, r)$,

$$\sup_{0 \le t \le T_A^{(\varepsilon)}}\left(\frac{M_t^{(\varepsilon)} - X_t^{(\varepsilon)}}{\varepsilon}\right) \xrightarrow{p} 0 \qquad \text{as } \varepsilon \to 0, \tag{2.11.12}$$

where

$$M_t^{(\varepsilon)} = \sup_{0 \le t \le T_A^\varepsilon} X_t^{(\varepsilon)} \tag{2.11.13}$$

is the record process of $X^{(\varepsilon)}$.

For a proof, see Genon-Catalot and Larédo (1987). For more details and other properties, see Genon-Catalot (1989) and Genon-Catalot and Larédo (1990).

2.11.1 Minimum Contrast Estimation

Let us consider the diffusion process $X^{(\varepsilon)}$ with drift $\mu(\varepsilon, u, \theta) = \mu(u, \theta)$, where θ is an unknown parameter, and with diffusion $\varepsilon\sigma(u)$. Suppose the process $X^{(\varepsilon)}$ is partially observed through its first hitting times $T_a^{(\varepsilon)}$ for $x \le a \le A$, where $X_0^{(\varepsilon)} = x$ and $A > x$ are given levels. We now consider the problem of estimation of the parameter θ.

We assume that the following conditions hold for μ and σ:

$$\mu : (\ell, r) \times \Theta \to (0, +\infty), \qquad \Theta \subset R^k, k \ge 1,$$

and

$$\sigma : (\ell, r) \to (0, +\infty), \qquad \ell < x < A < r.$$

Let θ_0 be the true value of the parameter and suppose that $\theta_0 \in \Theta^0$, the interior of θ.

We assume that the following conditions hold:

(AA0) Θ is a compact subset of R^k;
(AA1) $(u, \theta) \to \mu(u, \theta)$ is continuous on $(\ell, r) \times \Theta$ and C^2 on $(\ell, r) \times \Theta^0$;
(AA2) $\mu(\cdot, \theta) > 0$ for all $\theta \in \Theta$;
(AA3) $\mu(\cdot, \theta) \neq \mu(\cdot, \theta')$ for all $\theta \neq \theta'$;
(AA4) $\sigma \in C^2((\ell, r))$ and $\sigma(\cdot) > 0$;
(AA5) $S_\theta(\ell) = \int_\ell^x \exp(-2\int_x^v \frac{\mu(\xi,\theta)}{\varepsilon^2\sigma^2(\xi)}\, d\xi)\, dv = +\infty$ for all $\theta \in \Theta$;
(AA6) $\int_x^r (\int_x^u (\sigma^2(v) S'_\theta(v))^{-1}\, dv)\, dS_\theta(u) = +\infty$ for all $\theta \in \Theta$;
(AA7) the matrix $I(\theta_0)$ defined by

$$I(\theta_0) = \left(\left(\int_x^A \left[\left\{\frac{d\mu(u, \theta_0)}{d\theta^i}\frac{d\mu(u, \theta_0)}{d\theta^j}\right\}\Big/(\mu(u, \theta_0)\sigma^2(u))\right] du\right)\right)_{k\times k}$$

is positive definite. Here $\theta = (\theta^1, \theta^2, \ldots, \theta^k)$.

It is known that $\overline{\lim}_{t\to\zeta^\varepsilon} X_t^{(\varepsilon)} = r$ and $\zeta^{(\varepsilon)} = T_r^{(\varepsilon)}$ under condition (AA5) (cf. Ikeda and Watanabe 1981, pp. 361–362). In particular, it follows that, for all $x \leq a < r$, $T_a^{(\varepsilon)} < \infty$ a.s. Conditions (AA5) and (AA6) imply that $\zeta^{(\varepsilon)} = +\infty$ a.s. (cf. Ikeda and Watanabe 1981, pp. 361–362). Note that $\{X_t^{(\varepsilon)}\}$ can be identified as the solution

$$X_t = x + \int_0^t \mu(X_s, \theta)\, ds + \varepsilon \int_0^t \sigma(X_s)\, dW_s, \qquad t \geq 0. \tag{2.11.14}$$

Let us denote the solution corresponding to $\varepsilon = 0$ of (2.11.14) by $x(t, \theta)$ and its inverse funtion by $t(a, \theta) = \int_x^a du/\mu(u, \theta)$. Let $G(a, \theta)$ be the Gaussian process with mean 0 and

$$\operatorname{cov}(G(a, \theta), G(b, \theta)) = \int_x^{a\wedge b} \frac{\sigma^2(u)}{\mu^3(u, \theta)}\, du. \tag{2.11.15}$$

Consider the random function $U_A(\theta)$ defined by

$$\begin{aligned} U_A(\theta) = &\int_{[x,A)} [(\mu^2(u, \theta) - \mu^2(u, \theta_0))/2\sigma^2(u)]\, dT_u \\ &- \int_x^A [(\mu(u, \theta) - \mu(u, \theta_0))/\sigma^2(u)]\, du, \end{aligned} \tag{2.11.16}$$

where $T_u = \inf\{t : X_t \geq u\}$ as before. Note that the first integral on the right-hand side of (2.11.16) is a stochastic integral with respect to the increasing process $\{T_u\}$. Let $P_\theta^{(\varepsilon)}$ be the measure induced by the process $\{X_t\}$ on $(C, \mathcal{C})$, where C is the space of continuous functions on R^+ and $\mathcal{C} = \cup_t \mathcal{C}_t$ with $\mathcal{C}_t = \sigma(X_s, s \leq t)$.

Lemma 2.11.3 *If $\phi(u) \in C^1((\ell, r))$, then, under conditions (AA1), (AA2), (AA4) and (AA5),*

$$\int_{[x,A]} \phi(u)\, dT_u \xrightarrow{p} \int_x^A \frac{\theta(u)}{\mu(u, \theta_0)}\, du \text{ under } P_{\theta_0}^{(\varepsilon)}\text{-measure} \qquad \text{as } \varepsilon \to 0. \tag{2.11.17}$$

Proof. Applying the integration by parts formula to functions of bounded variation, it follows that

$$\int_{[x,A]} \phi(u)\,dT_u = \phi(A)T(A) - \int_x^A \phi'(u)T_u\,du. \tag{2.11.18}$$

Hence

$$\left| \int_{[x,A]} \phi(u)\,dT_u - \int_x^A \frac{\phi(u)}{\mu(u,\theta_0)}\,du \right| \leq |\phi(A)|\,|T_A - t(A,\theta_0)|$$

$$+ \sup_{x \leq a \leq A} |T_a - t(a,\theta_0)| \int_x^A |\phi'(u)|\,du.$$

The proof follows in view of Theorem 2.11.1.

Lemma 2.11.4 *Under the conditions (AA1)–(AA5),*

$$U_A(\theta) \xrightarrow{p} K(\theta_0,\theta) \text{ under } P_{\theta_0}^{(\varepsilon)}\text{-measure} \quad \text{as } \varepsilon \to 0, \tag{2.11.19}$$

where

$$K(\theta_0,\theta) = \int_x^A \{[\mu(u,\theta) - \mu(u,\theta_0)]^2/[2\sigma^2(u)\mu(u,\theta_0)]\}\,du \tag{2.11.20}$$

and the function $K(\theta_0,\theta)$ is nonnegative with strict minimum at $\theta = \theta_0$.

Proof. This lemma follows from Lemma 2.11.3 by choosing $\phi(u) = (\mu^2(u,\theta) - \mu^2(\mu,\theta_0))/2\sigma^2(u)$ and by noting that $K(\theta_0,\theta)$ is nonnegative with strict minimum at $\theta = \theta_0$ in view of (AA3).

The function $U_A(\theta)$ is a 'contrast' with respect to the 'contrast function' $K(\theta_0,\cdot)$ (cf. Dacunha-Castelle and Dufflo (1983, pp. 92–93)). Let us consider a minimum contrast estimator $\tilde{\theta}_\varepsilon$ for θ defined by

$$U_A(\tilde{\theta}_\varepsilon) = \inf\{U_A(\theta) : \theta \in \Theta\}. \tag{2.11.21}$$

Condition (AA0) ensures the existence of $\tilde{\theta}_\varepsilon$. It is clear from the definition of $U_A(\theta)$ (note that $dU_A(\theta)/d\theta^i$ does not depend on θ_0) that $\tilde{\theta}_\varepsilon$ does not depend on θ_0. We now study the properties of the estimator $\tilde{\theta}_\varepsilon$ for θ.

Note that the distributions $P_\theta^{(\varepsilon)}$ and $P_{\theta_0}^{(\varepsilon)}$ are equivalent on C_t for all $t \geq 0$ under (AA1)–(AA6) (cf. Liptser and Shiryayev 1977; or Basawa and Prakasa Rao 1980) and

$$L_t(\theta) = \left.\frac{dP^{(\varepsilon)}}{dP_{\theta_0}^{(\varepsilon)}}\right|_{C_t} = \exp\left\{\varepsilon^{-2}\left[\int_0^t \frac{\mu(X_s,\theta) - \mu(X_s,\theta_0)}{\sigma^2(X_s)}\,dX_s\right.\right.$$

$$\left.\left. - \int_0^t \frac{\mu^2(X_s,\theta) - \mu^2(X_s,\theta_0)}{2\sigma^2(X_s)}\,ds\right]\right\}. \tag{2.11.22}$$

Since $T_A < \infty$ a.s. $[P_\theta^{(\varepsilon)}]$ under (AA5) and (AA6) for all $\theta \in \Theta$, the distributions $P_\theta^{(\varepsilon)}$ and $P_{\theta_0}^{(\varepsilon)}$ are equivalent on C_{T_A}, with

$$\left.\frac{dP_\theta^{(\varepsilon)}}{dP_{\theta_0}^{(\varepsilon)}}\right|_{C_{T_A}} = L_{T_A}(\theta). \tag{2.11.23}$$

The following theorem shows the relation between the functions $L_{T_A}(\theta)$ and $U_A(\theta)$.

Theorem 2.11.5 *Under the conditions (AA1)–(AA6),*

$$-\varepsilon^2 \log L_{T_A}(\theta) = U_A(\theta) + V_A(\theta), \tag{2.11.24}$$

where $V_A(\theta) = o_P(\varepsilon)$ in $P_{\theta_0}^{(\varepsilon)}$-probability.

Proof. An application of Itô's formula in (2.11.22)–(2.11.23) shows that

$$\begin{aligned}-\varepsilon^2 \log L_{T_A}(\theta) &= \int_0^{T_A} \{(\mu^2(X_s,\theta) - \mu^2(X_s,\theta_0))/2\sigma^2(X_s)\}\, ds \\ &\quad + \varepsilon^2 \int_0^{T_A} h(X_s,\theta)\, ds - \int_x^A \{(\mu(u,\theta) - \mu(u,\theta_0))/\sigma^2(u)\}\, du,\end{aligned} \tag{2.11.25}$$

where

$$h(u,\theta) = \sigma^2(u) \frac{\partial}{\partial u} \left\{ \frac{\mu(u,\theta) - \mu(u,\theta_0)}{2\,\sigma^2(u)} \right\}. \tag{2.11.26}$$

Define

$$M_t = \sup_{0 \le s \le t} X_s. \tag{2.11.27}$$

Then, for all t and a satisfying $0 \le t \le T_A$ and $x \le a \le A$,

$$[M_t \ge a] = [T_a \le t]; \tag{2.11.28}$$

and hence, for all continuous functions ϕ,

$$\int_0^{T_A} \phi(M_s)\, ds = \int_{[x,A)} \phi(u)\, dT_u. \tag{2.11.29}$$

Choosing $\phi(u) = (\mu^2(u,\theta) - \mu^2(u,\theta_0))/2\sigma^2(u)$, together with (2.11.25), gives

$$-\varepsilon^2 \log L_{T_A}(\theta) = U_A(\theta) + V_A(\theta), \tag{2.11.30}$$

where

$$\begin{aligned}V_A(\theta) = \varepsilon^2 \int_0^{T_A} h(X_s,\theta)\, ds + \int_0^{T_A} &\{(\mu^2(X_s,\theta) - \mu^2(X_s,\theta_0))/2\sigma^2(X_s) \\ &- (\mu^2(M_s,\theta) - \mu^2(M_s,\theta_0))/2\sigma^2(M_s)\}\, ds.\end{aligned}$$

It is sufficient to prove that $\varepsilon^{-1} V_A(\theta) \xrightarrow{p} 0$ in $P_{\theta_0}^{(\varepsilon)}$-probability as $\varepsilon \to 0$ to complete the proof of Theorem 2.11.5. This can be shown by choosing the function $\phi(\cdot)$ suitably in the following lemma.

Lemma 2.11.6 *Suppose the conditions (AA1)–(AA6) hold. Let $\phi : (\ell, r) \to R$ be a Borel measurable function.*

(i) If ϕ is continuous, then

$$\int_0^{T_A} \phi(X_s) ds \xrightarrow{p} \int_0^{t(A,\theta_0)} \phi(x(s,\theta_0))\, ds = \int_x^A \frac{\phi(u)}{\mu(u,\theta_0)}\, du$$

in $P_{\theta_0}^{\varepsilon}$-probability as $\varepsilon \to 0$.

(ii) If ϕ is C^1, then

$$\varepsilon^{-1} \int_0^{T_A} (\phi(X_s) - \phi(M_s))\, ds \xrightarrow{p} 0 \text{ in } P_{\theta_0}^{(\varepsilon)}\text{-probability} \qquad \text{as } \varepsilon \to 0.$$

(Recall that $x(s, \theta)$ is the solution of the deterministic equation associated with the case $\varepsilon = 0$ in (2.11.14).)

Remark. This lemma is a consequence of Theorem 2.11.1. For a proof, see Genon-Catalot and Larédo (1987).

Note that the function $h(\cdot, \theta)$, defined by (2.11.26), is continuous and the function $u \to (\mu^2(u, \theta) - \mu^2(u, \theta_0))/2\sigma^2(u)$ is C^1. Applying Lemma 2.11.6, we can complete the proof of Theorem 2.11.5.

We now discuss the properties of consistency and the asymptotic normality of the MCE $\tilde{\theta}_\varepsilon$ defined by (2.11.21).

2.11.2 Consistency

Theorem 2.11.7 *Under the conditions (AA0)–(AA6),*

$$\lim_{\varepsilon \to 0} P_{\theta_0}^{\varepsilon}(\|\tilde{\theta}_\varepsilon - \theta_0\| > \delta) = 0 \text{ for all } \delta > 0. \tag{2.11.31}$$

Proof. Let $\delta > 0$ and $B = \{\|\theta - \theta_0\| < \delta\}$. Since $K(\theta_0, \theta)$ is continuous in θ, it is bounded from below on $\Theta \backslash B$ by a positive constant (say) 2α. Note that

$$[\tilde{\theta}_\varepsilon \notin B] \subset [\inf_{\Theta - B} U_A(\theta) \leq 0]. \tag{2.11.32}$$

Let B_i, $1 \leq i \leq m$, be m spheres with centers θ_i and radii smaller than η covering $\Theta - B$. Then

$$P_{\theta_0}^{(\varepsilon)}(\inf_{\Theta - B} U_A(\theta) \leq 0) \leq P_{\theta_0}^{(\varepsilon)}(\omega(U_A, \eta) > \alpha) + P_{\theta_0}^{(\varepsilon)}(\inf_{1 \leq i \leq m} U_A(\theta_i) \leq \alpha), \tag{2.11.33}$$

where

$$\omega(U_A, \eta) = \sup\{|U_A(\theta) - U_A(\theta')| : \|\theta - \theta'\| < \eta\}. \tag{2.11.34}$$

It is easy to see that

$$\inf_{1 \leq i \leq m} U_A(\theta_i) \xrightarrow{p} \inf_{1 \leq i \leq m} K(\theta_0, \theta_i) > 2\alpha \tag{2.11.35}$$

by Lemma 2.11.4. It suffices to prove that, for every $\alpha > 0$, there exists $\eta > 0$ such that

$$\lim_{\varepsilon \to 0} P_{\theta_0}^{(\varepsilon)}(\omega(U_A, \eta) > \alpha) = 0. \tag{2.11.36}$$

Since μ is uniformly continuous on the compact set $[x, A] \times \Theta$, the set of functions $\{\theta \to \mu(u, \theta), u \in [x, A]\}$ on Θ is uniformly equicontinuous. Hence

$$\beta(\eta) = \sup_{x \leq u \leq A} \sup_{\|\theta - \theta'\| \leq \eta} \left\{ \frac{|\mu(u, \theta) - \mu(u, \theta')|}{2\sigma^2(u)} \right\} \to 0, \qquad \text{as } \eta \to 0, \tag{2.11.37}$$

and

$$\gamma(\eta)) = \sup_{x \leq u \leq A} \sup_{\|\theta - \theta'\| < \eta} \left\{ \frac{|\mu^2(u, \theta) - \mu^2(u, \theta')|}{\sigma^2(u)} \right\} \to 0, \qquad \text{as } \eta \to 0. \tag{2.11.38}$$

But

$$0 \leq \omega(U_A, \eta) \leq T_A\gamma(\eta)/2 + (A - x)\beta(\eta))$$

from the definition of $U_A(\theta)$ given by (2.11.16). Relations (2.11.37) and (2.11.38), combined with Theorem 2.11.1, prove that

$$\omega(U_A, \eta) \xrightarrow{p} 0 \text{ in } P_{\theta_0}^{(\varepsilon)}\text{-probability} \quad \text{as } \varepsilon \to 0, \tag{2.11.39}$$

completing the proof.

2.11.3 Asymptotic Normality

Theorem 2.11.8 *Suppose that conditions (AA0)–(AA5) and (AA7) hold. Then $\tilde{\theta}_\varepsilon$ is asymptotically normal, that is,*

$$\varepsilon^{-1}(\tilde{\theta}_\varepsilon - \theta_0) \xrightarrow{\mathcal{L}} N_k(0, I(\theta_0)^{-1}) \text{ under } P_{\theta_0}^{(\varepsilon)}\text{-probability} \quad \text{as } \varepsilon \to 0. \tag{2.11.40}$$

Proof. Note that the mapping $\theta \to U_A(\theta)$ is in $C^2(\Theta^0)$. Let $\delta > 0$ such that the sphere $S(\theta_0, \delta) = \{\theta : \|\theta - \theta_0\| \leq \delta\} \subset \Theta^0$. Then, for $\tilde{\theta}_\varepsilon \in S(\theta_0, \delta)$, one has

$$\begin{aligned} 0 &= \varepsilon^{-1}\frac{\partial}{\partial\theta^i}U_A(\tilde{\theta}_\varepsilon) \\ &= \varepsilon^{-1}\frac{\partial}{\partial\theta^i}U_A(\theta_0) + \varepsilon^{-1}\sum_{j=1}^{k}(\tilde{\theta}_\varepsilon^j - \theta_0^j)\left\{\frac{\partial^2}{\partial\theta^i\,\partial\theta^j}U_A(\theta_0)\right. \\ &\quad \left. + \left[\int_0^1 \frac{\partial^2}{\partial\theta^i\,\partial\theta^j}U_A(\theta_0 + s(\tilde{\theta}_\varepsilon - \theta_0))\,ds - \frac{\partial^2}{\partial\theta^i\,\partial\theta^j}U_A(\theta_0)\right]\right\}. \end{aligned} \tag{2.11.41}$$

It is sufficient to show that:

$$\varepsilon^{-1}\nabla U_A(\theta_0) \xrightarrow{\mathcal{L}} N_k(0, I(\theta_0)) \text{ under } P_{\theta_0}^{(\varepsilon)}\text{-probability as } \varepsilon \to 0; \tag{2.11.42a}$$

$$\frac{\partial^2}{\partial\theta_i\,\partial\theta_j}U_A(\theta_0) \xrightarrow{p} I(\theta_0)_{ij} \text{ in } P_{\theta_0}^{(\varepsilon)}\text{-probability as } \varepsilon \to 0; \tag{2.11.42b}$$

$$I_{\{\tilde{\theta}_\varepsilon \in S(\theta_0, h)\}} \sup_{\|\alpha\| \leq \|\tilde{\theta}_\varepsilon - \theta_0\|} \left|\frac{\partial^2}{\partial\theta^i\,\partial\theta^j}U_A(\theta_0 + \alpha) - \frac{\partial^2}{\partial\theta^i\,\partial\theta^j}U_A(\theta_0)\right| \xrightarrow{p} 0 \text{ in } P_{\theta_0}^{(\varepsilon)}\text{-probability as } \varepsilon \to 0. \tag{2.11.42c}$$

From the definition of $U_A(\theta)$, it follows that

$$\varepsilon^{-1}\frac{\partial U_A(\theta_0)}{\partial\theta^i} = \int_{[x,A)}\left\{\frac{\mu(u, \theta_0)}{\sigma^2(u)}\frac{\partial}{\partial\theta^i}(\mu(u, \theta_0))\right\}\varepsilon^{-1}(dT_u - \{du/\mu(u, \theta_0)\}). \tag{2.11.43}$$

We will first show that

$$\varepsilon^{-1}\nabla U_A(\theta_0) \xrightarrow{\mathcal{L}} \left(\int_{[x,A)}(\mu(u, \theta_0)\frac{\partial}{\partial\theta^i}\mu(u, \theta_0)/\sigma^2(u))\,dG(u, \theta_0)\right)_{1 \leq i \leq k}, \tag{2.11.44}$$

where the integral is a stochastic integral with respect to the centered Gaussian martingale $\{G(a,\theta_0), x \le a < r\}$ whose covariance function is given by (2.11.15). Integrating by parts in (2.11.43), we have

$$\varepsilon^{-1}\frac{\partial}{\partial\theta^i}U_A(\theta_0) = (\mu(A,\theta_0)\frac{\partial}{\partial\theta^i}\mu(A,\theta_0)/\sigma^2(A))\varepsilon^{-1}(T_A - t(A,\theta_0))$$
$$- \int_x^A \frac{\partial}{\partial u}\{\mu(u,\theta_0)\frac{\partial}{\partial\theta^i}\mu(u,\theta_0)/\sigma^2(u)\}\varepsilon^{-1}(T_u - t(u,\theta_0))\,du. \tag{2.11.45}$$

Let

$$\alpha^i = \mu(A,\theta_0)\frac{\partial}{\partial\theta_i}\mu(A,\theta_0)/\sigma^2(A), \qquad 1 \le i \le k,\ \alpha = (\alpha^1,\dots,\alpha^k),$$

$$\phi_i(u) = -\frac{\partial}{\partial u}\left\{\mu(u,\theta_0)\frac{\partial}{\partial\theta^i}\mu(u,\theta_0)/\sigma^2(u)\right\}, \qquad 1 \le i \le k,\ \phi = (\phi_1,\dots,\phi_k),$$

and

$$\psi(g) = g(A)\alpha + \int_x^A g(u)\phi(u)\,du, \qquad g \in D,$$

where D is the space of real-valued functions on $[x,r)$ which are left-continuous with right limits endowed with the σ-algera $\mathcal{D}$ generated by the Skorokhod topology on each compact set. It can be shown that the mapping $\psi : D \to R^k$ is $\mathcal{D}$-measurable and has no discontinuity points in the set $C([x,r)) = \{g : [x,r) \to R, g \text{ continuous}\}$ (cf. Genon-Catalot and Larédo 1987, Lemma 3). An application of Theorem 2.11.2 proves that

$$\varepsilon^{-1}\nabla U_A(\theta_0) = \psi((\varepsilon^{-1}(T_a - t(a,\theta_0)), x \le a < r)) \xrightarrow{\mathcal{L}} \psi((G(a,\theta_0)), x \le a < r)$$

under $P_{\theta_0}^{(\varepsilon)}$-probability as $\varepsilon \to 0$. But $\psi = (\psi^1,\dots,\psi^k)$, where

$$\psi^i((G(a,\theta_0)), x \le a < A) = \left\{\mu(A,\theta_0)\frac{\partial}{\partial\theta^i}\mu(A,\theta_0)/\sigma^2(A)\right\}G(A,\theta_0)$$
$$- \int_x^A \frac{\partial}{\partial u}\left\{\mu(u,\theta_0)\frac{\partial}{\partial\theta^i}\mu(u,\theta_0)/\sigma^2(u)\right\}G(u,\theta_0)\,du$$
$$= \int_x^A \left\{\mu(u,\theta_0)\frac{\partial}{\partial\theta^i}\mu(u,\theta_0)/\sigma^2(u)\right\}dG(u,\theta_0),$$

by integrating by parts following Itô's lemma. This proves (2.11.44). Note that the Gaussian martingale $G(\cdot,\theta_0)$ has zero mean and has quadratic variation $\langle G(\cdot,\theta_0)\rangle = \int_x^{\cdot} \sigma^2(u)/\mu^3(u,\theta_0)\,du$ by (2.11.15). Hence the random vector $\psi(G(\cdot,\theta_0))$ is normal with mean zero and covariance matrix

$$\boldsymbol{I}(\theta_0) = \left(\left(\int_x^A \mu^2(u,\theta_0)\frac{\partial}{\partial\theta^i}\mu(u,\theta_0)\frac{\partial}{\partial\theta^j}\mu(u,\theta_0)/\sigma^4(u)d\langle G(\cdot,\theta_0)\rangle_u\right)\right)_{k\times k}. \tag{2.11.46}$$

Results (2.11.44) and (2.11.46) prove (2.11.42a). Next, (2.11.42b) follows by observing that

$$\frac{\partial^2}{\partial\theta^i\,\partial\theta^j}U_A(\theta_0) = \int_{[x,A)}\left\{\left[\frac{\partial}{\partial\theta^i}\mu(u,\theta_0)\frac{\partial}{\partial\theta^j}\mu(u,\theta_0)\right.\right.$$

$$+\mu(u,\theta_0)\frac{\partial^2}{\partial\theta^i\partial\theta^j}\mu(u,\theta_0)\Bigg]\Big/\sigma^2(u)\Bigg\}\,dT_u$$

$$-\int_x^A \frac{\partial^2}{\partial\theta^i\,\partial\theta^j}\left\{\frac{\mu(u,\theta_0)}{\sigma^2(u)}\right\}du, \tag{2.11.47}$$

and noting that Lemma 2.11.3 may be applied to the first integral (note that dT_u can be replaced by $dt(u,\theta_0)$ as $\varepsilon \to 0$). Let

$$\phi(u,\theta) = \frac{\partial^2}{\partial\theta^i\,\partial\theta^j}\frac{\mu(u,\theta)}{\sigma^2(u)} \tag{2.11.48}$$

and

$$\psi(u,\theta) = \left(\frac{\partial}{\partial\theta^i}\mu(u,\theta)\frac{\partial}{\partial\theta^j}\mu(u,\theta)\right)\Big/\sigma^2(u) + \mu(u,\theta)\frac{\partial^2}{\partial\theta^i\,\partial\theta^j}\left\{\frac{\mu(u,\theta)}{\sigma^2(u)}\right\}. \tag{2.11.49}$$

The set of functions $\theta \to \phi(u,\theta)$ and $\theta \to \psi(u,\theta)$ defined on $S(\theta_0,h)$ are uniformly equicontinuous for $u \in [x, A]$ and hence, for $0 < \eta \le h$,

$$C(\eta) = \sup_{x\le u\le A}\ \sup_{\|\alpha\|\le\eta} |\phi(u,\theta_0+\alpha) - \phi(u,\theta_0)| \to 0 \tag{2.11.50}$$

and

$$D(\eta) = \sup_{x\le u\le A}\ \sup_{\|\alpha\|\le\eta} |\psi(u,\theta_0+\alpha) - \psi(u,\theta_0)| \to 0. \tag{2.11.51}$$

Hence

$$I_{\{\tilde\theta_\varepsilon\in S(\theta_0,h)\}} \sup_{\|\alpha\|\le\|\tilde\theta_\varepsilon-\theta_0\|}\left|\frac{\partial^2}{\partial\theta^i\,\partial\theta^j}U_A(\theta_0+\alpha) - \frac{\partial^2}{\partial\theta^i\,\partial\theta^j}U_A(\theta_0)\right|$$
$$\le D(\|\tilde\theta_\varepsilon-\theta_0\|)T_A + C(\|\tilde\theta_\varepsilon-\theta_0\|)(A-x).$$

Since $\tilde\theta_\varepsilon$ is consistent and T_A converges in probability to $t(A,\theta_0)$ as $\varepsilon \to 0$ in $P^\varepsilon_{\theta_0}$-probability, we obtain (2.11.42c), following (2.11.50) and (2.11.51). Relations (2.11.44) and (2.11.43) prove the theorem.

Remark. In some cases, it is possible to prove the consistency and the asymptotic nomality of $\tilde\theta_\varepsilon$ directly without verifying conditions (AA0)–(AA6).

Example 2.11.1 (*Linear diffusion model*)
Suppose $\mu(u,\theta) = \theta\mu(u)$, with $\theta > 0$ and $\mu(\cdot) > 0$. Then

$$\tilde\theta_\varepsilon = \left(\int_x^A \frac{\mu(u)}{\sigma^2(u)}du\right)\left(\int_{[x,A)}\frac{\mu^2(u)}{\sigma^2(u)}dT_u\right)^{-1}$$

and

$$\varepsilon^{-1}(\tilde\theta_\varepsilon-\theta_0) \xrightarrow{\mathcal{L}} N\left(0,\theta_0\left(\int_x^A\frac{\mu(u)}{\sigma^2(u)}du\right)^{-1}\right) \text{ under } P^{(\varepsilon)}_{\theta_0}\text{-probability as } \varepsilon\to 0.$$

Example 2.11.2 (*Bilinear diffusion model*)
Here the state space is $(\ell, r) = (0, +\infty)$, $\sigma(u) = u$, $\mu(u, \theta) = \theta^1 u + \theta^2$, $\theta^1 > 0$, $\theta^2 > 0$ with $\theta = (\theta^1, \theta^2)$. Then

$$\tilde{\theta}_\varepsilon^1 = \frac{\log(A/x)\int_{[x,A)} u^{-2}\, dT_u - (x^{-1} - A^{-1})\int_{[x,A)} u^{-1}\, dT_u}{T_A \int_{[x,A)} u^{-2}\, dT_u - (\int_{[x,A)} u^{-1}\, dT_u)^2},$$

$$\tilde{\theta}_\varepsilon^2 = \frac{(x^{-1} - A^{-1})T_A - \log(A/x)\int_{[x,A)} u^{-1}\, dT_u}{T_A \int_{[x,A)} u^{-2}\, dT_u - (\int_{[x,A)} u^{-1}\, dT_u)^2}.$$

The denominators in the expressions for $\tilde{\theta}_\varepsilon^1$ and $\tilde{\theta}_\varepsilon^2$ are positive by the Cauchy–Schwarz inequality. Here the limiting covariance matrix $I(\theta) = ((I_{ij}(\theta)))_{2\times 2}$ is given by

$$I_{11}(\theta) = \frac{1}{\theta^1}\log\left(\frac{\theta^1 A + \theta^2}{\theta^1 x + \theta^2}\right),$$

$$I_{12}(\theta) = I_{21}(\theta) = \frac{1}{\theta^2}\left\{\log\left(\frac{A}{x}\right) - \log\left(\frac{\theta^1 A + \theta^2}{\theta^1 x + \theta^2}\right)\right\},$$

$$I_{22}(\theta) = \frac{1}{\theta^2}\left(\frac{1}{x} - \frac{1}{A}\right) + \frac{\theta^1}{(\theta^2)^2}\left\{\log\left(\frac{\theta^1 A + \theta^2}{\theta^1 x + \theta^2}\right) - \log\left(\frac{A}{x}\right)\right\}.$$

Example 2.11.3 (*Branching diffusion model*)
Here the state space is $(\ell, r) = (0, +\infty)$, $\sigma(u) = u^{1/2}$, $\mu(u, \theta) = \theta^1 u + \theta^2$ with $\theta^1 > 0$ and $\theta^2 > 0$. Then

$$\tilde{\theta}_\varepsilon^1 = \frac{\log(\frac{A}{x})\int_{[x,A)} u\, dT_u - T_A(A - x)}{(\int_{[x,A)} u^{-1}\, dT_u)(\int_{[x,A)} u\, dT_u) - T_A^2},$$

$$\tilde{\theta}_\varepsilon^2 = \frac{(A - x)\int_{[x,A)} u^{-1}\, dT_u - T_A\log(\frac{A}{x})}{(\int_{[x,A)} u^{-1}\, dT_u)(\int_{[x,A)} u\, dT_u) - T_A^2},$$

and the denominators are positive again by the Cauchy–Schwarz inequality. The covariance matrix $I(\theta) = ((I_{ij}(\theta)))_{2\times 2}$ is given by

$$I_{11}(\theta) = \frac{A - x}{\theta^1} - \frac{\theta^2}{(\theta^1)^2}\log\left(\frac{\theta^1 A + \theta^2}{\theta^1 x + \theta^2}\right),$$

$$I_{12}(\theta) = I_{21}(\theta) = \frac{1}{\theta^1}\log\left(\frac{\theta^1 A + \theta^2}{\theta^1 x + \theta^2}\right),$$

$$I_{22}(\theta) = \left\{\log\left(\frac{A}{x}\right) - \frac{1}{\theta^2}\log\left(\frac{\theta^1 A + \theta^2}{\theta^1 x + \theta^2}\right)\right\}.$$

Remark. (i) Let $\hat{\theta}_\varepsilon$ be the MLE defined under (AA0) by

$$\ell_A(\hat{\theta}_\varepsilon) = \inf\{\ell_A(\theta); \theta \in \Theta\},$$

where $\ell_A(\theta) = -\varepsilon^2 \log L_{T_A}(\theta)$ and $L_{T_A}(\theta)$ as defined by (2.11.25). Genon-Catalot and Larédo (1987) proved that the MLE $\hat{\theta}_\varepsilon$ and the MCE $\tilde{\theta}_\varepsilon$ are asymptotically equivalent; in fact,

$$\varepsilon^{-1}(\hat{\theta}_\varepsilon - \hat{\theta}_\varepsilon) \xrightarrow{P} 0 \text{ in } P_{\theta o}^{(\varepsilon)}\text{-probability} \qquad as \varepsilon \to 0,$$

under the conditions (AA0)–(AA7). Note that

$$\ell_{T_A}(\theta) = U_A(\theta) + o_p(\varepsilon)$$

from Theorem 2.11.5. We omit the details. In particular, it follows from earlier results that $\hat{\theta}_\varepsilon$ is consistent and asymptotically normal, generalizing the result of Ibragimov and Has'minskii (1981), Basawa and Prakasa Rao (1980), Kutoyants (1984a) and others for the MLE based on $X_t, 0 \le t \le T$, for fixed time T.

(ii) Here we have discussed the asymptotic properties of the MCE based on observation (Type I) of the first hitting-time process $T_a = \inf\{t \ge 0; X_t = a\}$ of levels a between two prescribed levels $x = X_0$ and $A > x$. Let $M_t = \sup\{X_s : 0 \le s \le t\}$. Then an observation of Type I is equivalent to an observation of M_t for $0 \le t \le T_A$. Let us consider an observation (Type II) of the values $(a, \Delta T_a)$ for all $a \in [x, A)$ such that the jump $\Delta T_a = T_{a+} - T_a > \eta$ for η positive and given. An observation of Type II is equivalent to the data consisting of flat stretches of M_t with length greater than η. Genon-Catalot and Larédo (1990) studied the asymptotic suffiency of such incomplete observation and obtained the asymptotic properties of the corresponding MCE. They prove that these estimators are asymptotically equivalent to the MLE $\hat{\theta}_\varepsilon$ under suitable conditions as $\eta = \varepsilon^\delta \to 0$ with $\varepsilon \to 0$ and $\delta > 4$. We omit the details.

(iii) Genon-Catalot (1987; 1989) considered the problem of testing the drift for a diffusion process on R^m, $m \ge 2$, with drift $\theta\mu(u)$ and diffusion $\varepsilon\sigma(u)$ based on diffusion obsevations on the first hitting times and positions of concentric spheres centered at $x = X_0$ with radii $r \le R$ for given R.

3

Parametric Inference for Diffusion Type Processes from Sampled Data

3.1 Introduction

We have assumed in Chapter 2 that the stochastic process under consideration can be observed continuously over a specified time period, and the statistical inference was based on either one or many realizations of the process over that time period. In practice, it is obviously impossible to observe a process continuously over any given time period – due, for example, to the limitations on the precision of the measuring instrument or to the unavailability of observations at every time point. In other words, statistical inference based on sampled data is of major importance in dealing with practical problems. For an earlier review, see Prakasa Rao (1988b; 1990a).

Sampling instants at which a process is observed may be of different types. They may be regularly or irregularly spaced. They may be deterministic or random. Random times of observation may be dependent on or independent of the process. Let $\{X_n, n \geq 1\}$ be a discrete-time stochastic process which is stationary and ergodic on a probability space $(\Omega, \mathcal{F}, P)$. If a complete realization of the process is available, then one can estimate all the finite-dimensional distributions of the process following the individual ergodic theorem. Suppose that it is possible only to observe a subsequence $\{X_{k_n}\}$ of $\{X_n\}$. The problem is to examine whether it is still possible to estimate the probability structure, in particular, all the finite-dimensional distributions of the original process $\{X_n\}$ from the sampled data. It is clear that it is in general impossible to obtain all the 'information' about $\{X_n\}$ from the subsequence $\{X_{k_n}\}$ of $\{X_n\}$. For instance if $k_n = kn$, for some $k > 1$, it may be impossible to estimate the joint distribution of (X_1, X_2). The process $\{X_{k_n}\}$ may not be stationary or ergodic even if the original process $\{X_n\}$ is stationary and ergodic. This problem was investigated by Blum and Rosenblatt (1964). Blum and Boyles (1981) discussed a similar problem for continuous-time stochastic processes where it is of interest since such processes are always observed via discrete sampling schemes in practice. In order to discuss whether the 'information' on the original process $\{X_t\}$ can be recovered from the 'information' obtained from the observed process (say) $\{X_{\tau_n}, n \geq 1\}$, it is important to know whether properties of the original process $\{X_t\}$ are inherited by the sampled process $\{X_{\tau_n}\}$. For instance, the concept of mixing for a stochastic process $\{X_t\}$ has to be strengthened further in order for the process $\{X_{\tau_n}\}$ to inherit the mixing property. These and related notions are investigated further in Prakasa Rao (1990b).

Let $(\Omega, \mathcal{F}, P)$ be a probability space. Let $\{\mathcal{F}_t, t \geq 0\}$ be an increasing flow of sub-σ-algebras of $\mathcal{F}$ and $\{\zeta_t, t \geq 0\}$ be a decreasing flow of sub-σ-algebras of $\mathcal{F}$. For any real-valued nonnegative random variable τ defined on $(\Omega, \mathcal{F}, P)$, let $\mathcal{F}_\tau$ be the σ-algebra generated by sets $A \in \mathcal{F}$ such that $A \cap [\tau \leq t] \in \mathcal{F}_t$ for every $t \geq 0$, and ζ_τ be the σ-algebra generated by sets $B \in \mathcal{F}$ such that $B \cap [\tau \geq s] \in \zeta_s, s \geq 0$.

Let $\{\tau_n, n \geq 1\}$ and $\{\sigma_n, n \geq 1\}$ be increasing sequences of nonnegative random variables defined on $(\Omega, \mathcal{F}, P)$. The increasing flow $\{\mathcal{F}_t\}$ is said to be *ϕ-mixing strongly* with the decreasing flow $\{\zeta_s\}$ with respect to the sequences $\{\tau_n\}$ and $\{\sigma_n\}$ if for every $A \in \mathcal{F}_{\tau_n}, n \geq 1$, and $B \in \zeta_{\sigma_m}, m \geq 1$,

$$|P(A \cap B) - P(A)P(B)| \leq E\{\phi(|\tau_n - \sigma_m|\}P(A), \tag{3.1.1}$$

where

$$E(\phi(|\tau_n - \sigma_m|)) \to 0 \text{ as } |\tau_n - \sigma_m| \xrightarrow{p} \infty \text{ as } m \to \infty.$$

If the above relation holds for every pair of sequences $\{\tau_n\}$ and $\{\sigma_n\}$ satisfying the conditions stated, then the increasing flow $\{\mathcal{F}_t\}$ is said to be ϕ-mixing strongly with the decreasing flow $\{\zeta_s\}$.

Examples of families of sub-σ-algebras which are ϕ-mixing strongly are discussed in Prakasa Rao (1990b). An example which is of interest in the present context is as follows. We first extend the notion of ϕ-mixing for stochastic processes (Ibragimov and Linnik 1971).

Suppose $\{X_t, t \geq 0\}$ is a progressively measurable stochastic process defined on a probability space $(\Omega, \mathcal{F}, P)$. Let $\{\tau_n, n \geq 1\}$ be an increasing sequence of nonnegative random variables defined on $(\Omega, \mathcal{F}, P)$. Let $\mathcal{F}_t^X$ be the σ-algebra generated by $\{X_u, 0 \leq u \leq t\}$ and ζ_s^X be the σ-algebra generated by $\{X_v, v \geq s\}$. Define $\mathcal{F}_{\tau_n}^X$ and $\zeta_{\sigma_{m,n}}^X$ as before, with $\sigma_{m,n} = \tau_{n+m}$. Then $\{X_t\}$ said to be ϕ-mixing strongly with respect to $\{\tau_n\}$ if (3.1.1) holds for $A \in \mathcal{F}_{\tau_n}^X$ and $B \in \zeta_{\sigma_{m,n}}^X$ for every $m \geq 1$ and $n \geq 1$.

Theorem 3.1.1 *Suppose $\{X_t, t \geq 0\}$ is a stationary ϕ-mixing progressively measurable stochastic process defined on a probability space $(\Omega, \mathcal{F}, P)$. Let $\{\tau_n, n \geq 1\}$ be an increasing sequence of nonnegative random variables defined on $(\Omega, \mathcal{F}, P)$ and independent on $\{X_t, t \geq 0\}$. Further suppose that $|\tau_n - \tau_{n+m}| \xrightarrow{p} \infty$ as $m \to \infty$ for every $n \geq 1$. Then $\{X_t\}$ is ϕ-mixing strongly with respect to $\{\tau_n\}$.*

For a proof, see Prakasa Rao (1990b).

Without going further into the discretization aspects and corresponding inference problems for general continuous-time processes, we shall consider the special case of diffusion type processes in this chapter.

As we have already seen, parametric inference for the diffusion processes should ideally be based on the likelihood function. We have discussed the likelihood theory for continuously observed diffusion processes in the previous chapter. For discretely observed diffusion processes, the likelihood function based on discrete-time observation is a product of the transition densities, which can be explicitly computed only in some special cases. One can either obtain the MLEs based on the exact likelihood function or calculate the estimators based on a good approximation to the likelihood function. Another approach is to approximate the continuous-time likelihood function and define the estimators based on this function. A third approch is via martingale estimating functions. As a further alternative, one can use the least-squares approach for the estimation of the parameters. For earlier work on this problem see Le Breton (1976), Prakasa Rao and Rubin (1981) and Prakasa Rao (1983b), as well as Basawa and Prakasa Rao (1980).

Unlike in the previous chapter, the problems of estimating the drift coefficient and the diffusion coefficient are both meaningful and important, as we shall see later, for instance in applications to modeling option prices in financial markets.

3.2 Estimation by the Least-squares Method

Consider a d-dimensional ($d \geq 1$) stationary stochastic process $\{X(t), t \geq 0\}$ satisfying the stochastic differential equation

$$dX(t) = a(\theta, X(t))\,dt + \Gamma\,dW(t), \qquad X(0) = X_0, t \geq 0, \tag{3.2.1}$$

where $E(X_0'X_0) < \infty$, $W(t) = (W_1(t), \ldots, W_d(t))$ is a d-dimensional standard Wiener process, Γ is a positive definite $d \times d$ matrix of the unknown diffusion coefficients (diffusion matrix), $\theta \in \Theta \subset R^\ell$ is an ℓ-dimensional vector of the unknown drift parameters, Θ is a compact convex subset of R^ℓ and $a(\cdot,\cdot)$ is a known R^d-valued function. Here α' denotes the transpose of the vector α.

Suppose the process is observed at the time points $0 = t_0 < t_1 < \cdots < t_n = T$ with $t_{i+1} - t_i = T/n$, $i = 0, 1, \ldots, n-1$. The problem is to estimate θ and $\Gamma\Gamma'$ based on the observation $\{X(t_i), 0 \leq i \leq n-1\}$.

Theorem 3.2.1 *Let Γ_n be defined by*

$$\Gamma_n\Gamma_n' = \frac{1}{T}\sum_{i=0}^{n-1}(X(t_{i+1}) - X(t_i))(X(t_{i+1}) - X(t_i))'.$$

Let $\theta = \theta_0$ be fixed and consider $a(\theta_0, x)$ as a function of x. Suppose that the following conditions hold:

(BB1) The process $\{X(t), t \geq 0\}$ is a stationary process satisfying (3.2.1) and for any function $g(\cdot) : R^d \to R^{d_1}$ ($d_1 \geq 1$) such that $E\|g(X(0))\| < \infty$,

$$\frac{1}{n}\sum_{k=1}^{n} g(X(t_k)) \xrightarrow{p} E(g(X(0)))$$

as $T \to \infty$ and $\frac{T}{n} \to 0$.

(BB2) $E_{\theta_0}\{a(\theta_0, X(0))a(\theta_0, X(0))'\}$ is a matrix with finite elements.

(BB3) (i) $\|a(\theta_0, x) - a(\theta_0, y)\| \leq L\|x - y\|, x, y \in R^d$;

(ii) $\|a(\theta_0, x)| \leq L(1 + \|x\|), x \in R^d$ for some constant $L > 0$.

(BB4) $T/\sqrt{n} \to 0$ as $T \to \infty$ (condition for a 'rapidly increasing experimental design').

Then

$$\sqrt{n}(\Gamma_n\Gamma_n' - \Gamma\Gamma') \xrightarrow{\mathcal{L}} N(0, (\Gamma \otimes \Gamma)A(\Gamma' \otimes \Gamma')) \text{ as } n \to \infty,$$

where A is a $d^2 \times d^2$ block matrix $A = ((A_{ij}))$ with

$$A_{ii} = \begin{pmatrix} 1 & 0 & \cdots & \vdots & \cdots & 0 \\ 0 & 1 & \cdots & \vdots & \cdots & 0 \\ \vdots & & \ddots & \vdots & & \\ 0 & \cdots & \cdots & 2 & & \\ \vdots & & & & \ddots & \\ 0 & 0 & \cdots & \cdots & \cdots & 1 \end{pmatrix} \quad A_{ij} = \begin{pmatrix} 0 & 0 & \cdots & \cdots & \cdots & 0 \\ \vdots & & & \vdots & & \\ 0 & \cdots & \cdots & 1 & \cdots & 0 \\ \vdots & & & & & \vdots \\ 0 & 0 & \cdots & \cdots & \cdots & 0 \end{pmatrix},$$

for $i \neq j, i, j = 1, 2, \ldots d$, and $\otimes$ denotes the Kronecker product of the matrices. Here '2' is the ith diagonal entry in A_{ii} and '1' is the (i, j)th entry in A_{ij}.

Remark. Note that the limiting normal distribution is singular if $d > 1$ since $\operatorname{rank}(\boldsymbol{A}) < d^2$. If $d = 1$, then the matrix $\boldsymbol{\Gamma}$ is a scalar $\sigma > 0$ (say) and

$$\sqrt{n}(\sigma_n^2 - \sigma^2) \xrightarrow{\mathcal{L}} N(0, 2\sigma^4) \qquad \text{as } n \to \infty.$$

Proof of Theorem 2.10.1. Let $\tau_n = T/n$ and denote

$$\boldsymbol{U}_k = \boldsymbol{X}(t_{k+1}) - \boldsymbol{X}(t_k) - \tau_n\, \boldsymbol{a}(\boldsymbol{X}(t_k)) = (U_{k1}, \ldots, U_{kd})' \tag{3.2.2}$$

and

$$\boldsymbol{W}_k = \boldsymbol{W}(t_{k+1}) - \boldsymbol{W}(t_k) = (W_{k1}, \ldots, W_{kd})'.$$

Here we have written $\boldsymbol{a}(\boldsymbol{x})$ for $\boldsymbol{a}(\theta_0, \boldsymbol{x})$ for simplicity. Then

$$\boldsymbol{U}_k = \boldsymbol{\Gamma}\boldsymbol{W}_k + \int_{t_k}^{t_{k+1}} (\boldsymbol{a}(\boldsymbol{X}(\tau)) - \boldsymbol{a}(\boldsymbol{X}(t_k)))\, d\tau.$$

In view of (BB3) and the properties of the diffusion processes,

$$\begin{aligned} E_{\theta_0}\left[\int_{t_k}^{t_{k+1}} \|\boldsymbol{a}(\boldsymbol{X}(\tau)) - \boldsymbol{a}(\boldsymbol{X}(t_k))\|^2\, d\tau\right] &\le L^2 E_{\theta_0}\left[\int_{t_k}^{t_{k+1}} \|\boldsymbol{X}(\tau) - \boldsymbol{X}(t_k)\|^2\, d\tau\right] \\ &\le L^2 M_1 \int_{t_k}^{t_{k+1}} (\tau - t_k)\, d\tau \le M_2 \tau_n^2 \end{aligned} \tag{3.2.3}$$

for some constants M_1 and M_2 finite.

Let $\boldsymbol{a} = (a_1, a_2, \ldots, a_d)'$. By the Cauchy–Schwarz inequality, it follows that

$$\begin{aligned} &E_{\theta_0} \int_{t_k}^{t_{k+1}} |\boldsymbol{a}_i(\boldsymbol{X}(\tau)) - \boldsymbol{a}_i(\boldsymbol{X}(t_k))|\, d\tau \\ &\le \tau_n^{1/2}\left\{E_{\theta_0} \int_{t_k}^{t_{k+1}} (a_i(\boldsymbol{X}(\tau)) - a_i(\boldsymbol{X}(t_k)))^2\, d\tau\right\}^{1/2} \\ &\le \tau_n^{1/2}\left\{E_{\theta_0} \int_{t_k}^{t_{k+1}} \|\boldsymbol{a}(\boldsymbol{X}(\tau)) - a(\boldsymbol{X}(t_k))\|^2\, d\tau\right\}^{1/2} \\ &\le M_2^{1/2} \tau_n^{3/2} \qquad \text{(from (3.2.3)).} \end{aligned} \tag{3.2.4}$$

Hence

$$\boldsymbol{U}_k = \boldsymbol{\Gamma}\boldsymbol{W}_k + O_p(\tau_n^{3/2}). \tag{3.2.5}$$

In view of (3.2.2), we have

$$\boldsymbol{X}(t_{k+1}) - \boldsymbol{X}(t_k) = \boldsymbol{U}_k + \tau_n \boldsymbol{a}(\boldsymbol{X}(t_k)),$$

and hence

$$\begin{aligned} \boldsymbol{\Gamma}_n\boldsymbol{\Gamma}_n' &= \frac{1}{T}\sum_{k=0}^{n-1}(\boldsymbol{U}_k\boldsymbol{U}_k' + \tau_n^2\, \boldsymbol{a}(\boldsymbol{X}(t_k))\boldsymbol{a}(\boldsymbol{X}(t_k))' + 2\tau_n \boldsymbol{U}_k \boldsymbol{a}(\boldsymbol{X}(t_k))') \\ &= \frac{1}{T}\sum_{k=0}^{n-1} \boldsymbol{\Gamma}\boldsymbol{W}_k\boldsymbol{W}_k'\boldsymbol{\Gamma}' + \frac{n}{T}O_p(\tau_n^3) + \frac{2}{T}\sum_{k=0}^{n-1} \boldsymbol{\Gamma}\boldsymbol{W}_k O_p(\tau_n^{3/2}) \\ &\quad + \frac{\tau_n^2}{T}\sum_{k=0}^{n-1} \boldsymbol{a}(\boldsymbol{X}(t_k))\boldsymbol{a}(\boldsymbol{X}(t_k))' + \frac{2\tau_n}{T}\sum_{k=0}^{n-1} \boldsymbol{U}_k\boldsymbol{a}(\boldsymbol{X}(t_k))' \end{aligned}$$

from (3.2.5). Therefore

$$\begin{aligned}\sqrt{n}(\boldsymbol{\Gamma}_n\boldsymbol{\Gamma}_n' - \boldsymbol{\Gamma}\boldsymbol{\Gamma}') &= \sqrt{n}\left(\frac{1}{T}\sum_{k=0}^{n-1}\boldsymbol{\Gamma}\boldsymbol{W}_k\boldsymbol{W}_k'\boldsymbol{\Gamma}' - \boldsymbol{\Gamma}\boldsymbol{\Gamma}'\right)\frac{n^{3/2}}{T}O_p(\tau_n^3)\\ &+\frac{2\sqrt{n}}{T}\sum_{k=0}^{n-1}(\boldsymbol{\Gamma}\boldsymbol{W}_k O_p(\tau_n^{3/2}))\frac{\sqrt{n}}{T}\tau_n^2\sum_{k=0}^{n-1}\boldsymbol{a}(\boldsymbol{X}(t_k))\boldsymbol{a}(\boldsymbol{X}(t_k))'\\ &+\frac{2\sqrt{n}\tau_n}{T}\sum_{k=0}^{n-1}U_k\boldsymbol{a}(\boldsymbol{X}(t_k))'. \end{aligned} \tag{3.2.6}$$

Let $\tilde{\boldsymbol{W}}_k = (n/T)^{1/2}\boldsymbol{W}_k$. Then $\tilde{W}_{ki}$ is an $N(0, 1)$ random variable for $i = 1, 2, \ldots d$, and it is easy to see that $E\tilde{W}_{ki}^4 = 3$, $E\tilde{W}_{ki}^2\tilde{W}_{kj}^2 = 1$, $i \neq j$, $E\tilde{W}_{ki}^3\tilde{W}_{kj} = 0$, $E\tilde{W}_{ki}\tilde{W}_{kj}\tilde{W}_{kl}\tilde{W}_{kq} = 0$ for different i, j, q, l where $k = 0, 1, \ldots, n-1$ and $i, j, q, l \in \{1, 2, \ldots, d\}$. From the central limit theorem, it follows that

$$\sqrt{n}\left(\frac{1}{n}\sum_{k=0}^{n-1}\tilde{\boldsymbol{W}}_k\tilde{\boldsymbol{W}}_k' - \boldsymbol{I}\right) \xrightarrow{\mathcal{L}} N(0, \boldsymbol{A}) \qquad \text{as } n \to \infty, \tag{3.2.7}$$

where $\boldsymbol{I}$ is the identity matrix. Hence

$$\begin{aligned}\sqrt{n}\left(\frac{1}{T}\sum_{k=0}^{n-1}\boldsymbol{\Gamma}\boldsymbol{W}_k\boldsymbol{W}_k'\boldsymbol{\Gamma}' - \boldsymbol{\Gamma}\boldsymbol{\Gamma}'\right) &= \sqrt{n}(\boldsymbol{\Gamma}\otimes\boldsymbol{\Gamma}')\left(\frac{1}{n}\sum_{k=0}^{n-1}\tilde{\boldsymbol{W}}_k\tilde{\boldsymbol{W}}_k' - \boldsymbol{I}\right)\\ &\xrightarrow{\mathcal{L}} N(0, (\boldsymbol{\Gamma}\otimes\boldsymbol{\Gamma})\boldsymbol{A}(\boldsymbol{\Gamma}'\otimes\boldsymbol{\Gamma}')) \qquad \text{as } n \to \infty,\end{aligned} \tag{3.2.8}$$

from the properties of the Kronecker product, namely that, if $\boldsymbol{A}$ is a matrix of order $k \times s$, $\boldsymbol{B}$ is a matrix of order $r \times \ell$ and $\boldsymbol{C}$ is a matrix of order $\ell \times k$, then

$$\boldsymbol{BCA} = (\boldsymbol{A}'\otimes\boldsymbol{B})\boldsymbol{C}, \qquad (\boldsymbol{A}\otimes\boldsymbol{B})' = \boldsymbol{A}'\otimes\boldsymbol{B}'.$$

Furthermore,

$$\frac{n^{3/2}}{T}O_p(\tau_n^3) = o_p(1) \qquad \text{(by condition (BB4))} \tag{3.2.9}$$

and

$$\begin{aligned}\frac{2\sqrt{n}}{T}\left\|\sum_{k=0}^{n-1}\boldsymbol{\Gamma}\boldsymbol{W}_k O_p(\tau_n^{3/2})\right\| &= \frac{2\sqrt{n}}{T}\left\|\sum_{k=0}^{n-1}(O_p(\tau_n^{3/2})\otimes\boldsymbol{\Gamma})\boldsymbol{W}_k\right\|\\ &= O_p\left(\frac{T}{\sqrt{n}}\right),\end{aligned} \tag{3.2.10}$$

since $\boldsymbol{W}_k = O_p(T/\sqrt{n})$. It follows by condition (BB1) that

$$\frac{1}{n}\sum_{k=0}^{n-1}\boldsymbol{a}(\boldsymbol{X}(t_k))\boldsymbol{a}(\boldsymbol{X}(t_k))' \xrightarrow{p} E\boldsymbol{a}(\boldsymbol{X}(0))\boldsymbol{a}(\boldsymbol{X}(0))' \qquad \text{as } n \to \infty.$$

Hence

$$\frac{\sqrt{n}}{T}\tau_n^2\sum_{k=0}^{n-1}\boldsymbol{a}(\boldsymbol{X}(t_k))\boldsymbol{a}(\boldsymbol{X}(t_k))' = \frac{T}{\sqrt{n}}\frac{1}{n}\sum_{k=0}^{n-1}\boldsymbol{a}(\boldsymbol{X}(t_k))\boldsymbol{a}(\boldsymbol{X}(t_k))' \xrightarrow{p} 0 \qquad \text{as } n \to \infty. \tag{3.2.11}$$

Furthermore,

$$\frac{2\sqrt{n}\tau_n}{T}\sum_{k=0}^{n-1} U_k a(X(t_k))' = \frac{2}{\sqrt{n}}\sum_{k=0}^{n-1}(U_k - \Gamma W_k)a(X(t_k))' + \frac{2}{\sqrt{n}}\Gamma\sum_{k=0}^{n-1} W_k a(X(t_k))'.$$

Note that W_k and $a(X(t_k))$ are independent random variables with $E[W_k] = 0$. Hence $E(\sum_{k=0}^{n-1} W_k a(X_{t_k})') = 0$. Furthermore, $\text{var}(W_{ki}) = T/n\text{var}(\tilde{W}_{ki}), 1 \le i \le d$. Hence each component $q_{ki}, i = 1, 2, \ldots, d$, of the vector $W_k a(X(t_k))'$ is $O_p((T/n)^{1/2})$. In fact

$$\text{var}\left(\frac{1}{\sqrt{n}}\sum_{k=0}^{n-1} q_{ki}\right) = \frac{1}{n}\sum_{k=0}^{n-1}\text{var}(q_{ki}) = O\left(\sqrt{\frac{T}{n}}\right).$$

Therefore,

$$\frac{2}{\sqrt{n}}\Gamma\sum_{k=0}^{n-1} W_k a(X(t_k))' = O_p\left(\sqrt{\frac{T}{n}}\right). \tag{3.2.12}$$

Consider now an arbitrary element S_{ij}^k of the matrix

$$S_k = (U_k - \Gamma W_k)a(X(t_k)).$$

Then

$$\begin{aligned}
&E|(U_{ki} - (\Gamma W)_{ki})a_j(X(t_k))| \\
&= E\left|\int_{t_k}^{t_{k+1}}[a_i(X(\tau)) - a_i(X(t_k))]\,d\tau\, a_j(X(t_k))\right| \\
&\le \{E[\int_{t_k}^{t_{k+1}}(a_i(X(\tau)) - a_i(X(t_k)))\,d\tau]^2 E[a_j(X(0))]^2\}^{1/2} \\
&\quad \text{(by the stationarity of the process } X(t)) \\
&\le \left\{E\int_{t_k}^{t_{k+1}}(a_i(X(\tau)) - a_i(X(t_k)))^2\,d\tau\ \tau_n\ E[a_j(X(0))]^2\right\}^{1/2} \\
&= O(\tau_n^{3/2}) \qquad \text{(by (3.2.3))}.
\end{aligned}$$

Hence

$$\frac{2}{\sqrt{n}}\sum_{k=0}^{n-1}(U_k - \Gamma W_k)a(X(t_k))' = n\frac{1}{\sqrt{n}}O_p\left(\frac{T^{3/2}}{n^{3/2}}\right) = O_p\left(\frac{T^{3/2}}{n}\right). \tag{3.2.13}$$

Relations (3.2.8) to (3.2.13), (3.2.6) and condition (BB4) prove that

$$\sqrt{n}(\Gamma_n\Gamma_n' - \Gamma\Gamma') \xrightarrow{\mathcal{L}} N_{d^2}(0, (\Gamma \otimes \Gamma)A(\Gamma' \otimes \Gamma')).$$

Remark. In view of (3.2.9) to (3.2.13), it is sufficient for the theorem to hold if T is fixed and $n \to \infty$.

Let us next consider the problem of estimating the parameter θ_0 in the model (3.2.1) given a positive definite matrix Γ. Consider the generalized least-squares estimator $\hat{\theta}_{nT}$ obtained by minimizing the functional

$$Q_n^T(\theta) = \sum_{k=0}^{n-1}\|X(t_{k+1}) - X(t_k) - \tau_n\, a(\theta, X(t_k))\|^2_{(\Gamma^{-1})'\Gamma^{-1}}, \tag{3.2.14}$$

where $\|Z\|_P^2 = Z'PZ$ for any $Z \in R^d$ and P is a $d \times d$ matrix.

Let us write $\tilde{X}(t_k) = \Gamma^{-1}X(t_k)$ and $\Gamma^{-1}a(\theta, X(t_k)) = \tilde{a}(\theta, \tilde{X}(t_k))$. Then (3.2.1) can be written in the form

$$d\tilde{X}(t) = \tilde{a}(\theta, \tilde{X}(t)) + dW(t), \qquad \tilde{X}(0) = \Gamma^{-1}X_0. \tag{3.2.15}$$

Let

$$\frac{\partial \tilde{a}(\theta, \tilde{X}(t))}{\partial \theta} = \Gamma^{-1}\begin{pmatrix} \frac{\partial a_1}{\partial \theta_1} & \cdots & \frac{\partial a_1}{\partial \theta_\ell} \\ \cdots & \cdots & \cdots \\ \cdots & \cdots & \cdots \\ \frac{\partial a_d}{\partial \theta_1} & \cdots & \frac{\partial a_d}{\partial \theta_\ell} \end{pmatrix} = \Gamma^{-1}\frac{\partial a(\theta, X(t))}{\partial \theta}.$$

Suppose an initial estimator $\hat{\theta}_{nT1}$ for the true parameter θ_0 is available. The estimator achieving the minimum of $Q_n^T(\theta)$ defined by (3.2.14) can be derived using the Gauss–Newton procedure iteratively. If the first iteration for θ_0 is $\hat{\theta}_{nT2}$, then

$$\Delta\theta_{nT} = \hat{\theta}_{nT2} - \hat{\theta}_{nT1}$$

is derived as the argument achieving the minimum of $\tilde{Q}_n^T(\Delta\theta)$ given by

$$\begin{aligned}
\tilde{Q}_n^T(\Delta\theta) = \sum_{k=0}^{n-1}\Bigg\{ & (X(t_{k+1})' - X(t_k)' - \tau_n a(\hat{\theta}_{nT1}, X(t_k))' \\
& -\tau_n \Delta\theta' \frac{\partial a}{\partial \theta}(\hat{\theta}_{nT1}, X(t_k))') \\
& \cdot (\Gamma^{-1})'\Gamma^{-1}(X(t_{k+1} - X(t_k) - \tau_n a(\hat{\theta}_{nT1}, X(t_k)) \\
& -\tau_n \frac{\partial a}{\partial \theta}(\hat{\theta}_{nT1}, X(t_k))\Delta\theta)\Bigg\} \\
= \sum_{k=0}^{n-1}\Bigg\{ & (\tilde{X}(t_{k+1})' - \tilde{X}(t_k)' - \tau_n \tilde{a}(\hat{\theta}_{nT1}, \tilde{X}(t_k))' \\
& -\tau_n \Delta\theta' \frac{\partial \tilde{a}}{\partial \theta}(\hat{\theta}_{nT1}, \tilde{X}(t_k))') \\
& \cdot (\tilde{X}(t_{k+1}) - \tilde{X}(t_k) - \tau_n \tilde{a}(\hat{\theta}_{nT1}, \tilde{X}(t_k)) \\
& -\tau_n \frac{\partial \tilde{a}}{\partial \theta}(\hat{\theta}_{nT1}, \tilde{X}(t_k)\Delta\theta)\Bigg\}.
\end{aligned} \tag{3.2.16}$$

Theorem 3.2.2 *Suppose that condition (BB1) holds. In addition, assume that the following conditions hold on* $a(\cdot,\cdot)$*:*

(BB2)′ $E_\theta \frac{\partial \tilde{a}}{\partial \theta}(\theta, \tilde{X}(0))' \frac{\partial \tilde{a}}{\partial \theta}(\theta, \tilde{X}(0))$ *is a matrix with finite elements.*

(BB3)′ *(i)* $\|a(\theta, x_1) - a(\theta, x_2)\| \le L(\theta)\|x_1 - x_2\|$ *for all* $\theta \in \Theta$ *and* $x_1, x_2 \in R^d$*, where* $\sup\{L(\theta), \theta \in \Theta\} < \infty$*;*

(ii) $\left|\frac{\partial a_i(\theta, x_1)}{\partial \theta_j} - \frac{\partial a_i(\theta, x_2)}{\partial \theta_j}\right| \le L_1(\theta)\|x_1 - x_2\|$ *for all* $\theta \in \Theta$ *and* $x_1, x_2 \in R^d$ *where* $L_1(\theta)$ *is continuous in* θ *and* $\sup\{L_1(\theta) : \theta \in \Theta\} < \infty$ *for* $1 \le i \le d$ *and* $1 \le j \le \ell$*;*

(iii) $\|\boldsymbol{a}(\theta, \boldsymbol{x}\| \leq L(\theta)(1 + \|\boldsymbol{x}\|)$ *for all* $\theta \in \Theta$;

(iv) $\left|\frac{\partial a_i}{\partial \theta_j}(\theta_1, \boldsymbol{x}) - \frac{\partial a_i}{\partial \theta_j}(\theta_2, \boldsymbol{x})\right| \leq \tau(\boldsymbol{x})\|\theta_1 - \theta_2\|$ *for all* $\theta_1, \theta_2 \in \Theta$ *and* $\boldsymbol{x} \in R^d$ *where* $\tau(\boldsymbol{x})$ *is continuous with* $E_\theta(\tau(\boldsymbol{X}(0))^{d+\chi}) < \infty$ *for some* $\chi > d$;

(BB4)′ $\hat{\theta}_{nT1} - \theta_0 = o_p(T^{-1/4})$.

(BB5)′ $T/\sqrt{n} \to 0$ *and* $T \to \infty$.

(BB6)′ *All partial derivatives of* $\boldsymbol{a}(\theta, \boldsymbol{x})$ *with respect to* θ *up to second order exist and are bounded.*

Then

$$\sqrt{T}(\hat{\theta}_{nT2} - \theta_0) \xrightarrow{\mathcal{L}} N_\ell(0, \{E_{\theta_0}\frac{\partial \boldsymbol{a}}{\partial \theta}(\theta_0, \boldsymbol{X}(0))'(\boldsymbol{\Gamma}^{-1})'\boldsymbol{\Gamma}\frac{\partial \boldsymbol{a}}{\partial \theta}(\theta_0, \boldsymbol{X}(0))\}^{-1}) \text{ as } n \to \infty.$$

Proof. It is easy to see that the solution of the minimization problem of the function defined by (3.2.16) is

$$\begin{aligned}\Delta\theta_{nT} = \frac{n}{T}&\left[\frac{1}{n}\sum_{k=0}^{n-1}\frac{\partial \tilde{\boldsymbol{a}}}{\partial \theta}(\hat{\theta}_{nT1}, \tilde{\boldsymbol{X}}(t_k))'\frac{\partial \tilde{\boldsymbol{a}}}{\partial \theta}(\hat{\theta}_{nT1}, \tilde{\boldsymbol{X}}(t_k))\right]^{-1}\\ &\times\frac{1}{n}\sum_{k=0}^{n-1}\frac{\partial \tilde{\boldsymbol{a}}}{\partial \theta}(\hat{\theta}_{nT1}, \tilde{\boldsymbol{X}}(t_k))'\Bigg[\tilde{\boldsymbol{X}}(t_{k+1}) - \tilde{\boldsymbol{X}}(t_k)\\ &-\frac{T}{n}\tilde{\boldsymbol{a}}(\hat{\theta}_{nT1}, \tilde{\boldsymbol{X}}(t_k))\Bigg].\end{aligned} \tag{3.2.17}$$

Note that

$$\tilde{\boldsymbol{X}}(t_{k+1}) - \tilde{\boldsymbol{X}}(t_k) = \boldsymbol{\Gamma}^{-1}\int_{t_k}^{t_{k+1}} \boldsymbol{a}(\theta_0, \boldsymbol{X}(s))\, ds + \boldsymbol{W}_k.$$

In view of (BB4)′ and (BB6)′, it follows that

$$\begin{aligned}\frac{T}{n}\tilde{\boldsymbol{a}}(\hat{\theta}_{nT1}, \tilde{\boldsymbol{X}}(t_k)) = &\frac{T}{n}\tilde{\boldsymbol{a}}(\theta_0, \tilde{\boldsymbol{X}}(t_k))\\ &+\frac{T}{n}\frac{\partial \tilde{\boldsymbol{a}}}{\partial \theta}(\theta_0, \tilde{\boldsymbol{X}}(t_k))(\hat{\theta}_{nT1} - \theta_0) + \frac{T}{n}o_p(T^{-1/2}).\end{aligned}$$

Hence

$$\begin{aligned}&\tilde{\boldsymbol{X}}(t_{k+1}) - \tilde{\boldsymbol{X}}(t_k) - \frac{T}{n}\tilde{\boldsymbol{a}}(\hat{\theta}_{nT1}, \tilde{\boldsymbol{X}}(t_k))\\ &= \int_{t_k}^{t_{k+1}}(\tilde{\boldsymbol{a}}(\theta_0, \tilde{\boldsymbol{X}}(s)) - \tilde{\boldsymbol{a}}(\theta_0, \tilde{\boldsymbol{X}}(t_k)))\, ds\\ &\quad -\tau_n\, o_p(T^{-1/2}) + \boldsymbol{W}_k - \tau_n\frac{\partial \tilde{\boldsymbol{a}}}{\partial \theta}(\theta_0, \tilde{\boldsymbol{X}}(t_k))(\hat{\theta}_{nT1} - \theta_0)\\ &= O_p(\tau_n^{3/2}) + \tau_n o_p(T^{-1/2}) + \boldsymbol{W}_k - \tau_n\frac{\partial \tilde{\boldsymbol{a}}}{\partial \theta}(\theta_0, \tilde{\boldsymbol{X}}(t_k))(\hat{\theta}_{nT1} - \theta_0).\end{aligned}$$

Therefore, using (BB4)′, we have

$$\begin{aligned}&\frac{1}{n}\sum_{k=0}^{n-1}\frac{\partial \tilde{\boldsymbol{a}}}{\partial \theta}(\hat{\theta}_{nT1}, \tilde{\boldsymbol{X}}(t_k))'\frac{\partial \tilde{\boldsymbol{a}}}{\partial \theta}(\hat{\theta}_{nT1}, \tilde{\boldsymbol{X}}(t_k))\\ &\quad = \frac{1}{n}\sum_{k=0}^{n-1}\frac{\partial \tilde{\boldsymbol{a}}}{\partial \theta}(\theta_0, \tilde{\boldsymbol{X}}(t_k))'\frac{\partial \tilde{\boldsymbol{a}}}{\partial \theta}(\theta_0, \tilde{\boldsymbol{X}}(t_k)) + o_p(T^{-1/4}).\end{aligned}$$

It can now be shown, by arguments in Fuller (1976), that

$$\left\{\frac{1}{n}\sum_{k=0}^{n-1}\frac{\partial\tilde{a}}{\partial\theta}(\hat{\theta}_{nT1},\tilde{X}(t_k))'\frac{\partial\tilde{a}}{\partial\theta}(\hat{\theta}_{nT1},\tilde{X}(t_k))\right\}^{-1}$$
$$=\left\{\frac{1}{n}\sum_{k=0}^{n-1}\frac{\partial\tilde{a}}{\partial\theta}(\theta_0,\tilde{X}(t_k))'\frac{\partial\tilde{a}}{\partial\theta}(\theta_0,\tilde{X}(t_k))\right\}^{-1}+o_p(T^{-1/4}).$$

Relation (3.2.17) now shows that

$$\begin{aligned}\Delta\theta_{nT}&=\frac{1}{T}\left(\frac{1}{n}\sum_{k=0}^{n-1}\frac{\partial\tilde{a}}{\partial\theta}(\theta_0,\tilde{X}(t_k))'\frac{\partial\tilde{a}}{\partial\theta}(\theta_0,\tilde{X}(t_k))\right)^{-1}\\&\quad\times\sum_{k=0}^{n-1}\frac{\partial\tilde{a}}{\partial\theta}(\hat{\theta}_{nT1},\tilde{X}(t_k))'W_k+o_p(T^{-1/2})\\&\quad+o_p(T^{-1/4})\frac{1}{T}\sum_{k=0}^{n-1}\frac{\partial\tilde{a}}{\partial\theta}(\hat{\theta}_{nT1},\tilde{X}(t_k))'W_k+O_p(\tau_n^{1/2})+\theta_0-\hat{\theta}_{nT1}.\end{aligned}\tag{3.2.18}$$

Let $D_T=\{\theta\in\Theta:\|\theta-\theta_0\|\leq T^{-1/4}\}$. Since $\hat{\theta}_{nT}\in D_T$ with probability approaching unity as $T\to\infty$ by (BB4)′, it follows that, with probability close to unity as $T\to\infty$,

$$\begin{aligned}&\frac{1}{T}\left\|\sum_{k=0}^{n-1}\frac{\partial\tilde{a}}{\partial\theta}(\hat{\theta}_{nT1},\tilde{X}(t_k))'W_k-\sum_{k=0}^{n-1}\frac{\partial\tilde{a}}{\partial\theta}(\theta_0,\tilde{X}(t_k))'W_k\right\|\\&\leq\sup_{\theta^*\in D_T}\frac{1}{T}\left\|\sum_{k=0}^{n-1}\frac{\partial\tilde{a}}{\partial\theta}(\theta^*,\tilde{X}(t_k))'W_k-\sum_{k=0}^{n-1}\frac{\partial\tilde{a}}{\partial\theta}(\theta_0,\tilde{X}(t_k))'W_k\right\|\\&=\sup_{\theta^*\in D_T}\frac{1}{T}\left\|\int_0^T(H^{(1)}(t,\theta^*)-H^{(1)}(t,\theta_0))\,dW(t)\right.\\&\quad\left.+\int_0^T\frac{\partial\tilde{a}}{\partial\theta}(\theta^*,X(t))'-\frac{\partial\tilde{a}}{\partial\theta}(\theta_0,X(t))'\,dW(t)\right\|,\end{aligned}\tag{3.2.19}$$

where

$$H^{(1)}(t,\theta)=\frac{\partial\tilde{a}}{\partial\theta}(\theta,\tilde{X}(t_k))'-\frac{\partial\tilde{a}}{\partial\theta}(\theta,\tilde{X}(t))'\qquad\text{for } t\in[t_k,t_{k+1}),\ 0\leq k\leq n-1.$$

Using (BB3)′ (ii), it can be shown by arguments similar to those in Lemma 3.1 of Prakasa Rao (1983b) that

$$\frac{1}{\sqrt{T}}\int_0^T H^{(1)}(t,\theta^*)dW(t)=o_P(1),\qquad\frac{1}{\sqrt{T}}\int_0^T H^{(1)}(t,\theta_0)dW(t)=o_P(1).\tag{3.2.20}$$

By (BB3)′ (iv) and Corollary 6.2 of Basu (1983a) (see Chapter 2), it follows that

$$P\left(\sup_{\theta^*\in D_T}\left\|\int_0^T\left(\frac{\partial\tilde{a}}{\partial\theta}(\theta^*,\tilde{X}(t)\right)-\frac{\partial\tilde{a}}{\partial\theta}(\theta_0,\tilde{X}(t))\right)dW(t)\right\|\geq\frac{2^\beta C}{T^{\beta/4}}\lambda^{1/d+\chi}\right)\leq C\frac{A_T}{\lambda}$$

where we choose (cf. Lemma 6.3 of Basu 1983a), $\gamma \in (2d, d+\chi)$ so that

$$\beta = \frac{\gamma - 2d}{d+\chi} > 0, \qquad d + \chi - \gamma > 0.$$

Here $A_T = \int_{D_T}\int_{D_T} \|\theta_1 - \theta_2\|^{d+\chi-\gamma}\, d\theta_2\, d\theta_1$ and $A_T \to 0$ as $T \to \infty$. Hence

$$\sup_{\theta^* \in D_T} \int_0^T \left(\frac{\partial \tilde{a}}{\partial \theta}(\theta^*, \tilde{X}(t))\right)' - \frac{\partial \tilde{a}}{\partial \theta}(\theta_0, \tilde{X}(t))')dW(t) = o_p(1). \tag{3.2.21}$$

Substituting the results in (3.2.20) and (3.2.21) into (3.2.19), we have

$$\frac{1}{T}\sum_{k=0}^{n-1} \frac{\partial \tilde{a}}{\partial \theta}(\hat{\theta}_{nT1}, \tilde{X}(t_k))' W_k = \frac{1}{T}\sum_{k=0}^{n-1} \frac{\partial \tilde{a}}{\partial \theta}(\theta_0, \tilde{X}(t_k))' W_k + o_p(T^{-1/2}).$$

Hence (3.2.18) can be written in the form

$$\begin{aligned}\Delta\theta_{nT} = {} & \frac{1}{T}\left(\frac{1}{n}\sum_{k=0}^{n-1} \frac{\partial \tilde{a}}{\partial \theta}(\theta_0, \tilde{X}(t_k))' \frac{\partial \tilde{a}}{\partial \theta}(\theta_0, \tilde{X}(t_k))\right)^{-1} \sum_{k=0}^{n-1} \frac{\partial \tilde{a}}{\partial \theta}(\theta_0, \tilde{X}(t_k))' W_k \\ & + o_p(T^{-1/2}) + o_p(T^{-1/4}) \cdot \frac{1}{T}\sum_{k=0}^{n-1} \frac{\partial \tilde{a}}{\partial \theta}(\theta_0, \tilde{X}(t_k))' W_k \\ & + \theta_0 - \hat{\theta}_{nT1} + O_p(\tau_n^{1/2}).\end{aligned} \tag{3.2.22}$$

Therefore

$$\begin{aligned}\sqrt{T}(\hat{\theta}_{nT2} - \theta_0) = {} & o_p(1) + \sqrt{T}\,O_p(\tau_n^{1/2}) \\ & + \left(\frac{1}{n}\sum_{k=0}^{n-1} \frac{\partial \tilde{a}}{\partial \theta}(\theta_0, \tilde{X}(t_k))' \frac{\partial \tilde{a}}{\partial \theta}(\theta_0, \tilde{X}(t_k))\right)^{-1} \\ & \times \frac{1}{\sqrt{T}}\sum_{k=0}^{n-1} \frac{\partial \tilde{a}}{\partial \theta}(\theta_0, \tilde{X}(t_k))' W_k \\ & + o_p(T^{-1/4}) \frac{1}{\sqrt{T}}\sum_{k=0}^{n-1} \frac{\partial \tilde{a}}{\partial \theta}(\theta_0, \tilde{X}(t_k))' W_k.\end{aligned} \tag{3.2.23}$$

But, from (3.2.20), we know that

$$\frac{1}{\sqrt{T}}\sum_{k=0}^{n-1} \frac{\partial \tilde{a}}{\partial \theta}(\theta_0, \tilde{X}(t_k))' W_k = \frac{1}{\sqrt{T}}\int_0^T \frac{\partial \tilde{a}}{\partial \theta}(\theta_0, \tilde{X}(t))'\, dW(t) + o_p(1).$$

On the other hand, using (BB1) and (BB2)$'$ and applying the central limit theorem for stochastic integrals (Basawa and Prakasa Rao 1980), it follows that

$$\frac{1}{\sqrt{T}}\int_0^T \frac{\partial \tilde{a}}{\partial \theta}(\theta_0, \tilde{X}(t))'\, dW(t) \xrightarrow{\mathcal{L}} N(0, E\frac{\partial \tilde{a}}{\partial \theta}(\theta_0, \tilde{X}(0))' \frac{\partial \tilde{a}}{\partial \theta}(\theta_0, \tilde{X}(0))). \tag{3.2.24}$$

Now, using (3.2.23), (BB5)$'$ and following (BB1)$'$, we have

$$\frac{1}{n}\sum_{k=0}^{n-1} \frac{\partial \tilde{a}}{\partial \theta}(\theta_0, \tilde{X}(t_k))' \frac{\partial \tilde{a}}{\partial \theta}(\theta_0, \tilde{X}(t_k)) \xrightarrow{P} E\left\{\frac{\partial a}{\partial \theta}(\theta_0, X(0))'(\Gamma^{-1})'\Gamma^{-1}\frac{\partial a}{\partial \theta}(\theta_0, X(0))\right\},$$

and this completes the proof of the theorem.

Remarks. (i) In Theorem 3.2.2, it was assumed that the matrix $\mathbf{\Gamma}$ is known. If $\mathbf{\Gamma}$ is unknown, one could use the estimator $\mathbf{\Gamma}_n$ discussed in Theorem 3.2.1 and it follows that

$$(\mathbf{\Gamma}_n\mathbf{\Gamma}_n')^{-1} = (\mathbf{\Gamma}\mathbf{\Gamma}')^{-1} + o_p(n^{-1/2}),$$

following arguments in Fuller (1976). We can replace $\Delta\theta_{nT}$ given by (3.2.17) with $\hat{\Delta\theta}_{nT}$ given by

$$\hat{\Delta\theta}_{nT} = \frac{n}{T}\left[\frac{1}{n}\sum_{k=0}^{n-1}\frac{\partial \boldsymbol{a}}{\partial\theta}(\hat{\theta}_{nT1}, \boldsymbol{X}(t_k))'(\mathbf{\Gamma}_n\mathbf{\Gamma}_n')^{-1}\frac{\partial \boldsymbol{a}}{\partial\theta}(\hat{\theta}_{nT1}, \boldsymbol{X}(t_k))\right]^{-1}$$
$$\times\frac{1}{n}\sum_{k=0}^{n-1}\frac{\partial \boldsymbol{a}}{\partial\theta}(\hat{\theta}_{nT1}, \boldsymbol{X}(t_k))'(\mathbf{\Gamma}_n\mathbf{\Gamma}_n')^{-1}$$
$$\times\left[\boldsymbol{X}(t_{k+1}) - \boldsymbol{X}(t_k) - \frac{T}{n}\boldsymbol{a}(\hat{\theta}_{nT1}, \boldsymbol{X}(t_k))\right],$$

and define

$$\hat{\hat{\theta}}_{nT2} = \hat{\theta}_{nT1} + \hat{\Delta\theta}_{nT}.$$

Note that $\hat{\Delta\theta}_{nT}$ can be computed from the data. The result in Theorem 3.2.2 holds if $\Delta\theta_{nT}$ is replaced by $\hat{\Delta\theta}_{nT}$ in the definition of $\hat{\theta}_{nT2}$ since

$$\hat{\Delta\theta}_{nT} = \Delta\theta_{nT} + o_p(T^{-3/4})$$

by the arguments similar to those given in Theorem 3.2.2. Hence

$$\sqrt{T}(\hat{\hat{\theta}}_{nT2} - \theta_0) = \sqrt{T}(\hat{\theta}_{nT1} + \hat{\Delta\theta}_{nT} - \theta_0)$$
$$= \sqrt{T}(\hat{\theta}_{nT1} + \Delta\theta_{nT} - \theta_0) + o_p(T^{-1/4})$$

The result now follows from Theorem 3.2.2.

In view of the above, the basic problem is to construct an initial estimator $\hat{\theta}_{nT1}$ satisfying (BB4)$'$. Penev (1985; 1986) gives some sufficient conditions for the existence of such estimators provided $a_i(\theta, \boldsymbol{x})$, $1 \le i \le d$, have smooth third-order partial derivatives and $T = o(n^{2/3})$. Note that the number of design points in the experimental design increases, more slowly than is needed for asymptotic normality, ($T = o(n^{1/2})$, but more rapidly than is needed for consistency ($T = o(n)$). Let $\hat{\theta}_{nT1}$ be the argument achieving the minimum of

$$\tilde{Q}_n^T(\theta) = \sum_{k=0}^{n-1}\|\boldsymbol{X}(t_{k+1}) - \boldsymbol{X}(t_k) - \tau_n\boldsymbol{a}(\theta, \boldsymbol{X}(t_k))\|^2_{(\mathbf{\Gamma}_n\mathbf{\Gamma}_n')^{-1}}$$

on Θ and suppose that $\hat{\theta}_{nT1} \xrightarrow{p} \theta_0$ as $T \to \infty$ and $T/n \to 0$. Then $\hat{\theta}_{nT1}$ satisfies (BB4)$'$.

The above results are generalizations and extensions to the multidimensional case following the work in Dorogovcev (1976) and Prakasa Rao (1983b).

(ii) Kasonga (1988) developed an alternate procedure for the estimation of the drift parameter of the stochastic differential equation

$$d\boldsymbol{X}_t = \boldsymbol{a}(\theta, \boldsymbol{X}_t)\,dt + d\boldsymbol{W}_t, \qquad \boldsymbol{X}_0 = \boldsymbol{x}_0, t \ge 0, \tag{3.2.25}$$

via the least-squares approach. Consider the deterministic analog of this equation, namely,

$$dU(\theta, t) = a(\theta, U(\theta, t))\, dt, \qquad U(\theta, t) = x_0. \tag{3.2.26}$$

Suppose the process X_t is observed at the time points

$$0 = t_0 < t_1 < \cdots < t_n = T$$

in $[0, T]$. Suppose the function $a(\theta, \cdot)$ satisfies Lipschitz and growth restrictions so that there exists a unique solution to (3.2.25) a.s. for each $\theta \in \Theta$. On each subinterval $[t_{k-1}, t_k)$, define a function $u(\theta, t)$ as a solution of the ordinary differential equation

$$\begin{aligned} \frac{dU_k}{dt} &= a(\theta, U_k(\theta, t)), & t \in [t_{k-1}, t_k), \\ U_k(\theta, t_{k-1}) &= X_{t_{k-1}}, & k = 1, 2, \ldots, n. \end{aligned} \tag{3.2.27}$$

Observe that, for $t_{k-1} < t \leq t_k$,

$$X(t) - X(t_{k-1}) = \int_{t_{k-1}}^{t} a(\theta, X_s)\, ds + W(t) - W(t_{k-1}) \tag{3.2.28}$$

and

$$U_k(\theta, t) = X(t_{k-1}) + \int_{t_{k-1}}^{t} a(\theta, U_k(\theta, s))\, ds. \tag{3.2.29}$$

Let

$$Q(\theta) = \sum_{k=1}^{n} \|X_{t_k} - U_k(\theta, t_k)\|^2. \tag{3.2.30}$$

Kasonga (1988) proved the following result.

Theorem 3.2.3 *Suppose that for every $\theta_1, \theta_2 \in \Theta$, $\theta_1 \neq \theta_2$,*

$$p - \lim_{n \to \infty} n^{-1} \sum_{k=1}^{n} |U_k(\theta_1, t_k) - U_k(\theta_2, t_k)|^2 > 0 \tag{3.2.31}$$

and $\Delta_n = \max_{1 \leq k \leq n} |t_k - t_{k-1}| \to 0$ as $n \to \infty$. Then the estimate of θ^*_{nT} obtained by minimizing $Q(\theta)$ defined by (3.2.30) is consistent as $n \to \infty$ and $T \to \infty$.

It was pointed out, on the basis of simulations, that the estimator obtained by minimizing

$$\sum_{k=1}^{n} \|X_{t_k} - X_{t_{k-1}} - a(\theta, X_{t_{k-1}})(t_k - t_{k-1})\|^2 \tag{3.2.32}$$

is severely biased as compared to the above estimator.

Kasonga (1988) proved that, for the linear stochastic differential equation

$$dX_t = \theta X_t\, dt + dW_t, \qquad X_0 = x_0,\ t \geq 0,\ \theta > 0,$$

the estimator θ^*_{nT} is strongly consistent provided

$$\begin{aligned} T &= O(n^{\alpha}), & 0 < \alpha \leq 1, \\ t_k &= \frac{kT}{n}, & 0 \leq k \leq n. \end{aligned}$$

Mishra and Bishwal (1995) studied approximate maximum likelihood estimation of the parameter in a linear drift coefficient of the stochastic differential equation by the 'trapezoidal rule' approximation by transforming the Itô type stochastic integral in the likelihood function to a Stratonovich type stochastic integral when the observations are made at regularly spaced, but dense, time points in the limit. Le Breton (1976) studied the case of a linear stochastic differential equation using the Riemann type approximation for the Itô stochastic integral and the ordinary integral present in the likelihood function (Basawa and Prakasa Rao 1980).

(iii) It was shown earlier in this section that the least-squares estimator obtained from the observations $X(t_k), 0 \le k \le n-1$, with $t_{i+1} - t_i = T/n = h$ is consistent and asymptotically normal provided $hn^{1/2} \to 0$ as $n \to \infty$ (condition for 'rapidly increasing experimental design') under some regularity conditions.

Florens-Zmirou (1989) considered the estimation problem with discrete observations for a one-dimensional diffusion process

$$dX_t = a(\theta, X_t)\,dt + \sigma\,dW_t, \qquad t \ge 0,$$

and showed that, under the condition $h^3 n \to 0$, one can construct a minimum contrast estimator of θ which is asymptotically normal and that, for a quadratic variation type estimator $\hat{\sigma}^2$ of σ^2, $n^{1/2}h^{1/2}(\hat{\sigma}^2 - \sigma^2)$ is asymptotically normal.

Consider the one-dimensional stochastic differential equation

$$dX_t = a(\theta, X_t)\,dt + \sigma\,dW_t, \qquad t \ge 0,$$

where θ and σ are both unknown. Let Δ_n be a sequence such that $\Delta_n \to 0$. Suppose the process is observed at the time points $k\Delta_n, 0 \le k \le n$. Let

$$Y_k^{(n)} = X_{k\Delta_n}$$

and

$$\begin{aligned} U_k^{(n)}(\theta) &= Y_k^{(n)} - Y_{k-1}^{(n)} - \Delta_n\, a(\theta, Y_{k-1}^{(n)}) \\ &= X_{k\Delta_n} - X_{(k-1)\Delta_n} - \Delta_n\, a(\theta, X_{(k-1)\Delta_n}). \end{aligned}$$

Define

$$H_n(\theta) = \frac{1}{n\Delta_n^2}\sum_{k=0}^{n-1}[U_k^{(n)}(\theta)]^2.$$

Let $\hat{\theta}_n$ be such that

$$H_n(\hat{\theta}_n) = \inf_{\theta\in\Theta} H_n(\theta).$$

Suppose that $\sigma = \sigma_0$ is known. Florens-Zmirou (1989) proved that $\hat{\theta}_n$ is consistent and asymptotically normal provided that $n\Delta_n \to \infty$ and $n\Delta_n^3 \to 0$ under some smoothness conditions on the function $a(\cdot, \cdot)$. In fact

$$\sqrt{n\Delta_n}(\hat{\theta}_n - \theta_0) \xrightarrow{\mathcal{L}} N\left(0, E_{\theta_0}\left(\frac{\dot{a}_{\theta_0}^2}{\sigma_0^2}\right)\right) \qquad \text{as } n \to \infty,$$

where $\dot{a}$ denotes the derivative of a with respect to θ. Furthermore, if σ is unknown, then

$$\hat{\sigma}_n^2 = \frac{1}{n\Delta_n}\sum_{k=0}^{n-1}(X_{k\Delta_n} - X_{(k-1)\Delta_n})^2 \xrightarrow{P} \sigma^2$$

if $n\Delta_n \to \infty$ and $\sqrt{n\Delta_n}(\hat{\sigma}_n^2 - \sigma_0^2)$ is asymptotically normal, and $\hat{\sigma}_n^2$ is asymptotically efficient if $n\Delta_n^3 \to 0$.

(iv) Consider the one-dimensional diffusion process satisfying the stochastic differential equation

$$dX_t = a(\theta, X_t)\,dt + \sigma\,dW_t, \qquad t \geq 0, \tag{3.2.33}$$

and suppose the process is observed at times $t_k = k\Delta, k \geq 0$, as in (iii). Dacunha-Castelle and Florens-Zmirou (1986) give a measure of the loss of information due to the discretization as a function of the sampling interval Δ. We will now briefly discuss their results.

Suppose the process X has an invariant probability measure with density $q(\cdot)$ with respect to the Lebesgue measure. Let $p_\Delta(\cdot)$ denote the transition density of (X_0, X_Δ), that is, the conditional density of X_Δ given X_0. Note that the joint density of (X_0, X_Δ) is $p_\Delta q$. The discrete chain $\{Y_k, k \geq 0\}$, $Y_k = X_{k\Delta}$, has $p_\Delta(\cdot)$ as its transition density and $q(\cdot)$ as its invariant density. It is known that, for some constant $k > 0$,

$$q(x) = k \exp[2G(x)/\sigma^2], \qquad G(x) = \int_0^x a(u)du, \tag{3.2.34}$$

where $a(\cdot)$ is the true drift function provided

$$\lim_{|x|\to\infty} \int_0^x \exp\left(\frac{-2G(u)}{\sigma^2}\right) du = \pm\infty \tag{3.2.35}$$

and

$$\int_{-\infty}^{\infty} \exp\left(\frac{2G(u)}{\sigma^2}\right) du < \infty. \tag{3.2.36}$$

Let $\alpha = (\theta, \sigma)$ and α_0 be the true value of α. Let

$$\ell_n(\alpha, \Delta) = \sum_{i=0}^{n-1} \log p_\Delta(Y_i, Y_{i+1}; \alpha). \tag{3.2.37}$$

The function $\ell_n(\alpha, \Delta)$ is the log-likelihood function associated with $(Y_0, \ldots, Y_n)$ when α is the true parameter. The Kullback–Leibler information associated with the Markov chain $\{Y_k\}$ is defined by

$$K(\alpha_0, \alpha; \Delta) = \lim_{n\to\infty} \frac{1}{n}[\ell_n(\alpha_0, \Delta) - \ell_n(\alpha, \Delta)], \tag{3.2.38}$$

assuming that the limit on the right-hand side exists. Under some regularity conditions on the diffusion process X, such as ergodic diffusion with smooth conditions on the drift and the diffusion, Dacunha-Castelle and Florens-Zmirou (1986) obtain an expansion of $p_\Delta(x, y)$ in terms of Δ and show that, as $\Delta \to 0$,

$$K(\alpha_0, \alpha; \Delta) = K(\sigma_0; \sigma) + \Delta K_1(\alpha_0; \alpha) + \Delta^2 K_2(\alpha_0; \alpha,) + O(\Delta^3), \tag{3.2.39}$$

where $K(\sigma_0; \sigma)$ is the Kullback–Leibler information between $N(0, \sigma_0^2)$ and $N(0, \sigma^2)$, and $K_1(\alpha_0; \alpha)$ is a function such that

$$\begin{aligned} K_1(\theta_0, \sigma_0; \theta, \sigma_0) &= \frac{\sigma_0^2}{2} E_q[a(X_0, \theta) - a(X_0, \theta_0)]^2 \\ &= K^0(\theta_0, \theta), \end{aligned} \tag{3.2.40}$$

K^0 being the Kullback–Leibler information from the continuous observations when $\sigma^2 = \sigma_0^2$. Furthermore, $K_2(\alpha_0; \alpha)$ has the property that

$$K_2(\theta_0, \sigma_0; \theta, \sigma_0) = 0. \tag{3.2.41}$$

Hence the Kullback–Leibler distance between the continuous model and the discrete model is $O(\Delta^0)$ if $\sigma_0^2 \neq \sigma^2$ and $O(\Delta^2)$ if $\sigma^2 = \sigma_0^2$. Using this result, they measure the loss of precision due to the discretization, precision being defined as the asymptotic variance of the maximum likelihood estimator of (θ, σ). If σ is known, then the loss is of the order $O(\Delta^2)$. If σ is unknown, then the loss is of higher order as $\Delta \to 0$. In other words, the fact that the diffusion coefficient is known or unknown plays a crucial role in obtaining asymptotically efficient parameter estimators from a discretized version of the process. Let $\hat{\alpha}_n$ be the MLE of $\alpha = (\theta, \sigma)$ based on $Y_i = X_{i\Delta_n}$, $0 \leq i \leq n$. Dacunha-Castelle and Florens-Zmirou (1986) study the properties of consistency and asymptotic normality of $\hat{\alpha}_n$.

(v) Making use of the results on the expansion of the transition density p_Δ, due to Dacunha-Castelle and Florens-Zmirou (1986), Dohnal (1987) investigated the problem of estimation of the diffusion coefficient. We now discuss these results briefly; the problem will be studied later in this chapter.

Let $\{X_t\}$ be the solution of the stochastic differential equation

$$dX_t = a(\theta, X_t)\,dt + \sigma(\theta, X_t)\,dW_t, \qquad 0 \leq t \leq 1 \tag{3.2.42}$$

where a and σ satisfy smooth conditions, $X_0 = x_0, \theta \in \Theta$ open in R. Suppose the process is observed at the time points $t_k = kT/n$, $0 \leq k \leq n$. Let $X_k^{(n)} = X(t_k)$ and $P_\theta^{(n)}$ be the probability measure generated by $\mathbf{X}^{(n)} = \{X_k^{(n)}, 0 \leq k \leq n\}$. Let θ_0 be the true parameter. It can be shown that the family $\{P_\theta^{(n)}, \theta \in \Theta\}$ is LAMN at $\theta_0 \in \Theta$ in the sense that

$$\log \frac{dP_{\theta_{n,h}}^{(n)}}{dP_{\theta_0}^{(n)}} - hJ_n\Gamma_n^{1/2} + \frac{1}{2}h^2\Gamma_n \to 0 \qquad \text{as } n \to \infty$$

under $P_{\theta_0}^{(n)}$, where the following conditions hold (Jeganathan 1982):

(CC1) $\theta_{n,h} = \theta_0 + d_n h$, $0 < d_n \to 0$, as $n \to \infty$;

(CC2) $(J_n, \Gamma_n) \xrightarrow{\mathcal{L}} (J, \Gamma)$ under $P_{\theta_0}^{(n)}$, as $n \to \infty$;

(CC3) J is $N(0, 1)$;

(CC4) $\Gamma = 2\int_0^1 \ell^2(X_t, \theta_0)\,dt > 0$ a.s., $\ell(x, \theta) = \frac{1}{\sigma(\theta,x)} \frac{\partial\sigma(\theta,x)}{\partial\theta}$;

(CC5) J and Γ are independent random variables.

Dohnal (1987) proved that any sequence of estimators $\hat{\theta}_n$ based on $\{X_k^{(n)}\}$ is asymptotically efficient in the sense of Hájek (1972) if

$$n^{1/2}(\hat{\theta}_n - \theta_0) - J_n\Gamma_n^{-1/2} \to 0 \text{ in } P_{\theta_0}^{(n)}\text{-probability} \qquad \text{as } n \to \infty. \tag{3.2.43}$$

Let us consider the following special case.

Example 3.2.1
Consider the stochastic differential equation given by

$$dX_t = a(X_t)\,dt + \sqrt{\theta}\sigma(X_t)\,dW_t, \qquad 0 \leq t \leq 1,\ \theta > 0, \tag{3.2.44}$$

where a and σ are known bounded functions with second-order continuous derivatives and $\sigma(x) > 0$ for all $x \in R$. Let θ_0 be the true parameter. It can be checked that $\{P_\theta^n\}$ is LAN at θ_0 and

$$\hat{\theta}_n = \sum_{k=0}^{n-1} \frac{[X_{k+1}^{(n)} - X_k^{(n)} - \frac{1}{n}\sigma(X_k^{(n)})]^2}{\sigma^2(X_k^{(n)})} \tag{3.2.45}$$

is a consistent and asymptotically efficient estimator of θ_0. Furthermore,

$$\frac{\hat{\theta}_n - \theta_0}{\theta_0\sqrt{2/n}} \xrightarrow{\mathcal{L}} N(0, 1) \qquad \text{as } n \to \infty. \tag{3.2.46}$$

Example 3.2.2
Consider the linear model

$$dX_t = \alpha X_t\, dt + \sqrt{\theta}\, dW_t, \qquad X_0 = x_0,\ 0 \le t \le 1, \tag{3.2.47}$$

where α is known. It can be checked that

$$\hat{\theta}_n = (1 - \bar{\rho}_n)\bar{s}_n, \qquad \bar{\rho}_n = \frac{1}{(n-1)\bar{s}_n^2}\sum_{k=1}^{n} \eta_k\eta_{k-1}, \qquad \bar{s}_n = \frac{1}{n}\sum_{k=1}^{n} \eta_k^2 \tag{3.2.48}$$

with $\eta_k = X_k^{(n)} - \frac{1}{n}\sum_{j=0}^{n} X_j^{(n)}$, is an estimator for θ (Arato 1982) for which

$$E_{\theta_0}(\bar{\theta}_n - \theta_0) = O\left(\frac{1}{n}\right); \text{ and } \operatorname{var}(\bar{\theta}_n) = \frac{2\theta_0^2}{n} + o\left(\frac{1}{n}\right).$$

Other alternate estimators are

$$\hat{\theta}_n = \frac{1}{n}\sum_{k=1}^{n}\left[X_k^{(n)}\left(2 + \frac{\alpha}{n}\right) - X_{k-1}^{(n)}\left(2 - \frac{\alpha}{n}\right)\right]^2 \tag{3.2.49}$$

and

$$\tilde{\theta}_n = \frac{2\alpha}{n(1-\rho^2)}\sum_{k=1}^{n}(X_k^{(n)} - \rho X_{k-1}^{(n)})^2, \qquad \rho = \exp\left(-\frac{\alpha}{n}\right). \tag{3.2.50}$$

It can be shown that

$$E_{\theta_0}(\hat{\theta}_n - \theta_0) = O\left(\frac{1}{n^2}\right), \qquad \operatorname{var}(\hat{\theta}_n) = \frac{2\theta_0^2}{n} + O\left(\frac{1}{n^3}\right), \tag{3.2.51}$$

and

$$E_{\theta_0}(\tilde{\theta}_n) = \theta_0, \qquad \operatorname{var}(\tilde{\theta}_n) = \frac{2\theta_o^2}{n}, \tag{3.2.52}$$

and $\tilde{\theta}_n$ is a minimum variance unbiased estimator for θ_0.

(vi) Yoshida (1992) considered the maximum likelihood estimation of the unknown parameter of a multidimensional diffusion process based on an approximate likelihood obtained from discrete observations on the process under a condition weaker than the condition of 'rapidly increasing experimental design'. He proved that the approximate MLE of the drift parameter θ and a suitable estimator of $\boldsymbol{\Gamma} = \boldsymbol{\sigma\sigma}'$, where $\boldsymbol{\sigma}$ is the diffusion matrix, have a joint limiting distribution after a suitable normalization under the condition $h^3 n \to 0$. We discuss these results briefly in the next section.

3.3 Estimation by the Maximum Likelihood Method

Consider the stochastic differential equation

$$dX_t = a(\theta, X_t) + b(X_t)\sigma\, dW_t, \qquad X_0 = x_0,\ t \geq 0, \tag{3.3.1}$$

where $\theta \in \bar{\Theta}$, Θ is a bounded convex set in R^m and $\sigma \in R^k \times R^r$ are unknown parameters, a is an R^d-valued function defined on $\bar{\Theta} \times R^d$, b is an $(R^d \times R^k)$-valued function defined on R^d and W is an r-dimensional standard Wiener process.

Suppose the process $\{X_t, t \geq 0\}$ is observed at the time instants $t_i = ih,\ h > 0,\ i = 0, 1, \ldots, N$. Let $\Gamma = \sigma\sigma'$. If Γ is known and the process $\{X_t\}$ can be observed continuously on $[0, Nh]$, then the likelihood function is given by

$$\exp\left\{\int_0^{Nh} a'\bar{B}\, dX_t - \frac{1}{2}\int_0^{Nh} a'\bar{B}a\, dt\right\}, \tag{3.3.2}$$

where $\bar{B} = b\sigma[\sigma'b'b\sigma]^{-2}\sigma'b'$. Here $[\sigma'b'b\sigma]^{-2}$ denotes the square of the Moore–Penrose generalized inverse of the matrix $\sigma'b'b\sigma$. Since $\bar{B}$ is invariant for σ satisfying $\Gamma = \sigma\sigma'$, we write $\bar{B} = \bar{B}(x, \Gamma)$. An approximation to the likelihood function, given by (3.3.2), is

$$Q_{h,N}(\Gamma, \theta) = \exp\left\{\sum_{i=1}^{N} a'_{i-1}(\theta)\bar{B}_{i-1}(\Gamma)\bar{\Delta}_i - \frac{h}{2}\sum_{i=1}^{N} a'_{i-1}(\theta)\bar{B}_{i-1}(\Gamma)a_{i-1}(\theta)\right\}, \tag{3.3.3}$$

where

$$a_i(\theta) = a(\theta, X_{t_i}), \qquad \bar{B}_i(\Gamma) = B(X_{t_i}, \Gamma), \qquad \bar{\Delta}_i = X_{t_i} - X_{t_{i-1}}. \tag{3.3.4}$$

An approximate MLE of θ is obtained by maximizing (3.3.3) over $\theta \in \bar{\Theta}$.

Since Γ is unknown in general, we first estimate Γ by some statistic $\hat{\Gamma}_0$ and then find the MLE, $\hat{\theta}_0$ say, by substituting $\hat{\Gamma}_0$ for Γ in $Q_{h,n}(\Gamma, \theta)$ given by (3.3.3). Using this estimator $\hat{\theta}_0$, we will construct a better estimator $\hat{\Gamma}$ for Γ. By maximizing $Q_{h,N}(\hat{\Gamma}, \theta)$, we obtain the MLE $\hat{\theta}$. It will be shown that $\hat{\theta}_0$ and $\hat{\theta}$ are both consistent estimators for θ but $\hat{\theta}$ is also an asymptotically efficient estimator for θ. Furthermore, $\hat{\theta}$ has an asymptotically normal distribution under the condition $h^3N \to 0$ after suitable normalization.

Let θ_0, σ_0 and Γ_0 denote the true values of θ, σ and Γ, respectively. Suppose that $\theta_0 \in \Theta$. For every matrix A, let $\|A\|^2$ denote the sum of squares of elements of A. Define

$$B(x) = (b'b)^{-1}b'(x), \qquad B_{i-1} = B(X_{t_{i-1}}), \tag{3.3.5}$$

$$\partial_i = \partial/\partial x^i,\ \partial = (\partial_1, \ldots, \partial_d),\ \delta_i = \partial/\partial\theta^i,\ \delta = (\delta_1, \ldots, \delta_m) \tag{3.3.6}$$

$$\Delta\bar{B}_i(\Gamma_2, \Gamma_1) = \bar{B}_i(\Gamma_2) - \bar{B}_i(\Gamma_1), \tag{3.3.7}$$

$$\Delta a_i(\theta_2, \theta_1) = a_i(\theta_2) - a_i(\theta_1), \tag{3.3.8}$$

$$\Delta_i(\theta) = X_{t_i} - X_{t_{i-1}} - h\, a(\theta, X_{t_{i-1}}), \tag{3.3.9}$$

and

$$L = \frac{1}{2}\sum_{\substack{i=1\\ j=1}}^{d} v^{ij}\partial_i\partial_j + \sum_{i=1}^{d} a^i\partial_i,\ v^{ij} = [b\Gamma b']^{ij} \qquad \text{for } \theta = \theta_0 \text{ and } \sigma = \sigma_0.$$

In the following discussion C denotes a generic positive constant independent of L, N and other variables in some cases.

We assume that the diffusion process $\boldsymbol{X}$ is ergodic with an invariant measure ν for $\theta = \theta_0$. Let

$$Y(\boldsymbol{\Gamma}, \theta) = \int_{R^d} \boldsymbol{a}'(\theta, \boldsymbol{x})\bar{\boldsymbol{B}}(\boldsymbol{x}, \boldsymbol{\Gamma})\{\boldsymbol{a}(\theta_0, \boldsymbol{x}) - \tfrac{1}{2}\boldsymbol{a}(\theta, \boldsymbol{x})\}\, \nu(d\boldsymbol{x}). \tag{3.3.10}$$

Suppose the following regularity conditions hold:

(DD1) There exists a constant L such that

$$\|\boldsymbol{a}(\theta_0, \boldsymbol{x})\| + \|\boldsymbol{b}(\boldsymbol{x})\| \leq L(1 + \|\boldsymbol{x}\|).$$

(DD2) There exists a constant L such that

$$\|\boldsymbol{a}(\boldsymbol{x}, \theta_0) - \boldsymbol{a}(\boldsymbol{y}, \theta_0)\| + \|\boldsymbol{b}(\boldsymbol{x}) - \boldsymbol{b}(\boldsymbol{y})\| \leq L\|\boldsymbol{x} - \boldsymbol{y}\|.$$

(DD3) $\inf_{\boldsymbol{x}} \det(\boldsymbol{b}\boldsymbol{b}')(\boldsymbol{x}) > 0$.

(DD4) $\sup_t E\|\boldsymbol{X}_t\|^p < \infty$ for each $p > 0$.

(DD5) The function $\boldsymbol{\Gamma} \to \bar{\boldsymbol{B}}(\boldsymbol{x}, \boldsymbol{\Gamma})$ is Hölder continuous in a neighborhood U of $\boldsymbol{\Gamma}_0$ in S_+^k, the set of $k \times k$ symmetric nonnegative definite matrices endowed with the relative topology, that is, there exist $\alpha > 0$ and $C > 0$ such that

$$\|\bar{\boldsymbol{B}}(\boldsymbol{x}, \boldsymbol{\Gamma}_2) - \bar{\boldsymbol{B}}(\boldsymbol{x}, \boldsymbol{\Gamma}_1)\| \leq C(1 + \|\boldsymbol{x}\|^C)\|\boldsymbol{\Gamma}_2 - \boldsymbol{\Gamma}_1\|^\alpha,$$

for all $\boldsymbol{\Gamma}_1, \boldsymbol{\Gamma}_2$ in U and all $\boldsymbol{x} \in R^d$.

(DD6) $\boldsymbol{a}(\theta, \boldsymbol{x})$ is twice differentiable in $\theta \in \bar{\Theta}$ and

$$\|\delta\boldsymbol{a}(\theta, \boldsymbol{x})\| + \|\delta^2\boldsymbol{a}(\theta, \boldsymbol{x})\| \leq C(1 + \|\boldsymbol{x}\|^C).$$

(DD7) The function $\theta \to Y(\boldsymbol{\Gamma}_0, \theta)$ has its unique minimum at $\theta = \theta_0$ in $\bar{\Theta}$.

(DD8) The functions $\boldsymbol{a}$, $\delta\boldsymbol{a}$, $\boldsymbol{b}$ and $\bar{\boldsymbol{B}}$ are smooth in $\boldsymbol{x}$ and their derivatives are of polynomial growth order in $\boldsymbol{x}$ uniformly in θ or $\boldsymbol{\Gamma}$.

(DD9) The matrix

$$\boldsymbol{\Phi} = \int_{R^d} \delta\boldsymbol{a}(\theta_0, \boldsymbol{x})'\bar{\boldsymbol{B}}(\boldsymbol{x}, \boldsymbol{\Gamma}_0)\delta\boldsymbol{a}(\theta_0, \boldsymbol{x})\, \nu(d\boldsymbol{x}) \tag{3.3.11}$$

is positive definite.

Consider the estimator $\hat{\boldsymbol{\Gamma}}_0$ of $\boldsymbol{\Gamma}$ defined by

$$\hat{\boldsymbol{\Gamma}}_0 = (hN)^{-1} \sum_{i=1}^{N} \boldsymbol{B}_{i-1}\bar{\boldsymbol{\Delta}}_i\bar{\boldsymbol{\Delta}}_i'\boldsymbol{B}_{i-1}'. \tag{3.3.12}$$

Yoshida (1992) proved the following results under conditions (DD1)–(DD4).

Lemma 3.3.1 *Under the conditions (DD1)–(DD4),*

$$E\|\hat{\boldsymbol{\Gamma}}_0 - \boldsymbol{\Gamma}_0\| \leq C(h^{1/2} + N^{-1/2}). \tag{3.3.13}$$

In particular, it follows that $\hat{\boldsymbol{\Gamma}}_0$ is consistent for $\boldsymbol{\Gamma}_0$ as $h \to 0$ and $N \to \infty$.

Lemma 3.3.2 *Under the conditions (DD1)–(DD7), $\hat{\theta}_0 \xrightarrow{p} \theta_0$ as $h \to 0$ and $N \to \infty$ such that $hN \to \infty$. Moreover, if $h^3N \to 0$, then $(hN)^{1/4}(\hat{\theta}_0 - \theta_0) \xrightarrow{p} 0$.*

Let

$$\hat{\boldsymbol{\Gamma}}_1 = \frac{1}{hN}\sum_{i=1}^{N} \boldsymbol{B}_{i-1}\boldsymbol{\Delta}_i(\hat{\theta}_0)\,\boldsymbol{\Delta}_i(\hat{\theta}_0)'\boldsymbol{B}'_{i-1} \tag{3.3.14}$$

and

$$\hat{\boldsymbol{\Gamma}} = \hat{\boldsymbol{\Gamma}}_1 - \frac{h}{2N}\sum_{i=1}^{N}[\boldsymbol{U}_{i-1} + \boldsymbol{V}_{i-1} + \boldsymbol{V}'_{i-1}], \tag{3.3.15}$$

where

$$\boldsymbol{U}_{i-1} = \boldsymbol{B}_{i-1}\boldsymbol{F}(\boldsymbol{X}_{t_{i-1}}, \hat{\boldsymbol{\Gamma}}_0, \hat{\theta}_0)\boldsymbol{B}'_{i-1},$$

$$\boldsymbol{V}_{i-1} = \boldsymbol{B}_{i-1}\partial\boldsymbol{a}(\boldsymbol{X}_{t_{i-1}}, \hat{\theta}_0)\boldsymbol{b}(\boldsymbol{X}_{t_{i-1}})\hat{\boldsymbol{\Gamma}}_0,$$

$$\boldsymbol{F}(\boldsymbol{x}, \boldsymbol{s}, \theta) = ((F_{\lambda\mu})),\ F_{\lambda\mu} = L_{\boldsymbol{\Gamma},\theta}((\boldsymbol{b}\boldsymbol{\Gamma}\boldsymbol{b}')_{\lambda\mu}), \qquad 1 \le \lambda, \mu \le d,$$

and $L_{\boldsymbol{\Gamma},\theta}$ is the generator corresponding to $\boldsymbol{\Gamma}$ and θ.

Yoshida (1992) studied the weak convergence of the likelihood ratio random field

$$Z_{h,N}(\boldsymbol{\Gamma}, \boldsymbol{u}) = Q_{h,N}(\boldsymbol{\Gamma}, \theta_0 + (hN)^{-1/2}\boldsymbol{u})/Q_{h,N}(\boldsymbol{\Gamma}, \theta_0), \tag{3.3.16}$$

where Q is as defined by (3.3.3). He proved the following theorem. We omit the proof in order to save space.

Theorem 3.3.3 *Under the conditions stated above, the sequence* $(N^{1/2}(\hat{\boldsymbol{\Gamma}} - \boldsymbol{\Gamma}_0), Z_{h,N}(\hat{\boldsymbol{\Gamma}}, \cdot))$ *converges to* $(\boldsymbol{H}, Z(\boldsymbol{\Gamma}_0, \cdot))$ *in distribution, where* $\boldsymbol{H} = ((H_{pq})) \in R^k \otimes R^k$ *is a multivariate normal random matrix with mean zero and*

$$\operatorname{cov}(H_{pq}, H_{st}) = \Gamma_{ps}\Gamma_{qt} + \Gamma_{pt}\Gamma_{qs},\ \boldsymbol{\Gamma}_0 = ((\Gamma_{ps})),$$

and where

$$Z(\boldsymbol{\Gamma}_0, \boldsymbol{u}) = \exp\{\boldsymbol{u}'\boldsymbol{\Delta} - \tfrac{1}{2}\boldsymbol{u}'\boldsymbol{\Phi}\boldsymbol{u}\} \tag{3.3.17}$$

in which $\boldsymbol{\Delta}$ *is* $N_m(0, \boldsymbol{\Phi})$ *independent of* $\boldsymbol{H}$ *and is given by (3.3.11) provided* $h \to 0$, $hN \to \infty$, *and* $h^3N = o(1)$.

Remark. Note that under the conditions stated in Theorem 3.3.3, for any continuous functional f on $(E_0, \mathcal{B})$, where $E_0 = (R^k \otimes R^k) \times C_0(R^m)$ and $\mathcal{B}$ is the Borel σ-algebra under the product topology on E_0,

$$E[f(N^{1/2}(\hat{\boldsymbol{\Gamma}} - \boldsymbol{\Gamma}_0), Z_{hN}(\hat{\boldsymbol{\Gamma}}, \cdot))] \to E[f(\boldsymbol{H}, Z(\boldsymbol{\Gamma}_0, \cdot))] \tag{3.3.18}$$

as $h \to 0, hN \to \infty$ and $h^3N = o(1)$. In particular, for the MLE $\hat{\theta}$ corresponding to $Q_{h,N}(\hat{\boldsymbol{\Gamma}}, \cdot)$,

$$(N^{1/2}(\hat{\boldsymbol{\Gamma}} - \boldsymbol{\Gamma}_0), (hN)^{1/2}(\hat{\theta} - \theta_0)) \xrightarrow{\mathcal{L}} (\boldsymbol{H}, \boldsymbol{\Phi}^{-1}\boldsymbol{\Delta}), \tag{3.3.19}$$

as $h \to 0, hN \to \infty$, with $h^3N = o(1)$.

Example 3.3.1

Let

$$\boldsymbol{X}_t = \begin{pmatrix} X_t \\ \dot{X}_t \end{pmatrix}, \qquad \boldsymbol{K} = \begin{pmatrix} 0 & -1 \\ w^2 & \alpha \end{pmatrix}, \qquad \boldsymbol{b} = \begin{pmatrix} 0 \\ 1 \end{pmatrix}$$

where α, w are positive constants such that $w^2 - \alpha^2/4 > 0$. Consider the two-dimensional diffusion

$$dX_t = -KX_t\,dt + b\sigma\,dW_t \tag{3.3.20}$$

where $\sigma > 0$. It is known that X_t is ergodic (Arnold and Kleimann 1987). Its invariant measure is the bivariate normal distribution with mean zero and covariance matrix diag$(\sigma^2/(2\alpha w^2)$, $\sigma^2/(2\alpha))$. The MLE $(\hat{\alpha}, \hat{w}, \hat{\sigma}^2)$ is asymptotically normal with covariance matrix diag$(2\alpha$, $w\alpha w^2, w\sigma^4)$.

Remarks. Let us again consider the problem of estimating the unknown parameter $\theta \in \Theta \subset R^p$ in the stochastic differential equation

$$dX_t = a(t, X_t; \theta)\,dt + \sigma(t, X_t; \theta)\,dW_t, \qquad X_0 = x_0,\ t \geq 0, \tag{3.3.21}$$

where W_t is an r-dimensional Wiener process, $a : [0, \infty) \times R^d \to R^d$ and $\sigma : [0, \infty) \times R^d \to m^{d \times r}$ the set of $d \times r$ matrices. Suppose the process X_t is observed at the discrete time points

$$0 = t_0 < t_1 < \cdots < t_n.$$

If the transition densities $p(s, x; t, y; \theta)$ are known, then θ can be estimated by the maximum likelihood method by maximizing the log-likelihood function

$$\ell_n(\theta) = \sum_{i=1}^{n} \log(p(t_{i-1}, X_{t_{i-1}}; t_i, X_{t_i}; \theta)).$$

Sufficient conditions for the consistency and the asymptotic normality of this maximum likelihood estimator $\hat{\theta}_n$ are known (Billingsley 1961; Prakasa Rao 1972, 1983b). However, the transition densities of the process X are usually unknown or difficult to compute. The approach is to approximate the transition densities p by p_N, calculate an approximate likelihood function $\ell_{n,N}(\theta) = \sum_{i=1}^{n} \log p_N(t_{i-1}, X_{t_{i-1}}; t_i X_{t_i}; \theta)$ and then maximize the approximate likelihood function $\ell_{n,N}(\theta)$ to obtain an estimator $\hat{\theta}_{nN}$ whose properties are as good as those of $\hat{\theta}_n$. We sketch briefly some results due to Pedersen (1995a; 1995b).

Consider the stochastic differential equation (3.3.21). Suppose the stochastic differential equation has a unique solution in law for all $x_0 \in R^d$ and $\theta \in \Theta$. Conditions that ensure this property are given in Rogers and Williams (1987) and Stroock and Varadhan (1979). Sufficient conditions are local Lipschitz and growth conditions for each $\theta \in \Theta$. We assume further that $\sigma(t, x; \theta)$ is positive definite for all $t \geq 0$, $x \in R^d$ and $\theta \in \Theta$. Let $\sigma(t, x; \theta)^{1/2}$ denote the positive definite square root of $\sigma(t, x; \theta)$. Under these assumptions, any solution of (3.3.21) is also a solution of the stochastic differential equation

$$dX_t = a(t, X_t; \theta)\,dt + \sigma(t, X_t; \theta)^{1/2}\,d\tilde{W}_t, \qquad X_0 = x_0,\ t \geq 0, \tag{3.3.22}$$

where

$$\tilde{W}_t = \int_0^t \sigma(s, X_s; \theta)^{-1/2}\, d\left(X_s - x_0 - \int_0^s a(u, X_u; \theta)\,du\right), \qquad t \geq 0,$$

is a d-dimensional Wiener process. Let P_θ be the probability measure induced by $\{X_t\}$ when θ is the parameter on the space $C([0, \infty), R^d)$ of continuous functions from $[0, \infty)$ to R^d endowed with its Borel σ-algebra $\mathcal{B}$. We assume that $P_{\theta_1} \equiv P_{\theta_2}$ implies $\theta_1 = \theta_2$. Due to the positive definiteness of $\sigma(t, x; \theta)$, a solution to (3.3.22) can be realized on the probability space $(C([0, \infty), R^d, \mathcal{B}, P_\theta)$ since

$$W_t^\theta = \int_0^t \sigma(s, X_s; \theta)^{-1/2} d\left(X_s - x_0 - \int_0^s a(u, X_u; \theta)\, du\right), \qquad t \geq 0,$$

is a d-dimensional Wiener process under P_θ and

$$X_t = x_0 + \int_0^t a(s, X_s; \theta)\, ds + \int_0^t \sigma(s, X_s; \theta)^{1/2}\, dW_s^\theta, \qquad t \geq 0.$$

In general, for each $\theta \in \Theta$, there exists a unique family of probability measures $\{P_{\theta,s,x}; s \geq 0, x \in R^d)$ on $(C([0, \infty), R^d), \mathcal{B})$ induced by the solution to (3.3.21) and (3.3.22) for $t \geq s$ with the initial condition $X_s = x$ (Friedman 1975; Stroock and Varadhan 1979) such that for each $s \geq 0, x \in R^d$,

$$P_{\theta,s,x}(X_u = x, 0 \leq u \leq s) = 1$$

and, under $P_{\theta,s,x}$,

$$X_t = x + \int_s^t a(u, X_u; \theta)\, du + \int_s^t \sigma(u, x_u; \theta)^{1/2}\, dW_u^{\theta,s}, \qquad t \geq s,$$

where

$$W_t^{\theta,s} = \int_s^t \sigma(u, X_u; \theta)^{-1/2} d\left(X_u - x - \int_s^u a(v, X_v; \theta)\, dv\right), \qquad t \geq s,$$

is a d-dimensional Wiener process after time s. Note that $P_{\theta,s,x}$ determines the transition function $P(s, x; t, A; \theta)$ of X under P_θ. For $0 \leq s < t, x \in R^d, \theta \in \Theta$ and $A \in \mathcal{B}(R^d)$,

$$P(s, x; t, A; \theta) = P_{\theta,s,x}(X_t \in A) = P_\theta(X_t \in A | X_s = x),$$

where $P_\theta(\cdot | X_s = x)$ is the conditional probability under P_θ given $X_s = x$.

Consider the following Euler–Maruyama approximation of X_t under $P_{\theta,s,x}$ (Kloeden and Platen 1992). For $k = 0, 1, \ldots, N$, define

$$\begin{aligned} \tau_k &= s + k\frac{t-s}{N}, \\ Y_{s,N} &= x, \\ Y_{\tau_k,N} &= Y_{\tau_{k-1},N} + \frac{t-s}{N} a(\tau_{k-1}, Y_{\tau_{k-1},N}; \theta) \\ &\quad + \sigma(\tau_{k-1}, Y_{\tau_{k-1},N}; \theta)^{1/2}(W_{\tau_k}^{\theta,s} - W_{\tau_{k-1}}^{\theta,s}). \end{aligned} \tag{3.3.23}$$

Under local Lipschitz and growth conditions, Kloden and Platen (1992) proved that

$$Y_{\tau_N,N} = Y_{t,N} \to X_t \tag{3.3.24}$$

in $L^1(P_{\theta,s,x})$ as $N \to \infty$. We define $y \mapsto p_N(s, x; t, y; \theta)$ to be the density (with respect to the Lebesgue measure on R^d) of the distribution of $Y_{t,N}$ under $P_{\theta,s,x}$ and take it as an approximation to the density $p(s, x; t, y; \theta)$.

Theorem 3.3.4 *For fixed* $0 \le s \le t$, $\boldsymbol{x} \in R^d$, $\theta \in \Theta$ *and* $N \ge 1$, *the distribution of* $\boldsymbol{Y}_{t,N}$ *under* $P_{\theta,s,\boldsymbol{x}}$ *has a density* $p_N(s, \boldsymbol{x}; t, \boldsymbol{y}; \theta)$ *with respect to the Lebesgue measure on* R^d. *For* $N = 1$,

$$
\begin{aligned}
p_1(s, \boldsymbol{x}; t, \boldsymbol{y}; \theta) = {} & (2\pi(t-s))^{-d/2} |\boldsymbol{\sigma}(s, \boldsymbol{x}; \theta)|^{-1/2} \\
& \times \exp\left\{ -\frac{1}{2(t-s)} [\boldsymbol{y} - \boldsymbol{x} - (t-s)\boldsymbol{a}(s, \boldsymbol{x}; \theta)]' \boldsymbol{\sigma}(s, \boldsymbol{x}; \theta)^{-1} \right. \\
& \left. \times [\boldsymbol{y} - \boldsymbol{x} - (t-s)\boldsymbol{a}(s, \boldsymbol{x}; \theta)] \right\}, \qquad (3.3.25)
\end{aligned}
$$

where $|\boldsymbol{\sigma}(s, \boldsymbol{x}; \theta)|$ *denotes the determinant of* $\boldsymbol{\sigma}(s, \boldsymbol{x}; \theta)$; *for* $N \ge 2$,

$$
\begin{aligned}
p_N(s, \boldsymbol{x}; t, \boldsymbol{y}; \theta) &= \int_{R^{d(N-1)}} \left\{ \prod_{k=1}^{N} p_1(\tau_{k-1}, \zeta_{k-1}; \tau_k, \zeta_k; \theta) \right\} d\zeta_1 \cdots d\zeta_{N-1} \\
&= E_{P_{\theta,s,\boldsymbol{x}}}(p_1(\tau_{N-1}, \boldsymbol{Y}_{\tau_{N-1},N}; t, \boldsymbol{y}; \theta)), \qquad (3.3.26)
\end{aligned}
$$

with $\zeta_0 = \boldsymbol{x}$ *and* $\zeta_N = \boldsymbol{y}$.

Remarks. We omit the proof of Theorem 3.3.4 (see Pedersen 1995a, 1995b). It is a consequence of the Chapman–Kolmogorov equation for the Markov chain $\{\boldsymbol{Y}_{\tau_{k,N}}, 0 \le k \le N\}$ under $P_{\theta,s,\boldsymbol{x}}$. Expression (3.3.26) gives a method for calculation of $\ell_{n,N}(\theta)$ by means of simulations of $\boldsymbol{Y}_{\tau_{N-1},N}$ under $P_{\theta,s,\boldsymbol{x}}$ (see Pedersen 1995a, 1995b).

In general it is difficult to prove pointwise convergence of $p_N(s, \boldsymbol{x}; t, \boldsymbol{y}; \theta)$ to $p(s, \boldsymbol{x}; t, \boldsymbol{y}; \theta)$ as $N \to \infty$. However, Pedersen (1995a, 1995b) has proved that if $\boldsymbol{\sigma}(t, \boldsymbol{x}, \theta) \equiv \boldsymbol{\sigma}(\theta)$ is independent of $t, \boldsymbol{x}$, then $p_N(s, \boldsymbol{x}; t, \boldsymbol{y}; \theta)$ converges to $p(s, \boldsymbol{x}; t, \boldsymbol{y}; \theta)$ in $L^1(R^d)$ if $\boldsymbol{a}(t, \boldsymbol{x}; \theta)$ is continuous in θ for every $(t, \boldsymbol{x}) \in R^+ \times R^d$ (in addition to the local Lipschitz and growth conditions) and $\boldsymbol{\sigma}(\theta)$ is positive definite. If $\boldsymbol{\sigma}(t, \boldsymbol{x}; \theta)$ is allowed to depend on t and (or) $\boldsymbol{x}$, then the result continues to hold under some stronger conditions; see Theorem 1 and Theorem 3 of Pedersen (1995a, 1995b). As a consequence we obtain that $\ell_{n,N}(\theta) \to \ell_n(\theta)$ in probability under P_{θ_0} as $N \to \infty$ for all $\theta \in \Theta$ and $n \ge 1$, where θ_0 is the true parameter.

Example 3.3.2
Consider the Ornstein–Uhlenbeck process satisfying the stochastic differential equation

$$
dX_t = \theta X_t\, dt + \sigma\, dW_t, \qquad X_0 = x_0,\ t \ge 0,
$$

with $(\theta, \sigma) \in (-\infty, 0) \times (0, \infty)$. Suppose the process is observed at the time instants $t_i = i\Delta$, $i = 0, 1, 2, \ldots$, for some $\Delta > 0$. It can be shown that the MLEs of θ and σ based on the Markov chain $X_0, \ldots, X_{n\Delta}$ are

$$
\hat{\theta}_n = \frac{1}{\Delta} \log \left(\frac{\sum_{i=1}^n X_{(i-1)\Delta} X_{i\Delta}}{\sum_{i=1}^n X_{(i-1)\Delta}^2} \right)
$$

and

$$
\hat{\sigma}_n^2 = -\frac{2\hat{\theta}_n}{n(1 - e^{-2\Delta\hat{\theta}_n})} \sum_{i=1}^n (X_{i\Delta} - X_{(i-1)\Delta} e^{\Delta\hat{\theta}_n})^2, \qquad (3.3.27)
$$

and they are consistent and asymptotically normal. Furthermore, $\ell_{n,N}(\theta)$ can be explicitly computed in this case, since

$$p_N(0, x; t, y; \theta) = \left(2\pi\sigma^2 \frac{t}{N} \sum_{k=1}^{N} \left(1 + \frac{t}{N}\theta\right)^{2(N-k)}\right)^{-1/2}$$
$$\times \exp\left(-\frac{[y - x(1 + \frac{t}{N}\theta)^N]^2}{2\sigma^2 \frac{t}{N} \sum_{k=1}^{N}(1 + \frac{t}{N}\theta)^{2(N-k)}}\right).$$

Pedersen (1995a, 1995b) compared the estimator $(\hat{\theta}_n, \hat{\sigma}_n^2)$ obtained above with the maximum likelihood estimators $\hat{\theta}_{n,N}$ and $\hat{\sigma}_{n,N}^2$ obtained from maximizing the likelihood function $\ell_{n,N}(\theta)$ based on $p_N(0, x; t, y; \theta)$. Let $\ell_n(\theta)$ be the likelihood function based on $(X_0, \ldots, X_{n\Delta})$. If $\ell_{n,N}(\theta) \to \ell_n(\theta)$ uniformly in θ in probability under P_{θ_0} as $N \to \infty$, then it follows that $\hat{\theta}_{n,N} \to \hat{\theta}_n$ in probability under P_{θ_0} as $N \to \infty$ and $\hat{\theta}_{\hat{n},N}$ will be consistent and asymptotically normal. In cases where there is no uniform convergence of $\ell_{n,N}(\theta)$ to $\ell_n(\theta)$ or where there is no convergence in probability of $\hat{\theta}_{n,N}$ to $\hat{\theta}_n$ or the latter convergence cannot be proved, one can still obtain consistency and asymptotic normality of $\hat{\theta}_{n,N}$ from general results proved in Pedersen (1995a, 1995b). We do not present the details here.

3.4 Other Methods of Estimation via Numerical Approximation Schemes

Let us again consider the stochastic differential equation

$$dX_t = a(X_t, \theta)\, dt + \sigma(X_t)\, dW_t, \qquad t \geq 0, \tag{3.4.1}$$

where $a(x, \cdot)$ and $\sigma(x)$ are twice continuously differentiable with respect to x.

For the estimation of the parameters by the maximum likelihood method, we have seen that there are two approches to the problem depending on the likelihood used. In the first approach, after discretizing an original continuous-time process, the likelihood function is derived from the discretized process. Let us call this approach the pseudo-likelihood approach. In the second approach, the likelihood function is the exact one derived from the original continuous-time process and the parameters are estimated so as to maximize the likelihood function. We call this the likelihood function approach.

3.4.1 Pseudo-likelihood Approach

Estimation Based on Discretization by the Euler Method

In this method, (3.4.1) is discretized under the assumption that the drift and diffusion are constant over the interval $[t, t + \Delta t)$. Then

$$X_{t+\Delta t} - X_t = a(X_t, \theta)\Delta t + \sigma(X_t)(W_{t+\Delta t} - W_t). \tag{3.4.2}$$

This discretized process is assumed to be a local approximation of the original process. Since $W_{t+\Delta t} - W_t$ is $N(0, \sigma^2(X_t)\Delta t)$, the transition density function $p(x_{t+\Delta t}|x_t)$ of the discretized process is

$$p(x_{t+\Delta t}|x_t) = \frac{1}{\sqrt{2\pi\sigma^2(x_t)\Delta t}} \exp\left\{-\frac{1}{2}\frac{(x_{t+\Delta t} - x_t - a(x_t, \theta)\Delta t)^2}{\sigma^2(x_t)\Delta t}\right\}. \tag{3.4.3}$$

Let $X_{(i)} = X_{t+i\Delta t}, 0 \le i \le N$. The joint density of $(X_{(0)}, \ldots, X_{(N)})$ is given by

$$p(x_{(0)}, \ldots, x_{(N)}) = \prod_{i=1}^{N} p(x_{(i)}|x_{(i-1)}p(x_{(0)}), \tag{3.4.4}$$

and the parameters of the original process may be estimated by maximizing the log-likelihood given by

$$\log p(x_{(0)}, \ldots, x_{(N)}) = -\frac{1}{2}\sum_{i=1}^{N}\left\{\frac{(x_{(i)} - x_{(i-1)} - a(x_{(i-1)}, \theta)\Delta t)^2}{\sigma^2(x_{(i-1)})\Delta t} + \log(2\pi\sigma^2(x_{(i-1)})\Delta t)\right\} + \log p(x_{(0)}), \tag{3.4.5}$$

where $t_i = t + i\Delta t$ and $x_{(i)} = x_{t+i\Delta t}$.

Consistency of the Euler estimator Let us consider the one-dimensional stochastic differential equation

$$dX_t = a(X_t; \theta)\, dt + \sigma\, dW_t, \qquad X_0 = X_0,\ t \ge 0, \tag{3.4.6}$$

where θ is a vector, $\sigma > 0$ is a constant and $\{W_t\}$ is the standard Wiener process.

Suppose the following conditions hold:

(EE1) Growth condition: there exists $L > 0$ and a positive constant m independent of θ such that $|a(x; \theta)| \le L(1 + |x|^m), \theta \in \Theta$.

(EE2) $\sup_t E|X_t|^p < \infty$ for every $p \ge 1$.

Suppose the process $\{X_t\}$ is observed at the discrete times

$$0 = t_0 < t_1 < \cdots < t_N, \qquad \text{with } \Delta t = t_i - t_{i-1}.$$

Let X_i denote $X_{t_i}, 0 \le i \le N$, for simplicity.

Consider the maximum likelihood estimator of the parameter θ based on the likelihood function. Then the likelihood

$$\exp\left[\sum_{i=1}^{N} a(X_{i-1}; \theta)\sigma^{-2}(X_i - X_{i-1}) - \tfrac{1}{2}\sum_{i=1}^{N} a(X_{i-1}; \theta)^2\sigma^{-2}\Delta t\right] \tag{3.4.7}$$

has to be maximized with respect to θ, or equivalently the function

$$\sum_{i=1}^{N} a(X_{i-1}; \theta)\sigma^{-2}(X_i - X_{i-1}) - \tfrac{1}{2}\sum_{i=1}^{N} a(X_{i-1}, \theta)^2\sigma^{-2}\Delta t \tag{3.4.8}$$

has to be maximized.

Since σ is a constant, the problem reduces to maximizing the discrete approximate likelihood function given by

$$\text{DLF}(\theta) \equiv \sum_{i-1}^{N} a(X_{i-1}; \theta)(X_i - X_{i-1}) - \tfrac{1}{2}\sum_{i=1}^{N} a(X_{i-1}; \theta)^2\Delta t. \tag{3.4.9}$$

Let

$$\hat{\theta}_{\mathrm{LF}} = \arg\max_{\theta\in\Theta} \mathrm{DLF}(\theta). \tag{3.4.10}$$

It was shown earlier (Section 3.3; or Yoshida 1992) that $\hat{\theta}_{\mathrm{LF}}$ has consistency and asymptotic normality under some conditions. In addition, Yoshida (1992) and Florens-Zmirou (1989) showed that

$$\hat{\sigma}^2_{\mathrm{LF}} = \frac{1}{N\Delta t}\sum_{i=1}^{N}(X_i - X_{i-1})^2 \tag{3.4.11}$$

is a consistent estimator for σ^2.

Let us now consider the least-squares estimator $\hat{\theta}_{\mathrm{LSQ}}$ of θ defined by

$$\mathrm{LSQ}(\theta) = \sum_{i=1}^{N}(X_i - X_{i-1} - a(X_{i-1};\theta)\Delta t)^2 \tag{3.4.12}$$

and

$$\hat{\theta}_{\mathrm{LSQ}} = \arg\max_{\theta\in\Theta} \mathrm{LSQ}(\theta). \tag{3.4.13}$$

The consistency and the asymptotic normality of this estimator are studied in Dorogovcev (1976) and Prakasa Rao (1983b).

Consider the Euler estimator $(\hat{\theta}, \hat{\sigma}^2)$ of (θ, σ^2) derived from the likelihood function of the process discretized by the Euler method. This estimator is defined by

$$\log \mathrm{Euler}(\theta, \sigma^2) \equiv -\frac{1}{2}\sum_{i=1}^{N}\left\{\frac{(X_i - X_{i-1} - a(X_{i-1};\theta)\Delta t)^2}{\sigma^2\Delta t} + \log(2\pi\sigma^2\Delta t)\right\} \tag{3.4.14}$$

and

$$(\hat{\theta}, \hat{\sigma}^2) = \arg\max_{(\theta,\sigma^2)\in\Theta\times\Sigma} \log \mathrm{Euler}(\theta, \sigma^2), \tag{3.4.15}$$

where $\Theta \times \Sigma$ is the parameter space. Observe that

$$\begin{aligned}
\hat{\theta}_{\mathrm{LF}} &= \arg\max_{\theta\in\Theta} \mathrm{DLF}(\theta)\\
&= \arg\max_{\theta\in\Theta}\{\mathrm{DLF}(\theta)\Delta t\}\\
&= \arg\max_{\theta\in\Theta}\left\{\mathrm{DLF}(\theta)\Delta t - \tfrac{1}{2}\sum_{i=1}^{n}(X_i - X_{i-1})^2\right\}\\
&= \arg\max_{\theta\in\Theta}\{-\tfrac{1}{2}\,\mathrm{LSQ}(\theta)\}\\
&= \arg\min_{\theta\in\Theta}\{\mathrm{LSQ}(\theta)\}\\
&= \hat{\theta}_{\mathrm{LSQ}}.
\end{aligned} \tag{3.4.16}$$

Let us now look at the relationship between $\hat{\theta}_{\mathrm{LSQ}}$ and $\hat{\theta}$. Let

$$\begin{aligned}
\sigma^2_*(\theta) &= \frac{1}{N\Delta t}\sum_{i=1}^{N}(X_i - X_{i-1} - a(X_{i-1};\theta)\Delta t)^2\\
&= \frac{1}{N\Delta t}\,\mathrm{LSQ}(\theta).
\end{aligned} \tag{3.4.17}$$

Then $(\hat{\theta}, \hat{\sigma}^2) = (\hat{\theta}, \sigma_*^2(\hat{\theta}))$ and

$$\begin{aligned}\hat{\theta} &= \arg\max_{\theta\in\Theta} \log \text{Euler}(\theta, \sigma_*^2(\theta)) \\ &= \arg\max_{\theta\in\Theta} \left\{ -\frac{1}{2} - \frac{N}{2} \log\left(\frac{2\pi\, \text{LSQ}(\theta)}{N} \right) \right\}. \end{aligned} \tag{3.4.18}$$

These relations prove that $\hat{\theta}_{\text{LSQ}}$ and $\hat{\theta}$ are the same. Hence all three estimators $\hat{\theta}_{\text{LSQ}}, \hat{\theta}$ and $\hat{\theta}_{\text{LF}}$ are the same. We now prove the consistency of the estimator

$$\hat{\sigma}^2 \equiv \frac{1}{N\Delta t} \sum_{i=1}^{N} (X_i - X_{i-1} - a(X_{i-1}; \hat{\theta})\Delta t)^2 \tag{3.4.19}$$

in the L_2 norm as $\Delta t \to 0$ and $N\Delta t \to \infty$.

Theorem 3.4.1 *If $\Delta t \to 0$ and $N\Delta t \to \infty$, then*

$$\hat{\sigma}^2 = \frac{1}{N\Delta t} \sum_{i=1}^{N} (X_i - X_{i-1} - a(X_{i-1}; \hat{\theta})\Delta t)^2 \to \sigma^2 \tag{3.4.20}$$

in L_2-mean.

We first prove a lemma.

Lemma 3.4.2 *Suppose that $\{X_t\}$ is a solution of the stochastic differential equation (3.4.6) and that conditions (EE1) and (EE2) hold. Then, for any integer $n \geq 1$,*

$$E|X_t - X_s|^{2n} = O(|t-s|^n). \tag{3.4.21}$$

Proof. Without loss of generality, suppose that $0 \leq s \leq t$. Then

$$\begin{aligned}|X_s - X_t|^{2n} &= \left| \int_s^t a(X_u; \theta)\, du + \int_s^t \sigma\, dW_u \right|^{2n} \\ &\leq 2^{2n-1} \left(\left| \int_s^t a(X_u; \theta)\, du \right|^{2n} + \left| \int_s^t \sigma\, dW_u \right|^{2n} \right) \\ &\leq 2^{2n-1} \left\{ (t-s)^{2n-1} \int_s^t |a(X_u; \theta)|^{2n}\, du + \sigma^{2n} \left| \int_s^t dW_u \right|^{2n} \right\}\end{aligned}$$

by Hölder's inequality. Since

$$|a(X_u; \theta)|^{2n} \leq L^{2n} 2^{2n-1} (1 + |X_u|^{2mn}),$$

and $E|X_t|^{2mn} < \infty$, it follows that

$$E\left[\int_s^t |a(X_u; \theta)|^{2n}\, du \right] = O(|t-s|).$$

It is well known that

$$E[|W_t - W_s|^{2n}] \leq C|t-s|^n.$$

for some constant $C > 0$. Hence

$$E|X_s - X_t|^{2n} = O(|t-s|^n).$$

Proof (of Theorem 3.4.1). Note that

$$\begin{aligned}
&(X_i - X_{i-1} - a(X_{i-1};\hat{\theta})\Delta t)^2 \\
&= (X_i - X_{i-1})^2 - 2(X_i - X_{i-1})\, a(X_{i-1};\hat{\theta})\Delta t + [a(X_{i-1};\hat{\theta})\Delta t]^2.
\end{aligned}$$

Following Itô's lemma,

$$\begin{aligned}
(X_i - X_{i-1})^2 = 2\int_{t_{i-1}}^{t_i} &(X_u - X_{i-1})a(X_u;\theta)\,du \\
+2\int_{t_{i-1}}^{t_i} &(X_u - X_{i-1})\sigma\, dW_u + \sigma^2 \Delta t.
\end{aligned}$$

Therefore

$$\begin{aligned}
|\hat{\sigma}^2 - \sigma^2| = \Bigg| &-\frac{2}{N\Delta t}\sum_{i=1}^{N}(X_i - X_{i-1})a(X_{i-1};\hat{\theta})\Delta t \\
&+\frac{1}{N\Delta t}\sum_{i=1}^{N}[a(X_{i-1};\hat{\theta})\Delta t]^2 \\
&+\frac{2}{N\Delta t}\sum_{i=1}^{N}\int_{t_{i-1}}^{t_i}(X_u - X_{i-1})a(X_u;\theta)\,du \\
&+\frac{2}{N\Delta t}\sum_{i=1}^{N}\int_{t_{i-1}}^{t_i}(X_u - X_{i-1})\sigma\, dW_u\Bigg|^2 \\
\le &\frac{16}{N(\Delta t)^2}\sum_{i=1}^{N}|(X_i - X_{i-1})a(X_{i-1};\hat{\theta})\Delta t|^2 \\
&+\frac{4}{N(\Delta t)^2}\sum_{i=1}^{N}|a(X_{i-1};\theta)\Delta t|^4 \\
&+\frac{16}{N(\Delta t)^2}\sum_{i=1}^{N}\left|\int_{t_{i-1}}^{t_i}(X_u - X_{i-1})a(X_u;\theta)\,du\right|^2 \\
&+\frac{16\sigma^2}{N(\Delta t)^2}\left|\sum_{i=1}^{N}\int_{t_{i-1}}^{t_i}(X_i - X_{i-1})\,dW_u\right|^2. \qquad (3.4.22)
\end{aligned}$$

Applying the Cauchy–Schwarz inequality, we have

$$\begin{aligned}
&E|(X_i - X_{i-1})a(X_{i-1};\hat{\theta})\Delta t|^2 \\
&\le (E|X_i - X_{i-1}|^4)^{1/2}(E|a(X_{i-1};\hat{\theta})\Delta t|^4)^{1/2}. \qquad (3.4.23)
\end{aligned}$$

Lemma 3.4.1 shows that

$$E|X_i - X_{i-1}|^4 = O((\Delta t)^2). \qquad (3.4.24)$$

From the growth condition on $a(\cdot,\theta)$ and the boundedness of $E|X_t|^p$ as a function of t, it follows that

$$E|a(X_t,\theta)|^4 \le 8L^4(1 + \sup_t E|X_t|^{4m}) < \infty. \qquad (3.4.25)$$

Relations (3.4.24) and (3.4.25) show that

$$E|(X_i - X_{i-1})a(X_{i-1}; \hat{\theta})\Delta t|^2 = O((\Delta t)^3). \tag{3.4.26}$$

An application of Hölder's inequality and the Cauchy–Schwarz inequality prove that

$$\begin{aligned}
&E\left|\int_{t_{i-1}}^{t_i} (X_u - X_{i-1})a(X_u; \theta)\, du\right|^2 \\
&\leq E\left[\Delta t \int_{t_{i-1}}^{t_i} |(X_u - X_{i-1})a(X_u; \theta)|^2\, du\right] \\
&\leq \Delta t \int_{t_{i-1}}^{t_i} (E|X_u - X_{i-1}|^4)^{1/2}(E|a(X_u, \theta)|^4)^{1/2}\, du \\
&= O((\Delta t)^3).
\end{aligned}$$

Let

$$M_t = \sum_{i=1}^{N} \int_{t\wedge t_{i-1}}^{t\wedge t_i} (X_u - X_{i-1})\, dW_u.$$

From the martingale moment inequality, there exists a constant C such that

$$E|M_t|^2 \leq CE\langle M\rangle_t$$

where $\langle M\rangle_t$ denotes the quadratic variation of M_t. Note that

$$\langle M\rangle_t = \sum_{i=1}^{N} \int_{t\wedge t_{i-1}}^{t\wedge t_i} (X_u - X_{i-1})^2\, du.$$

Since $E|X_u - X_{i-1}|^2 = O(|u - t_{i-1}|)$, it follows that

$$E\langle M\rangle_t = NO((\Delta t)^2)$$

and hence

$$E|\hat{\sigma}^2 - \sigma^2|^2 = O(\Delta t) + O\left(\frac{1}{N}\right). \tag{3.4.27}$$

Therefore $E|\hat{\sigma}^2 - \sigma^2| \to 0$ if $\Delta t \to 0$ and $N\Delta t \to \infty$ as $N \to \infty$. This implies that $\hat{\sigma}^2 \to \sigma^2$ in the L_2 norm.

Estimation by the (local linearization) Ozaki method (one-dimensional case)

The local linearization method consists in approximating the drift function of the stochastic differential equation (3.4.1) locally by a linear function. Since a linear stochastic differential equation with a constant coefficient is analytically tractable, it is convenient for estimation and simulation to approximate the original stochastic differntial equation by a linear one (cf. Basawa and Prakasa Rao 1980; Arato 1982). We first consider the stochastic differential equation

$$dX_t = a(X_t)\, dt + \sigma\, dW_t, \qquad d \geq 0, \tag{3.4.28}$$

where σ is constant. Let us consider the corresponding deterministic differential equation

$$\frac{dx_t}{dt} = a(x_t), \tag{3.4.29}$$

where x_t is a smooth function of t. Suppose x_t is differentiable twice with respect to t. Then we have

$$\frac{d^2x_t}{dt^2} = a'(x_t)\frac{dx_t}{dt}, \tag{3.4.30}$$

where $a'(x)$ denotes the derivative of $a(x)$ with respect to x. Suppose that $a'(x_t)$ is constant on a small time interval $[t, t+\Delta t)$. Integrating both sides from t to $u \in [t, t+\Delta t)$, we have

$$\left.\frac{dx_t}{dt}\right|_{t=u} = \frac{dx_t}{dt}\exp\{a'(x_t)(u-t)\}. \tag{3.4.31}$$

Integrating both sides of the above relation from t to $t+\Delta t$, we have

$$x_{t+\Delta t} = x_t + \frac{a(x_t)}{a'(x_t)}\{\exp(a'(x_t)\Delta t) - 1\}. \tag{3.4.32}$$

Suppose we approximate the drift function $a(x_t)$ in (3.4.28) by a linear function Lx_t, where L is a constant on $[t, t+\Delta t)$. A linear stochastic differential equation with a constant diffusion coefficient can be solved easily and

$$X_{t+\Delta t} = X_t\exp(L\Delta t) + \sigma\int_t^{t+\Delta t}\exp(L(t+\Delta t-u))\,dW_u. \tag{3.4.33}$$

In order to determine the coefficient L, suppose we impose the condition that the conditional expectation of $X_{t+\Delta t}$ given X_t at time t coincides with the trajectory given by (3.4.32). Then

$$\begin{aligned} E[X_{t+\Delta t}|X_t] &= X_t\exp(L\Delta t) \\ &= X_t + \frac{a(X_t)}{a'(X_t)}\{\exp(a'(X_t)\Delta t) - 1\}, \end{aligned} \tag{3.4.34}$$

using first (3.4.33) and then (3.4.32). Hence

$$L = \frac{1}{\Delta t}\log\left[1 + \frac{a(X_t)}{X_t\, a'(X_t)}\{\exp(a'(X_t)\Delta t) - 1\}\right].$$

Since L depends on t, let us denote it by L_t. The discretized process by the local linearization method is as follows:

$$X_{t+\Delta t} = X_t\exp(L_t\,\Delta t) + \sigma\int_t^{t+\Delta t}\exp(L_t(t+\Delta t-u))\,dW_u \tag{3.4.35}$$

from (3.4.33), where

$$L_t = \frac{1}{\Delta t}\log\left\{1 + \frac{a(X_t)}{X_t\, a'(X_t)}(\exp(a'(X_t)\Delta t) - 1)\right\}. \tag{3.4.36}$$

Since

$$\int_t^{t+\Delta t}\exp(L_t(t+\Delta t-u))\,dW_u$$

has the normal distribution with mean 0 and variance $(\exp(2L_t\Delta t) - 1)/(2L_t)$, the transition probability density function is given by

$$p(x_{(i)}|x_{(i-1)}) = \frac{1}{\sqrt{2\pi V_{i-1}}}\exp\left(-\frac{1}{2}\frac{(x_{(i)} - x_{(i-1)}\exp(L_{i-1}\Delta t))^2}{V_{i-1}}\right), \tag{3.4.37}$$

where

$$V_i = \sigma^2 \frac{\exp(2L_i \Delta t) - 1}{2L_i}, \tag{3.4.38}$$

$$L_i = \frac{1}{\Delta t} \log\left(1 + \frac{a(x_{(i)})}{x_i a'(x_{(i)})}(\exp(a'(x_i)\Delta t) - 1)\right) \tag{3.4.39}$$

and

$$x_{(i)} = x_{t+i\Delta t}. \tag{3.4.40}$$

As long as the piecewise linear approximation of the original stochastic differential equation is reasonable, the joint density of the process may be approximated by

$$p(x_{(0)}, \ldots, x_{(N)}) = \prod_{i=1}^{n} p(x_{(i)}|x_{(i-1)})p(x_{(0)})$$

and the parameters can be estimated by maximizing the log-likelihood given by

$$\log p(x_{(0)}, \ldots, x_{(N)}) = -\frac{1}{2}\sum_{i=1}^{N}\left\{\frac{x_{(i)} - x_{(i-1)}\exp(L_{i-1}\Delta t))^2}{V_{i-1}} + \log(2\pi V_{i-1})\right\}$$
$$+ \log(p(x_0)) \tag{3.4.41}$$

where V_i, L_i and $x_{(i)}$ are as defined above.

Remark. The local linearization method is expected to give a better approximation for the sample patterns of the original process since linear approximation is better than a constant approximation (Ozaki 1985, 1992, 1993) as in the Euler method.

Estimation by the (modified local linearization) Shoji–Ozaki method (one-dimensional case)

Consider a one-dimensional stochastic process X_t satisfying the stochastic differential equation

$$dX_t = a(X_t, t)\,dt + \sigma(X_t)\,dW_t, \qquad t \geq 0, \tag{3.4.42}$$

where $a(x,t)$ is twice continuously differentiable with respect to x and continuously differentiable with respect to t, and $\sigma(x)$ is a continuously differentiable function of x. Let us transform the differential equation by writing $Y_t = \phi(X_t)$. Then, by Itô's lemma,

$$dY_t = (a(X_t, t)\frac{d\phi}{dX_t} + \frac{\sigma^2(X_t)}{2}\frac{d^2\phi}{dX_t^2})\,dt + \tilde{\sigma}\,dW_t, \qquad t \geq 0, \tag{3.4.43}$$

where $\phi(x)$ satisfies the ordinary differential equation

$$\sigma(x)\frac{d\phi}{dx} = \tilde{\sigma},$$

with $\tilde{\sigma}$ constant. Thus, we can consider the stochastic differential equation

$$dX_t = a(X_t, t)\,dt + \sigma\,dW_t, \tag{3.4.44}$$

where $a(x,t)$ is twice continuously differentiable with respect to x and continuously differentiable with respect to t, and σ is a constant.

The local linearization method discussed earlier in this section is a method of approximation by which a drift function $a(x, t)$ in (3.4.44) not depending on t is locally approximated by a linear function of x. Let us now study the local behavior of $a(X_t, t)$. By Itô's formula,

$$da = \left(\frac{\sigma^2}{2}\frac{\partial^2 a}{\partial X_t^2} + \frac{\partial a}{\partial t}\right) dt + \frac{\partial a}{\partial X_t} dX_t. \tag{3.4.45}$$

In order to linearize a with respect to X_t and t, we assume that $\frac{\partial^2 a}{\partial x^2}(x_s, s)$, $\frac{\partial a}{\partial x}(x_s, s)$ and $\frac{\partial a}{\partial t}(x_s, s)$ are constant in a short interval $[s, s + \Delta s)$ for any s. Then (3.4.45) can be solved and we have

$$a(X_t, t) - a(X_s, s) = \left(\frac{\sigma^2}{2}\frac{\partial^2 a}{\partial X_t^2} + \frac{\partial a}{\partial t}\right)(t - s) + \frac{\partial a}{\partial X_t}(X_t - X_s). \tag{3.4.46}$$

Thus

$$a(X_t, t) = L_s X_t + M_s t + N_s, \tag{3.4.47}$$

where

$$L_s = \frac{\partial a}{\partial x}(X_s, s), \tag{3.4.48}$$

$$M_s = \frac{\sigma^2}{2}\frac{\partial^2 a}{\partial x^2}(X_s, s) + \frac{\partial a}{\partial t}(X_s, s) \tag{3.4.49}$$

and

$$N_s = a(X_s, s) - X_s \frac{\partial a}{\partial x}(X_s, s) - \left(\frac{\sigma^2}{2}\frac{\partial^2 a}{\partial x^2}(X_s, s) + \frac{\partial a}{\partial t}(X_s, s)\right) s. \tag{3.4.50}$$

Hence, instead of (3.4.44), we can consider the following stochastic differential equation on the interval $[s, s + \Delta s)$:

$$dX_t = (L_s X_t + M_s t + N_s)\, dt + \sigma\, dW_t, \qquad t \geq s, \tag{3.4.51}$$

as an approximation. By using Girsanov's theorem (see Theorem 1.2.1), (3.4.51) can be transformed into the equation

$$dX_t = L_s X_t + \sigma\, d\tilde{W}_t, \tag{3.4.52}$$

where

$$\tilde{W}_t = W_t - \int_s^t \gamma(u)\, du, \tag{3.4.53}$$

with

$$\gamma(u) = -\frac{1}{\sigma}(M_s u + N_s). \tag{3.4.54}$$

In other words,

$$d\tilde{W}_t = -\gamma(t)\, dt + dW_t = \frac{1}{\sigma}(M_t t + N_s)\, dt + dW_t. \tag{3.4.55}$$

Let $Y_t = e^{-L_s t} X_t$. Then the stochastic differential equation in Y_t can be solved and

$$\begin{aligned} Y_t &= Y_s + \sigma \int_s^t e^{-L_s u}\, d\tilde{W}_u \\ &= Y_s + \int_s^t (M_s u + N_s) e^{-L_s u}\, du + \sigma \int_s^t e^{-L_s u}\, dW_u. \end{aligned} \tag{3.4.56}$$

Hence a discretized process of X_t can be taken to be

$$X_t = X_s + \frac{a(X_s, s)}{L_s}(e^{L_s(t-s)} - 1) + \frac{M_s}{L_s^2}\{(e^{L_s(t-s)} - 1) - L_s(t-s)\} + \sigma \int_s^t e^{L_s(t-u)}\, dW_u, \qquad t \geq s, \tag{3.4.57}$$

where L_s and M_s are as defined in (3.4.48) and (3.4.49), respectively. Note that the conditional variance of X_t given X_s is

$$\operatorname{var}(X_t|X_s) = \sigma^2 \frac{e^{2L_s(t-s)} - 1)}{2L_s}. \tag{3.4.58}$$

The solution (3.4.57) leads to the following discrete-time model $\{X_t\}$ for the process:

$$X_{s+\Delta s} = A(X_s)X_s + B(X_s)N_{s+\Delta s}, \tag{3.4.59}$$

where

$$A(X_s) = 1 + \frac{a(X_s, s)}{X_s L_s}(\exp(L_s\Delta s) - 1) + \frac{M_s}{X_s L_s^2}(\exp(L_s \Delta s) - 1 - L_s \Delta s), \tag{3.4.60}$$

$$B(X_s) = \sigma \sqrt{\frac{\exp(2L_s \Delta s) - 1}{2L_s}}, \tag{3.4.61}$$

and $N_{s+\Delta s}$ is $N(0, 1)$.

Since the discretized process follows the normal distribution locally, the log-likelihood function can be easily calculated. Let $p(x_{(1)}, \ldots, x_{(N)})$ be the joint density function where $X_{(n)}$ is the nth equally spaced observation sample from the process $\{X_t\}$, Δt the time interval and N is the number of observations. By using the Markov property, the log-likelihood can be written as

$$\begin{aligned}
&\log(p(x_{(1)}, \ldots, x_{(N)})\\
&= \sum_{n=1}^{N-1} \log(p(x_{(n+1)}|x_{(n)})) + \log(p|x_{(1)})\\
&= -\frac{1}{2}\sum_{n=1}^{N-1}\left\{\frac{(x_{(n+1)} - E_n)^2}{V_n} + \log(2\pi V_n)\right\} + \log(p(x_{(1)})),
\end{aligned} \tag{3.4.62}$$

where E_n and V_n are the one-step-ahead conditional mean and conditional variance derived from the discretized process $\{X_{(i)}\}$. The new local linearization gives

$$E_n = x_{(n)} + \frac{a(x_{(n)}, t_n)}{L_n}(\exp(L_n\Delta t) - 1) + \frac{M_n}{L_n^2}(\exp(L_n\Delta t) - 1 - L_n\Delta t), \tag{3.4.63}$$

$$V_n = \frac{\exp(2L_n\Delta t) - 1}{2L_n}\sigma^2, \tag{3.4.64}$$

$$L_n = \frac{\partial a}{\partial x}(x_{(n)}, t_n) \tag{3.4.65}$$

and

$$M_n = \frac{\sigma^2}{2}\frac{\sigma^2 a}{\partial x^2}(x_{(n)}, t_n) + \frac{\partial a}{\partial t}(x_{(n)}, t_n). \tag{3.4.66}$$

Remarks. If the original stochastic differential equation contains a diffusion term which is not a constant but is a function of the process, we can transform the equation into a stochastic differential equation with diffusion coefficient as a constant. Let $p(x_{(1)}, \ldots, x_{(N)})$ and $p(y_{(1)}, \ldots, y_{(N)})$ be the joint density functions of $(x_{(1)}, \ldots, x_{(N)})$ and $(y_{(1)}, \ldots, y_{(N)})$, respectively. Then the transformed process $Y_t = \phi(X_t)$ satisfies the stochastic differential equation with a constant term as the diffusion coefficient. Since $p(x_{(1)}, \ldots, x_{(N)})$ and $p(y_{(1)}, \ldots, y_{(N)})$ are related by the equation

$$p(x_{(1)}, \ldots, x_{(N)}) = p(y_{(1)} \ldots, y_{(N)})\left|\frac{\partial(y_{(1)}, \ldots, y_{(N)})}{\partial(x_{(1)}, \ldots, y_{(N)})}\right| \tag{3.4.67}$$

where $|\frac{\partial(y_{(1)},\ldots,y_{(N)})}{\partial(x_{(1)},\ldots,x_{(N)})}|$ is the Jacobian which is equal to $\prod_{n=1}^{N} |\frac{d\phi}{dx}|_{x_{(n)}}$, it follows that

$$\begin{aligned}\log(p(x_{(1)}, \ldots, x_{(N)}) = &-\frac{1}{2}\sum_{n=1}^{N-1}\left\{\frac{(y_{(n+1)} - E_n)^2}{V_n} + \log(2\pi V_n)\right\} + \log(p(y_1)) \\ &+ \sum_{n=1}^{N} \log\left(\left|\frac{d\phi}{dx}\right|_{x=x_{(n)}}\right)\end{aligned} \tag{3.4.68}$$

and the estimation of the parameters can be done by the maximum likelihood method.

Estimation by the (modified local linearization) Shoji–Ozaki method (multi-dimensional case)

Consider the n-dimensional stochastic process $\boldsymbol{X}_t$ satisfying the stochastic differential equation

$$d\boldsymbol{X}_t = \boldsymbol{a}(\boldsymbol{X}_t)\,dt + \boldsymbol{\sigma}\,d\boldsymbol{W}_t, \qquad t \geq 0, \tag{3.4.69}$$

where $\boldsymbol{a}(\boldsymbol{X}_t)$ is an n-dimensional vector, $\boldsymbol{\sigma}$ is a constant $(n \times d)$ matrix and $\boldsymbol{W}_t$ is a d-dimensional standard Wiener process.

Let X_i denote the ith component of $\boldsymbol{X}_t$, suppressing the evolution at time t. Then

$$dX_i = a_i(\boldsymbol{X})\,dt + \sigma_i^j\,dW_j$$

where σ_i^j is the (i, j)th component of $\boldsymbol{\sigma}$ and, using the Einstein rule, the summation symbol is omitted. Here $a_i(\boldsymbol{X})$ is the ith component of $\boldsymbol{a}(\boldsymbol{X}_t)$. By Itô's formula

$$da_i = \frac{\partial a_i}{\partial X_j}dX_j + \frac{1}{2}\frac{\partial^2 a_i}{\partial X_k\,\partial X_\ell}d\langle X_k, X_\ell\rangle,$$

where $\langle\cdot, \cdot\rangle$ represents the quadratic variation. Since the components W_j are independent,

$$d\langle X_k, X_\ell\rangle = \sigma_k^p\sigma_\ell^q\,d\langle W_p, W_q\rangle = \delta_{pq}\sigma_k^p\sigma_\ell^q\,dt = \sigma_k^p\sigma_\ell^p\,dt.$$

Hence

$$da_i = \frac{\partial a_i}{\partial X_j}\, dX_j + \frac{1}{2}\frac{\partial^2 a_i}{\partial X_k\, \partial X_\ell}\sigma_k^p \sigma_\ell^q\, dt. \tag{3.4.70}$$

Let $\boldsymbol{H}_i$ be the Hessian matrix of a_i, that is,

$$\boldsymbol{H}_i = \left(\left(\frac{\partial^2 a_i}{\partial X_k\, \partial X_\ell}\right)\right)_{1\le k,\ell\le n}.$$

Note that $\sigma_k^p \sigma_\ell^p$ is the (k, ℓ)th element of the $n \times n$ matrix $\boldsymbol{\sigma\sigma}'$. For fixed k,

$$\frac{\partial^2 a_i}{\partial X_k\, \partial X_\ell}\sigma_k^p \sigma_\ell^q$$

is a kth diagonal element of $\boldsymbol{\sigma\sigma}'\boldsymbol{H}_i$. Hence

$$\frac{\partial^2 a_i}{\partial X_k\, \partial X_\ell}\sigma_k^p \sigma_\ell^q = \operatorname{tr}(\boldsymbol{\sigma\sigma}'\boldsymbol{H}_i)$$

where $\operatorname{tr}(\boldsymbol{A})$ denotes the trace of a square matrix $\boldsymbol{A}$. Hence (3.4.70) can be written in the form

$$da_i = \frac{\partial a_i}{\partial \boldsymbol{X}'}\, d\boldsymbol{X} + \frac{1}{2}\operatorname{tr}(\boldsymbol{\sigma\sigma}'\boldsymbol{H}_i)\, dt \tag{3.4.71}$$

where $\partial a_i/\partial \boldsymbol{X}'$ is the ith row of the Jacobian matrix

$$\boldsymbol{J} = \frac{\partial(a_1, \ldots, a_n)}{\partial(X_1, \ldots, X_n)}.$$

Let $\boldsymbol{M}$ be the n-dimensional vector with the ith component $\frac{1}{2}\operatorname{tr}(\boldsymbol{\sigma\sigma}'\boldsymbol{H}_i)$. Then (3.4.71) given rise to the differential of $\boldsymbol{a}$ of the form

$$d\boldsymbol{a} = \boldsymbol{J}\, d\boldsymbol{X} + \boldsymbol{M}\, dt. \tag{3.4.72}$$

In order to linearize $\boldsymbol{a}$ with respect to $\boldsymbol{x}$ and t, we assume that $\boldsymbol{J}$ and $\boldsymbol{M}$ are constant on a short time interval $[s, s + \Delta t)$. Then (3.4.72) can be written in the form

$$\boldsymbol{a}(\boldsymbol{X}_t) - \boldsymbol{a}(\boldsymbol{X}_s) = \boldsymbol{J}_s(\boldsymbol{X}_t - \boldsymbol{X}_s) + \boldsymbol{M}_s(t - s), \tag{3.4.73}$$

for $t \in [s, s + \Delta t)$. Thus, $\boldsymbol{a}$ is approximated by

$$\boldsymbol{a}(\boldsymbol{X}_t) = \boldsymbol{J}_s\boldsymbol{X}_t + \boldsymbol{M}_s t + \boldsymbol{a}(\boldsymbol{X}_s) - \boldsymbol{J}_s\boldsymbol{X}_s - \boldsymbol{M}_s s, \tag{3.4.74}$$

for $t \ge s$. Replacing $\boldsymbol{a}$ in (3.4.69) by the above approximation, we have

$$d\boldsymbol{X}_t = (\boldsymbol{J}_s\boldsymbol{X}_t + \boldsymbol{M}_s t + \boldsymbol{a}(\boldsymbol{X}_s) - \boldsymbol{J}_s\boldsymbol{X}_s - \boldsymbol{M}_s s)\, dt + \boldsymbol{\sigma}\, d\boldsymbol{W}_t, \qquad t \ge s. \tag{3.4.75}$$

Let $\boldsymbol{Y}_t = \exp(-\boldsymbol{J}_s t)$, where

$$\exp(\boldsymbol{A}) = \sum_{k=0}^{\infty} \frac{\boldsymbol{A}^k}{k!}. \tag{3.4.76}$$

Then, for $t \ge s$,

$$\begin{aligned} d\boldsymbol{Y}_t &= -\boldsymbol{J}_s \exp(-\boldsymbol{J}_s t)\boldsymbol{X}_t\, dt + \exp(-\boldsymbol{J}_s t)\, d\boldsymbol{X}_t \\ &= \exp(-\boldsymbol{J}_s t)\boldsymbol{M}_s t\, dt + \exp(-\boldsymbol{J}_s t)(\boldsymbol{a}(\boldsymbol{X}_s) - \boldsymbol{J}_s\boldsymbol{X}_s - \boldsymbol{M}_s s)\, dt \\ &\quad + \exp(-\boldsymbol{J}_s t)\boldsymbol{\sigma}\, d\boldsymbol{W}_t. \end{aligned} \tag{3.4.77}$$

Let us integrate both sides of this equation. The integral of the first term on the right-hand side is

$$\begin{aligned}
&\int_s^t \exp(-J_s u)M_s u\,du \\
&= [(-J_s)^{-1}\exp(-J_s u)M_s u]_s^t - \int_s^t (-J_s)^{-1}\exp(-J_s u)M_s\,du \\
&= [-J_s^{-1}\exp(-J_s u)M_s u - (J_s^{-1})^2\exp(-J_s u)M_s]_s^t \\
&= -J_s^{-1}(\exp(-J_s t)t - \exp(-J_s s)s)M_s \\
&\quad -(J_s^{-1})^2(\exp(-J_s t) - \exp(-J_s s))M_s.
\end{aligned}$$

Integration of the second term on the right-hand side of (3.4.77) gives

$$\begin{aligned}
&\int_s^t \exp(-J_s u)(a(X_s) - J_s X_s - M_s s)\,du \\
&= [(-J_s)^{-1}\exp(-J_s u)(a(X_s) - J_s X_s - M_s s)]_s^t \\
&= -J_s^{-1}(\exp(-J_s t) - \exp(-J_s s))(a(X_s) - J_s X_s - M_s s).
\end{aligned}$$

Hence, for $t \geq s$,

$$\begin{aligned}
Y_t - Y_s &= (\exp(-J_s s) - \exp(-J_s t))X_s \\
&\quad + J_s^{-1}(\exp(-J_s s) - \exp(-J_s t))a(X_s) \\
&\quad + (J_s^{-1})^2\{\exp(-J_s s) - \exp(-J_s t) - J_s\exp(-J_s t)(t-s)\}M_s \\
&\quad + \int_s^t \exp(-J_s u)\sigma\,dW_u.
\end{aligned}$$

Transforming Y_t to X_t again, we have

$$\begin{aligned}
X_t &= X_s + J_s^{-1}(\exp(J_s(t-s)) - I)a(X_s) \\
&\quad + (J_s^{-1})^2\{\exp(J_s(t-s)) - I - J_s(t-s)\}M_s \\
&\quad + \int_s^t \exp(J_s(t-u))\sigma\,dW_u
\end{aligned} \tag{3.4.78}$$

for $t \geq s$, where I is the $n \times n$ identity matrix. Note that the diffusion term is given by the covariance matrix V defined by

$$\begin{aligned}
V &= E_s\left[\int_s^t \exp(J_s(t-u))\sigma\,dW_u\,dW_u'\sigma'\exp(J_s'(t-u))\right] \\
&= \int_s^t \exp(J_s(t-u))\sigma\sigma'\exp(J_s'(t-u))\,du.
\end{aligned}$$

Explicit calculation of V is not easy. However, it can be calculated numerically following the method given below. Note that

$$\begin{aligned}
V &= \int_s^t \exp(J_s(t-u))\sigma\sigma'\exp(J_s'(t-u))\,du \\
&= \int_0^{t-s} \exp(J_s u)\sigma\sigma'\exp(J_s' u)\,du.
\end{aligned}$$

Integrating by parts, we have

$$\begin{aligned}
&\int_0^{t-s} \exp(J_s u)\sigma\sigma' \exp(J_s' u)\, du \\
&= [J_s^{-1} \exp(J_s u)\sigma\sigma' \exp(J_s' u)]_0^{t-s} \\
&\quad - \int_0^{t-s} J_s^{-1} \exp(J_s u)\sigma\sigma' \exp(J_s' u) J_s'\, du \\
&= [J_s^{-1} \exp(J_s u)\sigma\sigma' \exp(J_s' u)]_0^{t-s} - J_s^{-1} V J_s'.
\end{aligned}$$

Therefore

$$\begin{aligned}
J_s V + V J_s' &= [\exp(J_s u)\sigma\sigma' \exp(J_s' u)]_0^{t-s} \\
&= \exp(J_s(t-s))\sigma\sigma' \exp(J_s'(t-s)) - \sigma\sigma'.
\end{aligned} \tag{3.4.79}$$

If J_s is known, then V can be obtained by solving the above linear matrix equation. It can be shown that the above equation has a unique solution. If $(J_s V)' = J_s V$, then V can be calculated explicitly and, from (3.4.79), it follows that

$$V = \tfrac{1}{2} J_s^{-1} (\exp(J_s(t-s))\sigma\sigma' \exp(J_s'(t-s)) - \sigma\sigma'). \tag{3.4.80}$$

Parameter estimation from the discretized process now can be done by using the maximum likelihood techniques. Due to the local normality of the discretized process, the transition probability density function can be written as

$$p(X_{t+\Delta t}|X_t) = (2\pi|V|)^{-n/2} \exp[-\tfrac{1}{2}(X_{t+\Delta t} - X_t)' V^{-1} (X_{t+\Delta t} - X_t)], \tag{3.4.81}$$

where Δt is the discrete time interval. If the discrete observations sampled from the original process $\{X_t\}$ are $\{X_{t_i}, 0 \le i \le N\}$, then, by the Markov property of the process, the log-likelihood function is

$$\begin{aligned}
\log L &= \log p(x_{t_0}, \ldots, x_{t_N}) \\
&= \sum_{i=1}^{N} \log p(x_{t_i} | x_{t_{i-1}}) + \log p(x_{t_0})
\end{aligned}$$

and the paramaters can be estimated by maximizing $\log L$.

Example 3.4.1

Consider the two-dimensional stochastic process $(X(t), Y(t))$ characterizing the motion of a particle at $P = (x, y)$ which is forced towards the origin $O = (0, 0)$ with strength $|\vec{OP}| \cos\theta$ in the x-direction and $|\vec{OP}| \cos\theta$ in the y-direction, where θ is the angle between $\vec{OP}$ and the x-axis. Then the process can be modeled by the stochastic differential equations

$$\left.\begin{aligned}
dX(t) &= -\frac{X(t)}{\sqrt{X^2(t) + Y^2(t)}}\, dt + dW_1(t), \qquad t \ge 0, \\
dY(t) &= -\frac{Y(t)}{\sqrt{Y^2(t) + X^2(t)}}\, dt + dW_2(t), \qquad t \ge 0,
\end{aligned}\right\} \tag{3.4.82}$$

where $\boldsymbol{W} \equiv (W_1, W_2)$ is a two-dimensional standard Wiener process. After some calculations, it can be checked that

$$\boldsymbol{J} = \begin{pmatrix} -\frac{Y^2(t)}{Z^3(t)} & \frac{-X(t)Y(t)}{Z^3(t)} \\ \frac{-X(t)Y(t)}{Z^3(t)} & -\frac{X^2(t)}{Z^3(t)} \end{pmatrix},$$

$$\boldsymbol{M} = \begin{pmatrix} -\frac{X(t)}{2Z^3(t)} + \frac{3X(t)Y^2(t)}{Z^5(t)} \\ -\frac{Y(t)}{2Z^3(t)} + \frac{3X^2(t)Y(t)}{Z^5(t)} \end{pmatrix},$$

$$\boldsymbol{\sigma}\boldsymbol{\sigma}' = \boldsymbol{I},$$

where $\boldsymbol{I}$ is the 2×2 identity matrix and

$$Z(t) = [X^2(t) + Y^2(t)]^{1/2}.$$

For notational convenience, we write X_t for $X(t)$, Y_t for $Y(t)$ and Z_t for $Z(t)$. Following (3.4.78) the discretized process $\boldsymbol{X}_t = (X_t, Y_t)'$ is

$$\begin{aligned} \boldsymbol{X}_{t+\Delta t} = \boldsymbol{X}_t &+ \boldsymbol{J}_t^{-1}(\exp(\boldsymbol{J}_t\Delta t) - \boldsymbol{I})\boldsymbol{a}(\boldsymbol{X}_t) \\ &+ (\boldsymbol{J}_t^{-1})^2\{\exp(\boldsymbol{J}_t\Delta t) - \boldsymbol{I} - \boldsymbol{J}_t\Delta t\}\boldsymbol{M}_t \\ &+ \int_t^{t+\Delta t} \exp(\boldsymbol{J}_t(t+\Delta t - u))\boldsymbol{\sigma}\, d\boldsymbol{W}_u, \end{aligned} \tag{3.4.83}$$

where

$$\boldsymbol{a}(\boldsymbol{X}_t) = \begin{pmatrix} -\frac{X_t}{Z_t} \\ -\frac{Y_t}{Z_t} \end{pmatrix}, \qquad \boldsymbol{\sigma} = \boldsymbol{I}, \qquad \boldsymbol{W}_t = \begin{pmatrix} W_{1t} \\ W_{2t} \end{pmatrix}.$$

Here

$$\begin{aligned} (\boldsymbol{J}_t\boldsymbol{V}_t)' = \boldsymbol{V}_t'\boldsymbol{J}_t' &= \boldsymbol{V}_t\boldsymbol{J}_t \\ &= \int_0^{\Delta t} \exp(\boldsymbol{J}_t u)\boldsymbol{I}\exp(\boldsymbol{J}_t u)\, du\, \boldsymbol{J}_t \\ &= \int_0^{\Delta t} \exp(2\boldsymbol{J}_t u)\boldsymbol{J}_t\, du \\ &= \int_0^{\Delta t} \boldsymbol{J}_t \exp(2\boldsymbol{J}_t u)\, du \\ &= \boldsymbol{J}_t \int_0^{\Delta t} \exp(2\boldsymbol{J}_t u)\, du \\ &= \boldsymbol{J}_t\boldsymbol{V}_t, \end{aligned}$$

and hence

$$\boldsymbol{V}_t = \tfrac{1}{2}\boldsymbol{J}_t^{-1}(\exp(2\boldsymbol{J}_t\Delta t) - \boldsymbol{I}).$$

Example 3.4.2

Consider the system of stochastic differential equations

$$\left.\begin{aligned} dX_1 &= X_2\, dt, & t \geq 0, \\ dX_2 &= \{3(1 - X_1^2)X_2 - X_1\}\, dt + \sigma_1\, dW_t, & t \geq 0, \end{aligned}\right\} \tag{3.4.84}$$

where W is the standard Wiener process. If $\sigma_1 = 0$, then the second equation is known as the Van der Pol equation. The system (3.4.84) can be written in the form

$$d\boldsymbol{X}_t = \boldsymbol{a}(\boldsymbol{X}_t)\,dt + \boldsymbol{\sigma}\,d\boldsymbol{W}_t, \qquad t \geq 0 \tag{3.4.85}$$

where

$$\boldsymbol{X} = \begin{pmatrix} X_1 \\ X_2 \end{pmatrix}, \qquad \boldsymbol{a}(\boldsymbol{X}) = \begin{pmatrix} X_2 \\ 3(1 - X_1^2)X_2 - X_1 \end{pmatrix}, \qquad \boldsymbol{\sigma} = \begin{pmatrix} 0 \\ \sigma_1 \end{pmatrix}$$

and $\boldsymbol{W} = \binom{W_1}{W_2}$. Then, it can be checked that

$$\boldsymbol{J} = \begin{pmatrix} 0 & 1 \\ -6X_1X_2 - 1 & 3(1 - X_1^2) \end{pmatrix}, \qquad \boldsymbol{M} = \boldsymbol{0}$$

and

$$\boldsymbol{\sigma}\boldsymbol{\sigma}' = \begin{pmatrix} 0 & 0 \\ 0 & \sigma_1^2 \end{pmatrix}.$$

Since $\boldsymbol{M} = \boldsymbol{0}$, the discretized process is independent of the second derivatives of $\boldsymbol{a}$. The discretization by the Ozaki scheme and that by the Shoji–Ozaki scheme are the same.

A more general form of (3.4.84) is

$$\left.\begin{aligned} dX_1 &= X_2\,dt, \\ dX_2 &= -(\alpha(X_1)X_2 + \beta(X_1))\,dt + \sigma_1\,dW, \end{aligned}\right\} \tag{3.4.86}$$

where $\alpha(\cdot)$ and $\beta(\cdot)$ are linear or nonlinear functions. This system of stochastic differential equations can be written in the form (3.4.85) with

$$\boldsymbol{X} = \begin{pmatrix} X_1 \\ X_2 \end{pmatrix}, \qquad \boldsymbol{a}(\boldsymbol{X}) = \begin{pmatrix} X_2 \\ -\alpha(X_1)X_2 - \beta(X_1) \end{pmatrix}$$

and

$$\boldsymbol{\sigma}\boldsymbol{\sigma}' = \begin{pmatrix} 0 & 0 \\ 0 & \sigma_1^2 \end{pmatrix}.$$

Here $\boldsymbol{J}$ and $\boldsymbol{M}$ are given by

$$\boldsymbol{J} = \begin{pmatrix} 0 & 1 \\ -\frac{d\alpha}{dX_1}X_2 - \frac{d\beta}{dX_1} & -\alpha \end{pmatrix}, \qquad \boldsymbol{M} = \boldsymbol{0}.$$

The comments made earlier about the discretized version hold here.

Example 3.4.3

In modeling the dynamics of the interest rates in financial economics, short- and long-term interest rates have to be taken into account. They may be modeled by a two-dimensional stochastic process

$$d\boldsymbol{X} = (\boldsymbol{A}\boldsymbol{X} + \boldsymbol{b})\,dt + \boldsymbol{C}(\boldsymbol{X})\boldsymbol{\sigma}\,d\boldsymbol{W}, \qquad t \geq 0, \tag{3.4.87}$$

where

$$\boldsymbol{X} = \begin{pmatrix} X_1 \\ X_2 \end{pmatrix}, \qquad \boldsymbol{A} = ((a_{ij}))_{2\times 2}, \qquad \boldsymbol{b} = \begin{pmatrix} b_1 \\ b_2 \end{pmatrix},$$

$$\boldsymbol{C}(\boldsymbol{X}) = \begin{pmatrix} X_1 & 0 \\ 0 & X_2 \end{pmatrix}, \qquad \boldsymbol{\sigma} = ((\sigma_{ij}))_{2\times 2}$$

and

$$\boldsymbol{W} = \begin{pmatrix} W_1 \\ W_2 \end{pmatrix}.$$

Here X_1 and X_2 represent the short- and long-term interest rates respectively. This model implies that the dynamics of X_1 is affected by the long-term economic shocks as well as the short-term shocks, and so is the dynamics of X_2. Furthermore, the volatility of X_1 and X_2 is dependent on the levels of X_1 and X_2. Since the coefficient matrix of the diffusion term is not constant, it cannot be discretized by the Shoji–Ozaki scheme. However, one can transform the equation to an equation with a constant diffusion coefficient. Let

$$Y_i(t) = \log X_i(t), \qquad i = 1, 2.$$

Then the stochastic differential equation for $\boldsymbol{Y} = \binom{Y_1}{Y_2}$ is given by

$$d\boldsymbol{Y}_t = \boldsymbol{a}(\boldsymbol{Y}_t)\,dt + \boldsymbol{\sigma}\,d\boldsymbol{W}_t,$$

where

$$\boldsymbol{a}(y) = \begin{pmatrix} a_{11} + a_{12}e^{y_2-y_1} + b_1e^{-y_1} - \{(\sigma_{11}^2 + \sigma_{12}^2)/2\} \\ a_{21}e^{y_1-y_2} + a_{22} + b_2e^{-y_2} - \{(\sigma_{21}^2 + \sigma_{22}^2)/2\} \end{pmatrix}.$$

Applying the Shoji–Ozaki scheme, the Jacobian matrix $\boldsymbol{J}$ and the Hessian matrix $\boldsymbol{H}_i$, $i = 1, 2$, are given by

$$\boldsymbol{J} = \begin{pmatrix} -a_{12}e^{y_2-y_1} - b_1e^{-y_1} & a_{12}e^{y_2-y_1} \\ a_{21}e^{y_1-y_2} & -a_{21}e^{y_1-y_2} - b_2e^{-y_2} \end{pmatrix},$$

$$\boldsymbol{H}_1 = \begin{pmatrix} a_{12}e^{y_2-y_1} + b_1e^{-y_1} & -a_{12}e^{y_2-y_1} \\ -a_{21}e^{y_2-y_1} & a_{12}e^{y_2-y_1} \end{pmatrix},$$

$$\boldsymbol{H}_2 = \begin{pmatrix} a_{21}e^{y_1-y_2} & -a_{21}e^{y_1-y_2} \\ -a_{21}e^{y_1-y_2} & a_{21}e^{y_1-y_2} + b_2e^{-y_2} \end{pmatrix}.$$

Hence

$$\boldsymbol{M} = \begin{pmatrix} \mathrm{tr}(\boldsymbol{\sigma}\boldsymbol{\sigma}'\boldsymbol{H}_1)/2 \\ \mathrm{tr}(\boldsymbol{\sigma}\boldsymbol{\sigma}'\boldsymbol{H}_2)/2 \end{pmatrix}$$

$$= \frac{1}{2}\begin{pmatrix} a_{12}\{\sigma_{11}^2 + \sigma_{12}^2 + \sigma_{21}^2 + \sigma_{22}^2 - 2(\sigma_{11}\sigma_{21} + \sigma_{12}\sigma_{22})\}e^{y_2-y_1} + b_1\sigma_{11}e^{y_1} \\ a_{21}\{\sigma_{11}^2 + \sigma_{12}^2 + \sigma_{21}^2 + \sigma_{22}^2 - 2(\sigma_{11}\sigma_{21} + \sigma_{12}\sigma_{22})\}e^{y_1-y_2} + b_2\sigma_{22}e^{y_2} \end{pmatrix}.$$

Applying $\boldsymbol{J}$ and $\boldsymbol{M}$ computed above to the analog of (3.4.83), we get the discretized model for the short- and the long-term interest rates. The parameter estimation can be done using the MLE procedure as before.

3.4.2 Generalized Method of Moments

The generalized method of moments (GMM) due to Hansen (1982) has been used for the estimation of the parameters in linear and the nonlinear econometric models. This method does not use the likelihood function but requires specification of some moment conditions based on conditional as well as unconditional expectations. Let us again consider the stochastic differential equation

$$dX_t = a(X_t, t)\,dt + \sigma(X_t)\,dW_t, \qquad t \geq 0. \tag{3.4.88}$$

Suppose we discretize the process $\{X_t\}$ by the Euler method. Then

$$X_{t+\Delta t} = X_t + a(X_t, t)\Delta t + \sigma(X_t)(W_{t+\Delta t} - W_t). \tag{3.4.89}$$

Hence

$$E(X_{t+\Delta t}|X_t) = X_t + a(X_t, t)\Delta t \tag{3.4.90}$$

and

$$E[X_{t+\Delta t} - E(X_{t+\Delta t}|X_t)^2|X_t] = \sigma^2(X_t)\Delta t. \tag{3.4.91}$$

For convenience, we write $E_t(X_{t+\Delta t})$ for $E(X_{t+\Delta t}|X_t)$ and in general write $E_t(g(X_{t+\Delta t}))$ for $E(g(X(t+\Delta t))|X_t)$. Then we have

$$E_t(X_{t+\Delta t}) = X_t + a(X_t, t)\Delta t,$$

and

$$E_t[\{X_{t+\Delta t} - E_t(X_{t+\Delta t})\}^2] = \sigma^2(X_t)\Delta t.$$

The Δt-ahead forcasting errors are

$$\varepsilon^{(1)}_{t+\Delta t} = X_{t+\Delta t} - (X_t + a(X_t, t)\Delta t) \tag{3.4.92}$$

and

$$\varepsilon^{(2)}_{t+\Delta t} = (X_{t+\Delta t} - E_t(X_{t+\Delta t}))^2 - \sigma^2(X_t)\Delta t. \tag{3.4.93}$$

If we impose the additional conditions that $\varepsilon^{(1)}_t$ and $\varepsilon^{(2)}_t$ are orthogonal to 1 and X_t (instrumental variables in the language of econometrics), then it follows that

$$E_t(\boldsymbol{u}_{t+\Delta t}) = 0, \tag{3.4.94}$$

where

$$\boldsymbol{u}_{t+\Delta t} = (\varepsilon^{(1)}_{t+\Delta t}, \varepsilon^{(2)}_{t+\Delta t}, X_t\varepsilon^{(1)}_{t+\Delta t}, X_t\varepsilon^{(2)}_{t+\Delta})'. \tag{3.4.95}$$

The GMM estimators are obtained by minimizing the criterion

$$\boldsymbol{g}'_N \boldsymbol{D}_N \boldsymbol{g}_N, \tag{3.4.96}$$

where

$$\boldsymbol{g}_N = \frac{1}{N}\sum_{i=1}^{N} \boldsymbol{u}_i, \tag{3.4.97}$$

$$\boldsymbol{u}_i = (\varepsilon^{(1)}_{t_i}, \varepsilon^{(2)}_{t_i}, X_{t_{i-1}}\varepsilon^{(1)}_{t_i}, X_{t_{i-1}}\varepsilon^{(2)}_{t_i})', \tag{3.4.98}$$

$$t_i = t + i\Delta t \tag{3.4.99}$$

and $\boldsymbol{D}_N$ is a weight matrix which is symmetric positive definite and converges to a constant symmetric positive definite matrix $\boldsymbol{D}_0$. Note that $\boldsymbol{g}_N$ is a sample counterpart of $E_t(\boldsymbol{u}_{t+\Delta t})$ defined in (3.4.94). Hansen (1982) proved the strong consistency of the GMM estimator. It is clear that the GMM estimator depends on the choice of $\boldsymbol{D}_N$. Hansen (1982) proved that an asymptotically efficient estimator can be obtained when $\boldsymbol{D}_N \to \boldsymbol{D}_0$, where $\boldsymbol{D}_0 = \boldsymbol{S}_0^{-1}$ and $\boldsymbol{S}_0 = \sum_{j=-\infty}^{\infty} E(\boldsymbol{u}_i\boldsymbol{u}'_{i-j})$.

3.4.3 Comparison of Discretization Methods

Shoji and Ozaki (1997) carried out an extensive investigation of the computational aspects of the various methods described under the pseudo-likelihood approach in this section. We will briefly discuss some of them.

Example 3.4.4
Consider the stochastic differential equation

$$dX_t = -X_t^3\,dt + dW_t \qquad t \geq 0. \tag{3.4.100}$$

The Euler scheme gives the discrete time model

$$X_{t+\Delta t} = A(X_t)\,dt + B(X_t)N_{t+\Delta t}, \tag{3.4.101}$$

where

$$A(X_t) = 1 - X_t^2\Delta t,\ B(X_t) = \sqrt{\Delta t} \tag{3.4.102}$$

and $N_{t+\Delta t}$ is $N(0, 1)$.

The local linearization scheme of Ozaki gives the model (3.4.101) with

$$A(X_t) = \tfrac{2}{3} + \tfrac{1}{3}\exp(-3X_t^2\Delta t), \tag{3.4.103}$$

$$B(X_t) = \left\{\frac{\exp(2K_t\Delta t) - 1}{2K_t}\right\}^{1/2}, \tag{3.4.104}$$

$$K_t = \frac{1}{\Delta t}\log\left\{\frac{2}{3} + \frac{1}{3}\exp(-3X_t^2\Delta t)\right\}. \tag{3.4.105}$$

The modified local linearization scheme of Shoji and Ozaki yields

$$\begin{aligned} A(X_t) &= \tfrac{2}{3} + \tfrac{1}{3}\exp(-3X_t^2\Delta t) \\ &\quad - \frac{1}{3X_t^4}(\exp(-3X_t^2\Delta t) - 1 + 3X_t^2\Delta t), \end{aligned} \tag{3.4.106}$$

$$B(X_t) = \sqrt{\frac{\exp(-6X_t^2\Delta t) - 1}{-6X_t^2}}. \tag{3.4.107}$$

In this example, the difference in the schemes is in the behaviour of $A(X_t)$ as $X_t \to 0$ and $X_t \to \infty$. The term $A(X_t)$ in the Euler and Ozaki schemes goes to one as $X_t \to 0$ and, in the Shoji–Ozaki scheme, to $1 - (\frac{3}{2})(\Delta t)^2$. If $X_t \to \infty$, then $A(X_t) \to 2/3$ for the Ozaki and Shoji–Ozaki schemes but $A(X_t) \to \infty$ for the Euler scheme.

Example 3.4.5
A similar comparison may be made for the model given by the stochastic differential equation

$$dX_t = (1 - X_t^2)\,dt + X_t\,dW_t, \qquad t \geq 0, \tag{3.4.108}$$

which is used for modeling volatile financial time series. Here the diffusion term is not a constant but a function of x. Applying the transformation $Y_t = \log X_t$, the equation (3.4.108) can be transformed into

$$dY_t = -(\tfrac{1}{2} + 2\cosh\ (Y_t))\,dt + dW_t, \qquad t \geq 0.$$

For a comparison of the three schemes in this example, see Shoji (1995).

Shoji (1995) studied the performance of the estimators derived through the three schemes due to Euler, Ozaki and Shoji–Ozaki by simulation methods for the following general models generalizing Examples 3.4.4 and 3.4.5:

(i) $dX_t = (\alpha_1 X_t + \alpha_2 X_t^2 + \alpha_3 X_t^3)\, dt + \sigma\, dW_t, t \geq 0,$

(ii) $dX_t = (\alpha_1 X_t + \alpha_2 X_t^2 + \alpha_3 X_t^3 + \alpha_4 X_t^4 + \alpha_5 X_t^5)\, dt + \sigma\, dW_t, t \geq 0,$

(iii) $dX_t = (\alpha_0 + \alpha_1 X_t + \alpha_2 X_t^2 + \alpha_3 X_t^3)\, dt + \sigma X_t^\gamma\, dW_t, t \geq 0,$

where α_i, $1 \leq i \leq 5$ and $\gamma \neq 1$ are the parameters to be estimated. Models (i) and (ii) have a constant term as the diffusion coefficient, whereas model (iii) has an X-dependent diffusion which is used as a model for the movement of interest rates in financial economics. Model (iii) can be transformed into a model with a constant term as the diffusion coefficient by the transformation

$$Y_t = \phi(X_t) = X_t^{1-\gamma}/(1-\gamma).$$

Shoji (1995) has indicated that the Shoji–Ozaki scheme performs well compared to the Euler scheme and the Ozaki scheme, and the performance is considerably better if the sampling interval is large.

As a special case, Shoji and Ozaki (1997) made a comparative study of the methods of estimation described earlier by simulation techniques based on the following models:

(i) $dX_t = (\alpha + \beta X_t)\, dt + \sigma X_t\, dW_t, t \geq 0,$

(ii) $dX_t = \alpha X_t^3\, dt + \sigma\, dW_t, t \geq 0,$

when the true value of $(\alpha, \beta, \sigma^2)$ in (i) is $(1, -1, 1)$ and (α, σ^2) in (ii) is $(-1, 1)$.

In simulation studies for the first model, which is a stochastic differential equation with a linear drift and a state-dependent diffusion coefficient, it is observed that the differences among the methods except for the GMM are not large when the sampling interval is small. If the sampling interval is large, then the Shoji–Ozaki scheme is better than the others. It was also noted that, for the estimation of the diffusion coefficient, the dominance of Shoji–Ozaki is noticeable and that the other methods show large bias for large Δt. The GMM method is inferior to other methods.

For the second model, the Shoji–Ozaki scheme performed much better than the other schemes particularly for the estimation of the diffusion coefficient when the sampling interval is large. Unlike in the first model, GMM showed reasonable performance and the Ozaki scheme is somewhat inferior. For complete details, see Shoji (1995) and Shoji and Ozaki (1997).

3.5 Estimation via Martingale Estimating Functions

3.5.1 Linear Estimating Functions

Consider the stochastic differential equation

$$dX_t = a(X_t, \theta)\, dt + \sigma(X_t, \theta)\, dW_t, \qquad X_0 = x_0,\ t \geq 0,\ \theta \in R. \tag{3.5.1}$$

Suppose that the diffusion coefficient σ does not depend on θ. Then it is known that the measures corresponding to different values of θ are equivalent (Liptser and Shiryayev 1977; Basawa and Prakasa Rao 1980) under some additional conditions and the log-likelihood function is

$$\ell_t(\theta) = \int_0^t \frac{a(X_s, \theta)}{\sigma^2(X_t)}\, dX_s - \frac{1}{2}\int_0^t \frac{a^2(X_s, \theta)}{\sigma^2(X_s)}\, ds. \tag{3.5.2}$$

Approximating the integrals by Riemann sums and Itô sums, we have

$$\tilde{\ell}_n(\theta) = \sum_{i=1}^{n} \frac{a(X_{(i-1)\Delta}, \theta)}{\sigma^2(X_{(i-1)\Delta})}(X_{i\Delta} - X_{(i-1)\Delta}) - \frac{1}{2}\sum_{i=1}^{n} \frac{a^2(X_{(i-1)\Delta}, \theta)}{\sigma^2(X_{(i-1)\Delta})}\Delta$$

as an approximate likelihood function. Assuming the differentiability of $a(x, \theta)$ with respect to θ, we obtain

$$\begin{aligned}\dot{\tilde{\ell}}_n(\theta) &= \sum_{i=1}^{n} \frac{\dot{a}(X_{(i-1)\Delta}, \theta)}{\sigma^2(X_{(i-1)\Delta})}(X_{i\Delta} - X_{(i-1)\Delta}) \\ &\quad - \Delta \sum_{i=1}^{n} \frac{a(X_{(i-1), \Delta \theta)}}{\sigma^2(X_{(i-1)\Delta})}\dot{a}(X_{(i-1)\Delta}, \theta),\end{aligned}$$

where $\dot{a}(x, \theta)$ denotes the derivative of $a(x, \theta)$ with respect θ. Let us now consider the same function when σ also depends on θ by defining

$$\begin{aligned}\dot{\tilde{\ell}}_n(\theta) &\equiv \sum_{i=1}^{n} \frac{\dot{a}(X_{(i-1)\Delta}, \theta)}{\sigma^2(X_{(i-1)\Delta}, \theta)}(X_{i\Delta} - X_{(i-1)\Delta}) \\ &\quad - \Delta \sum_{i=1}^{n} \frac{a(X_{(i-1)\Delta}, \theta)\dot{a}(X_{(i-1)\Delta}, \theta)}{\sigma^2(X_{(i-1)\Delta}, \theta)} \qquad (3.5.3)\end{aligned}$$

Then $\{\dot{\tilde{\ell}}_n(\theta), n \geq 1\}$, as defined above, is not a martingale with respect to the filtration $\{\mathcal{F}_i\}$, where $\mathcal{F}_i$ is the σ-algebra generated by $\{X_0, X_\Delta, \ldots, X_{i\Delta}\}$. Let

$$F(x, \theta) = E_\theta(X_\Delta | X_0 = x). \qquad (3.5.4)$$

Then

$$\begin{aligned}\sum_{i=1}^{n} E_\theta[\dot{\tilde{\ell}}_i(\theta) - \dot{\tilde{\ell}}_{i-1}(\theta)|\mathcal{F}_{i-1}] &= \sum_{i=1}^{n} \frac{\dot{a}(X_{(i-1)\Delta}, \theta)}{\sigma^2(X_{(i-1)\Delta}, \theta)}\{F(X_{(i-1)\Delta}, \theta) - X_{(i-1)\Delta}\} \\ &\quad - \Delta \sum_{i=1}^{n} \frac{a(X_{(i-1)\Delta}, \theta)\dot{a}(X_{(i-1)\Delta}, \theta)}{\sigma^2(X_{(i-1)\Delta}, \theta)} \qquad (3.5.5)\end{aligned}$$

and hence

$$\begin{aligned}\tilde{G}_n(\theta) &\equiv \dot{\tilde{\ell}}_n(\theta) - \sum_{i=1}^{n} E_\theta[\dot{\tilde{\ell}}_i(\theta) - \dot{\tilde{\ell}}_{i-1}(\theta)|\mathcal{F}_{i-1}] \\ &= \sum_{i=1}^{n} \frac{\dot{a}(X_{i-1)\Delta}, \theta)}{\sigma^2(X_{(i-1)\Delta}, \theta)}\{-F(X_{(i-1)\Delta}, \theta) + X_{i\Delta}\}. \qquad (3.5.6)\end{aligned}$$

Note that $\{\tilde{G}_n(\theta), \mathcal{F}_n, n \geq 1\}$ is a a zero-mean P_θ-martingale whethere σ is a function of θ or not. The function $\tilde{G}_n(\theta)$ can be used as an estimating function for the estimation of the parameter θ in the sense of Godambe and Heyde (1987). See Chapter 5 of Prakasa Rao (1999).

Before we consider the properties of the estimator obtained from $\tilde{G}_n(\theta)$, let us look at general class ζ of estimating functions of the form

$$G_n(\theta) = \sum_{i=1}^{n} g_{i-1}(\theta)(X_{i\Delta} - F(X_{(i-1)\Delta}, \theta)), \qquad (3.5.7)$$

where g_{i-1} is a $\mathcal{F}_{i-1}$-measurable and continuously differentiable function of θ for $1 \leq i \leq n$. An estimating function is said to be *optimal* within the class ζ if it leads to a smallest asymptotic confidence interval around θ and yields an estimator with the smallest asymptotic variance; see Chapter 5 of Prakasa Rao (1999) or Godambe and Heyde (1987).

Note that the quadratic characteristic of $G(\theta)$ is

$$\langle G(\theta)\rangle_n = \sum_{i=1}^{n} g_{i-1}^2(\theta)\phi(X_{(i-1)\Delta}, \theta), \tag{3.5.8}$$

where

$$\phi(X_{(i-1)\Delta}, \theta) = E_\theta\{(X_{i\Delta} - F(X_{(i-1)\Delta}, \theta))^2 | X_{(i-1)\Delta}\}, \qquad 1 \leq i \leq n. \tag{3.5.9}$$

Furthermore,

$$\begin{aligned}\dot{G}_n(\theta) &= \sum_{i=1}^{n} \dot{g}_{i-1}(\theta)\{X_{i\Delta} - F(X_{(i-1)\Delta}, \theta)\} \\ &\quad - \sum_{i=1}^{n} g_{i-1}(\theta)\dot{F}(X_{(i-1)\Delta}, \theta),\end{aligned} \tag{3.5.10}$$

where F is assumed to be a differentiable function of θ. The first term on the right-hand side of (3.5.10) is a martingale and the compensator of $\{\dot{G}_n(\theta)\}$ is given by

$$\begin{aligned}\bar{G}_n(\theta) &\equiv \sum_{i=1}^{n} E\{\dot{G}_i(\theta) - \dot{G}_{i-1}(\theta) | \mathcal{F}_{i-1}\} \\ &= -\sum_{i=1}^{n} g_{i-1}(\theta)\dot{F}(X_{(i-1)\Delta}, \theta).\end{aligned} \tag{3.5.11}$$

An estimating function $G_n^*(\theta)$ is optimal within the class ζ if and only if

$$\frac{\langle G(\theta), G^*(\theta)\rangle_n}{\bar{G}_n(\theta)} = \frac{\langle G^*(\theta)\rangle_n}{\bar{G}_n^*(\theta)} \tag{3.5.12}$$

for all n and G in ζ by Heyde (1988); see also Chapter 5 of Prakasa Rao (1999). In our problem, (3.5.12) reduces to

$$\begin{aligned}&\frac{\sum_{i=1}^{n} g_{i-1}(\theta)g_{i-1}^*(\theta)\phi(X_{(i-1)\Delta}, \theta)}{-\sum_{i=1}^{n} g_{i-1}(\theta)\dot{F}(X_{(i-1)\Delta}, \theta)} \\ &= \frac{\sum_{i=1}^{n} g_{i-1}^{*2}(\theta)\phi(X_{(i-1)\Delta}, \theta)}{-\sum_{i=1}^{n} g_{i-1}^*(\theta)\dot{F}(X_{(i-1)\Delta}, \theta)},\end{aligned} \tag{3.5.13}$$

for all n and G in ζ. Equation (3.5.13) in turn holds if and only if

$$g_{i-1}^*(\theta) = \frac{\alpha(\theta)\dot{F}(X_{(i-1)\Delta}, \theta)}{\phi(X_{(i-1)\Delta}, \theta)}, \tag{3.5.14}$$

where $\alpha(\theta)$ is a nonrandom function of θ. Let us choose $\alpha(\theta) \equiv 1$. Then the optimal estimating function in ζ is of the form

$$G_n^*(\theta) = \sum_{i=1}^{n} \frac{\dot{F}(X_{(i-1)\Delta}, \theta)}{\phi(X_{(i-1)\Delta}, \theta)}\{X_{i\Delta} - F(X_{(i-1)\Delta}, \theta)\}. \tag{3.5.15}$$

Note that $\tilde{G}_n(\theta)$ defined by (3.5.6) can never be optimal since the equation

$$\frac{\dot{a}(x,\theta)}{\sigma^2(x,\theta)} = \frac{\dot{F}(x,\theta)}{\phi(x,\theta)}$$

does not hold in general. However, from the results in Florens-Zmirou (1984a, 1984b), it follows that

$$F(x,\theta) = x + \Delta\, a(x,\theta) + \tfrac{1}{2}\Delta^2\{a(x,\theta)a'(x,\theta) + \tfrac{1}{2}\sigma^2(x,\theta)a''(x,\theta)\} + o(\Delta^2) \tag{3.5.16}$$

and

$$\phi(x,\theta) = \Delta\,\sigma^2(x,\theta) + \Delta^2\{a(x,\theta)\sigma(x,\theta)\sigma'(x,\theta) + \frac{1}{2}\sigma^2(x,\theta)[2a'(x,\theta) + \sigma'(x,\theta)^2 + \sigma(x,\theta)\sigma''(x,\theta)]\} + O(\Delta^3), \tag{3.5.17}$$

where prime denotes differentiation with respect to x. Hence

$$\dot{F}(x,\theta) = \dot{a}(x,\theta)\Delta + O(\Delta^2),$$
$$\phi(x,\theta) = \sigma^2(x,\theta)\Delta + O(\Delta^2),$$

and, for small values of Δ, $\tilde{G}_n(\theta)$ is an approximation to the optimal function $G_n^*(\theta)$. Another approximately optimal estimating function can be obtained by using the expressions in (3.5.16) and (3.5.17) up to terms of third order in $G_n^*(\theta)$; see Bibby (1994) and Bibby and Sørensen (1995a, 1995c).

Let us now consider multidimensional version of the stochastic differential equation (3.5.1), where θ is k-dimensional, $\{X_t\}$ is d-dimensional, the drift coefficient $\boldsymbol{a}$ is d-dimensional, the diffusion coefficient $\boldsymbol{\sigma}$ is assumed to be a $d \times m$ matrix such that $\boldsymbol{\sigma}\boldsymbol{\sigma}^T$ is positive definite and the Wiener process is m-dimensional. Here $\boldsymbol{A}^T$ denotes the transpose of a matrix $\boldsymbol{A}$.

The $(k+1)$-dimensional estimating functions analogous to (3.5.6) and (3.5.15) are

$$\tilde{G}_n(\theta) = \sum_{i=1}^{n} \dot{\boldsymbol{a}}(\boldsymbol{X}_{(i-1)\Delta},\theta)^T\{\boldsymbol{\sigma}(\boldsymbol{X}_{(i-1)\Delta},\theta)\boldsymbol{\sigma}(\boldsymbol{X}_{(i-1)\Delta},\theta)^T\}^{-1}\{\boldsymbol{X}_{i\Delta} - \boldsymbol{F}(\boldsymbol{X}_{(i-1)\Delta},\theta)\} \tag{3.5.18}$$

and

$$G_n^*(\theta) = \sum_{i=1}^{n} \dot{\boldsymbol{F}}(\boldsymbol{X}_{(i-1)\Delta},\theta)^T\boldsymbol{\phi}(\boldsymbol{X}_{(i-1)\Delta},\theta)^{-1}\{\boldsymbol{X}_{i\Delta} - \boldsymbol{F}(\boldsymbol{X}_{(i-1)\Delta},\theta)\},$$

where $\dot{\boldsymbol{a}}$ and $\dot{\boldsymbol{F}}$ are the $d \times k$ matrices of partial derivatives of $\boldsymbol{a}$ and $\boldsymbol{F}$ with respect to the components of θ. Here we have assumed that $\boldsymbol{\phi}$ is positive definite.

Example 3.5.1 (*Ornstein–Uhlenbeck process*)
Consider the the stochastic differential equation

$$dX_t = \theta X_t\,dt + \sigma\,dW_t, \qquad X_0 = x_0,\ t \geq 0. \tag{3.5.19}$$

It is known in this case that the transition probability distribution is normal with mean $F(x,\theta) = xe^{\theta\Delta}$ and variance $\phi(\theta) = \sigma^2(e^{2\theta\Delta} - 1)/2\theta$. Hence

$$\tilde{G}_n(\theta) = \frac{1}{\sigma^2}\sum_{i=1}^{n}(X_{i\Delta} - X_{(i-1)\Delta}e^{\theta\Delta})X_{(i-1)\Delta}$$

from (3.5.6) and the equation $\tilde{G}_n(\theta) = 0$ leads to an estimator for θ of the form

$$\tilde{\theta}_n = \frac{1}{\Delta} \log \left\{ \frac{\sum_{i=1}^n X_{(i-1)\Delta} X_{i\Delta}}{\sum_{i=1}^n X_{(i-1)\Delta}^2} \right\} \tag{3.5.20}$$

provided that $\sum_{i=1}^n X_{(i-1)\Delta} X_{i\Delta} > 0$. It can be checked that G_n^* is proportional to $\tilde{G}_n$, leading to the same estimator. This is also the MLE when σ^2 is unknown.

Example 3.5.2 (*The mean-reverting process*)
Consider the class of diffusion processes satisfying the stochastic differential equation

$$dX_t = (\alpha + \theta X_t)\,dt + \sigma(X_t)\,dW_t, \qquad X_0 = X_0,\ t \geq 0, \tag{3.5.21}$$

where σ is a positive real-valued known function. The process satisfying (3.5.21) is referred to in the financial literature as the *mean-reverting process*. Here α and θ are both unknown. Let $f(t) = E_{\alpha,\theta}(X_t | X_0)$. Then

$$f'(t) = \alpha + \theta f(t)$$

from (3.5.21) and hence

$$F(x, \alpha, \theta) = x e^{\theta\Delta} + \frac{\alpha}{\theta}(e^{\theta\Delta} - 1).$$

This leads to the vector-valued estimating functions:

$$\tilde{G}_n(\alpha, \theta) = \left\{ \sum_{i=1}^n \frac{1}{\sigma^2(X_{(i-1)\Delta})} (X_{i\Delta} - X_{(i-1)\Delta} e^{\theta\Delta} + \frac{\alpha}{\theta}(1 - e^{\theta\Delta})), \right.$$
$$\left. \sum_{i=1}^n \frac{X_{(i-1)\Delta}}{\sigma^2(X_{(i-1)\Delta})} (X_{i\Delta} - X_{(i-1)\Delta} e^{\theta\Delta} + \frac{\alpha}{\theta}(1 - e^{\theta\Delta}) \right\}^T$$

and

$$G_n^*(\alpha, \theta) = \left\{ \sum_{i=1}^n \frac{e^{\theta\Delta} - 1}{\theta\, \phi(X_{(i-1)\Delta}, \alpha, \theta)} (X_{i\Delta} - X_{(i-1)\Delta} e^{\theta\Delta} + \frac{\alpha}{\theta}(1 - e^{\theta\Delta})), \right.$$
$$\sum_{i=1}^n \frac{\Delta e^{\theta\Delta}(X_{(i-1)\Delta} + \frac{\alpha}{\theta}) + \frac{\alpha}{\theta^2}(1 - e^{-\theta\Delta})}{\phi(X_{(i-1)\Delta}, \alpha, \theta)}$$
$$\left. \times (X_{i\Delta} - X_{(i-1)\Delta} e^{\theta\Delta} + \frac{\alpha}{\theta}(1 - e^{\theta\Delta})) \right\}^T.$$

In the case of $\tilde{G}_n$, we can solve the estimating equation

$$\tilde{G}_n(\alpha, \theta) = \begin{pmatrix} 0 \\ 0 \end{pmatrix}$$

directly and we obtain

$$e^{\tilde{\theta}_n \Delta} = \left[\left(\sum_{i=1}^n \frac{X_{(i-1)\Delta}}{\sigma^2(X_{(i-1)\Delta})} \right) \left(\sum_{i=1}^n \frac{X_{i\Delta}}{\sigma^2(X_{(i-1)\Delta})} \right) \right.$$

$$-\left(\sum_{i=1}^{n}\frac{X_{(i-1)\Delta}X_{i\Delta}}{\sigma^2(X_{(i-1)\Delta})}\right)\left(\sum_{i=1}^{n}\frac{1}{\sigma^2(X_{(i-1)\Delta})}\right)\Bigg]$$

$$\times\left[\left(\sum_{i=1}^{n}\frac{X_{(i-1)\Delta}}{\sigma^2(X_{(i-1)\Delta})}\right)^2-\left(\sum_{i=1}^{n}\frac{X^2_{(i-1)\Delta}}{\sigma^2(X_{(i-1)\Delta})}\right)\left(\sum_{i=1}^{n}\frac{1}{\sigma^2(X_{(i-1)\Delta})}\right)\right]^{-1} \tag{3.5.22}$$

and

$$\tilde{\alpha}_n=\frac{\tilde{\theta}_n}{1-e^{\tilde{\theta}_n\Delta}}\frac{\left(\sum_{i=1}^{n}\frac{X_{(i-1)\Delta}}{\sigma^2(X_{(i-1)\Delta})}\right)e^{\tilde{\theta}_n\Delta}-\left(\sum_{i=1}^{n}\frac{X_{i\Delta}}{\sigma^2(X_{(i-1)\Delta})}\right)}{\left(\sum_{i=1}^{n}\frac{1}{\sigma^2(X_{(i-1)\Delta})}\right)}.$$

If $\sigma(x)=\sigma\sqrt{x}$ (Cox *et al.* 1985), then the conditional second moment $g(t)\equiv E_{\alpha,\theta}(X_t^2|X_0)$ is a solution of the differential equation

$$g'(t)=(2\alpha+\sigma^2)f(t)+2\theta\, g(t)$$

and it can be checked that

$$\phi(x,\alpha,\theta)=\frac{\sigma^2}{2\theta^2}\{(\alpha+2\theta x)e^{2\theta\Delta}-2(\alpha+\theta x)e^{\theta\Delta}+\alpha\}.$$

If $\alpha=0$, θ is replaced by $-\theta$ in the drift and $\sigma(x)=\sqrt{\theta+x^2}$, where $\theta>0$, then the parameter θ of interest is involved in the diffusion coefficient in the mean-reverting process. In this case the conditional variance can be shown to be

$$\phi(x,\alpha,\theta)=x^2e^{-2\theta\Delta}(e^{\Delta}-1)+\frac{\theta}{2\theta-1}(1-e^{-(1-2\theta)\Delta});$$

see Bibby and Sørensen (1995a, 1995c).

Example 3.5.3 (*The hyperbolic diffusion process*)
Consider the stochastic differential equation

$$dX_t=\theta\frac{X_t}{\sqrt{1+X_t^2}}\,dt+\sigma\,dW_t,\qquad X_0=x_0,\ t\geq 0. \tag{3.5.23}$$

The solution of this differential equation is known as the *hyperbolic diffusion process* as it has a hyperbolic stationary distribution when $\theta<0$. The stationary density $g(x)$ is propotional to

$$\exp(\theta\sqrt{1+x^2}/\sigma^2).$$

As it is not possible to compute the conditional expectation explicitly, we can only write the estimating function $\tilde{G}_n(\theta)$ in the form

$$\tilde{G}_n(\theta)=\sum_{i=1}^{n}\frac{X_{(i-1)\Delta}}{\sigma^2\sqrt{1+X^2_{(i-1)\Delta}}}\{X_{i\Delta}-F(X_{(i-1)\Delta},\theta)\}. \tag{3.5.24}$$

3.5.2 Consistency and Asymptotic Normality

Let us consider observations at the discrete time points $i\Delta$, $i = 0, 1, \ldots n$, of a process $\{X_t\}$ which is a solution of (3.5.1). Suppose $\theta \in \Theta \subset R$, with Θ open. Let

$$s(x, \theta) = \exp\left(-2\int_0^x \frac{a(y, \theta)}{\sigma^2(y, \theta)}\,dy\right)$$

and suppose the following condition holds for all $\theta \in \Theta$:

(FF1) (a) $\int_0^\infty s(x, \theta)\,dx = \int_{-\infty}^0 s(x, \theta)\,dx = \infty$.

(b) $\int_{-\infty}^\infty \{s(x, \theta)\sigma^2(x, \theta)\}^{-1}\,dx = A(\theta) < \infty$.

(c) There exist constants $M > 0$ and $c > 0$ independent of θ such that

$$\left\{\frac{a(x, \theta)}{\sigma(x, \theta)} - \frac{1}{2}\sigma'(x, \theta)\right\}\operatorname{sgn}(x) \le -c \qquad \text{for } |x| > M.$$

Under conditions (FF1)(a) and (FF1)(b), the process X is ergodic with an invariant measure μ_θ having density

$$\frac{d\mu_\theta}{dx} = \{A(\theta)\sigma^2(x, \theta)s(x, \theta)\}^{-1}$$

with respect to the Lebesgue measure. Let P_θ be the measure corresponding to θ when $X_0 = x_0$. Define F and ϕ as in (3.5.4) and (3.5.9), respectively. Let us consider a general estimating function of the form

$$G_n(\theta) = \sum_{i=1}^n g(X_{(i-1)\Delta}, \theta)(X_{i\Delta} - F(X_{(i-1)\Delta}, \theta)).$$

Let $\theta_0 \in \Theta$ be the true parameter. In addition to (FF1), suppose the following condition holds for $G_n(\theta)$, for all $\theta \in \Theta$:

(FF2) (a) The functions $g(x, \theta)$ and $F(x, \theta)$ are continuously differentiable with respect to θ for all x.

(b) The function

$$h(\theta; x, y) = \dot{g}(x, \theta)\{y - F(x, \theta)\} - g(x, \theta)\dot{F}(x, \theta)$$

is locally dominated square-integrable with respect to $Q_\Delta^{\theta_0}$, where Q_t^θ is the measure defined by $\pi_t^\theta \otimes \mu_\theta$ with $\pi_t^\theta(dy, x) = P_\theta(X_t \in dy | X_0 = x)$; and

$$f(\theta_0) = Q_\Delta^{\theta_0}\{h(\theta_0)\} = -E_{\mu_{\theta_0}}(g(Z, \theta_0)\dot{F}(Z, \theta_0)) \neq 0,$$

where Z is a random variable with probability measure μ_{θ_0}.

(c) The function $g(x, \theta_0)(y - F(x, \theta_0))$ is in $L^2(Q_\Delta^{\theta_0})$ and

$$v(\theta_0) = E_{\mu_{\theta_0}}(g(Z, \theta_0)^2\phi(Z, \theta_0)\} > 0.$$

Following results in Florens-Zmirou (1989) and Barndorff-Nielsen and Sørensen (1994), Bibby and Sørensen (1995a) proved the following theorem giving the consistency and the asymptotic normality of the estimator obtained from the estimating equation $\tilde{G}_n(\theta) = 0$. We omit the details.

Theorem 3.5.1 *Under conditions (FF1) and (FF2), an estimator $\hat{\theta}_n$ exists for every n, which on a set C_n solves the equation*

$$G_n(\hat{\theta}_n) = 0, \tag{3.5.25}$$

where $P_{\theta_0}(C_n) \to 1$ as $n \to \infty$. Moreovr, as $n \to \infty$,

$$\hat{\theta}_n \to \theta_0 \text{ in probability under } P_{\theta_0} \tag{3.5.26}$$

and

$$\sqrt{n}(\hat{\theta}_n - \theta_0) \xrightarrow{\mathcal{L}} N(0, v(\theta_0)/f(\theta_0)^2) \text{ as } n \to \infty \text{ under } P_{\theta_0}. \tag{3.5.27}$$

Remarks. Bibby and Sørensen (1995a) investigate the practical problems involved in implementing the estimation procedures discussed here through simulation studies of the processes discussed in Examples 3.5.1 and 3.5.2. These studies show that the estimators discussed here give considerable improvement over estimators based on the discretized continuous-time likelihood function.

3.5.3 Quadratic Estimating Functions

In the previous subsection, we discussed optimal first-order polynomial (linear) martingale estimating functions closely related to the discretized continuous-time log-likelihood function. We now briefly discuss second-order polynomial (quadratic) martingale estimating functions and their applicability for estimation purposes.

Consider the stochastic differential equation

$$dX_t = a(X_t, \theta)\, dt + \sigma(X_t, \theta)\, dW_t, \qquad X_0 = x_0,\ t \geq 0 \tag{3.5.28}$$

as before, where $\sigma(\cdot, \cdot)$ is positive and a and σ are known and twice continuously differentiable with respect to both the arguments. The likelihood function for θ based on the observation $X_{i\Delta}, 1 \leq i \leq n$, is

$$L_n(\theta) = \prod_{i=1}^{n} p(\Delta, X_{(i-1)\Delta}, X_{i\Delta}; \theta), \tag{3.5.29}$$

where $y \to p(\Delta, x, y; \theta)$ is the density of X_Δ given $X_0 = x$ when θ is the true parameter. The transition density p is in general difficult to compute, but, for small Δ, it can be approximated by a normal distribution with mean

$$F(x, \theta) = E_\theta(X_\Delta | X_0 = x) \tag{3.5.30}$$

and variance

$$\phi(x, \theta) = \text{var}_\theta(X_\Delta | X_0 = x). \tag{3.5.31}$$

Hence an approximate log-likelihood function is

$$\ell_n^*(\theta) = -\tfrac{1}{2}\sum_{i=1}^{n} \log \phi(X_{(i-1)\Delta}, \theta) - \tfrac{1}{2}\sum_{i=1}^{n} \phi(X_{(i-1)\Delta}, \theta)^{-1}(X_{i\Delta} - F(X_{(i-1)\Delta}, \theta))^2$$

and an approximate score function (derivative of the log-likelihood function) is

$$\dot{\ell}_n^*(\theta) = \sum_{i=1}^{n} \left\{ \frac{\dot{F}(X_{(i-1)\Delta}, \theta)}{\phi(X_{(i-1)\Delta}, \theta)} (X_{i\Delta} - F(X_{(i-1)\Delta}, \theta)) + \frac{\dot{\phi}(X_{(i-1)\Delta}, \theta)}{2\phi(X_{(i-1)\Delta}, \theta)^2} [(X_{i\Delta} - F(X_{(i-1)\Delta}, \theta))^2 - \phi(X_{(i-1)\Delta}, \theta)] \right\}. \tag{3.5.32}$$

Note that $\dot{\ell}_n^*(\theta)$ is a martingale with respect to the filtration $\mathcal{F}_i = \sigma(X_\Delta, \ldots, X_{i\Delta}), i \geq 1$. It was seen that the estimator obtained by solving the equation $\dot{\ell}_n^*(\theta) = 0$ is consistent and asymptotically normal under some regularity conditions. However the estimating function $\dot{\ell}_n^*(\theta)$ need not be optimal in the class $\zeta^{(2)}$ of quadratic martingale estimating functions, that is, the estimating functions of the form

$$K_n(\theta) = \sum_{i=1}^{n} \{g_{i-1}(\theta)(X_{i\Delta} - F(X_{(i-1)\Delta}, \theta)) + h_{i-1}(\theta)[(X_{i\Delta} - F(X_{(i-1)\Delta}, \theta))^2 - \phi(X_{(i-1)\Delta}, \theta)]\}, \tag{3.5.33}$$

where $g_{i-1}(\theta)$ and $h_{i-1}(\theta)$ are $\mathcal{F}_{i-1}$-measurable. Following Heyde (1988) (see also Chapter 5 of Prakasa Rao 1999), it can be shown that the optimal estimating function is given by $K_n^*(\theta)$ with

$$g_{i-1}^*(\theta) = \frac{\dot{\phi}(X_{(i-1)\Delta}, \theta)\eta(X_{(i-1)\Delta}, \theta) - \dot{F}(X_{(i-1)\Delta}, \theta)\psi(X_{(i-1)\Delta}, \theta)}{\phi(X_{(i-1)\Delta}, \theta)\psi(X_{(i-1)\Delta}, \theta) - \eta(X_{(i-1)\Delta}, \theta)^2},$$

$$h_{i-1}^*(\theta) = \frac{\dot{F}(X_{(i-1)\Delta}, \theta)\eta(X_{(i-1)\Delta}, \theta) - \dot{\phi}(X_{(i-1)\Delta}, \theta)\phi(X_{(i-1)\Delta}, \theta)}{\phi(X_{(i-1)\Delta}, \theta)\psi(X_{(i-1)\Delta}, \theta) - \eta(X_{(i-1)\Delta}, \theta)^2}, \tag{3.5.34}$$

where

$$\eta(x, \theta) = E_\theta[(X_\Delta - F(x, \theta))^3 | X_0 = x]$$

and

$$\psi(x, \theta) = E_\theta[(X_\Delta - F(x, \theta))^4 | X_0 = x] - \phi(x, \theta)^2$$

(see Kessler 1995b, 1995c; Bibby and Sørensen 1995a). Note that the optimal estimating function $K_n^*(\theta)$ yields an estimator which has an asymptotic variance which is minimal within the class $\zeta^{(2)}$ and that the L_2 distance of K_n^* to $\dot{\ell}_n(\theta)$ is smaller than any other $K_n(\theta)$ in $\zeta^{(2)}$ to $\dot{\ell}_n(\theta)$, where $\dot{\ell}_n(\theta)$ is the derivative of $\ell_n(\theta) = \log L_n(\theta)$ defined by (3.5.29). Using the approximation $F(x, \theta) = x + a(x, \theta)\Delta$ and $\phi(x, \theta) = \sigma^2(x, \theta)\Delta$, one can obtain an approximately optimal estimating function $\tilde{K}_n(\theta)$ given by

$$\tilde{K}_n(\theta) = \sum_{i=1}^{n} \left\{ \frac{\dot{a}(X_{(i-1)\Delta}, \theta)}{\sigma^2(X_{(i-1)\Delta}, \theta)} (X_{i\Delta} - F(X_{(i-1)\Delta}, \theta)) + \frac{\dot{\sigma}(X_{(i-1)\Delta}; \theta)}{\sigma^3(X_{(i-1)\Delta}; \theta)\Delta} [(X_{i\Delta} - F(X_{(i-1)\Delta}, \theta)^2 - \phi(X_{(i-1)\Delta}, \theta)] \right\}. \tag{3.5.35}$$

Bibby and Sørensen (1995a) study the properties of estimators obtained from estimating functions of the form

$$K_n(\theta) = \sum_{i=1}^{n} k(\theta; X_{(i-1)\Delta}, X_{i\Delta}), \tag{3.5.36}$$

with $k : R^2 \to R$ of the form

$$k(\theta; x, y) = g(x, \theta)(y - F(x, \theta)) + h(x, \theta)((y - F(x, \theta))^2 - \phi(x, \theta)). \tag{3.5.37}$$

We omit the details.

Example 3.5.4 (*Cox–Ingersoll–Ross model (Example 3.5.2 continued)*)
Consider the diffusion process satisfying the stochastic differential equation

$$dX_t = (\alpha + \theta X_t)\,dt + \sigma\sqrt{X_t}\,dW_t, \qquad X_0 = x_0,\ \theta < 0,\ \sigma > 0, \tag{3.5.38}$$

which is widely used for modeling interest rates in mathematical finance (Cox *et al.* 1985). It can be checked that

$$F(x, \alpha, \theta) = \frac{1}{\theta}[(\alpha + \theta x)e^{\theta\Delta} - \alpha],$$

$$\phi(x, \alpha, \theta, \sigma) = \frac{\sigma^2}{2\theta^2}[(\alpha + 2\theta x)e^{2\theta\Delta} - 2(\alpha + \theta x)e^{\theta\Delta} + \alpha],$$

$$\eta(x, \alpha, \theta, \sigma) = \frac{\sigma^4}{2\theta^3}[(\alpha + 3\theta x)e^{3\theta\Delta} - 3(\alpha + 2\theta x)e^{2\theta\Delta} + 3(\alpha + \theta x)e^{\theta\Delta} - \alpha]$$

and

$$\begin{aligned}\psi(x, \alpha, \theta, \sigma) = {}& \frac{3\sigma^6}{4\theta^4}[(\alpha + 4\theta x)e^{4\theta\Delta} - 4(\alpha + 3\theta x)e^{3\theta\Delta} \\ & +6(\alpha + 2\theta x)e^{2\theta\Delta} - 4(\alpha + \theta x)e^{\theta\Delta} + \alpha] \\ & +2\phi^2(x, \alpha, \theta, \sigma).\end{aligned}$$

An explicit estimator for σ^2, can be obtained using the approximate optimal estimating function given by (3.5.35). The corresponding martingale estimating function has the solution

$$\tilde{\sigma}^2 = \frac{\sum_{i=1}^{n} X_{(i-1)\Delta}^{-1}(X_{i\Delta} - F(X_{(i-1)\Delta}, \alpha, \theta))^2}{\sum_{i=1}^{n} X_{(i-1)\Delta}^{-1}\tilde{\phi}(X_{(i-1)\Delta}, \alpha, \theta)}, \tag{3.5.39}$$

where $\tilde{\phi} = \phi/\sigma^2$ and ϕ is as defined above. For α and θ, initial estimates can be obtained by using the estimating function

$$G_n^*(\alpha, \theta) = \sum_{i=1}^{n} \frac{\dot{F}(X_{(i-1)\Delta}, \alpha, \theta)}{\phi(X_{(i-1)\Delta}, \alpha, \theta)}(X_{i\Delta} - F(X_{(i-1)\Delta}, \alpha, \theta)),$$

which is optimal in the class ζ as discussed earlier. Another estimator,

$$\hat{\sigma}^2 = \frac{1}{n}\sum_{i=1}^{n} \tilde{\phi}(X_{(i-1)\Delta}, \alpha, \theta)^{-1}(X_{i\Delta} - F(X_{(i-1)\Delta}, \alpha, \theta))^2, \tag{3.5.40}$$

for σ^2 can be obtained from the estimates of α and θ obtained by solving the equation $\dot{\ell}_n^*(\alpha, \theta) = 0$, where $\dot{\ell}_n^*(\alpha, \theta)$ is of the form (3.5.32).

3.5.4 Estimating Functions Based on Eigenfunctions

We have considered estimating functions of the type

$$G_n(\theta) = \sum_{i=1}^{n} g(X_{t_i}, X_{t_{i-1}}; \theta), \tag{3.5.41}$$

where $X_{t_0}, X_{t_1}, \ldots, X_{t_n}$ are the data observed at time points $t_0, t_1, \ldots, t_n$, respectively and $g(y, x; \theta)$ is a polynomial in y such that

$$E_\theta[g(X_{t_i}, X_{t_{i-1}}; \theta)|X_{t_{i-1}}] = 0, \tag{3.5.42}$$

for all $\theta \in \Theta$. Such polynomial estimating functions can be regarded as good approximations to the true score functions when the time interval between the observation times is small and the transition distribution can be approximated by a Gaussian law. If the time interval between the observation times is large and the Gaussian approximation is far from the true transition distribution, such polynomial estimating functions may not be the right choice.

Here we consider estimating functions based on the eigenfunctions for the generator of the diffusion model. This allows the estimating function to be closely tailored to the diffusion model under discussion. Karlin and Taylor (1981) gives an expansion for the transition density of a diffusion process in terms of the eigenfunctions of its generator.

Consider the stochastic differential equation

$$dX_t = a(X_t, \theta)\,dt + \sigma(X_t, \theta)\,dW_t, \qquad X_0 = x_0,\ t \geq 0, \tag{3.5.43}$$

where a and σ are known but for the parameter $\theta \in \Theta \subset R$. We assume as before that a and σ are smooth so that (3.5.43) has a unique solution denoted by P_θ for each θ. The problem is to estimate θ based on $X_{t_0}, X_{t_1}, \ldots, X_{t_n}$. For simplicity, we assume that $t_i - t_{i-1} = \Delta$.

The differential operator L_θ defined by

$$L_\theta = \frac{1}{2}\sigma^2(x, \theta)\frac{d^2}{dx^2} + a(x, \theta)\frac{d}{dx} \tag{3.5.44}$$

for all twice differentiable functions is the *generator* of the diffusion model (3.5.43). A twice continuously differentiable function $\phi(x, \theta)$ is called an eigenfunction for L_θ with the eigenvalue $\lambda(\theta)$ if

$$L_\theta\phi(x, \theta) = -\lambda(\theta)\phi(x, \theta) \tag{3.5.45}$$

for all x in the state space of X.

Under some regularity conditions, it can be shown that

$$E_\theta[\phi(X_{t_i}, \theta)|X_{t_{i-1}}] = e^{-\lambda(\theta)\Delta}\phi(X_{t_{i-1}}, \theta) \tag{3.5.46}$$

(Kessler and Sørensen 1995). Hence we can define a martingale estimating function $G_n(\theta)$ defined by (3.5.41) with

$$g(y, x; \theta) = \alpha(x, \theta)[\phi(y, \theta) - e^{-\lambda(\theta)\Delta}\phi(x, \theta)], \tag{3.5.47}$$

where $\alpha(x, \theta)$ is an arbitrary function. In general, if we have k eigenfunctions $\phi_1(x, \theta), \ldots, \phi_k(x, \theta)$ with distinct eigenvalues $\lambda_1(\theta), \ldots, \lambda_k(\theta)$ respectively, we can define a martingale estimating function $G_n(\theta)$ by (3.5.41), with

$$g(y, x; \theta) = \sum_{j=1}^{k} \alpha_j(x, \theta)[\phi_j(y, \theta) - e^{-\lambda_j(\theta)\Delta}\phi_j(x, \theta)], \tag{3.5.48}$$

where $\alpha_j(x,\theta)$, $1 \le j \le k$, are arbitrary functions. It is known from the spectral theory for diffusion processes that the set Λ_θ of all eigenvalues for L_θ is contained in $[0,\infty)$. The set Λ_θ is the *spectrum* of L_θ. For many diffusion models, the spectrum is a discrete set $\Lambda_\theta = \{\lambda_0(\theta), \lambda_1(\theta), \dots\}$ where

$$0 \le \lambda_0(\theta) < \lambda_1(\theta) < \cdots < \lambda_n(\theta) \uparrow \infty.$$

In this case, one can use the first k eigenfunctions to define $g(y,x;\theta)$ as in (3.5.48). If the spectrum is not discrete, then either one can choose k eigenfunctions associated with the eigenvalues $0 < \lambda_1(\theta) < \cdots < \lambda_k(\theta)$ and define $g(y,x;\theta)$ as in (3.5.48) or consider functions $g(y,x;\theta)$ of the type

$$g(y,x;\theta) = \int_{\Lambda_\theta} [\phi_\lambda(y,\theta) - e^{-\lambda\Delta}\phi_\lambda(x,\theta)]\, m_\theta(x, d\lambda) \tag{3.5.49}$$

where ϕ_λ is the eigenfunction associated with the eigenvalue λ and $m_\theta(x, d\lambda)$ is a measure on Λ_θ such that (3.5.49) exists.

Suppose the diffusion process is ergodic with the invariant measure μ_θ. Note that the eigenfunctions $\{\phi_i(x,\theta)\}$ have the following properties:

$$\phi_i(\cdot,\theta) \in L^2(\mu_\theta); \tag{3.5.50a}$$

$$\int \phi_i(y,\theta)\,\mu_\theta(dy) = 0; \tag{3.5.50b}$$

$$\int \phi_i(y,\theta)\phi_j(y,\theta)\,\mu_\theta(dy) = 0 \quad \text{for } i \ne j \tag{3.5.50c}$$

(Karlin and Taylor 1981).

Optimal choice

We now discuss how to construct an optimal martingale estimating function from k given eigenfunctions $\phi_j(x,\theta)$, $1 \le j \le k$, optimal in the sense of Godambe and Heyde (1987); see also Chapter 5 of Prakasa Rao (1999).

Let $\alpha_j(x,\theta)$, $1 \le j \le k$, be functions which are continuously differentiable with respect to θ. Consider the class $\zeta^{(k)}$ of functions $G_n(\theta)$ defined by

$$G_n(\theta) = \sum_{i=1}^{n}\sum_{j=1}^{k} \alpha_j(X_{t_{i-1}},\theta)(\phi_j(X_{t_i},\theta) - e^{-\lambda_j(\theta)\Delta}\phi_j(X_{t_{i-1}},\theta)). \tag{3.5.51}$$

Let $\pi_\Delta(\cdot, x;\theta)$ be the density, under P_θ, of X_Δ given $X_0 = x$. Define, for $1 \le i, j \le k$,

$$a_{ij}(x,\theta) = \int \phi_i(y,\theta)\phi_j(y,\theta)\pi_\Delta(y,x;\theta)\,dy - e^{-(\lambda_i(\theta)+\lambda_j(\theta))\Delta}\phi_i(x,\theta)\phi_j(x,\theta) \tag{3.5.52}$$

and, for $1 \le j \le k$,

$$b_j(x,\theta) = -\int \frac{\partial\phi_j(y,\theta)}{\partial\theta}\pi_\Delta(y,x;\theta)\,dy + \frac{\partial}{\partial\theta}\{e^{-\lambda_j(\theta)\Delta}\phi_j(x,\theta)\}. \tag{3.5.53}$$

Let $\boldsymbol{A} = ((a_{ij}))_{k\times k}$ and $\boldsymbol{B} = (b_1,\dots,b_k)^T$.

Kessler and Sørensen (1995) have proved that the optimal estimating function in the sense of Godambe and Heyde (1987) is given by $G_n^*(\theta)$ as defined by (3.5.51) with $\alpha_j = \alpha_j^*$, $1 \le j \le k$, satisfying the relation

$$A \begin{pmatrix} \alpha_1^* \\ \vdots \\ \alpha_k^* \end{pmatrix} = B. \tag{3.5.54}$$

We omit the proof.

Remarks. (i) If θ is vector-valued and $\theta \in \Theta \subset R^d$, then the optimal estimating function based on the eigenfunctions $\{\phi_j, 1 \le j \le k\}$ and the eigenvalues $\{\lambda_j, 1 \le j \le k\}$ is a d-dimensional vector the ith component of which is the projection in $L^2(\pi_\Delta(y, x; \theta)\, dy)$ of $\frac{\partial \pi_\Delta}{\partial \theta_j} \frac{1}{\pi_\Delta}$ onto the subspace $\gamma^{(k)}(x, \theta)$ defined by

$$\begin{aligned} \gamma^{(k)}(x, \theta) &= L^2(\pi_\Delta(y, x; \theta)\, dy) \cap \{g^{(k)}(\cdot, x; \theta) : g^{(k)}(y, x; \theta) \\ &= \sum_{j=1}^{k} \alpha_j(x, \theta)(\phi_j(y, \theta) - e^{-\lambda_j(\theta)\Delta}\phi_j(x, \theta)) \\ &\quad \text{for some } \alpha_i \text{ in } C, 1 \le i \le k\}. \end{aligned} \tag{3.5.55}$$

Here C is the space of $\Theta \times R \times R$ continuously differentiable functions with respect to θ. The estimating function is determined by a system of equations similar to (3.5.54) with $\partial/\partial\theta$ replaced by $\partial/\partial\theta_i$.

(ii) Unlike the linear or quadratic estimating functions, the estimating function derived from the eigenfunction is *invariant under a data transformation.* This can be seen from the fact that if u is a twice continuously differentiable bijection, then the diffusion $Y_t = u(X_t)$ has the eigenfunctions $\phi_j(u^{-1}(y), \theta)$ with the same eigenvalues $\lambda_j(\theta), j = 0, 1, 2, \ldots$, as the diffusion X and hence, by a change of variable approach, it follows that the estimating function based on the transformed $Y_{t_i} = u(X_{t_i}), 1 \le i \le n$, equals the one based on $X_{t_i}, 1 \le i \le n$. This property does not hold for the polynomial estimating functions.

(iii) For some diffusion processes, the eigenfunctions ϕ_j are polynomials of degree j and one can obtain an explicit optimal polynomial estimating function as given in the following examples.

Example 3.5.5 (*Cox–Ingersoll–Ross Model (Example 3.5.2 continued)*)
Consider the diffusion process satisfying the stochastic differential equation

$$dX_t = (\alpha + \theta X_t)\, dt + \sigma\sqrt{X_t}\, dW_t, \qquad X_0 = x_0,\ t \ge 0, \tag{3.5.56}$$

where $\alpha > 0, \theta < 0$ and $\sigma > 0$. For this process, the spectrum is

$$\Lambda_\theta = \{-n\theta : n = 0, 1, 2, \ldots\} \tag{3.5.57}$$

with corresponding eigenfunctions

$$\phi_i(x) = L_i^{(\nu)}(-2\theta x \sigma^{-2}), \tag{3.5.58}$$

where $L_i^{(\nu)}$ is the ith order Laguerre polynomial with the parameter $\nu = 2\beta\sigma^{-2} - 1$ (Karlin and Taylor 1981). The Laguerre polynomials are given by

$$L_i^{(\nu)}(x) = \sum_{m=0}^{i} (-1)^m \binom{i+\nu}{i-m} \frac{x^m}{m!} \tag{3.5.59}$$

and we have an example where polynomial estimating functions arise.

Example 3.5.6
Consider the diffusion process satisfying the stochastic differential equation

$$dX_t = -\theta \tan(X_t)\, dt + dW_t, \qquad t \geq 0, \tag{3.5.60}$$

where $\theta \geq 1/2$. Then this process is an ergodic diffusion process on the interval $(-\pi/2, \pi/2)$ and the invariant measure has a density proportional to $(\cos x)^{2\theta}$. The spectrum of the above model is

$$\Lambda_\theta = \{n(\theta + n/2) : n = 0, 1, 2, \ldots\}, \tag{3.5.61}$$

with corresponding eigenfunctions given by

$$\phi_i(x, \theta) = C_i^\theta(\sin x), \tag{3.5.62}$$

where C_i^θ is a Gegenbauer polynomial of order i. Note that the Gegenbauer polynomial of order i is a solution of the differential equation

$$f''(y) + \frac{(2\theta + 1)y}{y^2 - 1} f'(y) - \frac{i(2\theta + i)}{y^2 - 1} f(y) = 0,$$

so that $\phi_i(x, \theta)$ is a solution of the differential equation

$$\frac{1}{2}\phi_i''(x, \theta) - \theta(\tan x)\phi_i'(x, \theta) = -i\left(\theta + \frac{i}{2}\right)\phi_i(x, \theta).$$

Here ϕ_i is the eigenfunction for the generator of the model corresponding to the eigenvalue $i(\theta + i/2)$. It follows from Gradshteyn and Ryzhik (1965) that

$$\phi_i(x, \theta) = \sum_{m=0}^{i} \binom{\theta - 1 + m}{m} \binom{\theta - 1 + i - m}{i - m} \cos[(2m - i)(\pi/2 - x)]. \tag{3.5.63}$$

For $i = 1$, the eigenfunction is $c \sin x$ for some constant c corresponding to the eigenvalue $\theta + \frac{1}{2}$. From the martingale estimating function

$$\sum_{i=1}^{n} \sin X_{t_i} [\sin X_{t_i} - e^{-(\theta + \frac{1}{2})\Delta} \sin X_{t_{i-1}}], \tag{3.5.64}$$

one can obtain the estimator

$$\hat{\theta}_n = -\frac{1}{\Delta} \log\left(\frac{\sum_{i=1}^{n} \sin X_{t_{i-1}} \sin X_{t_i}}{\sum_{i=1}^{n} \sin^2 X_{t_{i-1}}}\right) - \frac{1}{2}. \tag{3.5.65}$$

This is not necessarily optimal. However it is $\sqrt{n}$-consistent and asymptotically normal.

Example 3.5.7 (*Generalized logistic diffusion*)
Consider a diffusion process with drift coefficient

$$a(x,\theta_1,\theta_2) = [(\theta_1-\theta_2)\cosh(x/2) - (\theta_1+\theta_2)\sinh(x/2)]\cosh(x/2), \tag{3.5.66}$$

where $\theta_1 > 0$ and $\theta_2 > 0$ and diffusion coefficient

$$\sigma(x) = 2\cosh(x/2). \tag{3.5.67}$$

This process is an ergodic diffusion process on the real line with the invariant measure having the generalized logistic density

$$B(\theta_1+1,\theta_2+1)e^{(\theta_1+1)x}(1+e^x)^{-(\theta_1+\theta_2+2)}.$$

Here $B(\cdot,\cdot)$ is the beta function. This distribution was studied in Barndorff-Nielsen *et al.* (1982). Such models are of use in mathematical finance and the theory of turbulence where proceses with marginal distributions with log-linear tails are used for modeling. This diffusion process has a discrete spectrum given by

$$\Lambda_{\theta_1,\theta_2} = \left\{\frac{n}{2}(n+\theta_1+\theta_2+1) : n = 0,1,\ldots\right\} \tag{3.5.68}$$

and the associated eigenfunctions are

$$\phi_i(x,\theta_1,\theta_2) = P_i^{\theta_2,\theta_1}(\tanh(x/2)), \tag{3.5.69}$$

where $P_i^{\theta_2,\theta_1}$ denotes the Jacobi polynomial of order i given by

$$P_i^{\theta_2,\theta_1}(y) = 2^{-1}\sum_{m=0}^{i}\binom{i+\theta_2}{m}\binom{i+\theta_1}{i-m}(y-1)^{i-m}(y+1)^m. \tag{3.5.70}$$

Note that the Jacobi polynomial $P_i^{\theta_2,\theta_1}(y)$ is a solution of the differential equation

$$(1-y^2)f''(y) + [\theta_1-\theta_2-(\theta_1+\theta_2+2)y]f'(y) + i(i+\theta_1+\theta_2+1)f(y) = 0$$

and, from this equation, it follows that $\phi_i(x,\theta_1,\theta_2)$ is a solution of the equation

$$\begin{aligned}
&2\cosh^2(x/2)\phi_i''(x,\theta_1,\theta_2)\\
&\quad +[(\theta_1-\theta_2)\cosh(x/2) - (\theta_1+\theta_2)\sinh(x/2)]\cosh(x/2)\phi_i'(x,\theta_1,\theta_2)\\
&= -\tfrac{1}{2}i(i+\theta_1+\theta_2+1)\phi_i(x,\theta_1,\theta_2).
\end{aligned}$$

and that $\phi_i(x,\theta_1,\theta_2)$ is an eigenfunction of the generator of the model with eigenvalue $\frac{1}{2}i(i+\theta_1+\theta_2+1)$.

Example 3.5.8 (*Stopped Bessel process*)
Consider the Bessel process defined by the equation

$$dX_t = \theta X_t^{-1}\,dt + dW_t, \qquad t \geq 0, \tag{3.5.71}$$

where the state space is $(0, \infty)$. Suppose the process $\{X_t\}$ has started at a point in the interval $(0, 1)$ and stopped as soon as the process hits unity. Further suppose that $\theta \geq 1/2$. Then the eigenfunctions are

$$\phi_i(x, \theta) = x^{-\theta+1/2} J_{\theta-1/2}(x\, \zeta_i(\theta)), \qquad i \geq 1, \tag{3.5.72}$$

where J_ν is a Bessel function and $\{\zeta_i(\theta); i = 1, 2, \ldots\}$ is the sequence of positive zeros of the function $J_{\theta-1/2}$ (Karlin and Taylor 1981). The spectrum is

$$\{2\zeta_i^2(\theta), i = 1, 2, \ldots\}.$$

Example 3.5.9 (*Radial Ornstein–Uhlenbeck process*)
The class of processes satisfying the stochastic differential equation

$$dX_t = (\theta X_t^{-1} - X_t)\, dt + dW_t, \qquad t \geq 0, \tag{3.5.73}$$

where $\theta > 0$, is called the radial Ornstein–Uhlenbeck process. Here the state space is $(0, \infty)$ and the process is ergodic if $\theta \geq 1/2$. The eigenfunctions are $\phi_i(x, \theta) = L_i^{(\theta-1/2)}(x^2)$, where $L_i^{(\nu)}$ is the ith-order Laguerre polynomial with parameter ν (see Example 3.5.5). The associated eigenvalues are $\{2i; i = 1, 2, \ldots\}$. It is easy to check that the optimal estimating function based on the first eigenfunction is

$$G_n(\theta) = \sum_{i=1}^{n} \frac{X_{t_i}^2 - e^{-2\Delta} X_{t_{i-1}}^2 - (1 - e^{-2\Delta})(\theta + 1/2)}{2X_{t_{i-1}}^2 + (\theta + 1/2)(e^{2\Delta} - 1)}.$$

One can obtain an approximate optimal estimating function for Δ small by omitting the last term in the numerator. This leads to an estimating function which is linear in θ and an explicit solution can be computed.

Remarks. Let

$$G_n(\theta) = \sum_{i=1}^{n} \sum_{j=1}^{k} \alpha_j(X_{t_{i-1}}, \theta)(\phi_j(X_{t_i}, \theta) - e^{-\lambda_j(\theta)\Delta} \phi_j(X_{t_{i-1}}, \theta)).$$

Under conditions similar to (FF1) and (FF2), one can prove that there exists a solution $\hat{\theta}_n$ for the equation $G_n(\theta) = 0$ with probability tending to one as $n \to \infty$ under P_{θ_0}. Moreover, as $n \to \infty$, $\hat{\theta}_n \to \theta_0$ in probability under P_{θ_0} with $\sqrt{n}(\hat{\theta}_n - \theta_0)$ asymptotically normal with mean 0 and variance $v(\theta_0)/f^2(\theta_0)$, where

$$v(\theta_0) = \sum_{r=1}^{k} \sum_{s=1}^{k} \int \alpha_r(x, \theta_0) \alpha_s(x, \theta_0) a_{r,s}(x, \theta_0) \mu_{\theta_0}(x)\, dx$$

and

$$f(\theta_0) = \sum_{j=1}^{k} Q_\Delta^{\theta_0}\left(\left. \frac{\partial g_j}{\partial \theta} \right|_{\theta=\theta_0} \right).$$

Here $Q_\Delta^\theta(x, y) = \mu_\theta(x) \times \pi_\Delta(x, y; \theta)$, $Q_\Delta^\theta(g)$ denotes the integral of g under the measure Q_Δ^θ and $a_{rs}(x, \theta)$ is as defined by (3.5.52). It is assumed that $f(\theta_0) \neq 0$. For a proof, see Kessler and Sørensen (1995). For the optimal estimating function, $v(\theta_0) = f(\theta_0)$ and the asymptotic variance is equal to $1/v(\theta_0)$.

Example 3.5.10 (*Example 3.5.6 continued*)
Consider the process defined by (3.5.60) and the estimating function given by (3.5.64). It can be checked that $\sqrt{n}(\hat{\theta}_n - \theta_0)$ is asymptotically normal with mean zero and variance

$$\frac{v(\theta_0)}{f^2(\theta_0)} = \frac{1 - e^{-2(\theta_0+1)\Delta}}{\Delta^2 e^{-2(\theta_0+1/2)\Delta}} + 3\frac{(\theta_0+1)(e^{-\Delta}-1)}{(\theta_0+2)\Delta^2},$$

where $\hat{\theta}_n$ is defined by (3.5.65).

Example 3.5.11 (*Bibby–Sørensen model*)
It is observed that the logarithm of stock prices is a process with nearly uncorrelated increments but that the increments are not independent (Taylor 1986). Bibby and Sørensen (1995b) suggested a model for the stock price S_t of the form

$$S_t = \exp(kt + X_t), \qquad t \geq 0, \tag{3.5.74}$$

where

$$X_t = X_0 + \int_0^t v(X_s)\,dW_s, \qquad t \geq 0. \tag{3.5.75}$$

Applying Itô's formula, it follows that

$$dS_t = S_t\{[k + \tfrac{1}{2}v^2(\log S_t - kt)]\,dt + v(\log S_t - kt)\,dW_t\}, \qquad t \geq 0. \tag{3.5.76}$$

It $v(x)$ is a constant, then the model for S_t is the geometric Brownian motion used in the derivation of the Black–Scholes formula. A more general version of the model can be obtained if $\log S_t$ has a general deterministic drift. It is known that (3.5.75) with a given initial distribution has a weak solution if and only if $v^{-2}(x)$ is locally integrable (Engelbert and Schmidt 1985). Furthermore, the solution is unique in distribution if and only if $v^2(x) > 0$ for all $x \in R$. If $v(x)$ is chosen so that

$$\int_{-\infty}^{\infty} v^{-2}(x)\,dx < \infty,$$

then the diffusion process defined by (3.5.75) is ergodic and X_t converges in law to a probability measure with density proportinal to $1/v^2(x)$ (Skorokhod 1989).

Let $\{W_t, t \geq 0\}$ be a standard Wiener process. Define

$$C_t = \int_0^t v^{-2}(W_s)\,ds \tag{3.5.77}$$

and

$$T_t = \inf\{s \geq 0 : C_s > t\}. \tag{3.5.78}$$

Then C_t is strictly increasing and $C_\infty = \infty$ a.s. (Engelbert and Schmidt 1981), and T_t defines a stochastic time transformation. It follows from Engelbert and Schmidt (1985) that

$$X_t = W_{T_t} \tag{3.5.79}$$

is a weak solution of (3.5.75).

Let $v(x)$ be defined by

$$v^2(x) = \sigma^2 \exp\Big[\alpha\sqrt{\delta^2 + (x-\mu)^2} - \beta(x-\mu)\Big]. \tag{3.5.80}$$

Note that $v^{-2}(x)$ is proportional to the density of the hyperbolic distribution discussed by Barndorff-Nielsen (1977).

Suppose that the price of a particular stock at time $t > 0$ is S_t and that the observations are $S_{t_1}, \ldots, S_{t_n}$ at times $t_1, t_2, \ldots, t_n$, respectively. Note that $t_1, t_2, \ldots, t_n$ are the trading days for the stock and that there can be more than one day between two observation times. Consider the log-price of a stock adjusted for the trend, that is,

$$X_{t_i} = \log S_{t_i} - kt_i, \qquad 1 \le i \le n.$$

For each stock, k may be determined by the ordinary linear regression. Bibby and Sørensen (1995b) modeled the adjusted log-prices of the shares using the hyperbolic diffusion process defined by

$$dX_t = \sigma \exp\left\{\tfrac{1}{2}\sigma\sqrt{\delta^2 + (X_t - \mu)^2} - \beta(X_t - \mu)\right\} dW_t, \qquad X_0 = x_0,\ t \ge 0. \quad (3.5.81)$$

In order to determine whether the shape of the diffuson coefficient assumed is a good fit for the variability in the data observed, one can consider the nonparametric estimator of the squared diffuson coefficient $v^2(x)$ proposed by Florens-Zmirou (1993), namely

$$V_n(x) = \frac{\sum_{i=1}^n I_{\{|X_{t_{i-1}}-x|<h\}}(X_{t_i} - X_{t_{i-1}})^2}{\sum_{i=1}^n I_{\{|X_{t_{i-1}}-x|<h\}}(t_i - t_{i-1})}, \quad (3.5.82)$$

where h is the bandwidth to be chosen, for the model

$$dX_t = v(X_t)\, dW_t, \qquad X_0 = x_0,\ t \ge 0$$

(we will return to this later in this chapter). For an application to real data based on daily stock prices, see Bibby and Sørensen (1995b).

3.6 Estimation of the Diffusion Parameter

3.6.1 Deterministic Sampling

Let $\boldsymbol{X}_t \equiv (X_t^1, \ldots, X_t^d)$ be a d-dimensional stochastic process satisfying the stochastic differential equation

$$d\boldsymbol{X}_t = \boldsymbol{a}(t, \boldsymbol{X})\, dt + \boldsymbol{b}(\theta, t, \boldsymbol{X}_t)\, d\boldsymbol{W}_t, \qquad \mathcal{L}(\boldsymbol{X}_0) = \nu,\ 0 \le t \le 1, \quad (3.6.1)$$

where ν is a probability measure on R^d possibly completely unknown or known but for an unknown parameter θ, $\boldsymbol{a}(\cdot, \cdot)$ is a nonanticipative R^d-valued function possibly unknown or known but depending on the unknown parameter θ, and $\boldsymbol{b}(\cdot, \cdot, \cdot)$ is a known function mapping $\Theta \times [0, 1] \times R^d$ to $R^d \otimes R^d$. Further suppose that $\boldsymbol{W}$ is an m-dimensional Wiener process and Θ is a compact subset of R^q. Here $\mathcal{L}(\boldsymbol{X}_0)$ denotes the probability measure of the random vector $\boldsymbol{X}_0$.

We have seen the role played by the diffusion coefficient $\boldsymbol{\Gamma} = \boldsymbol{b}\boldsymbol{b}^T$ earlier in the estimation of the drift coefficient $\boldsymbol{a}$. Note that $\boldsymbol{\Gamma}$ is a function from $\Theta \times [0, 1] \times R^d$ into the set of all symmetric nonnegative definite $d \times d$ matrices. We assume the following:

(GG1) (i) $\boldsymbol{a}(t, \boldsymbol{x})$ is continuous in t.

(ii) The partial derivatives $\nabla_{\boldsymbol{x}}^2 \boldsymbol{b} = \frac{\partial^2 \boldsymbol{b}}{\partial \boldsymbol{x}^2}$, $\nabla_\theta^2 \boldsymbol{b} = \frac{\partial^2 \boldsymbol{b}}{\partial \theta^2}$, $\nabla_{\boldsymbol{x}} \nabla_\theta \boldsymbol{b} = \frac{\partial^2 \boldsymbol{b}}{\partial \boldsymbol{x}\, \partial \theta}$ and $\nabla_{\boldsymbol{x}} \frac{\partial \boldsymbol{b}}{\partial t} = \frac{\partial^2 \boldsymbol{b}}{\partial \boldsymbol{x}\, \partial t}$ exist and are continuous on $\Theta \times [0, 1] \times R^d$.

(iii) Equation (3.6.1) has a nonexploding strong solution on [0, 1] (strong in the sense that the solution is adapted to the filtration generated by the Wiener process $\boldsymbol{W}$ and the initial value $\boldsymbol{X}_0$).

Suppose the process $\{\boldsymbol{X}_t\}$ is observed at the times

$$0 = t_{n,0} < t_{n,1} < \cdots < t_{n,n}$$

in (0, 1]. We assume that the sequence of sampling times has the following property:

(GG2) The probability measure

$$\mu_n = \frac{1}{n} \sum_{1 \le i \le n} \varepsilon_{t_{ni}}$$

on [0, 1] converges weakly to a limiting measure μ as $n \to \infty$, where ε_δ denotes the probability measure degenerate at δ.

In addition, let us assume that the following identifiability assumption holds:

(GG3) For all $\theta \in \Theta$, for $P^\theta_{\nu,a}$-a.s. for all $\zeta \ne \theta$, the functions $t \to \boldsymbol{\Gamma}(\theta, t, \boldsymbol{X}_t(\omega))$ and $t \to \boldsymbol{\Gamma}(\zeta, t, \boldsymbol{X}_t(\omega))$ are not equal to μ-a.s.

(Here $P^\theta_{\nu,a}$ is the probability measure induced by the process $\boldsymbol{X}_t$ on $C([0, 1], R^d)$ with the usual σ-algebra generated by the filtration $(\mathcal{F}_t, t \in [0, 1])$).

If $\mathcal{F}^{(n)}$ denotes the σ-algebra generated by $\{\boldsymbol{X}_{t_{ni}}, 1 \le i \le n\}$ at stage n, then condition (GG3) amounts to saying that, under (GG2), for all $\zeta \ne \theta$, the two sequences $\{P^\theta_{\nu,a}$ on $(\Omega, \mathcal{F}^{(n)}), n \ge 1\}$ and $\{P^\zeta_{\nu,a}$ on $(\Omega, \mathcal{F}^{(n)}), n \ge 1\}$ are entirely separated (Prakasa Rao 1987) which is necessary for the existence of a consistent estimator.

Let us write t_i for t_{ni} and δ_i for $\delta_{ni} = t_i - t_{i-1}$ for simplicity. Let

$$\boldsymbol{X}_i^{(n)} = \frac{1}{\sqrt{\delta_i}} (\boldsymbol{X}_{t_i} - \boldsymbol{X}_{t_{i-1}}), \qquad 1 \le i \le n.$$

The problem is to estimate the parameter θ from the observations $\boldsymbol{X}_i^{(n)}, 1 \le i \le n$ or equivalently from the σ-algebra $\mathcal{F}^{(n)}$ generated by $\{\boldsymbol{X}_{t_i}, 1 \le i \le n\}$. Since the Radon–Nikodym derivative $dP^\theta_{\nu,a}/dP^\theta_{\nu',a'}$ restricted to $\mathcal{F}^{(n)}$ is unknown in general, we consider the method of minimum contrast estimation.

If $\boldsymbol{a} = 0$, $\nu = \varepsilon_{x_0}$ and $\boldsymbol{\Gamma}(\theta, t, \boldsymbol{x}) = \boldsymbol{\Gamma}(\theta, t)$ is invertible, then the log-likelihood of the family $\{\boldsymbol{X}_{t_{ni}}, 1 \le i \le n\}$ with respect to the Lebesgue measure on $(R^d)^n$ is

$$-\frac{d}{2}\log(2\pi) - \frac{1}{2}\sum_{i=1}^{n}[\log\det \boldsymbol{\Gamma}(\theta)_i^n + (\boldsymbol{X}_{t_i} - \boldsymbol{X}_{t_{i-1}})^T(\boldsymbol{\Gamma}(\theta)_i^n)^{-1}(\boldsymbol{X}_{t_i} - \boldsymbol{X}_{t_{i-1}})],$$

where $\boldsymbol{\Gamma}(\theta)_i^n = \int_{t_{i-1}}^{t_i} \boldsymbol{\Gamma}(\theta, s)\, ds$ since $\boldsymbol{X}$ is a continuous Gaussian process with independent increments and the MLE for θ is one which maximizes the above expression over $\theta \in \Theta$. Motivated by the above, we can consider the contrast function

$$\frac{1}{n}\sum_{i=1}^{n}[\log\det \boldsymbol{\Gamma}(\theta, t_{i-1}, \boldsymbol{X}_{t_{i-1}}) + (\boldsymbol{X}_i^n)^T \boldsymbol{\Gamma}(\theta, t_{i-1}, \boldsymbol{X}_{t_{i-1}})^{-1}(\boldsymbol{X}_i^n)]. \tag{3.6.2}$$

In general, let us consider a contrast function of the form

$$U_n(\theta) = \frac{1}{n}\sum_{i=1}^{n} f(\boldsymbol{\Gamma}(\theta, t_{n,i-1}, \boldsymbol{X}_{t_{n,i-1}}), \boldsymbol{X}_i^n), \tag{3.6.3}$$

where $f = f(\boldsymbol{G}, \boldsymbol{x})$ is a function from $\mathcal{M} \times R^d$ to R. Here $\mathcal{M} \subset \mathcal{M}_d^+$, the set of symmetric nonnegative definite matrices of order $d \times d$ such that the set $\cup_{t,\zeta}\{\omega : \Gamma(\zeta, t, X_t(\omega)) \notin \zeta\}$ is $P_{\nu,\boldsymbol{a}}^{\theta}$-null for all $\theta \in \Theta$.

We assume the following:

(HH1) $\mathcal{M} = \cup \mathcal{M}_\ell$, where $\{\mathcal{M}_\ell\}$ is an increasing sequence of compact subsets of $\mathcal{M}_d^+$ with

$$\lim_{\ell\to\infty} P_{\nu,\boldsymbol{a}}^{\theta}\{\boldsymbol{\Gamma}(\zeta, t, \boldsymbol{X}_t) \in \mathcal{M}_\ell \text{ for all } \zeta, t\} = 1$$

for all θ. In addition assume that for $i, j = 0, 1, 2$, the partial derivatives $\nabla_{\boldsymbol{G}}^i \nabla_{\boldsymbol{x}}^j f$ are continuous and satisfy $|\nabla_{\boldsymbol{G}}^i \nabla_{\boldsymbol{x}}^j f| \le \gamma(\boldsymbol{G})(1 + \|\boldsymbol{x}\|^{\gamma(\boldsymbol{G})})$ for some continuous function γ on $\mathcal{M}$.

(HH2) f is even in its second argument.

(HH3) Let $\rho_{\boldsymbol{G}}$ denote the d-dimensional Gaussian distribution with the mean zero and with the covariance matrix $\boldsymbol{G}$. Suppose the function

$$\boldsymbol{G}' \to \int f(\boldsymbol{G}', \boldsymbol{x})\, \rho_{\boldsymbol{G}}(d\boldsymbol{x})$$

has a unique minimum at $\boldsymbol{G}' = \boldsymbol{G}$.

Let

$$\rho_{\theta,t} = \rho_{\boldsymbol{\Gamma}(\theta,t,\boldsymbol{X}_t)}, \qquad (3.6.4)$$
$$F(\theta, t, \boldsymbol{x}) = f(\boldsymbol{\Gamma}(\theta, t, \boldsymbol{X}_t), \boldsymbol{x}),$$

and

$$U_t(\theta, \zeta) = \int F(\zeta, t, \boldsymbol{x})\, \rho_{\theta,t}(d\boldsymbol{x}).$$

Genon-Catalot and Jacod (1993) proved the following results.

Theorem 3.6.1 *Under the conditions (GG1), (GG2) and (HH1), for each $\theta \in \Theta$, $U_n(\zeta)$ converges uniformly in $\zeta \in \Theta$ in $P_{\nu,\boldsymbol{a}}^{\theta}$-measure to the random variable*

$$U(\theta, \zeta) = \int U_t(\theta, \zeta)\, \mu(dt).$$

Let $\hat{\theta}_n$ be a minimum contrast estimator associated with $U_n(\theta)$, that is,

$$U_n(\hat{\theta}_n) = \inf_{\theta\in\Theta} U_n(\theta), \qquad (3.6.5)$$

where $U_n(\theta)$ is as defined by (3.6.3).

Consistency

Theorem 3.6.2 *Under conditions (GG1)–(GG3), (HH1) and (HH2),*

$$\hat{\theta}_n \xrightarrow{p} \theta \text{ under } P_{\mu,\boldsymbol{a}}^{\theta}\text{-measure.}$$

Asymptotic Mixed Normality

Let

$$\begin{aligned}
\boldsymbol{B}(\theta)_t &= \int_{R^d} \nabla_\theta^2 F(\theta, t, \boldsymbol{x})\, \rho_{\theta,t}(d\boldsymbol{x}), \\
\boldsymbol{D}(\theta)_t &= \int_{R^d} \nabla_\theta F(\theta, t, \boldsymbol{x}) \nabla_\theta F(\theta, t, \boldsymbol{x})^T\, \rho_{\theta,t}(d\boldsymbol{x}), \\
\boldsymbol{B}(\theta) &= \int_{[0,1]} \boldsymbol{B}(\theta)_t\, \mu(dt), \quad \text{and} \\
\boldsymbol{D}(\theta) &= \int_{[0,1]} \boldsymbol{D}(\theta)_t\, \mu(dt).
\end{aligned} \tag{3.6.6}$$

Note that $\boldsymbol{B}(\theta)_t$ and $\boldsymbol{D}(\theta)_t$ are processes taking values in the set of symmetric $q \times q$ matrices and $\boldsymbol{D}(\theta)_t$ is nonnegative definite.

Theorem 3.6.3 *Suppose that the conditions (GG1)–(GG3) and (HH1)–(HH3) hold. Let $\theta \in \Theta^0$ such that $\boldsymbol{B}(\theta)$ is $P^\theta_{\nu,\boldsymbol{a}}$-a.s. invertible. Then $\boldsymbol{S}_n = \sqrt{n}(\hat{\theta}_n - \theta)$ converges in law under $P^\theta_{\nu,\boldsymbol{a}}$ to a 'mixed normal' random vector $\boldsymbol{S}$ which, conditionally on $\mathcal{F}_1$, is centered Gaussian with covariance matrix*

$$\boldsymbol{\Gamma}(\theta) = \boldsymbol{B}(\theta)^{-1} \boldsymbol{D}(\theta) \boldsymbol{B}(\theta)^{-1}. \tag{3.6.7}$$

It can be shown that the contrast (3.6.2) is the best among all the contrasts of the form (3.6.3) in the sense that the asymptotic conditional variance of the minimum contrast estimator is minimal for f given by

$$f(\boldsymbol{G}, \boldsymbol{x}) = \log \det \boldsymbol{G} + \boldsymbol{x}^T \boldsymbol{G}^{-1} \boldsymbol{x}, \tag{3.6.8}$$

where $\boldsymbol{G} = \boldsymbol{\Gamma}(\theta)$ as defined earlier. For proofs of Theorems 3.6.1 to 3.6.3, see Genon-Catalot and Jacod (1993). Suppose that:

(GG4) For all θ outside a $P^\theta_{\nu,\boldsymbol{a}}$-null set, the matrices $\boldsymbol{\Gamma}(\zeta, t, \boldsymbol{x}_t)$ are invertible for all $t \in [0, 1]$ and $\zeta \in \Theta$.

Under conditions (GG1)–(GG4) and (HH1)–(HH3), it can be shown that, for the contrast f defined by (3.6.8) with $\boldsymbol{G} = \boldsymbol{\Gamma}(\theta)$,

$$\boldsymbol{D}(\theta)_t = 2\boldsymbol{B}(\theta)_t, \qquad \boldsymbol{B}(\theta)_t^{ij} = \operatorname{tr}\left[\left(\frac{\partial \boldsymbol{\Gamma}}{\partial \theta_i} \boldsymbol{\Gamma}^{-1} \frac{\partial \boldsymbol{\Gamma}}{\partial \theta_j}\right)(\theta, t, \boldsymbol{X}_t)\right], \tag{3.6.9}$$

and if $\boldsymbol{B}(\theta)$ is $P^\theta_{\nu,\boldsymbol{a}}$-a.s. invertible, then the asymptotic conditional covariance matrix of the associated minimum contrast estimator is

$$\boldsymbol{\Gamma}(\theta) = 2\boldsymbol{B}(\theta)^{-1}. \tag{3.6.10}$$

If $\boldsymbol{I}(\theta, t, \boldsymbol{x})$ is the Fisher information matrix at θ for the model $(R^d, (P_{\Gamma(\theta,t,\boldsymbol{x})})_{\theta \in \Theta})$, then $\boldsymbol{B}(\theta)_t = 2\boldsymbol{I}(\theta, t, \boldsymbol{X}_t)$ and hence

$$\boldsymbol{\Gamma}(\theta) = \left[\int_{[0,1]} \boldsymbol{I}(\theta, t, \boldsymbol{X}_t)\, \mu(dt)\right]^{-1}. \tag{3.6.11}$$

For details, see Genon-Catalot and Jacod (1993).

Example 3.6.1 (*One-dimensional case*)
Suppose $d = m = 1$. Further assume that $q = 1$. Let $\dot{\Gamma}$ denote the partial derivative of Γ with respect to θ. If (GG4) holds, then we can use the contrast function

$$f(G, x) = \log G + \frac{x^2}{G}, \tag{3.6.12}$$

in which case

$$D(\theta)_t = 2\,B(\theta)_t, \qquad B(\theta)_t = \left(\frac{\dot{\Gamma}}{\Gamma}\right)^2 (\theta, t, X_t). \tag{3.6.13}$$

If (GG4) does not hold, then we can use the contrast function

$$f_0(G, x) = (x^2 - G)^2. \tag{3.6.14}$$

It can be checked that

$$D_0(\theta)_t = 8(\Gamma\dot{\Gamma})^2(\theta, t, X_t), \qquad B_0(\theta)_t = 2\dot{\Gamma}^2(\theta, t, X_t). \tag{3.6.15}$$

If μ in (GG2) is a Dirac measure and if (GG4) holds, then the asymptotic conditional variances $J(\theta)$ and $J_0(\theta)$ of the estimators $\hat{\theta}_n$ and θ_n, based on the contrasts f and f_0 respectively, are equal.

Example 3.6.2 (*One-dimensional linear case*)
Suppose that $d = m = q = 1$ and $b(\theta, t, x) = \sigma(\theta, t)\, x$. Further suppose that $\nu = \varepsilon_{x_0}$ for $x_0 \neq 0$. In other words, consider the equation

$$dX_t = a(t, X)\, dt + \sigma(\theta, t) X_t\, dW_t, \qquad X_0 = x_0 \neq 0,\ t \geq 0. \tag{3.6.16}$$

Then $\Gamma(\zeta, t, x) = C(\zeta, t)x^2$, where $C = \sigma^2$. Suppose that $\sigma \neq 0$. Then the measures $P^\theta_{\nu,a}$ and $P^\theta_{\nu,0}$ are equivalent and, under $P^\theta_{\nu,0}$, the solution X_t never hits zero a.s. Hence this is also true under $P^\theta_{\nu,a}$ and (GG4) holds. The optimal contrast associated with f in (3.6.8) is

$$f(G, x) = \log G + \frac{x^2}{G} \tag{3.6.17}$$

and (3.6.13) shows that

$$B(\theta)_t = \left(\frac{\dot{C}}{C}\right)(\theta, t)^2, \qquad \Gamma(\theta) = 2\left\{\int B(\theta)_t\, \mu(dt)\right\}^{-1} \tag{3.6.18}$$

which are nonrandom in this case. Hence the estimators based on these contrasts are asymptotically normal.

Since we would like to have estimators with asymptotic variance $\Gamma(\theta)$ as small as possible, the best choice of sampling times t_{ni} are those leading to a limiting measure μ which maximizes $\int B(\theta)_t\, \mu(dt)$. Since θ is unknown, it is not possible to find this in general.

If $\sigma(\theta, t) = \sigma(\theta)$ does not depend on t, then $B(\theta)_t$ does not depend on t and all the sampling schemes are equivalent. If $\sigma(\theta, t) = \sigma(\theta)$ and $a \equiv 0$, then we have a multiplicative model.

Example 3.6.3 (*Noninvertible case*)
Suppose $d = m = q = 1$. Consider the equation

$$dX_t = a\,dt + \sigma(\theta, X_t)\,dW_t, \qquad \mathcal{L}(X_0) = \nu,\ t \geq 0, \tag{3.6.19}$$

where $a > 0$ and $\sigma(\theta, x) \geq 0$. Further suppose that $\sigma(\theta, x) = 0$ if $x \geq \alpha$ and $x_0 < \alpha$. Then $\Gamma = \sigma^2 > 0$ on a set that does not depend on θ.

Consider the contrast f_0 defined by (3.6.14). If $\dot\Gamma(\theta, 0, x_0) \neq 0$ and if 0 belongs to the support of μ, then $B'(\theta)$ is a.s. nonzero and the asymptotic conditional variance of the estimator is

$$J(\theta) = 2\frac{\int(\dot\Gamma\Gamma)^2(\theta, t, X_t)\,\mu(dt)}{(\int(\dot\Gamma)^2(\theta, t, X_t)\,\mu(dt))^2}. \tag{3.6.20}$$

The integration can be restricted to the subset of $[0, 1]$ where $X_t < \alpha$.

Example 3.6.4 (*Two-dimensional case*)
Here $d = 2$, $m = 1$ and $q = 1$. Suppose ν is a Dirac measure here. Consider the equation

$$\begin{aligned} dX_t^{(1)} &= a^{(1)}(t, X)\,dt + \sigma(\theta, X_t)\,dW_t, \qquad X_0^{(1)} = x_0^{(1)},\ t \geq 0,\\ dX_t^{(2)} &= a^{(2)}(t, X)\,dt, \qquad X_0^{(2)} = x_0^{(2)},\ t \geq 0, \end{aligned} \tag{3.6.21}$$

where $\alpha > 0$. Let $\beta = \sigma^2$ and $S = \begin{bmatrix} \beta & 0 \\ 0 & 0 \end{bmatrix}$. Here (GG4) does not hold. Consider the set $\mathcal{M}$ of matrices $G = \begin{bmatrix} g & 0 \\ 0 & 0 \end{bmatrix}$ with $g > 0$. Set

$$f(G, x) = \log g + (x^{(1)})^2/g \qquad \text{for } G = \begin{bmatrix} g & 0 \\ 0 & 0 \end{bmatrix}. \tag{3.6.22}$$

Then the estimator $\hat\theta_n$ minimizing the contrast based on f has asymptotic conditional variance

$$J(\theta) = 2B(\theta), \qquad B(\theta) = \int B(\theta)_t\,\mu(dt), \qquad B(\theta)_t = \left(\frac{\dot\beta}{\beta}\right)(\theta, X_t). \tag{3.6.23}$$

It can be shown that these estimators are optimal in the sense discussed earlier. This can be seen by noting that the asymptotic variance does not depend on the drift a so that one can take $a^{(1)} = a^{(2)} = 0$, in which case $X_t^{(2)} = x_0^{(2)}$ for all t and the problem reduces to a one-dimensional problem.

Example 3.6.5 (*Mathematical finance modeling*)
Let $m = d = 2$ and consider the stochastic differential system

$$\begin{aligned} dX_t^{(1)} &= \gamma(X_t^{(2)} - X_t^{(1)})\,dt + \beta X_t^{(1)}\,dW_t^{(1)}, \qquad t \geq 0,\\ dX_t^{(2)} &= \alpha X_t^{(2)}\,dt + \rho\sigma X_t^{(2)}\,dW_t^{(1)} + \sigma X_t^{(2)}\,dW_t^{(2)}, \qquad t \geq 0 \end{aligned} \tag{3.6.24}$$

where $\alpha, \beta, \gamma, \sigma, \rho$ are unknown and $W_t^{(1)}, W_t^{(2)}$ are independent standard Wiener processes (Courtadon 1985). The problem of interest is the estimation of $\theta = (\beta, \sigma, \rho)$. Here $\Theta = (0, \infty) \times (0, \infty) \times R$. It can be shown that (GG4) holds in this example as in Example 3.6.2. Consider the contrast function f given by

$$f(G, x) = \log\det G + x^T G^{-1} x. \tag{3.6.25}$$

The associated contrast at stage n is

$$U_n(\theta) = \log(\beta^2\sigma^2) + \frac{1+\rho^2}{\beta^2}\Big(\sum_{11}^{n}\Big) - \frac{2\rho}{\beta\sigma}\Big(\sum_{12}^{n}\Big) + \frac{1}{\sigma^2}\Big(\sum_{22}^{n}\Big)$$
$$+\frac{1}{n}\log\{(X_{t_i}^{(1)} - X_{t_{i-1}}^{(1)})^2(X_{t_i}^{(2)} - X_{t_{i-1}}^{(2)})^2\}, \tag{3.6.26}$$

where

$$\sum_{jk}^{n} = \frac{1}{n}\sum_{i=1}^{n}\frac{(X_{t_i}^{(j)} - X_{t_{i-1}}^{(j)})(X_{t_i}^{(k)} - X_{t_{i-1}}^{(k)})}{\delta_i X_{t_{i-1}}^{(j)} X_{t_{i-1}}^{(k)}}, \qquad \delta_i = t_i - t_{i-1}. \tag{3.6.27}$$

Then the minimum contrast estimator is $\hat{\theta}_n = (\hat{\beta}_n, \hat{\rho}_n, \hat{\sigma}_n)$ where

$$\hat{\beta}_n = \Big(\sum_{11}^{n}\Big)^{1/2},$$
$$\hat{\rho}_n = \Big(\sum_{12}^{n}\Big)\Big(\sum_{11}^{n}\sum_{22}^{n} - \Big(\sum_{12}^{n}\Big)^2\Big)^{1/2}, \tag{3.6.28}$$
$$\hat{\sigma}_n = \Big\{\Big(\sum_{11}^{n}\sum_{22}^{n} - \Big(\sum_{12}^{n}\Big)^2\Big)/\sum_{11}^{n}\Big\}^{1/2}$$

and the asymptotic conditional covariance matrix for these estimators at the point $\theta = (\beta, \rho, \sigma)$ is

$$J(\theta) = \begin{pmatrix} \beta^2/2 & \rho\beta/2 & 0 \\ \rho\beta/2 & 1+\rho^2 & -\rho\sigma/2 \\ 0 & -\rho\sigma/2 & \sigma^2/2 \end{pmatrix}. \tag{3.6.29}$$

Note that $J(\theta)$ is nonrandom and the limiting distribution is normal. Further $J(\theta)$ does not depend on μ and hence all the sampling schemes are equivalent.

3.6.2 Random Sampling

Let us now consider 'random sampling schemes' for the estimation of a parameter θ based on the observations at random times of the process $\{X_t\}$ solving the stochastic differential equation

$$dX_t = \boldsymbol{a}(\theta, X_t)\,dt + \boldsymbol{b}(\theta, X_t)\,dW_t, \qquad X_0 = \boldsymbol{x}_0,\ 0 \le t \le 1, \tag{3.6.30}$$

where $\boldsymbol{x}_0$ is fixed and the coefficients $\boldsymbol{a}, \boldsymbol{b}$ are known functions of $(\theta, \boldsymbol{x})$. We assume that the process X is d-dimensional and $\theta \in \Theta = [\theta_0, \theta_1] \subset R$. We further assume that $\boldsymbol{\Gamma} = \boldsymbol{b}\boldsymbol{b}^T$ is nondegenerate and the functions $\boldsymbol{a}$ and $\boldsymbol{b}$ are smooth.

A *random sampling scheme* is a family T_{ni}, $1 \le i \le n$, of successive sampling times such that: (i) T_{n1} is deterministic; (ii) T_{ni}, $i \ge 2$, is measurable with respect to the σ-algebra generated by $X_{T_{nj}}$, $1 \le j < i$; and (iii) $T_{ni} \ne T_{nj}$ for $i \ne j$.

Among all the random sampling schemes, let us further consider a class of *restricted* schemes, that is, those for which T_{ni} is increasing in i. Then T_{ni} are stopping times and this class contains all the deterministic schemes as discussed earlier. Jacod (1993) discussed the case of random sampling schemes for continuous Gaussian processes with independent increments (which is the case when $\boldsymbol{a}$ and $\boldsymbol{b}$ in (3.6.30) depend on t only and not on $\boldsymbol{x}$).

The problem is to find an 'optimal' scheme which can give an estimator with the smallest asymptotic variance for estimating the parameter θ (Prakasa Rao 1988b).

The concept of local asymptotic mixed normality (Jeganathan 1982; or Le Cam and Yang 1990) is discussed in Prakasa Rao (1999). A sequence $\mathcal{E}_n = (\Omega, \mathcal{F}_n, (P_\theta)_{\theta \in \Theta})$ of statistical models has the LAMN property at θ with the normalizing factor $\sqrt{n}$ and conditional variance $1/W(\theta)$ if the following hold:

(i) $W(\theta) > 0$,

(ii) $\log Z_n(\theta, \theta + \zeta/\sqrt{n}) = \zeta Y_n \sqrt{W(\theta)} - \frac{\zeta^2}{2} W(\theta) + R_n(\theta, \zeta)$,

where $R_n(\theta, \zeta) \xrightarrow{P} 0$ in P_θ-measure, $(Y_n, W(\theta)) \xrightarrow{\mathcal{L}} (Y, W(\theta))$ under P_θ-measure where Y is $N(0, 1)$ independent of $W(\theta)$. Here $Z_n(\theta, \phi) = dP_\phi / dP_\theta|_{\mathcal{F}_n}$. If the LAMN property holds, then one can find an estimator $\hat{\theta}_n$ such that $\sqrt{n}(\hat{\theta}_n - \theta) \xrightarrow{\mathcal{L}} Y/\sqrt{W(\theta)}$; furthermore, for any other regular sequence θ_n^* of estimators, if $\sqrt{n}(\theta_n^* - \theta)$ converges in limit, then this limit is more spread out than $Y/\sqrt{W(\theta)}$. In other words $1/W(\theta)$ is the lowest bound for the (conditional) asymptotic variance of the estimator for the sequence $\{\mathcal{E}_n\}$ of stochastic models. The following result is proved by Genon-Catalot and Jacod (1994) in the case where $d = 1$ and $\{W_t\}$ is one-dimensional.

Theorem 3.6.4 *Suppose the following regularity conditions hold:*

(i) $\inf_{\theta,x} |b(\theta, x)| > 0$;

(ii) $\nabla_\theta^i \nabla_x^j a$ *exists and is bounded for* $i \leq 3$, $j \leq 5$; *and*

(iii) $\nabla_\theta^i \nabla_x^j b$ *exists and is bounded for* $i \leq 3$, $j \leq 6$.

If the LAMN property is satisfied at some θ *for some sequence of random samplings, then the corresponding conditional variance* $1/W(\theta)$ *satisfies*

$$W(\theta) \leq M(\theta) \equiv \tfrac{1}{2} \sup_t B(\theta)_t,$$

where

$$B(\theta)_t = \left[\left(\frac{\dot{\Gamma}}{\Gamma} \right)^2 (\theta, X_t) \right], \qquad \Gamma = b^2.$$

In addition, suppose that the following conditions holds:

(iv) (Identifiability.) For all θ, *for* P_θ*-almost surely, for all* $\zeta \neq \theta$, *there is a* $t = t(\omega, \theta, \zeta)$ *with* $\Gamma(\zeta, X_t(\omega)) \neq \Gamma(\theta, X_t(\omega))$.

Then there is a sequence of random sampling times such that the LAMN property holds for each value of θ *and there is a sequence of estimators* $\hat{\theta}_n$ *such that* $\sqrt{n}(\hat{\theta}_n - \theta)$ *is* P_θ*-asymptotically mixed normal with the asymptotic conditional variance* $1/M(\theta)$ *for all* θ.

Remarks. For a more general result relaxing some of the assumptions stated above, see Genon-Catalot and Jacod (1994). They discuss the problem for $d = 1$ and for $d \geq 2$ under a special structure for $\boldsymbol{a}$ and $\boldsymbol{b}$. The case $d \geq 2$ in general is open.

Optimal random sampling scheme

Let us now describe the optimal random sampling scheme mentioned in the above theorem.

Let $\tau_n = \{T_{ni}, 1 \le i \le n\}$ be a sequence of random sampling schemes and ζ_i^n be the σ-algebra generated by $X_{T_{nj}}, 1 \le j \le i$. Let $\mathcal{F}^{(n)} = \zeta_n^n$ be the σ-algebra generated by $X_{T_{nj}}$, $1 \le j \le n$. We assume that the following conditions hold:

(JJ1) (a) The solution X_t of the equation (3.6.30) satisfies the following conditions: (i) under each P_θ, the process X_t has $P_\theta(X_0 = x_0) = 1$ and is a homogeneous Markov process with nonvanishing transition densities $y \to p_t^\theta(x, y)$ with respect to the Lebesgue measure; (ii) for each θ, the function $p_t^\theta(x, y)$ and $1/p_t^\theta(x, y)$ are locally bounded in (t, x, y) and continuous in t on $(0, 1] \times E \times E$, where E is an open subset.

(b) The matrix $\Gamma(\theta, x)$ is invertible for all $\theta \in \Theta, x \in E$ (recall that $\Gamma = bb^T$).

(c) The derivatives $\nabla_\theta^2 \Gamma$ and $\nabla_\theta \nabla_x \Gamma$ exist and are continuous on $\Theta \times E$.

Let

$$\gamma(\theta, x) = \tfrac{1}{2} \operatorname{tr}[(\dot{\Gamma}\Gamma^{-1}\dot{\Gamma}\Gamma^{-1})(\theta, x)] \tag{3.6.31}$$

and

$$\tilde{M}(\theta) = \sup_{t \in [0,1]} \gamma(\theta, X_t). \tag{3.6.32}$$

Let ρ_G be the d-dimensional centered Gaussian distribution with covariance matrix G.

In addition to (JJ1), suppose the following condition holds.

(JJ2) (Identifiability.) For all $\theta \in \Theta$, for P_θ-almost all ω, for all $\zeta \ne 0$ there exists a time $t = t(\omega, \theta, \zeta)$ such that

$$\Gamma(\zeta, X_t(\omega)) \ne \Gamma(\theta, X_t(\omega)).$$

Let n be fixed. We now describe a scheme which will give an optimal random sampling scheme under some additional regularity conditions on the transition density functions.

1. Let $m = [\sqrt{n}]$ and $T_{ni} = i/m, 1 \le i \le m$. In other words, the first m points are equally spaced. We find a consistent estimator θ_n^* based on observations at these time points under P_θ-measure; for instance, we can choose the optimal estimator which minimizes the contrast

$$\begin{aligned} U_n(\zeta) = \frac{1}{m} \sum_{i=1}^{m-1} &[\log \det(\Gamma(\zeta, X_{(i/m)})) \\ &+ (X_{((i+1)/m)} - X_{(i/m)})^T \Gamma(\zeta, X_{(i/m)})^{-1} (X_{((i+1)/m)} - X_{(i/m)})]. \end{aligned} \tag{3.6.33}$$

2. Let V_n be one of the time points $t = i/m, 0 \le i \le m - 1$, which maximize the function $\gamma(\theta_n^*, X_{(i/m)})$, where $\gamma(\theta, x)$ is as defined by (3.6.31). Choose the remaining $N = n - m$ sampling times at

$$T_{n,m+i} = V_{ni} = V_n + \frac{i}{n^2}. \tag{3.6.34}$$

3. Consider the following contrast, where we write V_i for V_{ni} for simplicity:

$$U_n(\zeta) = \frac{1}{N}\sum_{i=1}^{N}[\log\det(\boldsymbol{\Gamma}(\zeta, \boldsymbol{X}_{V_{i-1}})) \\ + n^2(\boldsymbol{X}_{V_i} - \boldsymbol{X}_{V_{i-1}})^T \boldsymbol{\Gamma}(\zeta, \boldsymbol{X}_{V_{i-1}})^{-1}(\boldsymbol{X}_{V_i} - \boldsymbol{X}_{V_{i-1}})]. \tag{3.6.35}$$

Let $\hat{\theta}_n$ be the argument of the local minimum of $U_n(\zeta)$ which is closest to θ_n^* obtained in step 1.

Under some additional technical conditions involving the moments of the transition densities, Genon-Catalot and Jacod (1993) show that if $\theta \in \Theta^0$ and $\tilde{M}(\theta) > 0$ P_θ-a.s., then $\sqrt{n}(\hat{\theta}_n - \theta)$ converges in law under P_θ to a centered mixed normal random variable with asymptotic conditional variance $1/\tilde{M}(\theta)$.

In addition to the conditions (JJ1) and (JJ2), suppose that the following conditions hold:

(JJ3) There exists a function $\gamma^*(\theta, \boldsymbol{x})$ on $\Theta \times E$ locally Lipschitz in the second variable and a constant C_p such that

$$|I_t^\theta(\boldsymbol{x}) - \gamma^*(\theta, \boldsymbol{x})| \le C_p\sqrt{t} \qquad \text{for all } \boldsymbol{x} \in K_p,$$

where $\{K_p\}$ is an increasing sequence of compact subsets of E such that there exist times $T_p = \inf\{t : \boldsymbol{X}_t \notin K_p\}$ satisfying

$$\lim_{p\to\infty} \sup_\theta P_\theta(T_p < \infty) = 0,$$

and

$$I_t^\theta(\boldsymbol{x}) = \int \frac{p_t^\theta(\boldsymbol{x}, \boldsymbol{y})}{\bar{p}_t^\theta(\boldsymbol{x}, \boldsymbol{y})^2}\, d\boldsymbol{y}, \qquad \bar{p}_t^\theta = \dot{p}_t^\theta / p_t^\theta.$$

Define

$$J_n(\theta) = \frac{1}{n}\sum_{i=1}^{n}\gamma^*(\theta, \boldsymbol{X}_{T_{ni}})$$

for any arbitrary random sampling scheme T_{ni}, $1 \le i \le n$. Suppose that $J_n(\theta) \xrightarrow{p} W(\theta)$ as $n \to \infty$ with $W(\theta) > 0$ P_θ-a.s. Let

$$M(\theta) = \sup_{t\in[0,1]} \gamma^*(\theta, \boldsymbol{X}_t).$$

It is obvious that $W(\theta) \le M(\theta)$ in general. Let

$$\mu_n = \frac{1}{n}\sup_{1\le i\le n} \varepsilon_{T_{ni}}.$$

Assume also that:

(JJ4) μ_n converges weakly in P_θ-probability to a (possibly random) probability measure μ.

It can be shown that

$$J_n(\theta) \xrightarrow{p} W(\theta) = \int_0^1 \gamma^*(\theta, \boldsymbol{X}_t)\,\mu(dt) \qquad \text{as } n \to \infty,$$

and that LAMN property holds at θ under some technical conditions too many to state here (Genon-Catalot and Jacod (1993)). In general $\gamma \le \gamma^*$. If $\gamma = \gamma^*$ and $M = \tilde{M}$, P_θ-a.s., then under conditions (JJ1)–(JJ4) and additional technical conditions, the sequence of random sampling schemes constructed above is optimal in the sense that the LAMN property holds with $W(\theta) = M(\theta)$. Furthermore, the estimator $\hat{\theta}_n$ is asymptotically optimal among all the estimators based on all possible random sampling schemes satisfying the conditions stated above. For a proof, see Genon-Catalot and Jacod (1994). The optimal sampling scheme proposed above is not restricted in the sense described earlier. Recall that a random sampling scheme $\{T_{ni}, 1 \leq i \leq n\}$ is restricted if T_{ni} is increasing in i. It is not clear whether there exists a restricted random sampling scheme which is optimal.

Example 3.6.6 (*Perturbed Black–Scholes model*)
Consider the one-dimensional equation

$$dY_t = \beta Y_t\,dt + Y_t(\alpha + \theta f(Y_t))\,dW_t, \qquad Y_0 = y_0 > 0,\ t \geq 0. \tag{3.6.36}$$

Suppose that $\alpha > 0, \beta > 0$ are known and f is a positive known C^∞ function with compact support. Suppose that $\Theta = [0, 1]$. The process $\{Y_t\}$ takes values in $(0, \infty)$ and $X_t = \log Y_t$ satisfies

$$dX_t = [\beta - \tfrac{1}{2}(\alpha + \theta\, f(e^{X_t}))^2]\,dt + (\alpha + \theta\, f(e^{X_t}))\,dW_t, \qquad X_0 = \log y_0,\ t \geq 0. \tag{3.6.37}$$

This is a stochastic differential equation of the type (3.6.30) satisfying the conditions of Theorem 3.6.4. Here

$$B(\theta)_t = \frac{4f(Y_t)^2}{(\alpha + \theta f(Y_t))^2}$$

and

$$M(\theta) = \frac{2Z^2}{(\alpha + \theta Z)^2}$$

where Z denotes the maximum of $f(Y_t)$ over $[0, 1]$. All the above comments on the optimal sampling schemes are valid here.

Example 3.6.7
Consider the equation

$$dY_t = \beta Y_t\,dt + Y_t(\alpha + g(Y_t) + \theta f(Y_t))\,dW_t, \qquad Y_0 = y_0 > 0,\ t \geq 0,$$

where α, β, f are as in the Example 3.6.6 and g is another C^∞ function. Let $X_t = \log Y_t$. Then the equation reduces to the form (3.6.30). Here

$$B(\theta)_t = \frac{4f(Y_t)^2}{(\alpha + g(Y_t) + \theta f(Y_t))^2}.$$

The maximum of $t \to B(\theta)_t$ is attained at a random point which depends on the path of Y and also on θ.

3.6.3 Estimation by Approximation of the Transition Function

Approximation of the transition function by a suitable Gaussian density

Consider the stochastic differential equation

$$dX_t = a(X_t, \theta)\,dt + S(X_t, \sigma)\,dW_t, \qquad X_0 = x_0,\ t \geq 0, \tag{3.6.38}$$

where $\alpha = (\theta, \sigma) \in H$, a compact subset of R^2, W_t is the standard Wiener process and a and S are known real-valued functions defined on R^2. We assume that sufficient conditions hold on a and S so that there exists a unique solution in law for (3.6.38). Let P_α denote this distribution.

Suppose the process is observed at the discrete time points, $t_i^n = ih_n, 0 < i \le n$, and h_n is called the discretization step. We assume that $h_n \to 0$ and $nh_n \to \infty$ and that the process $\{X_t\}$ has the ergodic property. Let

$$m(\alpha, X_{t_{i-1}^n}) = E_\alpha[X_{t_i^n} \mid X_{t_{i-1}^n}] \tag{3.6.39}$$

and

$$m_2(\alpha, X_{t_{i-1}}^n) = E_\alpha[(X_{t_i^n} - m(\alpha, X_{t_{i-1}^n}))^2 \mid X_{t_{i-1}^n}]. \tag{3.6.40}$$

Let us approximate the transition density by a Gaussian density within mean $m(\alpha, x)$ and variance $m_2(\alpha, x)$. Then the approximate log-likelihood function of $\{X_{t_i^n}, 0 \le i \le n\}$ can be written in the form

$$\sum_{i=1}^{n}\left\{\frac{(X_{t_i^n} - m(\alpha, X_{t_{i-1}^n}))^2}{m_2(\alpha, X_{t_{i-1}^n})} + \log m_2(\alpha, X_{t_{i-1}^n})\right\} \tag{3.6.41}$$

and one can estimate α by minimizing (3.6.41). If $S(\cdot, \sigma) \equiv \sigma$, then $m(\vartheta, x) \simeq x + h_n a(x, \theta)$ and $m_2(\vartheta, x) \simeq h_n S^2(x, \sigma) = h_n \sigma^2$ and the above approximation can be chosen to be

$$\left\{\sum_{i=1}^{n}\frac{(X_{t_i^n} - X_{t_{i-1}^n} - h_n a(X_{t_{i-1}^n}, \theta))^2}{\sigma^2} + n \log \sigma^2\right\}. \tag{3.6.42}$$

Prakasa Rao (1983b) studied the estimation of θ by minimization of (3.6.42) when σ is known. In general there are no closed-form expressions for $m(\theta, x)$ and $m_2(\theta, x)$ and hence these functions cannot be written down explicitly. We will try to give close approximations of them.

Let $\alpha_0 = (\theta_0, \sigma_0)$ be the true value of the parameter $\alpha = (\theta, \sigma)$. For any real sequence $\{u_n\}$, let J denote a function from $H \times R^2$ to R for which there exists a constant c such that

$$J(\alpha, u_n, x) \le u_n c(1 + |x|)^c, \tag{3.6.43}$$

for all α, n and x. Suppose the following conditions hold.

(KK1) There exists a constant C such that

$$|S(x, \sigma_0) - S(y, \sigma_0)| + |a(x, \theta_0) - a(y, \theta_0)| \le C|x - y|, \tag{3.6.44}$$

where C is a constant independent of x and y.

(KK2) The process X is ergodic for $\alpha = \alpha_0$. Let μ_0 be the invariant probability measure. Suppose μ_0 has all moments finite.

(KK3) $\inf_{x,\sigma} S^2(x, \sigma) > 0$.

(KK4) For all $p \ge 0$, $\sup_t E|X_t|^p < \infty$.

(KK5) $[k]$ (i) The functions S and a are continuously differentiable with respect to x up to order k for all θ and σ, and their derivatives up to order k are of polynomial growth in x uniformly in α.

(ii) $S(x, \cdot)$ (or $a(x, \cdot)$) and all its partial derivatives with respect to x up to order k are three times differentiable with respect to σ (or θ) for all x in R. Moreover, these derivatives up to the third order with respect to σ (or θ) are of polynomial growth in x uniformly in α.

Let L_α denote the infinitesimal generator of the diffusion process defined by (3.6.38). For any $f \in C^2(R)$,

$$L_\alpha f(x) = a(x,\theta)\frac{\partial f}{\partial x} + \frac{S^2(x,\sigma)}{2}\frac{\partial^2 f}{\partial x^2}, \qquad L_0 = L_{\alpha_0}. \tag{3.6.45}$$

Let L_α^k denote the kth iterate of L_α. This is well defined under assumption (KK5)[$2(k-1)$](i) and its domain is $C^{2k}(R)$. We set L_α^0 to be the identity operator.

We further assume that the following identification condition holds.

(KK6) (i) $a(x,\theta) = a(x,\theta_0)$, $[\mu_0]$-a.s. for all $x \Rightarrow \theta = \theta_0$;

(ii) $S(x,\sigma) = S(x,\sigma_0)$, $[\mu_0]$-a.s. for all $x \Rightarrow \sigma = \sigma_0$.

For $\ell \geq 0$, under assumption (KK5)[$2(\ell-1)$](i), let $f(y) = y$ and

$$r_\ell(h_n, x, \alpha) = \sum_{i=0}^{\ell} \frac{h_n^i}{i!} L_\alpha^i f(x). \tag{3.6.46}$$

Let

$$g_{h_n,x,\alpha,\ell}(y) = (y - r_\ell(h_n, x, \alpha))^2. \tag{3.6.47}$$

Note that for fixed x, y and α, $g_{h_n,x,\alpha,\ell}$ is a polynomial in h_n of degree 2ℓ. Let $\bar{g}_{h_n,x,\alpha,\ell}$ be the sum of its terms up to degree ℓ. Then

$$\bar{g}_{h_n,x,\alpha,\ell} = \sum_{j=0}^{\ell} h_n^j \bar{g}_{x,\alpha}^{(j)}(y), \tag{3.6.48}$$

where

$$\bar{g}_{x,\alpha}^{(0)}(y) = (y-x)^2, \tag{3.6.49}$$

$$\bar{g}_{x,\alpha}^{(j)}(y) = -2(y-x)\frac{L_\alpha^j f(x)}{j!} + \sum_{\substack{r,s\geq 1\\ r+s=j}} \frac{L_\alpha^r f(x)}{r!}\frac{L_\alpha^s f(x)}{s!}, \tag{3.6.50}$$

for $i \leq j \leq \ell$. From (3.6.50), we can see that $L_\alpha^r \bar{g}_{x,\alpha}^{(j)}$ is well defined for $r + j = \ell$ under the assumption (KK5)[$2(\ell-1)$](i), and we define

$$\Gamma_\ell(h_n, x, \alpha) = \sum_{j=0}^{\ell} h_n^j \sum_{r=0}^{\ell-j} \frac{h_n^r}{r!} L_\alpha^r \bar{g}_{x,\alpha}^{(j)}(x). \tag{3.6.51}$$

The function Γ_ℓ can be written as

$$\Gamma_\ell(h_n, x, \alpha) = \sum_{i=0}^{\ell} h_n^i \gamma_i(x, \alpha). \tag{3.6.52}$$

It can be checked that

$$\gamma_0(x,\alpha) = 0, \qquad \gamma_1(x,\alpha) = S^2(x,\sigma). \tag{3.6.53}$$

Define

$$\bar{\Gamma}_\ell(h_n, x, \alpha) = \frac{\sum_{i=2}^{\ell} h_n^i \gamma_i(x,\alpha)}{h_n S^2(x,\sigma)} \tag{3.6.54}$$

so that

$$\Gamma_\ell(h_n, x, \alpha) = h_n S^2(x, \sigma)\{1 + \bar{\Gamma}_\ell(h_n, x, \alpha)\}. \tag{3.6.55}$$

The following lemma can be proved for the expansion of $m(\alpha, x)$ and $m_2(\alpha, x)$ using Lemma 1 from Florens-Zmirou (1989).

Lemma 3.6.5 (*Kessler 1997*) *Under the assumptions stated above*

$$m(\alpha, , X_{t_{i-1}^n}) = r_\ell(h_n, X_{t_{i-1}^n}, \alpha) + J(\alpha, h_n^{\ell+1}, X_{t_{i-1}^n}), \tag{3.6.56}$$

where J is a function as defined earlier in (3.6.43), and

$$\begin{aligned} &E_\alpha[(X_{t_i^n} - r_\ell(h_n, X_{t_{i-1}}^n, \alpha))^2 \mid X_{t_{i-1}^n}] \\ &= h_n c_{i-1}(\sigma_0)(1 + \bar{\Gamma}_\ell(h_n, X_{t_{i-1}^n}, \alpha)) + J(\alpha, h_n^{\ell+1}, X_{t_{i-1}^n}), \end{aligned} \tag{3.6.57}$$

where

$$c_{i-1}(\sigma) = \frac{\partial}{\partial\sigma} S^2(X_{t_{i-1}^n}, \sigma). \tag{3.6.58}$$

Note that r_ℓ and $\bar{\Gamma}_\ell$ have closed forms and they involve only the coefficients of the diffusion (3.6.38) and their derivatives with respect to x. For further simplification, we consider the expansions of Γ_ℓ^{-1} and $\log \Gamma_\ell$ in h_n.

For the condition $nh_n^p \to 0$, we set $k_0 = [p/2]$ and assume that (KK5)[$2k_0$](i) holds. Let d_j (or e_j) be the coefficient of h_n^j in the Taylor expansion of $(1 + \bar{\Gamma}_{k_0+1}(h_n, X_{t_{i-1}^n}, \alpha))^{-1}$ (or $\log(1 + \bar{\Gamma}_{k_0+1}(h_n, X_{t_{i-1}^n}, \alpha))$. Define the function

$$\begin{aligned} I_{p,n}(\alpha) = \sum_{i=1}^{n} \Bigg[&\frac{(X_{t_i^n} - r_{k_0}(h_n, X_{t_{i-1}^n}, \alpha))^2}{h_n c_{i-1}(\sigma)} \left\{1 + \sum_{j=1}^{k_0} h_n^j d_j(X_{t_{i-1}^n}, \alpha)\right\} \\ &+ \sum_{i=1}^{n} \left\{\log c_{i-1}(\sigma) + \sum_{j=1}^{k_0} h_n^j e_j(X_{t_{i-1}^n}, \alpha)\right\}\Bigg]. \end{aligned} \tag{3.6.59}$$

Note that $I_{p,n}$ is explicit. It is easy to see that, for all $j \le k_0$, d_j and e_j are three times differentiable with respect to α. Moreover, all the derivatives with respect to α are of polynomial growth in x uniformly in α. Furthermore,

$$d_1(x, \alpha) = -e_1(x, \alpha) = -\frac{\gamma_2(x, \alpha)}{S^2(x, \sigma)}. \tag{3.6.60}$$

Let $\widehat{\alpha}_{p,n} = (\hat{\theta}_{p,n}, \hat{\sigma}_{p,n})$ be any of the solutions to

$$I_{p,n}(\widehat{\alpha}_{p,n}) = \inf_\alpha I_{p,n}(\alpha). \tag{3.6.61}$$

We now state the following main result proved by Kessler (1997).

Theorem 3.6.6 *Let p be an integer and $k_0 = [p/2]$. Under assumptions (KK1)–(KK4), (KK5)[$2k_o$] and (KK6), if $h_n \to 0$ and $nh_n \to \infty$, then*

$$\alpha \hat{}_{p,n} \xrightarrow{p} \alpha_0 \quad \textit{as } n \to \infty \textit{ under } P_{\alpha_0}. \tag{3.6.62}$$

If, in addition, $nh_n^p \to 0$ and $\alpha_0 \in H^0$, the interior of H, then

$$(\sqrt{nh_n}(\widehat{\theta}_{p,n} - \theta_0), \sqrt{n}(\widehat{\sigma}_{p,n} - \sigma_0)) \xrightarrow{\mathcal{L}} N_2(0, K_0), \tag{3.6.63}$$

where

$$K_0 = \begin{bmatrix} (\int \frac{1}{c}(\frac{\partial}{\partial\theta}a)^2(x,\alpha_0)\,\mu_0(dx))^{-1} & 0 \\ 0 & 2(\int(\frac{1}{c}\frac{\partial}{\partial\sigma}c)^2(x,\alpha_0)\,\mu(dx))^{-1} \end{bmatrix} \tag{3.6.64}$$

and $c = S^2$.

Remark. The estimator $\widehat{\theta}_{p,n}$ is asymptotically efficient since

$$\int \frac{1}{c}\left(\frac{\partial}{\partial\theta}a\right)^2 (x,\alpha_0)\,\mu(dx)$$

is the asymptotic Fisher information (Dacunha-Castelle and Florens-Zmirou 1986). Let π_{h_n} be the transition density of the process X and define

$$I_{h_n}^{\alpha}(x) = \int \left(\frac{1}{\pi_{h_n}}\frac{\partial \pi_{h_n}}{\partial\alpha}\right)^2 \pi_{h_n}(\alpha, x, y)\,dy.$$

Then it can be shown that

$$\lim_{n\to\infty}\frac{1}{n}\sum_{i=1}^{n} I_{h_n}^{\alpha_0}(X_{t_{i-1}^n}) = 2\int\left(\frac{1}{c}\frac{\partial c}{\partial\sigma}\right)^2 (x,\alpha_0)\,\mu_0(dx).$$

Alternate schemes for obtaining an asymptotically efficient estimator for the parameter (θ, σ) are given in Kessler (1995a).

Approximation of the transition function by Gaussian and close to Gaussian densities

Consider the stochastic differential equation

$$dX_t = a(X_t,\theta)\,dt + \sigma(X_t,\theta)\,dW_t, \qquad t \geq 0,$$

where a and σ are known functions and θ is unknown with $\theta \in \Theta$, an open bounded subset of R^k. Let $p_X(\Delta, x|x_0;\theta)$ denote the conditional density of $X_{t+\Delta}$ given $X_t = x_0$. Suppose the process is observed at the time points $\{t = i\Delta, 0 \leq i \leq n\}$ where Δ is fixed. The log-likelihood function of the observations is proportional to

$$\ell_n(\theta) = n^{-1}\sum_{i=1}^{n}\log\{p_X(\Delta, X_{i,\Delta}|X_{(i-1)}\Delta;\theta)\}. \tag{3.6.65}$$

Following Ait-Sahalia (1997), we construct a sequence $\ell_n^{(J)}$, $J \geq 1$, of approximations in a closed form to the log-likelihood function ℓ_n such that $\ell_n^{(J)}$ converges to ℓ_n as J increases. Let

$$\hat{\theta}_n^{(J)} = \arg\max_{\theta\in\Theta}\ell_n^{(J)}(\theta) \tag{3.6.66}$$

and

$$\hat{\theta}_n = \arg\max_{\theta \in \Theta} \ell_n(\theta). \tag{3.6.67}$$

We will give sufficient conditions which ensure that $\hat{\theta}_n^{(J)}$ tends to $\hat{\theta}_n$ as J gets larger and larger. Empirical evidence shows that for models that are relevant in finance $J = 4$ is adequate. Since the expression for $\ell_n^{(J)}$ is explicit, standard maximum likelihood estimation can be carried through for computation of $\hat{\theta}_n^{(J)}$.

Let $D_X = (\underline{x}, \overline{x})$ be the domain of the diffusion X. We assume that D_X is either $(-\infty, +\infty)$ or $(0, \infty)$, and that the following conditions hold.

(LL1) (Smoothness of the coefficients.) The functions $a(x; \theta)$ and $\sigma(x; \theta)$ are infinitely differentiable in x on D_X and twice continuously differentiable in θ in $\Theta \subset R^k$, Θ being an open bounded set.

(LL2) (Nondegeneracy of the diffusion.) (i) If $D_X = (-\infty, +\infty)$, then there exists a constant c such that $\sigma(x; \theta) > c > 0$ for all $x \in D_X$ and for all $\theta \in \Theta$.

(ii) If $D_X = (0, \infty)$, then the function σ can have local degeneracy at $x = 0$. However, if $\sigma(0; \theta) = 0$, then there exist constants $\xi_0 \geq 0$, $c \geq 0$ and ρ such that $\sigma(x; \theta) \geq cx^\rho$ for all $0 < x \leq \xi_0$ and $\theta \in \Theta$. Away from $x = 0$, σ is nondegenerate, that is, for each $\xi > 0$, there exists a constant c_ξ such that $\sigma(x; \theta) \geq c_\xi > 0$ for all $x \in [\xi, \infty)$ and $\theta \in \Theta$.

Let

$$Y_t \equiv \gamma(X_t; \theta) = \int^{X_t} \frac{du}{\sigma(u; \theta)} \tag{3.6.68}$$

where any primitive of the function $1/\sigma$ may be chosen. Since $\sigma > 0$ on D_X, the function γ is increasing and invertible. It maps D_X into $D_Y = (\underline{y}, \overline{y})$, where $\underline{y} = \lim_{x \to \underline{x}} \gamma(x, \theta)$ and $\overline{y} = \lim_{x \to \overline{x}} \gamma(x, \theta)$. We assume that the domain D_Y is independent of $\theta \in \Theta$. Applying Itô's lemma, we have

$$dY_t = a_Y(Y_t; \theta)\, dt + dW_t, \tag{3.6.69}$$

where

$$a_Y(y; \theta) = \frac{a(\gamma^{-1}(y; \theta); \theta)}{\sigma(\gamma^{-1}(y; \theta); \theta)} - \frac{1}{2}\frac{\partial \sigma}{\partial x}(\gamma^{-1}(y; \theta); \theta). \tag{3.6.70}$$

An infinitely differentiable function f is said to have at most exponential growth if $f(y) = O(e^{\lambda |y|})$ for some $\lambda \geq 0$.

Suppose the following additional condition holds.

(LL3) (Boundary behaviour.) For all $\theta \in \Theta$, $a_Y(y; \theta)$, $\frac{\partial a_Y}{\partial y}(y; \theta)$ and $\frac{\partial^2 a_Y}{\partial y^2}(y; \theta)$ have at most exponential growth near the infinity boundaries, and

$$\lim_{y \to \overline{y} \text{ or } \underline{y}} g(y; \theta) < \infty$$

where

$$g(y; \theta) \equiv -\frac{1}{2}\left(a_Y^2(y; \theta) + \frac{\partial}{\partial y} a_Y(y; \theta)\right). \tag{3.6.71}$$

Furthermore,

(i) if $\underline{y} = 0^+$, then there exist constants ε_0, k, α such that for all $0 < y \leq \varepsilon_0$ and $\theta \in \Theta$,

$$a_Y(y, ; \theta) \geq k y^{-\alpha},$$

where either $\alpha > 1$ and $k > 0$ or $\alpha = 1$ and $k \geq \frac{1}{2}$;

(ii) if $\underline{y} = -\infty$, then there exist constants $E_0 > 0$ and $K > 0$ such that for all $0 < y \leq E_0$ and $\theta \in \Theta$,

$$a_Y(y; \theta) \geq K y;$$

(iii) if $\bar{y} = +\infty$, then there exist constants $E_0 > 0$ and $K > 0$ such that for all $y \geq E_0$ and $\theta \in \Theta$,

$$a_Y(y; \theta) \leq K y;$$

(iv) and if $\overline{y} = 0^{-\sigma}$, then there exist constants ε_0, k, α such that for all $0 > y \geq -\varepsilon_0$ and $\theta \in \Theta$,

$$a_Y(y; \theta) \leq -k|y|^{-\alpha},$$

where either $\alpha > 1$ and $k > 0$ or $\alpha = 1$ and $k \geq \frac{1}{2}$.

For given $\Delta > 0$, $\theta \in \Theta$ and $y_0 \in R$, define the 'pseudo-normalized' increment of Y as

$$Z_t \equiv Z(\Delta, Y_t | y_0; \theta) = \Delta^{-1/2}(Y_t - y_0 - a_Y(y_0; \theta)\Delta). \tag{3.6.72}$$

Let $p_Y(\Delta, y|y_0; \theta)$ denote the conditional density of $Y_{t+\Delta}$ given Y_t and define

$$p_Z(\Delta, z|y_0; \theta) = \Delta^{1/2} p_Y(\Delta, \Delta^{1/2} z + y_0 + a_Y(y_0; \theta)\Delta | y_0; \theta). \tag{3.6.73}$$

It is easy to see that

$$p_Y(\Delta, y|y_0; \theta) \equiv \Delta^{-1/2} p_Z(\Delta, \Delta^{-1/2}(y - y_0 - a_Y(y_0, \theta)\Delta)|y_0; \theta). \tag{3.6.74}$$

Note that p_X and p_Y are connected by the relations

$$p_X(\Delta, x|x_0; \theta) = \sigma(x; \theta)^{-1} p_Y(\Delta, \gamma(x; \theta)|\gamma(x_0, \theta); \theta) \tag{3.6.75}$$

and

$$p_Y(\Delta, y|y_0; \theta) = \sigma(\gamma^{-1}(y; \theta); \theta) p_X(\Delta, \gamma^{-1}(y; \theta)|\gamma^{-1}(y_0; \theta); \theta). \tag{3.6.76}$$

Note that the random variable Z_t is constructed so that its density p_Z is close to a standard normal density. We will now approximate the density function p_Z by a Hermite series expansion. Let

$$H_j(z) \equiv e^{z^2/2} \frac{d^j}{dz^j}(e^{-z^2/2}), \quad j \geq 1, \qquad \Phi(z) = \frac{1}{\sqrt{2\pi}} e^{-z^2/2}. \tag{3.6.77}$$

Let

$$p_Z^{(J)}(\Delta, z|y_0; \theta) \equiv \Phi(z) \sum_{j=0}^{J} \eta_j(\Delta, y_0; \theta) H_j(z) \tag{3.6.78}$$

be the Hermite series expansion of the density function $p_Z(\Delta, z|y_0; \theta)$ up to the Jth term for fixed Δ, y_0 and θ. The coefficients η_j are defined by

$$\eta_j(\Delta, y_0; \theta) = \frac{1}{j!} \int_{-\infty}^{\infty} H_j(z) p_Z(\Delta, z|y_0; \theta)\, dz. \tag{3.6.79}$$

Following (3.6.73), we construct a sequence of approximations to p_Y by

$$p_Y^{(J)}(\Delta, y|y_0;\theta) \equiv \Delta^{-1/2} p_Z^{(J)}(\Delta, \Delta^{-1/2}(y - y_0 - a_Y(y_0;\theta)\Delta)|y_0;\theta) \tag{3.6.80}$$

and to p_X by

$$p_X^{(J)}(\Delta, x \mid x_0;\theta) \equiv \sigma(x,\theta)^{-1} p_Y^{(J)}(\Delta, \gamma(x;\theta)|\gamma(x_0,\theta);\theta). \tag{3.6.81}$$

The following theorem due to Ait-Sahalia (1997) proves that the density of the random variable Z_t is close to the standard normal density and the approximation $p_X^{(J)}(\Delta, x|x_0;\theta)$ is close to the true transition density function $p_X(\Delta, x|x_0,\theta)$. We note that the sampling interval Δ remains fixed and it is not assumed that $\Delta \to 0$ but $J \to \infty$, where J is the order of the approximation.

Theorem 3.6.7 *Under the assumptions (LL1)–(LL3), there exists $\bar{\Delta} > 0$ such that for every $\Delta \in (0, \bar{\Delta})$, $\theta \in \Theta$ and $(x, x_0) \in D_X^2$,*

$$p_X^{(J)}(\Delta, x|x_0;\theta) \to p_X(\Delta, x|x_0;\theta) \qquad \text{as } J \to \infty \tag{3.6.82}$$

uniformly for $\theta \in \Theta$ and x_0 over compact subsets of D_X. If σ is nondegenerate, then the convergence is uniform in x over D_X. If σ is degenerate at zero, then the convergence is uniform in x in each interval of the form $[\varepsilon, \infty)$, $\varepsilon > 0$.

For a proof, see Ait-Sahalia (1997). With the above sequence of approximations to the transition function, consider

$$\ell_n^{(J)}(\theta) \equiv n^{-1} \sum_{i=1}^{n} \log\{p_X^{(J)}(\Delta, X_{i\Delta}|X_{(i-1)\Delta};\theta)\}, \tag{3.6.83}$$

which can be taken as an approximation to the function $\ell_n(\theta)$ defined by (3.6.65). Let $\hat{\theta}_n^{(J)}$ be a maximizer of $\ell_n^{(J)}(\theta)$ over $\theta \in \Theta$. We will now give sufficient conditions under which $\hat{\theta}_n^{(J)}$ converges to $\hat{\theta}_n$ as $J \to \infty$ and further prove that $\hat{\theta}_n^{(J_n)} \to \theta_0$, the true parameter, provided $J_n \to \infty$ as $n \to \infty$. This is a consequence of the convergence of the approximating transition density $p_X^{(J_n)}$ to p_X uniformly in x and θ.

In the definition of the log-likelihood function in (3.6.83), we have ignored the term $\log \pi(x_0;\theta)$ corresponding to the initial density (Billingsley 1961; Basawa and Prakasa Rao 1980). Let

$$L_i(\theta) \equiv \log p_X(\Delta, X_{i\Delta}|X_{(i-1)\Delta};\theta).$$

It is easy to see that $p_X(\Delta, x|x_0,\theta)$ admits two continuous derivatives with respect to θ in Θ and this also holds for its approximation $p_X^{(J)}$ of any order J under (LL1). Let $\dot{L}_i(\theta) \equiv \partial L_i(\theta)/\partial\theta$ and $\ddot{L}_i(\theta) \equiv \partial^2 L_i(\theta)/\partial\theta\, \partial\theta^T$, where T denotes transposition. Define

$$I_n(\theta) \equiv \sum_{i=1}^{n} \operatorname{diag} E_\theta[\dot{L}_i(\theta)\dot{L}_i(\theta)^T|X_{(i-1)\Delta}]$$

and

$$i_n(\theta) = E_\theta[I_n(\theta)].$$

Define $H_n(\theta) = -\sum_{i=1}^n \ddot{L}_i(\theta)$. It can be checked that $E_\theta[H_n(\theta)] = E_\theta[I_n(\theta)] = i_n(\theta)$. In addition to conditions (LL1)–(LL3), suppose that the following conditions hold.

(LL4) The true parameter $\theta_0 \in \Theta$ and $i_n^{-1}(\theta) \overset{\text{a.s.}}{\to} 0$ as $n \to \infty$ uniformly in $\theta \in \Theta$.

(LL5) There exists a matrix $G(\theta)$ possibly random, almost surely finite and positive definite, such that $G_n(\theta) \equiv i_n^{-1/2}(\theta)H_n(\theta)i_n^{-1/2}(\theta) \overset{p}{\to} G(\theta)$ uniformly over compact subsets of Θ.

(LL6) In assumptions (LL2)(ii) and (LL3)(i), if $\alpha = 1$, then either $\rho \geq 1$ with no restriction on k or $k \geq 2\rho/(1-\rho)$ if $0 \leq \rho < 1$. If $\alpha > 1$, no restriction is imposed.

The following theorem, proved by Ait-Sahalia, gives the asymptotic properties of the estimator $\hat{\theta}_n^{(J_n)}$.

Theorem 3.6.8 *Suppose that conditions (LL1)–(LL6) hold and $\Delta \in (0, \bar{\Delta})$. Then*

(i) $\hat{\theta}_n^{(J)} \overset{p}{\to} \hat{\theta}_n$ under P_{θ_0} as $J \to \infty$ for any fixed n; and

(ii) as $n \to \infty$, there exists a sequence $\bar{J}_n \to \infty$ such that for any $J_n \geq \bar{J}_n$, $\hat{\theta}_n^{(J_n)} \overset{p}{\to} \theta_0$ and $i_n^{1/2}(\theta_0)(\hat{\theta}_n^{(J_n)} - \theta_0) \overset{\mathcal{L}}{\to} G^{-1/2}(\theta_0)Z$ under P_{θ_0},

where Z is $N(0, Id)$ is independent of $G^{-1/2}(\theta_0)$ and Id is the least asymptotic variance achievable by $i_n^{1/2}(\theta_0)$-consistent and asymptotically normal estimators of θ_0.

We now give a method of computation of the terms in the expression of $p_X^{(J)}$. Theorem 3.6.8 shows that

$$p_Z(\Delta, z \mid y_0; \theta) = \Phi(z) \sum_{j=0}^{\infty} \eta_j(\Delta, y_0; \theta) H_j(z). \tag{3.6.84}$$

Note that $p_Z^{(J)}(\Delta, z|y_0; \theta)$ denotes the partial sum in (3.6.84) up to $j = J$. Relation (3.6.78) implies that

$$\begin{aligned}
\eta_j(\Delta, y_0; \theta) &= \frac{1}{j!}\int_{-\infty}^{\infty} H_j(z) p_Z(\Delta, z|y_0; \theta)\, dz \\
&= \frac{1}{j!}\int_{-\infty}^{\infty} H_j(z)\Delta^{1/2} p_Y(\Delta, \Delta^{1/2}z + y_0 + a_Y(y_0; \theta)\Delta|y_0; \theta)\, dz \\
&= \frac{1}{j!}\int_{-\infty}^{\infty} H_j(\Delta^{-1/2}(y - y_0 - a_Y(y_0; \theta)\Delta)) p_Y(\Delta, y|y_0, \theta)\, dy \\
&= \frac{1}{j!} E[H_j(\Delta^{-1/2}(Y_{t+\Delta} - y_0 - a_Y(y_0; \theta)\Delta))|Y_t = y_0; \theta].
\end{aligned} \tag{3.6.85}$$

In order to calculate the coefficients explicitly, we need to calculate the conditional moments. The following lemma, proved by Ait-Sahalia (1997), will be needed.

Lemma 3.6.9 *Suppose that conditions (LL1)–(LL3) hold. Let f be a function such that f and all its derivatives have at most exponential growth. Then, for $\Delta \in (0, \bar{\Delta})$, $y_0 \in R$ and $\theta \in \Theta$, there exists δ in $(0, \Delta]$ such that*

$$E[f(Y_{t+\Delta})|Y_t = y_0] = \sum_{j=1}^{J} A^j(\theta) \bullet f(y_0)\frac{\Delta^j}{j!} + E[A^{J+1}(\theta) \bullet f(Y_{t+\delta})|Y_t = y_0]\frac{\Delta^{J+1}}{(J+1)!}, \tag{3.6.86}$$

where $A(\theta)$ is the infinitesimal generator of the diffusion defined by

$$A(\theta)f = a_Y(\cdot,\theta)\frac{\partial f}{\partial y}(\cdot) + \frac{1}{2}\frac{\partial^2 f}{\partial y^2}(\cdot) \tag{3.6.87}$$

and $A^j(\theta) \bullet f(y_0)$ means that the operator $A(\theta)$ is applied j times to the function $f(y)$ and evaluated at $y = y_0$. Furthermore, there exists a constant K_J depending on J, but independent of f and δ, such that

$$|E(A^{J+1}(\theta) \bullet f(Y_{t+\delta})|Y_t = y_0)| \le K_J. \tag{3.6.88}$$

We do not require the remainder term in the Taylor series expansion given in (3.6.86) to converge to zero as $J \to \infty$ for fixed $\Delta \in (0, \bar{\Delta})$. The convergence of the Taylor series expansion in (3.6.86) is interpreted in the sense that, for any given J,

$$\lim_{\Delta\to 0} \Delta^{-(J+1)}\left\{E[f(Y_{t+\Delta})\Big|Y_t = y_0] - \sum_{j=1}^{J} A^j(\theta) \bullet f(y_0)\frac{\Delta^j}{j!}\right\}$$
$$= \frac{A^{J+1}(\theta) \bullet f(y_0)}{(J+1)!} \tag{3.6.89}$$

and the expansion in (3.6.89) is the Taylor series expansion of the conditional expectation operator in Δ.

Let $\tilde{p}_Z^{(J)}$ denote the approximation to p_Z obtained by retaining in the expansion for $p_Z^{(J)}$ all the terms in η_j, $j = 0, \ldots, J$ of order smaller than or equal to $\Delta^{J/2}$ in $p_Z^{(J)}$. It can be shown that η_j is of the order $\Delta^{J/2}$, so that

$$\Phi(z) \sum_{j=J+1}^{\infty} \eta_j(\Delta, y_0; \theta)H_j(z)$$

is of order $O(\Delta^{j/2})$. Hence, the expression $\tilde{p}_Z^{(J)}$ does retain all the terms up to $\Delta^{j/2}$ in the expansion for p_Z and not only in the expansion for $p_Z^{(J)}$.

The first four terms of the sequence $\tilde{p}_Z^{(J)}(\Delta, z|y_0; \theta)$ are

$$\tilde{p}_Z^{(1)} = \Phi, \tag{3.6.90}$$
$$\tilde{p}_Z^{(2)} = \tilde{p}_Z^{(1)} + \Phi[H_2 a_Y^{[1]}/2]\Delta, \tag{3.6.91}$$
$$\tilde{p}_Z^{(3)} = \tilde{p}_Z^{(2)} - \Phi[H_1\{a_Y a_Y^{[1]}/2 + a_Y^{[2]}/4\} + H_3 a_Y^{[2]}/6]\Delta^{3/2}, \tag{3.6.92}$$
$$\tilde{p}_Z^{(4)} = \tilde{p}_Z^{(3)} + \Phi[H_2\{a_Y^{[1]2}/3 + a_Y a_Y^{[2]}/3 + a_Y^{[3]}/6\}$$
$$+ H_4\{a_Y^{[1]2}/8 + a_Y^{[3]}/24\}]\Delta^2, \tag{3.6.93}$$

where Φ is the standard normal density, H_j is the jth Hermite polynomial and $a_Y^{[k]m}$ stands for $\left[\frac{\partial^k a_Y(y;\theta)}{\partial y^k}\Big|_{y=y_0}\right]^m$. Note that $H_1(z) \equiv -z$, $H_2(z) = z^2 - 1$, $H_3(z) = -z^3 + 3z$ and $H_4(z) = z^4 - 6z^2 + 3$.

The corresponding expression for $\tilde{p}_X^{(J)}(\Delta, x|x_0; \theta)$ is given by

$$\begin{aligned}\tilde{p}_X^{(J)}(\Delta, x|x_0, \theta) = \sigma(x, \theta)^{-1}\Delta^{-1/2} \\ \times \tilde{p}_Z^{(J)}(\Delta, \Delta^{-1/2}(\gamma(x;\theta) - \gamma(x_0;\theta) \\ - a_Y(\gamma(x_0;\theta);\theta)\Delta) | \gamma(x_0;\theta);\theta),\end{aligned} \tag{3.6.94}$$

and then replacing γ and a_Y by their expressions in (3.6.68) and (3.6.70). It can be checked that

$$\tilde{p}_X^{(1)}(\Delta, x|x_0;\theta) = \frac{1}{\sqrt{2\pi\Delta}\sigma(x;\theta)} \exp[-(\gamma(x;\theta) - \gamma(x_0;\theta) - a_Y(\gamma(x_0,\theta);\theta)\Delta)^2/2\Delta]$$

and

$$\begin{aligned}\tilde{p}_X^{(2)}(\Delta, x|x_0;\theta) &= \tilde{p}_X^{(1)}(\Delta, x|x_0;\theta) \\ &\times \left\{1 + H_2\left(\frac{\gamma(x;\theta) - \gamma(x_0;\theta) - a_Y(\gamma(x_0;\theta);\theta)\Delta}{\Delta^{1/2}}\right)\right. \\ &\left.\times \frac{a_Y^{(1)}((\gamma(x_0;\theta);\theta)}{2}\Delta\right\}\end{aligned}$$

etc. One can compute the approximate log-likelihood functions for Z which are given by

$$\tilde{\ell}_{Z,n}^{(1)}(\theta) = n^{-1}\sum_{i=1}^{n} \log \Phi(Z_{i\Delta}),$$

$$\tilde{\ell}_{Z,n}^{(2)}(\theta) = \tilde{\ell}_{Z,n}^{(1)}(\theta) + n^{-1}\sum_{i=1}^{n}[H_2 a_Y^{[1]}/2]\Delta,$$

$$\tilde{\ell}_{Z,n}^{(3)}(\theta) = \tilde{\ell}_{Z,n}^{(2)}(\theta) - n^{-1}\sum_{i=1}^{n}[H_1\{a_Y a_Y^{[1]}/2 + a_Y^{[2]}/4\} + H_3 a_Y^{[2]}/6]\Delta^{3/2}$$

etc.; the expansion $\tilde{\ell}_n^{(J)}(\theta)$ is obtained from $\tilde{\ell}_{Z,n}^{(J)}(\theta)$ just as $\tilde{p}_X^{(J)}$ from $\tilde{p}_Z^{(J)}$. It is sufficient to include the first three terms in the expansion, and maximizing $\tilde{\ell}_n^{(3)}(\theta)$ is more than adequate in practical problems.

The following examples illustrate the size of the approximation when replacing p_X by $p_X^{(J)}$ and indicate how fast the error decreases as more and more terms are added.

Example 3.6.8 (*Vasicek model*)
Consider the Ornstein–Uhlenbeck process

$$dX_t = \beta(\bar{\alpha} - X_t)\,dt + \sigma\,dW_t, \qquad t \geq 0,$$

proposed by Vasicek (1997) for the short-term interest rate. Here the transition density can be explicitly computed and it is given by

$$\begin{aligned}&p_X(\Delta, x|x_0;\theta) \\ &= (\pi\sigma^2(1 - e^{-2\beta\Delta})/\beta)^{-1/2} \\ &\quad \times \exp\{-(x - \bar{\alpha} - (x_0 - \bar{\alpha})\bar{e}^{\beta\Delta})^2\beta/(\sigma^2(1 - e^{-2\beta\Delta}))\},\end{aligned}$$

with $\theta = (\bar{\alpha}, \beta, \sigma)$. Here $Y_t = \gamma(X_t;\theta) = \sigma^{-1}X_t$ and $a_Y(y;\theta) = \beta\bar{\alpha}\sigma^{-1} - \beta y$.

Example 3.6.9 (*Cox–Ingersoll–Ross model*)
Consider the process

$$dX_t = \beta(\bar{\alpha} - X_t)\,dt + \sigma\sqrt{X_t}\,dW_t, \qquad t \geq 0,$$

with $\theta = (\bar{\alpha}, \beta, \sigma)$ all positive, proposed as a model by Cox *et al.* (1985) for the short-term interest rate. Here X_t is distributed on $D_X = (0, +\infty)$ provided $q = (2\beta\bar{\alpha}/\sigma^2) - 1 \geq 0$. Its transition density is the noncentral chi-squared density given by

$$p_X(\Delta, x|x_0; \theta) = ce^{-u-v}\left(\frac{v}{u}\right)^{q/2} I_q(2(uv)^{1/2}),$$

where $c = 2\beta/(\sigma^2(1 - e^{-\beta\Delta}))$, $u \equiv cx_0e^{-\beta\Delta}$, $v \equiv cx$ and I_q is the modified Bessel function of the first kind of order q. Here $Y_t = \gamma(X_t; \theta) = 2\sqrt{X_t}/\sigma$ and $a_Y(y; \theta) = (q + \frac{1}{2})/y - \beta y/2$. It can be checked that the conditions (LL1)–(LL3) hold for $q \geq 3/2$.

Example 3.6.10 (*CEV model*)
Consider the model

$$dX_t = \beta(\bar{\alpha} - X_t)\,dt + \sigma X_t^{\rho}\,dW_t, \qquad t \geq 0,$$

with $\theta = (\bar{\alpha}, \beta, \sigma, \rho)$ and X_t distributed over $(0, +\infty)$, where $\bar{\alpha} > 0$ $\beta > 0$, and $\rho > 1/2$. This model does not have a closed form for the transition density (Cox 1996; and Chan *et al.* 1992). For $1/2 < \rho < 1$, the transformation from X to Y is given by

$$Y_t = \gamma(X_t; \theta) = X_t^{1-\rho}/\{\sigma(1 - \rho)\}$$

and

$$a_Y(y; \theta) = \phi_0 y^{-\rho/(1-\rho)} - \psi y^{-1} - \phi_1 y,$$

where $\phi_0 \equiv \beta\bar{\alpha}\sigma^{-1/(1-\rho)}(1 - \rho)^{-\rho/(1-\rho)}$, $\psi \equiv -\rho/\{2(1 - \rho)\}$ and $\phi_1 = \beta(1 - \rho)$.

Example 3.6.11 (*Nonlinear mean reversion*)
Consider the model

$$dX_t = \left(\frac{\alpha - 1}{X_t} + \alpha_0 + \alpha_1 X_t + \alpha_2 X_t^2\right) dt + \sigma X_t^{\rho}\,dW_t, \qquad t \geq 0,$$

introduced by Ait-Sahalia (1996a), Conley *et al.* (1997) and Tauchen (1997) to produce short mean reversion while interest rate values remain in the middle part of their domain and strong nonlinear mean reversion at the end of the domain. Here $\theta = (\alpha_1, \alpha_0, \alpha_1, \alpha_2, \sigma, \rho)$ and $Y_t = \gamma(X_t, \theta) = X_t^{1-\rho}/\{\sigma(1 - \rho)\}$ for $\rho \neq 1$ and $Y_t = \sigma \log X_t$ for $\rho = 1$. Furthermore, for $\rho \neq 1$,

$$a_Y(y; \theta) = \varphi_{-1} y^{-(1+\rho)/(1-\rho)} + \varphi_0 y^{-\rho/(1-\rho)} + \psi y^{-1} + \varphi_1 y + \varphi_2 y^{(2-\rho)/(1-\rho)}$$

where

$$\begin{aligned}
\varphi_{-1} &\equiv \alpha_{-1}\sigma^{-2/(1-\rho)}(1 - \rho)^{-(1+\rho)/(1-\rho)} \\
\varphi_0 &\equiv \alpha_0\sigma^{-1/(1-\rho)}(1 - \rho)^{-\rho/(1-\rho)}, \\
\psi &\equiv -\rho/\{2(1 - \rho)\} \\
\varphi_1 &\equiv \alpha_1(1 - \rho) \\
\varphi_2 &\equiv \alpha_2\sigma^{1/(1-\rho)}(1 - \rho)^{(1-\rho)/(1+\rho)}.
\end{aligned}$$

Example 3.6.12 (*Black–Scholes–Merton model*)
Consider the geometric Brownian motion

$$dX_t = \mu X_t\, dt + \sigma X_t\, dW_t, \qquad t \geq 0,$$

which is distributed on $D_X = (0, \infty)$. Its transition density is log-normal:

$$p_X(\Delta, x|x_0; \theta) = (2\pi\Delta\sigma^2 x^2)^{-1/2} \exp\left\{ -\left(\log\left(\frac{x}{x_0}\right) - \left(\mu - \frac{\sigma^2}{2}\right)\Delta \right)^2 \Big/ (2\Delta\sigma^2) \right\},$$

with $\theta = (\mu, \sigma)$. Here $Y_t = \gamma(x_t; \theta) = \sigma \log X_t$ so that $D_Y = (-\infty, \infty)$ and $a_Y(y; \theta) = (\mu/\sigma) - (\sigma/2)$. Furthermore, $p_X^{(J)} = p_X$ for all $J \geq 1$ in this example.

Ait-Sahalia (1997) shows that, in all the examples discussed above, the terms of order 2 provide an approximation to p_X which is better by a factor of roughly 10 than the term of order 1 and that each additional order produces additional improvements by a factor of roughly 10. It is generally not necessary to go beyond $J = 3$ in the above examples to estimate the true density with a high degree of precision.

The discussion and results given above are due to Ait-Sahalia (1997).

3.7 Simulation-Based Estimation Methods

Simulation methods are widely used in econometrics to examine properties of estimators. These methods are useful when the computation of the exact theoretical estimators is impossible or difficult. They give approximations of these estimators based on the simulation of the true data generating process.

Consider the dynamic model

$$Y_t = \psi(Y_{t-1}, Y_{t-2}, \ldots, Y_{t-q}, \varepsilon_t, \theta)$$

where θ is a p-dimensional parameter. Suppose that we have discrete time observation $Y_T(\theta) = (Y_0, Y_1, \ldots, Y_T)$. From the initial value Y_0 and a simulation of the process $\tilde{\varepsilon}_t$ it is possible to derive, from the model, simulated paths $\tilde{Y}(\theta)$, namely,

$$\tilde{Y}_t(\theta) = \psi(\tilde{Y}_{t-1}, \tilde{Y}_{t-2}, \ldots, \tilde{Y}_{t-q}, \tilde{\varepsilon}_t, \theta), \qquad t = 1, \ldots, N = KT,$$

for $K \geq 1$. If the true parameter θ_0 can be idettified by the expectation of a given function g, we can choose $\hat{\theta}$ as that value which minimizes the distance between the empirical moment

$$G_T(Y) = \frac{1}{T} \sum_{t=r+1}^{T} g(Y_t, Y_{t-1}, \ldots, Y_{t-r})$$

and

$$G_N(\tilde{Y}(\theta)) = \frac{1}{N} \sum_{t=r+1}^{N} g(\tilde{Y}_t(\theta) Y_{t-1}^{\sim}(\theta), \ldots, Y_{t-r}^{\sim}(\theta)).$$

The simulated moments estimator is defined as

$$\hat{\theta_T} = \arg\min_{\theta \in \Theta} [G_T(Y) - G_N(\tilde{Y}(\theta))]' \hat{\Omega_T} [G_T(Y) - G_N(\tilde{Y}(\theta))],$$

where $\hat{\Omega_T}$ is a symmetric positive definite matrix converging to a deterministic positive definite matrix Ω, when $T \to \infty$ with $K = N/T$ fixed, the simulated moments estimator is consistent and is asymptotically normal under some regularity conditions (cf. Gouriéroux and Monfort 1993). Let us consider the stochastic differential equation

$$dY_t = a(Y_t, \theta)\, dt + \sigma(Y_t, \theta)\, dW_t, \qquad t \geq 0, \tag{3.7.1}$$

where $\{Y_t, t \geq 0\}$ is an R^d-valued process.

Suppose the following conditions holds:

(MM0) There exists a unique ergodic stationary process satisfying the stochastic differential equation given by (3.7.1). The process can be observed at T fixed but equally space sampling times.

Let τ be the sampling interval and

$$Y_T = (Y_\tau, Y_{2\tau}, \ldots, Y_{T\tau}) \tag{3.7.2}$$

be the observation. Let $\theta_0 \in \Theta^0$. Consider a function S from $\Omega \times \Theta$ to R of the form $S(\omega; \theta) = S_T(Y_T, \theta)$ with the property that $S(\cdot, \theta)$ is measurable for all $\theta \in \Theta$ and $S(\omega; \cdot)$ is continuous in θ for all $\omega \in \Omega$. Let $\hat{\theta}^*_T(\theta_0)$ denote any measurable solution of the equation

$$\hat{\theta}^*_T(\theta_0) = \arg\max_{\theta^* \in \Theta} S_T(Y_T, \theta^*) \tag{3.7.3}$$

Suppose also that the following conditions hold:

(MM1) For all $\theta_0 \in \Theta$, there exists a nonrandom limit $S_\infty(\theta_0, \theta^*)$ such that

$$P_{\theta_0}[\lim_{T\to\infty} S_T(Y_T, \theta^*) = S_\infty(\theta_0, \theta^*)] = 1$$

uniformly in $\theta^* \in \Theta$. Furthermore, for all $\theta \in \Theta$, the function $S_\infty(\theta_0, \cdot)$ admits a unique maximum,

$$\theta^*_0(\theta_0) = \arg\max_{\theta^* \in \Theta} S_\infty(\theta_0, \theta^*). \tag{3.7.4}$$

(MM2) The criterion function $S_T(Y_T, \theta^*)$ is twice continuously differentiable with respect to θ^* and

(i) for all $\theta_0 \in \Theta$, there exists a nonrandom matrix $\ddot{S}_\infty(\theta_0, \theta^*)$ such that

$$P_{\theta_0}\left[\lim_{T\to\infty} \frac{\partial^2 S_T(Y_T, \theta^*)}{\partial\theta^*\, \partial\theta^{*\prime}} = \ddot{S}_\infty(\theta_0, \theta^*)\right] = 1$$

uniformly in $\theta^* \in \Theta$;

(ii) $\ddot{S}_\infty(\theta_0, \theta^*_0(\theta_0))$ is invertible;

(iii) the sequence $[\sqrt{T}\frac{\partial S_T(Y_T, \theta^*)}{\partial\theta^*}]_{\theta^* = \theta^*_0(\theta_0)}$ converges in law to $N(0, J^*(\theta_0))$.

Under assumptions, (MM0), (MM1) and (MM2), and it can be shown that

$$\sqrt{T}(\hat{\theta}^*_T(\theta_0) - \theta^*_0(\theta_0)) \xrightarrow{\mathcal{L}} N(0, \Gamma^*_0) \qquad \text{as } T \to \infty,$$

where

$$\Gamma^*_0 = [\ddot{S}_\infty(\theta_0, \theta^*_0(\theta_0))]^{-1} J^*(\theta_0) [\ddot{S}_\infty(\theta_0, \theta^*_0(\theta_0))]^{-1}$$

(see Gallant and White 1988; Gouriéroux and Monfort 1989). In general $\theta^*_0(\theta_0) \neq \theta_0$ and hence the estimator $\hat{\theta}^*_T(\theta_0)$ is inconsistent (see Lo 1988).

Example 3.7.1 (*Ornstein–Uhlenbeck process*)
Consider the stochastic differential equation

$$dX_t = (\alpha - \mu X_t)\,dt + \sigma\,dw_t, \tag{3.7.5}$$

where $\{Y_t, t \geq 0\}$ is a real-valued process with $Y_0 > 0$ and the problem is to estimate $\theta = (\alpha, \mu, \sigma^2)$. We assume that $\mu > 0$ and $\sigma^2 > 0$ so that the equation given above has a stationary ergodic solution. Euler discretization gives

$$X_{t+\tau} - X_t = (\alpha^* - \mu^* X_t)\tau + \sigma^*(W_{t+\tau} - W_t) \tag{3.7.6}$$

while the exact discretization is given by

$$X_{t+\tau} - X_t = \frac{\alpha}{\mu}(1 - e^{-\mu\tau}) - (1 - e^{-\mu\tau})X_t + \sigma e^{-\mu t}\int_t^{t+\tau} e^{\mu s}\,dW_s. \tag{3.7.7}$$

The asymptotic properties of the MLE $\hat{\theta}_T^*(\theta_0)$ can be obtained. The MLE converges to

$$\theta_0^*(\theta_0) = \left(\frac{\alpha_0}{\mu_0\tau}(1 - e^{-\mu_0\tau}), \frac{1 - e^{-\mu\tau}}{\tau}, \frac{\sigma_0^2}{2\mu_0\tau}(1 - e^{-2\mu_0\tau})\right).$$

Hence all the components are biased for fixed sampling interval τ and the bias goes to zero as τ decreases and it is of order τ in the neighborhood of continuous time sampling. For details, see Broze *et al.* (1995).

Example 3.7.2
Consider the geometric Brownian motion $\{X_t\}$ satisfying the equation

$$dX_t = \mu X_t\,dt + \sigma X_t\,dW_t, \qquad t \geq 0,\ X_0 > 0. \tag{3.7.8}$$

It is easy to check that

$$\log X_t = \log X_0 + (\mu - \tfrac{1}{2}\sigma^2)t + \sigma W_t. \tag{3.7.9}$$

Applying Euler discretization to (3.7.9) yields the relation

$$X_{t+\tau} - X_t = \mu^* X_t \tau + \sigma^* X_t(W_{t+\tau} - W_t) \tag{3.7.10}$$

It can be shown that the MLE $\hat{\theta}_T^*(\theta_0)$ converges to

$$\theta_0^*(\theta_0) = \left(\frac{e^{\mu_0\tau} - 1}{\tau}, \frac{e^{2\mu_0\tau}}{\tau}(e^{\sigma_0^2\tau} - 1)\right) \tag{3.7.11}$$

in this example. For details, see Broze *et al.* (1995).

3.7.1 Indirect Inference for Diffusion Processes

In view of the asymptotic bias present in the above estimators, an indirect inference procedure has been suggested by Smith (1990), Gallant and Tauchen (1996) and Gouriéroux *et al.* (1993). We will now describe this procedure. Let $S_T(\cdot, \cdot)$ be an auxiliary criterion depending on the observations $\boldsymbol{Y}_T$ and on an auxiliary parameter β. Let

$$\hat{\beta}_T = \arg\max_{\beta \in B} S_T(\boldsymbol{Y}_T; \beta).$$

Suppose the model

$$\tilde{Y}_t^{i}(\theta) = \psi(\tilde{Y}_{t-1}^i, \tilde{Y}_{t-2}^i, \ldots, \tilde{Y}_{t-q}^i(\theta), \tilde{\varepsilon}_t^i, \theta), \qquad t = 1, \ldots, T, \ i = 1, \ldots, K,$$

is simulated and define

$$\hat{\beta}_T^i(\theta) = \arg\max_{\beta \in B} S_T(\tilde{Y}_T^i(\theta); \beta)$$

for each value of θ and i, that is, for each simulated path $\tilde{Y}_T^i(\theta)$. The indirect estimator of θ can be chosen to be

$$\hat{\theta}_T^{(1)} = \arg\min_{\theta \in \Theta} \left[\hat{\beta}_T - \frac{1}{K} \sum_{i=1}^{K} \hat{\beta}_T^i(\theta) \right]' \hat{\Omega}_T \left[\hat{\beta}_T - \frac{1}{K} \sum_{i=1}^{K} \hat{\beta}_T^i(\theta) \right],$$

where $\hat{\Omega}_T$ is a symmetric positive definite matrix converging to a deterministic positive definite matrix Ω. Other indirect estimators can also be obtained. For instance, define

$$\hat{\beta}_T^*(\theta) = \arg\max_{\beta \in B} \frac{1}{K} \sum_{i=1}^{K} S_T(Y_T^i(\theta); \beta)$$

and

$$\hat{\theta}_T^{(2)} = \arg\min_{\theta \in \Theta} [\hat{\beta}_T - \hat{\beta}_T^*(\theta)]' \hat{\Omega}_T [\hat{\beta}_T - \hat{\beta}_T^*(\theta)].$$

Another possibility consists in simulating only one path of length KT, which leads to

$$\hat{\beta}_{KT}(\theta) = \arg\max_{\beta \in B} S_{KT}(\hat{Y}_{KT}(\theta); \beta)$$

and

$$\hat{\theta}_T^{(3)} = \arg\min_{\theta \in \Theta} [\hat{\beta}_T - \hat{\beta}_{KT}(\theta)]' \hat{\Omega}_T [\hat{\beta}_T - \hat{\beta}_{KT}(\theta)].$$

Finally, another estimator based on the score function derived from the criterion can be defined by

$$\hat{\theta}_T^{(4)} = \arg\min_{\theta \in \Theta} \left[\sum_{i=1}^{K} \frac{\partial S_T}{\partial \beta}(Y_T^i(\theta); \hat{\beta}_T) \right]' \hat{\Omega}_T \left[\sum_{i=1}^{K} \frac{\partial S_T}{\partial \beta}(Y_T^i(\theta); \hat{\beta}_T) \right]$$

(Gallant and Tauchen 1996). However, all the estimators $\hat{\theta}_T^{(i)}$, $1 \le i \le 4$, have the same asymptotic properties (Gouriéroux *et al.* 1993).

3.7.2 Quasi-indirect Inference for Diffusion Processes

The application of the indirect inference procedure to avoid asymptotic bias due to the fixed sampling interval consists of the following:

(i) creation of simulated data sets of size T and time interval τ using (3.7.1) as the data generating process for all values of θ;
(ii) maximization of a given criterion $S_T(\cdot, \theta^*)$ applied to the observation;
(iii) maximization of the same criterion using the simulated data; and
(iv) calibration with respect to θ in order to minimize the distance between the outputs of steps (ii) and (iii)

(Gouriéroux *et al.* 1993).

Since the transition densities of a diffusion process are not easily computable the statement in (i) cannot be executed. An obvious approach is to use a discretization of (i) with a *fixed discretization step h smaller than the observation step* τ as the generating process to obtain simulated data. Since the true model is not used for simulation this method of indirect inference is termed *quasi-indirect inference*.

Consider the following sequence of discrete-time models indexed by the time unit h and depending on the parameter θ defined in (3.7.1):

$$Y^{(h)}_{(n+1)h} = F_h(Y^{(h)}_{nh}; \varepsilon^{(h)}_{(n+1)h}; \theta), \tag{3.7.12}$$

where $\{\varepsilon^{(h)}_{(n+1)h}, n \geq 1\}$ is a sequence of $R^{(d)}$-valued standard Gaussian independent random variables. Special cases of the model (3.7.12) are the Euler discretization of (3.7.1) given by

$$Y^{(h)}_{(n+1)h} = Y^{(h)}_{nh} + a(Y^{(h)}_{nh}; \theta)h, +\sigma(Y^{(h)}_{nh}; \theta)h^{1/2}\varepsilon^{(h)}_{(n+1)h}, \tag{3.7.13}$$

and the Milstein (1976) discretization, for $d = 1$,

$$\begin{aligned} Y^{(h)}_{(n+1)h} &= Y^{(h)}_{nh} + [a(Y^{(h)}_{nh}; \theta) - \tfrac{1}{2}\sigma(Y^{(h)}_{nh}; \theta)\frac{\partial \sigma}{\partial Y}(Y^{(h)}_{nh}; \theta)]h \\ &\quad +\sigma(Y^{(h)}_{nh}; \theta)h^{1/2}\varepsilon^{(h)}_{(n+1)h} + \frac{1}{2}\sigma(Y^{(h)}_{nh}; \theta)\frac{\partial \sigma}{\partial Y}(Y^{(h)}_{nh}; \theta)h\varepsilon^{(h)2}_{(n+1)h}. \end{aligned} \tag{3.7.14}$$

We make the following assumption:

(MM0)′ The solution process of the model given by (3.7.13) is unique and it is stationary and ergodic.

Let $\tau/\mathcal{N} = \{h \in R : \tau/h \in \mathcal{N}\}$. For any $h \in \tau/\mathcal{N}$ and $\theta \in \Theta$, let

$$\tilde{Y}^{(h)}_T(\theta) = (\tilde{Y}^{(h)}_\tau(\theta), \tilde{Y}^{(h)}_{2\tau}(\theta), \ldots, \tilde{Y}^{(h)}_{T\tau}(\theta))$$

denote a set of simulated variables drawn from the model given by (3.7.13). We consider the maximization of the criterion S_T as discussed earlier applied to the simulated data set.

Let $\hat{\theta}^*_T(\theta, h)$ denote any measurable solution of the maximization

$$\hat{\theta}^*_T(\theta, h) = \arg\max_{\theta^* \in \Theta} S_T(\tilde{Y}^{(h)}_T(\theta), \theta^*).$$

Suppose the following condition (MM0)″ and analogs of the conditions (MM1) and (MM2) hold:

(MM0)″ For all $h \in \tau/\mathcal{N}$, there exists a unique solution $\theta_0(\theta_0, h)$ to the equation

$$\hat{\theta}^*_0(\theta, h) = \theta^*_0(\theta_0),$$

where $\theta^*_0(\theta_0)$ is as defined by (3.7.4) and $\hat{\theta}^*_0(\theta, h)$ is as as defined in (MM1)′ given below.

(MM1)$'$ For all $\theta \in \Theta$ and all $h \in \tau/\mathcal{N}$, there exists a nonrandom limit $S_\infty^{(h)}(\theta, \theta^*)$ such that

$$P_\theta\left[\lim_{T\to\infty} S_T(\bar{Y}_T^{(h)}(\theta); \theta^*) = S_\infty^{(h)}(\theta, \theta^*)\right] = 1$$

uniformly in $\theta^* \in \Theta$. Furthermore, for all $\theta \in \Theta$, the function $S_\infty^{(h)}(\theta, \cdot)$ admits a unique maximum

$$\tilde{\theta}_0^*(\theta, h) = \arg\max_{\theta^*\in\Theta} S_\infty^{(h)}(\theta, \theta^*).$$

(MM2)$'$ The criterion function $S_T(Y_T^{(h)}(\theta), \theta^*)$ is twice continuously differentiable with respect to θ^* and

(i) for all $\theta \in \Theta$, there exists a nonrandom matrix $\ddot{S}_\infty^{(h)}(\theta, \theta^*)$ such that

$$P_\theta\left[\lim_{T\to\infty} \frac{\partial^2 S_T(\bar{Y}_T^{(h)}(\theta), \theta^*)}{\partial\theta^*\partial\theta^{*\prime}} = \ddot{S}_\infty^{(h)}(\theta, \theta^*)\right] = 1$$

uniformly in $\theta^* \in \Theta$;

(ii) $\ddot{S}_\infty^{(h)}(\theta, \tilde{\theta}_0^*(\theta, h))$ is invertible for all $\theta \in \Theta$; and

(iii) the sequence $\left[\sqrt{T}\frac{\partial S_T(\bar{Y}_T^{(h)}(\theta),\theta^*)}{\partial\theta^*}\right]_{\theta^*=\tilde{\theta}_0^*(\theta,h)}$ converges in distribution to $N(0, J^*(\theta, h))$.

Under conditions (MM0)$'$–(MM2)$'$, it can be shown that the simulation-based estimator is asymptotically normal, that is,

$$\sqrt{T}[\hat{\theta}_T^*(\theta, h) - \tilde{\theta}_0^*(\theta, h)] \xrightarrow{\mathcal{L}} N(0, \tilde{\Gamma}^{(h)}(\theta)) \qquad \text{as } T \to \infty,$$

where

$$\tilde{\Gamma}^{(h)}(\theta) = [\ddot{S}_\infty^{(h)}(\theta, \tilde{\theta}_0^*(\theta, h))]^{-1} J^*(\theta, h)[\ddot{S}_\infty^{(h)}(\theta, \tilde{\theta}_0^*(\theta, h))]^{-1}.$$

A *quasi-indirect estimator* of θ is a sequence $\hat{\theta}_T(\theta_0, h)$, $h \in \tau/\mathcal{N}$, where $\hat{\theta}_T(\theta_0, h)$ is any measurable solution of

$$\hat{\theta}_T(\theta_0, h) = \arg\min_{\theta\in\Theta} \|\hat{\theta}_T^*(\theta, h) - \hat{\theta}_T^*(\theta_0)\|$$

and $\|\cdot\|$ denotes any norm in R^p.

Suppose there exists a nonrandom matrix $V_0^{(h)}$ such that

$$\left.\frac{\partial\hat{\theta}_T^{*\prime}(\theta, h)}{\partial\theta}\right|_{\theta=\theta_0(\theta_0,h)} \to V_0^{(h)} \text{ a.s.} \qquad \text{as } T \to \infty,$$

where $V_0^{(h)\prime} V_0^{(h)}$ is nonsingular and $\lim_{h\to 0} V_0^{(h)} = V_0$. Here $\theta_0(\theta_0, h)$ is as defined by (MM0)$''$. Under some additional conditions, Broze *et al.* (1995) prove that

$$\sqrt{T}(\hat{\theta}_T(\theta_0, h) - \theta_0) \xrightarrow{\mathcal{L}} N(0, \Omega_0) \qquad \text{as } T \to \infty \text{ and } h \to 0,$$

where

$$\Omega_0 = 2[V_0' V_0]^{-1} V_0' \Gamma_0^* [V_0' V_0]^{-1}.$$

If $h = h(T) = T^{-\delta}$, where $\frac{1}{4} < \delta \leq \frac{1}{2}$, then

$$\sqrt{T}\left(\hat{\theta}^*_T(\theta_0, h) - \theta_0 - h\left[\frac{\partial\theta_0(\theta_0, h)}{\partial h}\right]_{h=0}\right) \xrightarrow{\mathcal{L}} N(0, \Omega_0)$$

as $T \to \infty$, and if $\delta > \frac{1}{2}$, then

$$\sqrt{T}(\hat{\theta}^*_T(\theta_0, h) - \theta_0) \xrightarrow{\mathcal{L}} N(0, \Omega_0) \quad \text{as } T \to \infty.$$

For additional remarks based on Monte Carlo experiments, see Broze *et al.* (1995). For related results and comments on quasi-indirect estimation, see DeWinne (1995). An extensive survey of the application of the methods described above for the study of stochastic volatility is given in Ghysels *et al.* (1995).

3.7.3 Estimation by Simulated Moment Methods

Suppose we observe a discrete-time process $\{Y_{t_i}, 0 \leq i \leq n\}$ with a time interval equal to Δ between consecutive times of observations, and these observations are a realization of a process $\{Y_t\}$ satisfying (3.7.1).

In order to describe this method, we consider a function $f : R^2 \to R^p$ and consider the empirical moment

$$F_n = \frac{1}{n}\sum_{i=0}^{n-1} f(Y_{t_i}, Y_{t_{i+1}})$$

obtained from the true observation of the process $\{Y_t\}$. If the process $\{Y_t\}$ is ergodic, then

$$\begin{aligned} F_n \xrightarrow{\text{a.s.}} F(\theta_0) &= \int\int f(x, y) p^0_\Delta(x, y)\pi^0(x)\, dx\, dy \\ &= E_{p^0_\Delta \pi^0} f, \end{aligned}$$

where $p^0_\Delta(x, y)$ and $\pi^0(x)$ denote the densities of the transition probability function $P^0_\Delta(x, \cdot)$ and the invariant measure Π^0, respectively.

Since the transition probability function of the discrete-time process $\{Y_{t_i}\}$ is not easy to calculate, the limit function F is in general intractable amd the idea of this method of estimation is to replace F by its sample counterpart calculated with simulated observations.

Suppose that we can simulate the diffusion process Y_t for any parameter value θ from a Wiener process W^s independent of W and from an initial value $Y_0^{\theta,s}$. We assume that $Y_0^{\theta,s}$ is independent of W^s and follows a distribution ν^θ. Let $\{Y_t^{\theta,s}\}$ be the simulated process solution of the stochastic differential equation

$$dY_t^{\theta,s} = a(Y_t^{\theta,s}, \theta)\, dt + \sigma(Y_t^{\theta,s}, \theta)\, dW_t^s,$$

with $Y_0^{\theta,s} \sim \nu^\theta$. Let $P^\theta_\Delta(x, \cdot)$ be the transition probability function, Π^θ the invariant measure, $p^\theta_\Delta(x, y)$ and $\pi^\theta(x)$ their densities and

$$F(\theta) = \int\int f(x, y) p^\theta_\Delta(x, y)\pi^\theta(x)\, dx\, dy = E_{p^\theta_\Delta \pi^\theta} f.$$

From simulated observations $Y_t^{\theta,s}$, we can define the empirical moment

$$F_N(\theta) = \frac{1}{N}\sum_{i=0}^{N-1} f(Y_{t_i}^{\theta,s}, Y_{t_{i+1}}^{\theta,s}),$$

where N denotes the number of simulated observations. If we simulate a finite number of trajactries of the diffusion process for a finite number of parameter values, a simulated moment estimator will be given by a calibration between F_n and $F_N(\theta)$ as θ varies over Θ_m, a finite subset of Θ. The simulated moment estimator is defined as a solution of the minimization problem

$$\hat{\theta}_{n,N}^m = \arg\min_{\theta\in\Theta_m} \|F_n - F_N(\theta)\|,$$

where $\|\cdot\|$ denotes the Euclidean norm in R^p. Note that $\hat{\theta}_{n,N}^m$ belongs to Θ_m by definition.

We assumed earlier that we could simulate discrete-time observations $Y_t^{\theta,s}$ of the process which is a solution of (3.7.1). This assumption is not realistic as the transition probability functions are generally unknown. In practice, we will simulate a process $Y_{t,h}^{\theta,s}$ from a discretization scheme, for example the Euler discretization scheme, with a discretization step equal to Δ/h, that is,

$$Y_{t+\frac{\Delta}{h},h}^{\theta,s} - Y_{t,h}^{\theta,s} = a(Y_{t,h}^{\theta,s},\theta)\frac{\Delta}{h} + \sigma(Y_{t,h}^{\theta,s},\theta)(W_{t+\frac{\Delta}{h}}^s - W_t^s),$$

with $Y_{0,h}^{\theta,s}$ having the distribution ν^θ. We consider the estimator based on this approximated process, namely,

$$\hat{\theta}_{n,N}^{m,h} = \arg\min_{\theta\in\Theta_m} \|F_n - F_N^h(\theta)\|$$

where

$$F_N^h(\theta) = \frac{1}{N}\sum_{i=0}^{N-1} f(Y_{t_ih}^{\theta,s}, Y_{t_{i+1},h}^{\theta,s}).$$

Let the moment estimator $\hat{\theta}_n$ be defined by

$$\hat{\theta}_n = \arg\min_{\theta\in\Theta} \|F_n - F(\theta)\|.$$

Clement (1997) studied the rate of divergence of the number of simulated observations N, the cardinality of the finite subset Θ_m and the inverse of the simulation step h as functions of the number of observations in order for the simulated estimators $\hat{\theta}_{n,N}^m$ and $\hat{\theta}_{n,N}^{m,h}$ to inherit the asymptotic properties of the moment estimator $\hat{\theta}_n$. The following theorem is proved by Clement (1997).

Theorem 3.7.1 *Suppose the following conditions hold:*

(NN1) *Θ is compact and $\theta_0 \in \Theta^0$, the interior of Θ.*

(NN2) *The functions $a(\theta, x)$ and $\sigma(\theta, x)$ are C^∞ functions for all $\theta \in \Theta$ and all their derivatives are bounded, with*

(i) *$0 < \underline{\sigma} \le \sigma(\theta, x) \le \overline{\sigma} < \infty$ for all $(\theta, x) \in \Theta \times R$, and*

(ii) *$-\alpha x \le a(\theta, x) \le -\gamma x$ for all $(\theta, x) \in \Theta \times R$, where $\alpha \ge \gamma > 0$.*

(NN3) *f is a C^∞ function with at most polynomial growth.*

(NN4) F is of the class C' and the differential of F, $\mathrm{DF}(\theta)$, is invertible for all $\theta \in \Theta^0$.
(NN5) $F(\theta) = F(\theta')$ if and only if $\theta = \theta'$.
Let $m(n)$, $h(n)$ and $N(n)$ be sequences such that

(i) $\lim_{n\to\infty} \frac{\sqrt{n}}{m(n)} = 0$ and $\lim_{n\to\infty} \frac{\sqrt{n}n^{\varepsilon}}{m(n)} = \infty$ for all $\varepsilon > 0$;

(ii) $\lim_{n\to\infty} \frac{m(n)}{h(n)} = 0$ and $\lim_{n\to\infty} \frac{\sqrt{n}n^{\varepsilon}}{h(n)} = \infty$ for all $\varepsilon > 0$; and

(iii) there exists $\delta > 0$ such that

$$\lim_{n\to\infty} n^{1+\frac{1}{2}(\frac{\overline{\sigma}}{\underline{\sigma}})^2+\delta} N(n)^{-1} = 1.$$

Then

$$\lim_{n\to\infty} \hat{\theta}_{n,N(n)}^{m(n),h(n)} = \theta_0 \qquad \textit{in probability}$$

and

$$\sqrt{n}(\hat{\theta}_{n,N(n)}^{m(n),h(n)} - \theta_0) \xrightarrow{\mathcal{L}} N(0, \mathrm{DF}(\theta_0)^{-1}\overset{0}{\sum}(f)\,\mathrm{DF}(\theta_0)^{-1}),$$

where

$$\overset{0}{\sum}(f) = \int\int (f(x,y) - F(\theta_0))(f(x,y) - F(\theta_0))' p_{\Delta}^{0}(x,y)\pi^0(x)\,dx\,dy$$
$$+2\int\left(\int (f(x,y) - F(\theta_0))p_{\Delta}^{0}(x,y)\,dy\right)G'(y)\pi^0(x)\,dx$$

and

$$G(x) = \sum_{k\geq 1}\int (f(x,y) - F(\theta_0))p_{k\Delta}^{0}\,dy.$$

3.7.4 Estimation by the Method of Minimum Chi-squared

Consider the stochastic differential equation

$$dX_t = a(X_t, \rho)\,dt + \sigma(X_t, \rho)\,dW_t, \qquad t \geq 0,$$

where $\rho \in D \subset R^d$. Suppose that the system is observed at equally spaced time intervals $t = 0, 1, \ldots$ and that selected characteristics

$$Y_t = S(X_{t+L}), \qquad t = -L, -L+1, \ldots,$$

of the state are recorded. Further suppose that the process X_t is stationary and ergodic. Then Y_t is in general also stationary and ergodic under reasonable conditions on $S(\cdot)$. Suppose the stationary distribution of Y_t is absolutely continuous. Let us denote the data observed by $\tilde{Y}_t$, the simulation by $\hat{Y}_t$ and the random variables to which they correspond by Y_t. From the ergodic theorem, it follows that

$$\lim_{N\to\infty} \frac{1}{N}\sum_{t=0}^{N} g(\hat{Y}_{t-L}, \ldots, \hat{Y}_t)$$
$$= \int \cdots \int g(y_{-L}, \ldots, y_0)p(y_{-L}, \ldots, y_0; \rho)\,dy_{-L}\cdots dy_0$$

almost surely, where $\{\hat{Y}_t, t = -L, \ldots, N\}$ is a realization of the process $\{Y_t, t = -L, \ldots, N\}$. Here $p(y_{-L}, \ldots, y_0; \rho)$ is the time-invariant density. Suppose that the distribution of the process is correctly described both by the density $p(y_{-L}, \ldots, y_0; \rho)$ induced by the stochastic differential equation given above and by another time-invariant density $f(y_{-L}, \ldots, y_0; \theta)$ which we shall call the *auxiliary model*. The process $\{Y_t\}$ is a Markov process of order L. Let

$$f(y_0|x_{-1}, \theta) = \frac{f(y_{-L}, \ldots, y_0; \theta)}{\int f(y_{-L}, \ldots, y_0; \theta)\, dy_0},$$

where $x_{-1} = (y_{-L}, \ldots, y_{-1})$. Under some regularity conditions, the maximum likelihood estimate $\tilde{\theta}_n$ of the parameter θ in the auxiliary model is a sufficient statistic. Furthermore, there exists a mapping $\theta = g(\rho)$ between the parameter ρ of the stochastic differential equation and the parameter θ of the auxiliary model. If $\tilde{\mathcal{I}}_n$ denotes the estimated information matrix of the auxiliary model, then the minumum chi-squared estimator

$$\hat{\rho} = \arg\min_{\rho \in \mathcal{R}} \{\tilde{\theta}_n - g(\rho)\}' \tilde{\mathcal{I}}_n \{\tilde{\theta}_n - g(\rho)\}$$

is asymptotically efficient. Under some regularity conditions, it can be checked that

$$0 \equiv \int \cdots \int \frac{\partial}{\partial \theta} \log f(y_0|x_{-1}, g(\rho)) p(y_{-L}, \ldots, y_0; \rho)\, dy_{-L} \cdots dy_0,$$

$$\begin{aligned}\mathcal{I} &= \int \cdots \int \left[\frac{\partial}{\partial \theta} \log f(y_0|x_{-1}, g(\rho))\right] \\ &\quad \times \left[\frac{\partial}{\partial \theta} \log f(y_0|x_{-1}, g(\rho))\right]' p(y_{-L}, \ldots, y_0; \rho)\, dy_{-L} \cdots dy_0 \\ &= -\int \cdots \int \frac{\partial^2}{\partial \theta\, \partial \theta'} \log f(y_0|x_{-1}, g(\rho)) p(y_{-L}, \ldots, y_0; \rho)\, dy_{-L} \cdots dy_0,\end{aligned}$$

and

$$\begin{aligned}\frac{\partial}{\partial \theta} \log f(y_0 \mid x_{-1}, \theta) &\simeq \frac{\partial}{\partial \theta} \log f(y_0 \mid x_{-1}, g(\rho)) \\ &\quad + \frac{\partial^2}{\partial \theta\, \partial \theta'} \log f(y|x_{-1}, g(\rho))(\theta - g(\rho)).\end{aligned}$$

By Taylor's expansion, the above relations imply that

$$\begin{aligned}m(\rho, \theta) &\equiv \int \cdots \int \frac{\partial}{\partial \theta} \log f(y_0|x_{-1}, \theta) p(y_{-L}, \ldots, y_0; \rho)\, dy_{-L} \cdots dy_0 \\ &\simeq -\mathcal{I}(\theta - g(\rho)).\end{aligned}$$

Hence, to the first order, minimizing $\{\tilde{\theta}_n - g(\rho)\}' \tilde{\mathcal{I}}_n \{\hat{\theta}_n - g(\rho)\}$ is the same as minimizing $m(\rho, \tilde{\theta}_n)'(\tilde{\mathcal{I}}_n)^{-1} m(\rho, \tilde{\theta}_n)$. Minimizing this function is possible because $m(\rho, \theta)$ can be computed by averaging over a long simulation:

$$m(\rho, \theta) \simeq \frac{1}{N} \sum_{t=0}^{N} \frac{\partial}{\partial \theta} \log\{f(\hat{Y}_t | \hat{Y}_{t-L}, \ldots, \hat{Y}_{t-1}; \theta)\}.$$

This leads to the consideration of the minimum chi-squared estimator

$$\hat{\rho}_n = \arg\min_{\rho \in \mathcal{R}} m(\rho, \tilde{\theta}_n)'(\tilde{\mathcal{I}}_n)^{-1} m(\rho, \tilde{\theta}_n).$$

Gallant and Tauchen (1996) give sufficient conditions for the above intuitive argument to hold and show that $\hat{\rho}_n$ is asymptotically efficient. An alternate procedure of choosing an auxiliary model has been suggested by Gallant and Long (1997). Suppose $\mathcal{H}$ is a normed space containing the densities $\{p(\cdot|\rho), \rho \in R\}$ and the classes of densities $\mathcal{H}_k = \{f_k(\cdot|\theta), \theta \in R^{p_k}\}$ that expand in the sense that $\mathcal{H}_k \subset \mathcal{H}_{k+1}$ and are dense so that $\mathcal{H} = \bigcup_{k=0}^{\infty} \mathcal{H}_k$. Then the density $f_k(\cdot|\theta)$ can be used as the auxiliary model and the estimator $\hat{\rho}_n$ will be nearly as efficient as the maximum likelihood estimator for k large. For details, see Gallant and Long (1997).

3.8 Bayesian Inference

Let us consider the stochastic differential equation

$$dX_t = a(X_t, \theta)\, dt + b(X_t)\sigma\, dW_t, \qquad X_0 = x_0,\ t \geq 0, \tag{3.8.1}$$

where $\theta \in \Theta$, Θ compact in R, and $\sigma \in R^+$ are unknown. Suppose the process $\{X_t\}$ is observed at the times $t_i = ih$, $0 \leq i \leq n$, $h > 0$. We have discussed various methods of estimation of the parameters θ and σ based on the discrete data

$$\mathbf{X}_0^{n,h} \equiv \{X_{t_i}, t_i = ih, h > 0, 0 \leq i \leq n\}. \tag{3.8.2}$$

Suppose a continuous observation of the process $\{X_t\}$ over $[0, nh]$ is available. Then the likelihood function of θ is given by

$$L(\sigma^2, \theta) = \exp\left\{\int_0^{nh} \frac{a(X_t, \theta)}{b^2(X_t)\sigma^2}\, dX_t - \frac{1}{2}\int_0^{nh} \frac{a^2(X_t, \theta)}{b^2(X_t)\sigma^2}\, dt\right\}. \tag{3.8.3}$$

An approximation for $L(\sigma^2, \theta)$ is given by

$$L_{nh}(\sigma^2, \theta) = \exp\left\{\sum_{i=1}^{n} \frac{a(X_{t_{i-1}}, \theta)}{b^2(X_{t_{i-1}})\sigma^2}[X_{t_i} - X_{t_{i-1}}] - \frac{1}{2}\sum_{i=1}^{n} \frac{a^2(X_{t_{i-1}}, \theta)}{b^2(X_{t_{i-1}})\sigma^2}(t_i - t_{i-1})\right\}. \tag{3.8.4}$$

Let (θ_0, σ_0) denote the true parameter.

Suppose that Λ is a prior probability measure on $(\Theta, \mathcal{B})$, where $\mathcal{B}$ is the σ-algebra of Borel subsets of Θ. Further assume that, with respect to the Lebesgue measure, Λ has density $\lambda(\cdot)$ which is continuous, positive and bounded with continuous, bounded first and second derivatives.

We define a *psuedo-Bayes estimator* $\tilde{\theta}_{nh}$ as an estimator which minimizes

$$B_{nh}(\phi) = \int_{\Theta} M_{nh}(\theta, \phi)\, p_{nh}(\theta|\mathbf{X}_0^{n,h})\, d\theta, \tag{3.8.5}$$

where $M_{nh} : \Theta \times \Theta \to R$, $n \geq 1$, $h > 0$ are loss functions. Assume that the loss function $M_{nh}(\theta, \phi) = (\sqrt{nh}\,)^{\alpha}|\theta - \phi|^{\alpha}$ (more generally of the type $\ell(x)$, where ℓ is convex and even, with $\ell(0) = 0$). Here

$$p_{nh}(\theta|\mathbf{X}_0^{n,h}) = \frac{\lambda(\theta)L_{nh}(\sigma^2, \theta)}{\int_{\Theta} \lambda(\theta)L_{nh}(\sigma^2, \theta)\, d\theta} \tag{3.8.6}$$

is the psuedo-posterior density. Note that this is not the posterior density given the discrete data $\mathbf{X}_0^{n,h}$. It is based on an approximation $L_{nh}(\sigma^2, \theta)$ to the likelihood $L(\sigma^2, \theta)$.

Let C be a generic constant in the following independent of h, n and other variables. Suppose the following conditions hold.

(PP1) The diffusion process X is ergodic with the invariant measure ν for $\theta = \theta_0$.

(PP2) There exists a constant K such that

$$|a(x,\theta_0)| + |b(x)| \le K(1+|x|)$$

and

$$|a(x,\theta_0) - a(y,\theta_0)| + |b(x) - b(y)| \le K|x-y|.$$

(PP3) $\inf_x b^2(x) > 0$.

(PP4) $\sup_t E|X_t|^\beta < \infty$ for some $\beta > 0$.

(PP5) The function $\sigma^2 \to (b^2(x)\sigma^2)^{-1}$ is Hölder continuous in a neighborhood U of σ_0^2, that is, there exist $\gamma > 0$ and $c > 0$ such that

$$|(b^2(x)\sigma_1^2)^{-1} - (b^2(x)\sigma_2^2)^{-1}| \le C(1+|x|^c)|\sigma_1^2 - \sigma_2^2|^\gamma$$

for σ_1^2, σ_2^2 in U and all $x \in R$.

(PP6) The function $a(x,\theta)$ is twice differentiable in $\theta \in \Theta$ and

$$|a'(x,\theta)| + |a''(x,\theta)| \le C(1+|x|^c),$$

where a' and a'' denote the first and second derivatives of a with respect to θ.

(PP7) Let

$$J(\sigma^2,\theta) = \int_R \frac{a(x,\theta)\{a(x,\theta_0) - \frac{1}{2}a(x,\theta)\}}{b^2(x)\sigma^2}\nu(dx).$$

The function $y \to J(\sigma_0^2,\theta)$ has its unique maximum at $\theta = \theta_0$ in Θ.

(PP8) The functions a, a', b and $(b^2(\cdot)\sigma^2)^{-1}$ are smooth in x and their derivatives are of polynomial growth order in x uniformly in θ or σ^2.

(PP9) $I(\theta_0) = \int_R \frac{a'^2(x,\theta_0)}{b^2(x)\sigma_0^2}\nu(dx) > 0$.

Since $\tilde{\theta}_{nh}$ depends on the possibly unknown σ^2, this cannot be used as an estimator for θ. Let

$$\hat{\sigma}^2_{n,h,0} = \frac{1}{nh}\sum_{i=1}^{n} \frac{(X_{t_i} - X_{t_{i-1}})^2}{b^2(X_{t_{i-1}})}.$$

Lemma 3.8.1 *Under conditions (PP1)–(PP5),*

$$E|\hat{\sigma}^2_{n,h,0} - \sigma_0^2| \le C(h^{1/2} + n^{-1/2}).$$

Let $\hat{\hat{\theta}}_{nh}$ be an estimtor of θ defined by

$$L_{nh}(\hat{\sigma}^2_{n,h,0}, \hat{\hat{\theta}}_{nh}) = \sup_{\theta\in\Theta} L_{nh}(\hat{\sigma}^2_{n,h,0}, \theta).$$

Theorem 3.8.2 *Under conditions (PP1)–(PP5),*

$$\hat{\hat{\theta}}_{nh} \xrightarrow{P} \theta_0 \quad \textit{as } h \to 0,\ n \to \infty \textit{ and } nh \to \infty.$$

Furthermore, if $h^3 n \to 0$*, then* $(hn)^{1/4}(\hat{\hat{\theta}}_{nh} - \theta_0) \xrightarrow{P} 0$.

Lemma 3.8.1 and Theorem 3.8.2 follow from Propositions 1 and 2 in Yoshida (1992) (Section 3.3 above). Substituting $\hat{\hat{\theta}}_{nh}$ for θ, let us consider

$$\hat{\sigma}^2_{n,h,1} = \frac{1}{nh}\sum_{i=1}^{n}\frac{[X_{t_i} - X_{t_{i-1}} - h\,a(X_{t_{i-1}}, \hat{\hat{\theta}}_{nh})]^2}{b^2(X_{t_{i-1}})},$$

$$\tilde{\sigma}^2_{n,h,1} = \frac{1}{nh}\sum_{i=1}^{n}\frac{[X_{t_i} - X_{t_{i-1}} - h\,a(X_{t_{i-1}}, \theta_0)]^2}{b^2(X_{t_{i-1}})},$$

$$\hat{\sigma}^2_{n,h} = \hat{\sigma}^2_{n,h,1} - \frac{h}{2n}\sum_{i=1}^{n}(U_{i+1} + 2V_{i-1}),$$

where

$$U_{i-1} = \frac{F(X_{t_{i-1}}, \hat{\sigma}^2_{n,ho}, \hat{\hat{\theta}}_{n,h})}{b^2(X_{t_{i-1}})},$$

$$V_{i-1} = \frac{a'(X_{t_{i-1}}, \hat{\hat{\theta}}_{n,h})\hat{\sigma}^2_{n,h,0}}{b(X_{t_{i-1}})},$$

$$F(x, \sigma^2, \theta) = \mathcal{L}_{\sigma^2,\theta}(b^2\sigma^2).$$

Here $\mathcal{L}_{\sigma^2,\theta}$ is the generator corresponding to σ^2 and θ. Furthermore, let

$$\tilde{\sigma}^2_{n,h} = \tilde{\sigma}^2_{n,h,1} - \frac{1}{nh}\sum_{i=1}^{n}\frac{h^2F(X_{t_{i-1}}, \sigma_0^2, \theta_0)}{2b^2(X_{t_{i-1}})}$$

$$-\frac{2}{nh}\sum_{i=1}^{n}\int_{t_{i-1}}^{t_i}\int_{t_{i-1}}^{t}\frac{\sigma_0^2 a'(X_u, \theta_0)b(X_u)}{b(X_{t_{i-1}})}\,du\,dt.$$

Using the consistent estimators $\hat{\hat{\theta}}_{n,h}$ and $\hat{\sigma}^2_{n,h,0}$ one can obtain a better estimator for σ_0^2 having a faster rate of convergence.

Note that

$$\tilde{\sigma}^2_{n,h} = \sigma_0^2 + 2w + \rho,$$

where

$$w = \frac{1}{hn}\sum_{i=1}^{n}\left\{\int_{t_{i-1}}^{t_i}\sigma_0\,dW_t\left(\int_{t_{i-1}}^{t}\sigma_0\,dW_s\right)\right\}$$

$$= \frac{1}{hn}\sigma_0^2\sum_{i=1}^{n}\int_{t_{i-1}}^{t_i}[W_t - W_{t_{i-1}}]\,dW_t$$

and

$$n^{1/2}E|\rho| \le C(h^2 + h^{1/2} + h^{3/2}n^{1/2} + h^2n^{1/2})$$

under the conditions stated above.

Theorem 3.8.3 *Under the conditions stated above,*

$$\sqrt{n}(\hat{\sigma}^2_{n,h} - \sigma_0^2) \xrightarrow{\mathcal{L}} N(0, 2\sigma_0^4) \qquad \text{if } h \to 0,\ nh \to \infty,\ nh^3 \to 0$$

and

$$\sqrt{n}(\hat{\sigma}^2_{n,h,1} - \sigma_0^2) \xrightarrow{\mathcal{L}} N(0, 2\sigma_0^4) \qquad \text{if } h \to 0,\ nh \to \infty,\ nh^2 \to 0.$$

Theorem 3.8.3 follows from Yoshida (1992).

Let

$$Z_{nh}(\sigma^2, u) = \frac{L_{nh}(\sigma^2, \theta_0 + (nh)^{-1/2}u)}{L_{nh}(\sigma^2, \theta_0)}$$

and

$$B_{c,h,n} = \{u \in R : |u| \leq c, \theta_0 + (nh)^{-1/2}u \in \Theta\},$$

for $c > 0$. The following result proving the weak convergence of the random field $Z_{nh}(\hat{\sigma}^2_{nh}, \cdot)$ follows from arguments similar to those in Yoshida (1992).

Theorem 3.8.4 *If $h \to 0$, $nh \to \infty$ and $nh^3 = o(1)$, then for each $u \in R$,*

$$\log Z_{nh}(\hat{\sigma}^2_{n,h}, u) = u\Delta_{n,h} - \tfrac{1}{2}u^2 I + \rho_{n,h}(u),$$

where

$$(\Delta_{n,h}, \sqrt{n}(\hat{\sigma}^2_{n,h} - \sigma_0^2)) \xrightarrow{\mathcal{L}} (\Delta, H) \qquad \text{as } n \to \infty$$

in which $\Delta \simeq N(0, I)$ independent of $H \simeq N(0, 2\sigma_0^4)$, $I = I(\theta_0)$ is as defined by (PP9) and $\rho_{n,h} \xrightarrow{p} 0$.

Following arguments in Yoshida (1992), the following result holds.

Theorem 3.8.5 *Supose that conditions (PP1)–(PP9) hold. Then the sequence $\{\sqrt{n}(\hat{\sigma}^2_{n,h} - \sigma_0^2), Z_{nh}(\hat{\sigma}^2_{n,h}, \cdot)\}$ converges to $\{H, Z(\sigma_0^2, \cdot)\}$ in law, where*

$$Z(\sigma_0^2, u) = \exp\{u\Delta - \tfrac{1}{2}u^2 I\}.$$

In view of Theorem 3.8.5, it follows that for any bounded continuous functional f on $(E, \mathcal{H})$, where $E = (R \otimes R) \times C_0(R)$ is endowed with the product topology and $\mathcal{H}$ the corresponding Borel σ-algebra,

$$E\{f(\sqrt{n}(\hat{\sigma}^2_{n,h} - \sigma_0^2), Z_{n,h}(\hat{\sigma}^2_{n,h}, \cdot))\} \to E\{f(H, Z(\sigma_0^2, \cdot))\}$$

as $h \to 0$, $nh \to \infty$ and $nh^3 = o(1)$.

Following again the arguments in the proof of Theorem III.2.1 of Ibragimov and Has'minskii (1981) or Theorem 3.4.2 of Kutoyants (1984a), it follows that the pseudo-Bayes estimator $\tilde{\theta}_{n,h}$ satisfies the property

$$\{\sqrt{n}(\hat{\sigma}^2_{n,h} - \sigma_0^2), (nh)^{1/2}(\tilde{\theta}_{nh} - \theta_0)\} \xrightarrow{\mathcal{L}} (H, I^{-1}\Delta) \qquad \text{as } n \to \infty$$

under the conditions stated earlier (Basawa and Prakasa Rao 1980).

4

Nonparametric Inference for Diffusion Type Processes from Continuous Sample Paths

4.1 Introduction

In the previous two chapters, we have discussed the estimation of parameters in the drift as well as the diffusion coefficient in a stochastic differential equation when their functional forms are known but for the parameters involved. It is often the case that even the forms of these functionals are unknown. Misspecification of the model by a specific functional form might lead to erroneous conclusions if used for inference purposes. It is now accepted that one should resort to methods of inference which allow the data to speak for themselves. The methods of nonparametric functional estimation are extensively discussed, for instance, in Prakasa Rao (1983a). We will now discuss the methods of nonparametric inference for diffusion type processes when a complete path of the process can be observed. As mentioned earlier, the estimation of the drift coefficient is the major problem in this case as the diffusion coefficient can be estimated almost surely using the quadratic variation of the process. In the following discussion, we consider three different types of asymptotics in the study of properties of the estimators: (i) the process is observed over the interval $[0, T]$ and the time T tends to infinity; (ii) the variance of the noise tends to zero keeping the time T of observation fixed; and (iii) the number n of independent realizations of the process observed over a fixed time period $[0, T]$ tends to infinity. The exact nature of asymptotics under consideration will be clear from the discussion.

4.2 Stationary Diffusion Model

We first discuss some general results for the stationary Markov processes.

Let $\{X_t, t \geq 0\}$ be a stationary Markov process with initial density $f(\cdot)$. For each $t \geq 0$, define the transition probability operator

$$(T_t g)(x) = E(g(X_t)|X_0 = x), \qquad x \in R, \tag{4.2.1}$$

for any bounded Borel measurable function g on R. We say that $\{X_t, t \geq 0\}$ satisfies *condition* G_p, $1 \leq p \leq \infty$, if there exists $s > 0$ such that

$$\langle T_s \rangle_p \equiv \sup_{g \perp 1} \frac{\|T_s g\|_p}{\|g\|_p} \leq \alpha < 1, \tag{4.2.2}$$

where the supremum is over the functions g which are bounded and Borel measurable on R. Here $\|g\|_p = (E|g(X_0)|^p)^{1/p}$, and $g \perp 1$ denotes the set of all g such that $E(g(X_0)) = 0$.

The process $\{X_t \geq 0\}$ is said to satisfy *condition* L_p, $1 < p < \infty$, if

$$\lim_{t\to\infty} \langle T_t \rangle_p = 0. \tag{4.2.3}$$

The process $\{X_t, t \geq 0\}$ is said to be *asymptotically uncorrelated* if

$$\lim_{t\to\infty} \sup_{g_1 \perp 1, g_2 \perp 1} \frac{E[g_1(X_0)g_2(X_t)]}{\|g_1\|_2 \|g_2\|_2} = 0. \tag{4.2.4}$$

The process $\{X_t, t \geq 0\}$ is said to be *strong-mixing* with mixing coefficient $\alpha(t)$ if

$$\sup_{A\in\mathcal{B}_0^s, B\in\mathcal{B}_{t+s}^\infty} |P(A\cap B) - P(A)P(B)| \leq \alpha(t), \tag{4.2.5}$$

where $\alpha(t) \to 0$ as $t \to \infty$. Here $\mathcal{B}_0^s$ and $\mathcal{B}_{t+s}^\infty$ denote the σ-algebras generated by $\{X_t, 0 \leq t \leq s\}$ and $\{X_q, q \geq t+s\}$, respectively.

Hereafter we assume that the process $\{X_t, t \geq 0\}$ is a stationary Markov process satisfying the condition G_2. It is known that such a process satisfies condition L_2 and is also asymptotically uncorrelated (see Banon and Nguyen 1978).

Since the process $\{X_t, t \geq 0\}$ is a stationary Markov process, the transition probability operator T_t satisfies the semigroup property, i.e. $T_{s+t} = T_sT_t = T_tT_s$ (Wong 1971, p. 183) and the operator T_t is a contraction under the condition G_2. It can be shown that

$$\|T_t\|_2 < \frac{\beta^t}{\alpha}, \tag{4.2.6}$$

with $0 < \beta = \alpha^{1/s} < 1$ (Banon 1977, p. 79; Prakasa Rao 1983a). Furthermore, for $t < s$, there exists a constant C independent of t and s such that

$$\begin{aligned}\sup_{A\in\mathcal{B}_0^t, B\in\mathcal{B}_s^\infty} |P(A\cup B) - P(A)P(B)| &\leq C \sup_{g\perp 1} \frac{\|T_{s-t}g\|_1}{\|g\|_\infty} \\ &\leq C \sup \frac{\|T_{s-t}g\|_2}{\|g\|_2} \\ &\leq \frac{C}{\alpha}\beta^{s-t} = C_0\beta^{s-t}.\end{aligned} \tag{4.2.7}$$

Hence $\{X_t, t \geq 0\}$ is a strong-mixing process with the mixing coefficient $O(\beta^t)$, where $0 < \beta < 1$. This in turn implies that, for $t < s$,

$$\sup_{\substack{\eta_1\in m(\mathcal{B}_0^t),\\ \eta_2\in m(\mathcal{B}_s^\infty),\\ |\eta_1|\leq 1, |\eta_2|\leq 1}} |E(\eta_1\eta_2) - E(\eta_1)E(\eta_2)| \leq C_1\beta^{s-t} \tag{4.2.8}$$

for some constant $C_1 > 0$ independent of s and t, where $m(\mathcal{B}_0^t)$ and $m(\mathcal{B}_s^\infty)$ denote the family of functions η_1 and η_2 measurable with respect to the σ-algebras $\mathcal{B}_0^t$ and $\mathcal{B}_s^\infty$, respectively. Let m_s^t be the family of functions η measurable with respect to the σ-algebra $\mathcal{B}_s^t$ generated by $\{X_u, s \leq u \leq t\}$. If $\eta_k \in m(\mathcal{B}_{s_k}^{t_k})$, $|\eta_k| \leq 1$, $1 \leq k \leq n$, where $s_1 \leq t_1 < s_2 < \cdots < s_n \leq t_n$, $s_{k+1} - t_k > \gamma$, $1 \leq k \leq n$, then

$$\left| E(\eta_1\eta_2\cdots\eta_n) - \prod_{i=1}^n E(\eta_i)\right| \leq C_1(n-1)\beta^\gamma. \tag{4.2.9}$$

This is a consequence of Lemma 11.4 of Rozanov (1967).

4.2.1 Central Limit Theorem

Let $\{X_t, t \geq 0\}$ be a stationary Markov process satisfying the condition G_2 as discussed earlier. Let $\{f_t(x), t \geq 0\}$ be a family of uniformly bounded Borel measurable functions defined on the state space $\mathcal{X}$ of $\{X_t\}$. Define

$$H(t, s) = \int_s^t f_u(X_u)\, du. \tag{4.2.10}$$

The function $H(t, s)$ is well defined in the sense of convergence in quadratic mean, that is, as an L_2-stochastic integral provided

$$\int_t^s \int_t^s \operatorname{cov}[f_u(X_u), f_v(X_v)]\, du\, dv < \infty \tag{4.2.11}$$

(Loeve 1963, p. 472). Suppose that $E[f_u(X_u)] = 0$, $u \in R_+$. Then $E[H(t, s)] = 0$ for all t and s. Let $H(t) \equiv H(0, t)$ and

$$g(t) = \operatorname{var}[H(t)] = \int_t^s \int_t^s \operatorname{cov}[f_u(X_u), f_v(X_v)]\, du\, dv.$$

Theorem 4.2.1 *Under the assumptions stated above, if $g(t) \to \infty$ as $t \to \infty$, then*

$$\frac{1}{\sqrt{g(t)}} \int_0^t f_u(X_u)\, du \xrightarrow{\mathcal{L}} N(0, 1) \qquad \text{as } t \to \infty. \tag{4.2.12}$$

For a proof, see Prakasa Rao (1979).

4.2.2 Estimation by the Method of Delta Families

Definition. A family of nonnegative functions $\{\delta_h, h > 0\}$ is called a *delta family of positive type* α if:

$$\text{there exist constants } A > 0, B > 0 \text{ such that } \left|1 - \int_{-A}^{B} \delta_h(x)\, dx\right| = O(h^\alpha),$$

$$\sup\{|\delta_h(x)| : |x| \geq h^\alpha\} = O(h^\alpha), \quad \text{and} \quad \|\delta_h\|_\infty \simeq h^{-1} \tag{4.2.13}$$

as $h \to 0$.

Let $h(t)$ be a nonnegative function such that:

$$h(t) \downarrow 0 \qquad \text{as } t \to \infty;$$

$$h(t) \text{ is locally integrable, that is, } \int_0^t h(s)\, ds < \infty, \qquad t > 0;$$

$$\gamma(t) = \int_0^t h(s)\, ds \to \infty \qquad \text{as } t \to \infty. \tag{4.2.14}$$

Estimation of the Marginal Density for Stationary Markov Processes

Let $\{X_t\}$ be a stationary Markov process with the initial density $f(x)$. Define

$$f_t(x) = \frac{1}{\gamma(t)} \int_0^t h(s)\delta_{h(s)}(x - X_s)\,ds \tag{4.2.15}$$

where $\{\delta_h\}$ is a delta family as defined above. Suppose the stationary Markov process is observed over $[0, t]$. Prakasa Rao (1979) studied the properties of the estimator $f_t(x)$ as an estimator of the initial density $f(x)$.

Theorem 4.2.2 *Let $\{\delta_h, h > 0\}$ be a delta family of positive type α. Further suppose that the initial density $f \in \mathrm{Lip}(\lambda)$ for some $0 < \lambda \leq 1$, that is,*

$$|f(x) - f(y)| \leq C|x - y|^\lambda, \qquad x, y \in R. \tag{4.2.16}$$

Then

$$\sup_x E[f_t(x) - f(x)]^2 = O\left(\frac{1}{\gamma(t)} + \left\{\frac{1}{\gamma(t)} \int_0^t h(s)^{1+\alpha\lambda}\,ds\right\}^2\right). \tag{4.2.17}$$

For a proof, see Prakasa Rao (1978a; 1979). In particular, if $h(s) \simeq s^{-\tau}, 0 < \tau < 1$, then $\gamma(t) \simeq s^{-\tau+1}$ and

$$\sup_x E[f_t(x) - f(x)]^2 = O(t^{\tau-1} + t^{-2\tau\alpha\lambda}). \tag{4.2.18}$$

If $\tau = (1 + 2\alpha\lambda)^{-1}$, then

$$\sup_x E[f_t(x) - f(x)]^2 = O(t^{-2\alpha\lambda/(1+2\alpha\lambda)}). \tag{4.2.19}$$

It is easy to see that, for every $x \in R$, the estimator $f_t(x)$ is asymptotically unbiased and consistent in quadratic mean for $f(x)$. Fix $x \in R$. Let

$$g_s(y) = h(s)\delta_{h(s)}(x - y) - E[h(s)\delta_{h(s)}(x - X_0)]. \tag{4.2.20}$$

Then

$$\frac{f_t(x) - Ef_t(x)}{\sqrt{\mathrm{var}\, f_t(x)}} = \frac{1}{\sqrt{\mathrm{var}\, H(t)}} \int_0^t g_s(X_s)\,ds, \tag{4.2.21}$$

where

$$H(t) = \int_0^t g_s(X_s)\,ds. \tag{4.2.22}$$

Applying the central limit theorem stated in Theorem 4.2.1, it can be proved that if $\mathrm{var}[H(t)] \to \infty$ as $t \to \infty$, then

$$\frac{f_t(x) - Ef_t(x)}{\sqrt{\mathrm{var}\, f_t(x)}} \xrightarrow{\mathcal{L}} N(0, 1) \qquad \text{as } t \to \infty. \tag{4.2.23}$$

Suppose the initial density f of the stationary Markov process is differentiable. Let f' denote the derivative of f. We now briefly describe the estimation of the derivative f'.

In addition to conditions (4.2.13), suppose that the delta family $\{\delta_h\}$ also satisfies the following additional conditions:

$$\begin{aligned} &\delta_h(\cdot) \text{ is differentiable for every } h > 0; \\ &\|\delta'_h\|_\infty \simeq h^{-1}. \end{aligned} \tag{4.2.24}$$

Let

$$\psi_t(x) = f_t'(x)$$

be chosen as the estimator of $f'(x)$. Note that for any differentiable function $g(\cdot)$ with $g(t) \to 0$ as $t \to \pm\infty$,

$$\begin{aligned}(g' * f)(x) &\equiv \int_{-\infty}^{\infty} g'(x-t) f(t)\, dt \\ &= \int_{-\infty}^{\infty} g(x-t)\, f'(t)\, dt \\ &= (g * f')(x).\end{aligned}$$

Using these relations, the following theorem can be proved.

Theorem 4.2.3 *Let $\{\delta_h, h > 0\}$ be a delta family of positive type satisfying the condition (4.2.24). Further suppose that the derivative $f' \in \operatorname{Lip}(\mu)$ for some $0 < \mu \leq 1$ and $f' \in L_1(R)$. Then*

$$\sup_x E[\psi_t(x) - f'(x)]^2 = O\left(\frac{1}{\gamma(t)} + \left\{\frac{1}{\gamma(t)} \int_0^t h(s)^{1+\alpha\mu}\, ds\right\}^2\right). \tag{4.2.25}$$

In particular,

$$\sup_x E[\psi_t(x) - f'(x)]^2 \to 0 \qquad \text{as } t \to \infty. \tag{4.2.26}$$

For other properties of the estimator $\psi_t(x)$ and related results, see Prakasa Rao (1983a).

Estimation of the Drift for Diffusion Processes

Let $\{X_t, t \geq 0\}$ be a diffusion process satisfying the stochastic differential equation

$$dX_t = a(X_t)\, dt + \sigma(X_t)\, dW_t, \qquad X_0 = x, t \geq 0. \tag{4.2.27}$$

Sufficient conditions are known for the existence and uniqueness of the solution $\{X_t\}$ of (4.2.27) which is a stationary Markov process with stationary transition density $f_{X_t|X_0=x}(\cdot)$. Conditions are also known under which $f_{X_t|X_0=x}(\cdot)$ must tend to a limiting density $f(\cdot)$ as $t \to \infty$ and the process satisfies the condition G_2 (Banon 1978, p. 390). Suppose the initial density of the process is f and that the process $\{X_t\}$ is a stationary Markov process. Under some regularity conditions, it is known that a, σ and f are related by the equation

$$a(x) = \frac{1}{2}\left[\frac{d\sigma^2(x)}{dx} + \sigma^2(x)\frac{f'(x)}{f(x)}\right],$$

whenever $f(x)$ is not identically equal to zero (Banon 1978, p. 387). We assume that $\sigma(x)$ is known. Our interest is in the estimation of the drift (trend) coefficient $a(\cdot)$ by estimating $f'(\cdot)$ and $f(\cdot)$ by the method described earlier. The problem reduces to the estimation of the function $f'(x)/f(x)$.

Following the methods of estimation of $f(x)$ and $f'(x)$ described earlier, let

$$q_t(x) = \frac{\int_0^t h(s)\delta'_{h(s)}(x - X_s)\, ds}{\int_0^t h(s)\delta_{h(s)}(x - X_s)\, ds}$$

for x for which $f(x) \neq 0$, where $\{\delta_h\}$ is a delta family of positive type satisfying conditions (4.2.13) and (4.2.24). Let $q(x) = f'(x)/f(x)$ for x for which $f(x) \neq 0$. It is easy to see that the estimator $q_t(x)$ is consistent for $q(x)$ as $t \to \infty$ under the conditions stated in Theorems 4.2.2 and 4.2.3. Let

$$a_t(x) = \frac{1}{2}\left[\frac{d\sigma^2(x)}{dx} + \sigma^2(x)q_t(x)\right].$$

Then

$$a_t(x) \xrightarrow{p} a(x) \quad \text{as } t \to \infty.$$

Remarks. (i) If $\sigma^2(x)$ is unknown but a constant, then $a(x) = \frac{1}{2}\sigma^2 q_t(x)$ and an estimator for $a(x)$ is $a_t(x) = \frac{1}{2}\hat{\sigma}_t^2 q_t(x)$, where $\hat{\sigma}_t$ is any consistent estimator for σ.

(ii) Banon (1978) and Banon and Nguyen (1978) discussed the method of kernels for the estimation of $a(\cdot)$ (Prakasa Rao 1983a). Note that if $K(\cdot)$ is a smooth kernel, then the function $\delta_h(x) = \frac{1}{h}K(\frac{x}{h})$ satisfies the properties of a delta family and the method of delta sequences generalizes the method of kernels (Walter and Blum 1979).

4.2.3 Estimation of the Drift for Diffusion Processes by Other Methods

Using the regression structure

Consider the stochastic differential equation

$$d\boldsymbol{X}_t = \boldsymbol{a}(\boldsymbol{X}_t)\,dt + \sigma(\boldsymbol{X}_t)\,d\boldsymbol{W}_t, \qquad t \geq 0, \tag{4.2.28}$$

where we assume that $\boldsymbol{X}_0$ has an initial density $f(\cdot)$ and $\{\boldsymbol{X}_t\}$ is an R^d-valued stationary Markov process. The problem is to estimate $\boldsymbol{a}(\boldsymbol{x})$ based on $\{\boldsymbol{X}_t\}$ observed over $[0, t]$ for each $\boldsymbol{x}$ for which $f(\boldsymbol{x}) > 0$.

Suppose the Markov process $\{\boldsymbol{X}_t, t \geq 0\}$ is sampled at the times $t = j\Delta$, $j \geq 1$, $\Delta > 0$. Let

$$\boldsymbol{\alpha}_\Delta(\boldsymbol{x}) = E(\boldsymbol{X}_{t+\Delta}|\boldsymbol{X}_t = \boldsymbol{x}) \tag{4.2.29}$$

and

$$\boldsymbol{\beta}_\Delta(\boldsymbol{x}) = [\operatorname{cov}(\boldsymbol{X}_{t+\Delta}|\boldsymbol{X}_t = \boldsymbol{x})]^{1/2}, \tag{4.2.30}$$

that is, $\boldsymbol{\beta}_\Delta(\boldsymbol{x})$ is the positive square root of the conditional covariance matrix of $\boldsymbol{X}_{t+\Delta}$ given that $\boldsymbol{X}_t = \boldsymbol{x}$. Suppose that $\boldsymbol{\alpha}_\Delta(\boldsymbol{x})$ and $\boldsymbol{\beta}_\Delta(\boldsymbol{x})$ exist for all $\boldsymbol{x}$. Then

$$\begin{aligned}\boldsymbol{X}_{t+\Delta} &= \boldsymbol{\alpha}_\Delta(\boldsymbol{X}_t) + \boldsymbol{\beta}_\Delta(\boldsymbol{X}_t)\boldsymbol{\varepsilon}_{t+\Delta}\\ &= \boldsymbol{X}_t + \boldsymbol{a}_\Delta(\boldsymbol{X}_t) + \boldsymbol{\beta}_\Delta(\boldsymbol{X}_t)\boldsymbol{\varepsilon}_{t+\Delta},\end{aligned} \tag{4.2.31}$$

where $E(\boldsymbol{\varepsilon}_{t+\Delta}|\boldsymbol{X}_s, s \leq t) = 0$ and $E(\boldsymbol{\varepsilon}_{t+\Delta}\boldsymbol{\varepsilon}'_{t+\Delta}|\boldsymbol{X}_s, s \leq t) = \boldsymbol{I}$. We will call $\boldsymbol{a}_\Delta(\cdot)$ and $\boldsymbol{\beta}_\Delta(\cdot)$ the discrete drift and diffusion coefficients, respectively. The function $\boldsymbol{a}_\Delta(\cdot)/\Delta$ can be considered as an approximation for $\boldsymbol{a}(\cdot)$ for Δ small from (4.2.31).

Suppose $\boldsymbol{a}_\Delta(\cdot)$ is a constant in a neighborhood V of $\boldsymbol{x}$. Let $S = \{j : \boldsymbol{X}_{j\Delta} \in V, 1 \leq j \leq n\}$. Then

$$\boldsymbol{X}_{j\Delta+\Delta} - \boldsymbol{X}_{j\Delta} = \boldsymbol{a}_\Delta(\boldsymbol{x}) + \boldsymbol{\beta}_\Delta(\boldsymbol{X}_{j\Delta})\boldsymbol{\varepsilon}_{j\Delta+\Delta}, \qquad j \in S. \tag{4.2.32}$$

Hence a natural estimator for $\boldsymbol{a}_\Delta(\boldsymbol{x})$ is

$$\frac{1}{\operatorname{card}(S)}\sum_{j\in S}(\boldsymbol{X}_{j\Delta+\Delta} - \boldsymbol{X}_{j\Delta}) = \frac{\sum_{j=1}^n(\boldsymbol{X}_{j\Delta+\Delta} - \boldsymbol{X}_{j\Delta})I(\boldsymbol{X}_{j\Delta} \in V)}{\sum_{j=1}^n I(\boldsymbol{X}_{j\Delta} \in V)}. \tag{4.2.33}$$

If $a_\Delta(\cdot)$ is not a constant, then it will be almost a constant if V is small (say) $V = V_n = \{y : (y - x)/h_n \in U\}$, where U is some neighborhood of zero, $h_n \downarrow 0$, so that V_n shrinks to $\{x\}$ as $n \to \infty$. Then the estimator given above can be written in the form

$$\hat{a}_{\Delta,n}(x) = \frac{\sum_{j=1}^{n} K\left(\frac{X_{j\Delta}-x}{h_n}\right)(X_{j\Delta+\Delta} - X_{j\Delta})}{\sum_{j=1}^{n} K\left(\frac{X_{j\Delta}-x}{h_n}\right)}, \tag{4.2.34}$$

where $K(\cdot) = I_V(\cdot)$. This is the analog of the naive estimator for a density function (Prakasa Rao 1983a). If $K(\cdot)$ is any other probability density, then $\hat{a}_{\Delta,n}(x)$ defined by the formula (4.2.34) can be considered as an estimator for $a(x)$. However, the estimator given by (4.2.34) is not recursive. Following Deheuvels (1973), Prakasa Rao (1978b) and Nguyen (1979), one can consider a recursive estimator which is more easily computable and can be updated as and when more observations are available. Consider

$$a_{\Delta,n}(x) = \frac{\sum_{j=1}^{n} H_{j\Delta} K\left(\frac{X_{j\Delta}-x}{h_{j\Delta}}\right)(X_{j\Delta+\Delta} - X_{j\Delta})}{\sum_{j=1}^{n} H_{j\Delta} K\left(\frac{X_{j\Delta}-x}{h_{j\Delta}}\right)}, \tag{4.2.35}$$

where $h_s > 0$, $H_s > 0$ with $h_s \downarrow 0$ as $s \to \infty$ and

$$g_n = \sum_{j=1}^{n} h_{j\Delta}^d H_{j\Delta} \to \infty. \tag{4.2.36}$$

(For more details on sequential and recursive estimation of probability density, see Prakasa Rao 1983a.) Since the process is observed continuously, we can take a limiting version of the estimator $a_{\Delta,n}(x)$, namely $a_T(x)$ given by

$$a_T(x) = \frac{\int_0^T H_t K\left(\frac{X_t-x}{h_t}\right) dX_t}{\int_0^T H_t K\left(\frac{X_t-x}{h_t}\right) dt}, \tag{4.2.37}$$

where h_t and H_t are as given above such that the integral

$$g_T = \int_0^T h_t^d H_t \, dt \to \infty \qquad \text{as } T \to \infty, \tag{4.2.38}$$

as an estimator for $a(x)$. Note that the numerator in the definition of $a_T(x)$ is a stochastic integral. Asymptotic properties of $a_T(x)$ are investigated in Pham Dinh Tuan (1981) when the process $\{X_t\}$ satisfies the condition G_2. We omit the details.

The kernel method

Consider the class of diffusion processes $X_\varepsilon = \{X_\varepsilon(t), 0 \le t \le T\}$ satisfying the stochastic differential equation

$$dX_\varepsilon(t) = a(X_\varepsilon(t))\, dt + \varepsilon\, dW(t), \qquad X_\varepsilon(0) = x_0, \; 0 \le t \le T, \; \varepsilon \ge 0; \tag{4.2.39}$$

the problem is to estimate the function $a(x)$ from a realization of X_ε on $[0, T]$. Let us consider the corresponding ordinary differential equation

$$\frac{dX_0(t)}{dt} = a(X_0(t)), \qquad X_0(0) = x_0, \; 0 \le t \le T. \tag{4.2.40}$$

Note that the above differential equation corresponds to the case $\varepsilon = 0$ in (4.2.39). Let

$$\alpha = \min_{0 \le t \le T} X_0(t), \qquad \delta = \max_{0 \le t \le T} X_0(t).$$

It is easy to check that the function $a(x)$ for $x \in [\alpha, \delta]$ can be recovered from $\{X_0(t), 0 \le t \le T\}$ and the problem of estimation of the function $\{a(x), x \in [\alpha, \delta]\}$ is equivalent to the problem of estimation of the function $\{f_0(t) = a(X_0(t)), 0 \le t \le T\}$. Note that the functions $\{f_0(t), X_0(t), 0 \le t \le T\}$ completely determine the values of the function $a(x)$ for $x \in [\alpha, \delta]$.

We now give a method of kernel type for the estimation of the functions $a(x)$ and $f_0(t)$. We assume that the following conditions hold:

(QQ1) The function $a(x)$ satisfies the conditions

$$|a(x) - a(y)| \le L|x - |y,$$
$$|a(x)| \le L(1 + |x|). \qquad (4.2.41)$$

(QQ2) The function $a(x)$ has k bounded derivatives on $[\alpha, \delta]$ and the kth derivative is Lipschitzian of order τ, that is,

$$|a^{(k)}(x) - a^{(k)}(y)| \le L_\tau |x - y|^\tau \qquad (4.2.42)$$

for some constant L_τ.

Under condition (QQ1), it is known that

$$|X_\varepsilon(t) - X_0(t)| \le \varepsilon|W(t)| + \varepsilon L \int_0^t e^{L(t-s)}|W(s)|\, ds \qquad (4.2.43)$$

and hence

$$\sup_{0 \le t \le T} |X_\varepsilon(t) - X_0(t)| \le \varepsilon e^{Lt} \sup_{0 \le t \le T} |W(t)|. \qquad (4.2.44)$$

(Kutoyants 1984a, p. 89).

Estimation of the function $f_0(t)$: Motivated by the kernel type method for the estimation of a density in nonparametric inference (Prakasa Rao 1983a), we choose

$$\hat{f}_\varepsilon(t) = \gamma_\varepsilon^{-1} \int_0^T G\left(\frac{\tau - t}{\gamma_\varepsilon}\right) dX_\varepsilon(\tau), \qquad (4.2.45)$$

where G is a bounded kernel with

$$\int_{-\infty}^{\infty} G(u)\, du = 1 \qquad (4.2.46)$$

and $\gamma_\varepsilon \to 0$ as $\varepsilon \to 0$, as an estimator for $f_0(t)$. Note that

$$X_\varepsilon(t) = x_0 + \int_0^t a(X_\varepsilon(\tau))\, d\tau + \varepsilon W(t), \qquad 0 \le t \le T. \qquad (4.2.47)$$

We now show that $\hat{f}_\varepsilon(t)$ is a uniformly consistent estimator of $f_0(t)$ under some conditions. Let $\theta(L)$ be the class of real-valued functions which are uniformly bounded and satisfy the Lipschitz condition (4.2.42). Suppose that the bounded kernel G is such that $G(u) = 0$ for u not in $[A, B]$, where $A < 0$ and $B > 0$. Let $[c, d]$ be an arbitrary interval contained in $(0, T)$.

Theorem 4.2.4 (*Mean square consistency*) *Suppose* $a(\cdot) \in \theta(L), \gamma_\varepsilon > 0, \gamma_\varepsilon \to 0$ *and* $\varepsilon^2\gamma_\varepsilon^{-1} \to 0$ *as* $\varepsilon \to 0$. *Then*

$$\lim_{\varepsilon\to 0} \sup_{a(\cdot)\in\theta(L)} \sup_{c\le t\le d} E|\hat{f}_\varepsilon(t) - f_0(t)|^2 = 0. \tag{4.2.48}$$

Proof. It is easy to see that

$$(a+b+c)^2 \le 4(a^2+b^2+c^2).$$

Hence

$$\begin{aligned} E|\hat{f}_\varepsilon(t) - f_0(t)|^2 &= E\left\{\gamma_\varepsilon^{-1}\int_0^T G\left(\frac{\tau-t}{\gamma_\varepsilon}\right)[a(X_\varepsilon(\tau)) - a(X_0(\tau))]\,d\tau\right. \\ &\quad +\gamma_\epsilon^{-1}\int_0^T G\left(\frac{\tau-t}{\gamma_\varepsilon}\right)a(X_0(\tau))\,d\tau - a(X_0(t)) \\ &\quad \left. +\varepsilon\gamma_\varepsilon^{-1}\int_0^T G\left(\frac{\tau-t}{\gamma_\varepsilon}\right)dW(\tau)\right\}^2 \\ &\le 4E\left\{\gamma_\varepsilon^{-1}\int_0^T G\left(\frac{\tau-t}{\gamma_\varepsilon}\right)[a(X_\varepsilon(\tau)) - a(X_0(\tau))]d\tau\right\}^2 \\ &\quad +4\left\{\gamma_\varepsilon^{-1}\int_0^T G\left(\frac{\tau-t}{\gamma_\varepsilon}\right)a(X_0(\tau))\,d\tau - a(X_0(t))\right\}^2 \\ &\quad +4\frac{\varepsilon^2}{\gamma_\varepsilon^2}\int_0^T \left[G\left(\frac{\tau-t}{\gamma_\varepsilon}\right)\right]^2 d\tau. \end{aligned} \tag{4.2.49}$$

Let $\varepsilon_1 = \min(\varepsilon', \varepsilon'')$, where

$$\varepsilon' = \sup\left\{\varepsilon : \gamma_\varepsilon \le -\frac{c}{A}\right\}$$

and

$$\varepsilon'' = \sup\left\{\varepsilon : \gamma_\varepsilon \le \frac{T-d}{B}\right\}.$$

Then, for $0 < \varepsilon < \varepsilon_1$,

$$\begin{aligned} \frac{\varepsilon^2}{\gamma_\varepsilon^2}\int_0^T \left[G\left(\frac{\tau-t}{\gamma_\varepsilon}\right)\right]^2 d\tau &= \frac{\varepsilon^2}{\gamma_\varepsilon}\int_0^T G(u)^2\,du \\ &= \frac{\varepsilon^2}{\gamma_\varepsilon}\sigma^2 \text{ (say)}. \end{aligned} \tag{4.2.50}$$

Furthermore,

$$\left|\gamma_\varepsilon^{-1}\int_0^T G\left(\frac{\tau-t}{\gamma_\varepsilon}\right)[a(X_0(\tau))\,d\tau - a(X_0(t))\right|$$

$$= \left| \int_{-\infty}^{\infty} G(u)[a(X_0(t+u\gamma_\varepsilon)) - a(X_0(t))]\, du \right|$$
$$\leq L \int_{-\infty}^{\infty} |G(u)||X_0(t+u\gamma_\varepsilon) - X_0(t)|\, du$$
$$= LL_0\gamma_\varepsilon \int_{-\infty}^{\infty} |uG(u)|\, du. \tag{4.2.51}$$

This follows from the observation that

$$|X_0(t+\delta) - X_0(t)| = \left| \int_0^{\delta} a(X_0(t+u))\, du \right| \leq L_0\delta,$$

since $|a(x)| \leq L_0 < \infty$ for some constant L_0. Applying (4.2.44), we obtain that

$$E|X_\varepsilon(t) - X_0(t)|^2 \leq 2T\varepsilon^2\varepsilon^{2LT}, \qquad 0 \leq t \leq T, \tag{4.2.52}$$

from the properties of the Wiener process which implies that

$$E\left\{\gamma_\varepsilon^{-1} \int_0^T G\left(\frac{\tau - t}{\gamma_\varepsilon}\right)[a(X_\varepsilon(\tau)) - a(X_0(\tau))]\, d\tau\right\}^2$$
$$= E\left\{\int_{-\infty}^{\infty} G(u)[a(X_\varepsilon(t+u\gamma_\varepsilon)) - a(X_0(t+u\gamma_\varepsilon))]\, du\right\}^2$$
$$\leq (B-A)L^2 \int_{-\infty}^{\infty} G(u)^2 E|X_\varepsilon(t+u\gamma_\varepsilon) - X_0(t+u\gamma_\varepsilon)|^2\, du$$
$$\leq 2(B-A)L^2\sigma^2T\varepsilon^2e^{2LT}. \tag{4.2.53}$$

Combining (4.2.50), (4.2.51) and (4.2.53), we obtain that

$$\sup_{0\leq t\leq T} E|\hat{f}_\varepsilon(t) - f_0(t)|^2 \leq C_1\varepsilon^2 + C_2\gamma_\varepsilon + C_3\frac{\varepsilon^2}{\gamma_\varepsilon}, \tag{4.2.54}$$

where C_1, C_2, C_3 depend on A, B, T and the kernel G. The proof of the theorem is now complete in view of (4.2.54) under the condition $\varepsilon \to 0$ since $\gamma_\varepsilon \to 0$ and $\frac{\varepsilon^2}{\gamma_\varepsilon} \to 0$.

Let Θ_β, $\beta = k + r$, be the set of k-times differentiable functions $Q(x)$, $x \in [\alpha, \delta]$, such that there exists $L_r > 0$ such that

$$|Q^{(k)}(x_2) - Q^{(k)}(x_1)| \leq L_r|x_2 - x_1|^r, \tag{4.2.55}$$

where $Q^{(k)}(\cdot)$ denotes the kth derivative of $Q(\cdot)$. Further suppose that the kernel $G(\cdot)$ satisfies the condition

$$\int_{-\infty}^{\infty} u^j G(u)\, du = 0, \qquad 1 \leq j \leq k. \tag{4.2.56}$$

Then the following theorem can be proved by arguments analogous to those given above.

Theorem 4.2.5 *Let $a(\cdot) \in \Theta_\beta$. Then*

$$\lim_{\varepsilon\to 0} \sup_{a(\cdot)\in\Theta_\beta} \sup_{c\leq t\leq d} E|\hat{f}_\varepsilon(t) - f_0(t)|^2\varepsilon^{-4\beta/2\beta+1} < \infty. \tag{4.2.57}$$

This can be seen from the fact that the function $f_0(t)$ is k times differentiable under the conditions of the Theorem with the kth derivative Liptschitzian of order r. Note that

$$f_0^{(2)}(t) = a^{(2)}(X_0(t))[a(X_0(t))^2 + a^{(1)}(X_0(t))^2 a(X_0(t))], \tag{4.2.58}$$

where $a^{(i)}(\cdot)$ denotes the ith derivative of $a(\cdot)$. By Taylor's expansion, it follows that

$$f_0(t+h) = f_0(t) + \sum_{j=1}^{k-1} \frac{f_0^{(j)}(t)}{j!} h^j + \frac{h^k}{k!}[f_0^{(k)}(t+vh) - f_0^{(k)}(t)], \tag{4.2.59}$$

for $0 < v < 1$, and hence, by (4.2.56), for $\varepsilon < \varepsilon_1$,

$$\begin{aligned}
&\left| \gamma_\varepsilon^{-1} \int_0^T G\left(\frac{\tau - t}{\gamma_\varepsilon}\right)[a(X_0(\tau))\,d\tau - a(X_0(t))\right| \\
&= \left| \int_{-\infty}^{\infty} G(u)[f_0(t+u\gamma_\varepsilon) - f_0(t)]\,du \right| \\
&\leq \frac{L_r'}{k!} \gamma_\varepsilon^{k+r} \int_{-\infty}^{\infty} |u|^k |G(u)|\,du
\end{aligned} \tag{4.2.60}$$

for some constant $L_r' > 0$. Applying other estimates derived in the proof of Theorem 4.2.4, we obtain that there exist constants $C_i > 0$, $1 \leq i \leq 3$, such that

$$\sup_{c \leq t \leq d} E|\hat{f}_\varepsilon(t) - f_0(t)|^2 \leq C_1 \varepsilon^2 + C_2 \gamma_\varepsilon^{2\beta} + C_3 \varepsilon^2 \gamma_\varepsilon^{-1}. \tag{4.2.61}$$

Choosing $\gamma_\varepsilon = \varepsilon^{2/2\beta+1}$, we obtain Theorem 4.2.5.

Asymptotic normality for the estimator $\hat{f}_\varepsilon(t)$ For the estimator $\hat{f}_\varepsilon(t)$ given by (4.2.45), the following representation holds for sufficiently small ε:

$$\begin{aligned}
\hat{f}_\varepsilon(t) &= \int_{-\infty}^{\infty} a(X_0(t+u\gamma_\varepsilon))G(u)\,du + \varepsilon \gamma_\varepsilon^{-1/2} \int_{-\infty}^{\infty} G(u)\,d\tilde{W}(u) \\
&\quad + \int_{-\infty}^{\infty} [a(X_\varepsilon(t+u\gamma_\varepsilon)) - a(X_0(t+u\gamma_\varepsilon))]G(u)\,du \\
&= \Phi(t,\gamma_\varepsilon) + \varepsilon^{2\beta/[2\beta+1]}\xi + \varepsilon\zeta
\end{aligned} \tag{4.2.62}$$

where $\tilde{W}(u) = \gamma_\varepsilon^{-1/2}[W(t+u\gamma_\varepsilon) - W(t)]$ is a Wiener process, ζ is a random variable bounded in probability and ξ is a Gaussian random variable with parameters 0 and σ^2. Relation (4.2.59) shows that

$$\varepsilon^{-2\beta/(2\beta+1)}[\Phi(t,\gamma_\varepsilon) - f_0(t)] = \frac{1}{k!} \int_{-\infty}^{\infty} u^{k+r} G(u) \frac{[f_0^{(k)}(t+u\gamma_\varepsilon) - f_0^{(k)}(t)]}{(u\gamma_\varepsilon)^r}\,du. \tag{4.2.63}$$

The integrand is bounded by the condition that $a(\cdot) \in \Theta_\beta$. Hence, if $m(t)$ denotes the limit of the expression on the right-hand side of (4.2.63) as $\varepsilon \to 0$, then

$$[\hat{f}_\varepsilon(t) - f_0(t)]\varepsilon^{-\frac{2\beta}{2\beta+1}} \xrightarrow{\mathcal{L}} N(m(t),\sigma^2) \quad \text{as } \varepsilon \to 0. \tag{4.2.64}$$

Estimation of the drift function $a(x)$ Let us consider again the stochastic differential equation

$$dX_\varepsilon(t) = a(X_\varepsilon(t))\,dt + \varepsilon\,dW(t), \qquad X_\varepsilon(0) = x_0, 0 \le t \le T. \tag{4.2.65}$$

We assume that $a(\cdot) \in \Theta'_L$, satisfying the conditions

$$a(x) \ge k > 0, \qquad x \in [\alpha, \delta],$$
$$|a(x) - a(y)| \le |x - y|, \qquad x, y \in [\alpha, \delta] \tag{4.2.66}$$

and

$$|a(x)| \le L(1 + |x|). \tag{4.2.67}$$

Let $x \in [\alpha, \delta]$. Let $\tau_\varepsilon = \inf\{t : X_\varepsilon(t) = x\}$. Define $\tau_\varepsilon = T$ if the set $\{t : X_\varepsilon(t) = x, t \in [0, t]\}$ is empty. Let $\tau = \inf\{t : X_0(t) = x\}$. Let $[r.q] \subset (\alpha, \delta)$. Consider

$$\hat{a}_\varepsilon(x) = \gamma_\varepsilon^{-1} \int_0^T G\left(\frac{t - \tau_\varepsilon}{\gamma_\varepsilon}\right) dX_\varepsilon(t). \tag{4.2.68}$$

Kutoyants (1985a) proved the following theorem.

Theorem 4.2.6 *Suppose* $a(\cdot) \in \Theta'_L, \gamma_\varepsilon \to 0$ *and* $\varepsilon^2\gamma_\varepsilon^{-1} \to 0$ *as* $\varepsilon \to 0$. *Then*

$$\hat{a}_\varepsilon(x) = \gamma_\varepsilon^{-1} \int_0^T G\left(\frac{t - \tau_\varepsilon}{\gamma_\varepsilon}\right) dX_\varepsilon(t) \tag{4.2.69}$$

is uniformly consistent for $a(\cdot)$, *and* $x \in [r, q]$, *that is,*

$$\lim_{\varepsilon \to 0} \sup_{a \in \Theta'_L} \sup_{r \le x \le q} E|\hat{a}_\varepsilon(x) - a(x)|^2 = 0.$$

Suppose that $a(\cdot) \in \Theta'_\beta = \Theta_\beta \cap \Theta'_L$, where Θ_β is as defined earlier. Further suppose that the kernel $G(\cdot)$ satisfies (4.2.56). Then the following Theorem holds.

Theorem 4.2.7 *Let* $a(\cdot) \in \Theta'_\beta$. *Then*

$$\lim_{\varepsilon \to 0} \sup_{a \in \Theta'_\beta} \sup_{r \le x \le q} E|\hat{a}_\varepsilon(x) - a(x)|^2 \varepsilon^{-4\beta/(2\beta+1)} < \infty. \tag{4.2.70}$$

We omit the proofs of Theorems 4.2.6 and 4.2.7 as they are analogous to those of Theorems 4.2.4 and 4.2.5 respectively, except for minor modifications. For details, see Kutoyants (1985a). Kutoyants (1985a) has also proved that the rate of convergence in Theorem 4.2.7 is the best possible. In other words, if $a(\cdot) \in \Theta'_\beta$, then for any $x \in (\alpha, \delta)$,

$$\liminf_{\varepsilon \to 0} \inf_{a_\varepsilon^*(x)} \sup_{a(\cdot) \in \Theta'_\beta} E|a_\varepsilon^*(x) - a(x)|^2 \varepsilon^{-4\beta/(2\beta+1)} > 0. \tag{4.2.71}$$

This was proved following the techniques for similar results in the problem of density estimation (Prakasa Rao 1983a; Ibragimov and Has'minskii 1981).

Remarks. Techniques similar to those used above can be employed for the estimation of functions of the type $f^*(t) = a(t, X_0(t))$ for stochastic differential equations of the type

$$dX_\varepsilon(t) = a(t, X_\varepsilon(t))\,dt + \varepsilon\,dW(t), \qquad X_\varepsilon(0) = x_0,\ 0 \le t \le T,$$

or

$$dX_\varepsilon(t) = a(t, X_\varepsilon(t))\,dt + \varepsilon b(t, X_\varepsilon(t))\,dW(t), \qquad X_\varepsilon(0) = x_0,\ 0 \le t \le T,$$

when the function $b(t, x)$ is known (Kutoyants 1985a).

The method of sieves

Consider again the stochastic differential equation

$$dX_t = a(X_t)\,dt + dW_t, \qquad t \geq 0,\ X_0 = x_0. \tag{4.2.72}$$

The problem is to estimate $a(x)$ for $x \in [-\lambda, \lambda]$ for some $\lambda > 0$. Let

$$u(x) = \int_0^x \exp\left(-2\int_0^y a(z)\,dz\right) dy.$$

Suppose that

$$\lim_{x\to+\infty} u(x) = \infty, \qquad \lim_{x\to-\infty} u(x) = -\infty.$$

Then the process $\{X_t\}$ will be recurrent following Chapter 9 of Friedman (1975). Let

$$e_1 = \inf\{t \geq 0, |X_t| = \lambda\}, \qquad R_1 = \inf\{t \geq e_1, X_t = 0\}$$

and, in general,

$$e_{i+1} = \inf\{t \geq R_i, |X_t| = \lambda\}, \qquad R_{i+1} = \inf\{t \geq e_{i+1}, X_t = 0\}, \tag{4.2.73}$$

for $i \geq 1$. Let

$$X_t^{(1)} = I_{[0,e_1]}(t)X_t,$$

and

$$X_t^{(i)} = I_{[0,e_i - R_{i-1}]}(t)X_{R_{i-1}+t}, \qquad i \geq 2. \tag{4.2.74}$$

Then $\{X_t^{(i)}\}, i \geq 1$ are i.i.d. random processes. Let Θ be the space of Lipschitz continuous functions, that is, $\Theta = \{a(x) : a(\cdot)$ is continuous and there exists L such that $|a(x) - a(y)| \leq L|x - y|, x, y \in R\}$. For any α_1 and α_2 in Θ, define

$$d(\alpha_1, \alpha_2) = \left(E\int_0^{e_1} |\alpha_1(X_s) - \alpha_2(X_s)|^2\,ds\right)^{1/2}. \tag{4.2.75}$$

Let

$$S_m = \left\{\sum_{k=-m}^{m} a_k e^{ik\frac{\pi}{\lambda}x} : \sum_{k=-m}^{m} |a_k| \leq K\log m,\ a_k = \bar{a}_{-k} \text{ for each } k\right\} \tag{4.2.76}$$

for some constant K. Then $\cup_{m=1}^{\infty} S_m$ is dense in Θ since any $a(\cdot) \in \Theta$ can be uniformly approximated by the trigonometric polynomials on $[-\lambda, \lambda]$. The Radon–Nikodym derivative of the measure generated by the process X with respect to the Wiener process given that the process is observed up to time e_1 is

$$f(x, a) \equiv \exp\left(\int_0^{e_1} a(X_s)\,dX_s - \frac{1}{2}\int_0^{e_1} a^2(X_s)\,ds\right), \tag{4.2.77}$$

where e_1 is the first exit time for the process X_t for $[-\lambda, \lambda]$. Define

$$L_n(a) = \prod f(x_t^{(i)}, a) \tag{4.2.78}$$

and

$$M_m^{(n)} = \left\{ \alpha \in S_m \,\middle|\, L_n(\alpha) = \sup_{\beta \in S_m} L_n(\beta) \right\}. \tag{4.2.79}$$

Note that $M_m^{(n)}$ is the set of all the maximum likelihood estimators in S_m of α given the processes $X_t^{(i)}, 1 \leq i \leq n$. Geman and Hwang (1982) proved the following theorem.

Theorem 4.2.8 *If $m_n \simeq n^{1-\epsilon}$ for some $\epsilon > 0$, then*

$$M_{m_n}^{(n)} \to a \text{ a.s.} \qquad \text{as } n \to \infty. \tag{4.2.80}$$

Remark. The method proposed above is known as 'the method of sieves' and was suggested by Grenander (1981). Here the 'sieve' refers to to the subspace over which the likelihood is maximized. Properties of the estimators so obtained depend on the growth of the sieve as compared to the growth of the sample size.

Using observations related to the hitting times

Genon-Catalot and Larédo (1986) considered the problem of nonparametric estimation of the drift function $a(\cdot)$ when the diffusion process X satisfying the stochastic differential equation

$$dX_t = a(X_t)\,dt + \sigma\,dW_t, \qquad t \geq 0, X_0 = x,$$

is partially observed in the following sense. It is assumed that $\sigma > 0$ is known. Let

$$M_t = \sup_{0 \leq s \leq t} X_s.$$

The process M_t is called the *record process* of the process X_t. Suppose the data consist of lengths of time intervals greater than a given $\varepsilon > 0$ where M_t is constant and of the values of M_t on each of these intervals.

Let $T_u = \inf\{t \geq 0 : X_t = u\}$. Suppose the record process M_t is partially observed up to the hitting time $T_{\bar{\alpha}}$ of a given fixed level $\bar{\alpha} > x = X_0$. The time interval, where $M_t, 0 \leq t \leq T_{\bar{a}}$, remains constant and is equal to $u \in [x, \bar{\alpha})$ is $[T_u, T_{u+0}]$. Here T_{v+0} denotes the right limit of T_v at v. Thus the data are given by $(u, \Delta T_u)$ for all $u \in [x, \bar{\alpha})$ where $\Delta T_u = T_{u+0} - T_u \geq \varepsilon > 0$. Genon-Catalot and Larédo (1986) obtained an estimator for

$$G(y) - G(x) = \int_x^y a(u)\,du, \qquad x \leq y \leq \bar{\alpha}, \tag{4.2.81}$$

based on the statistic

$$\hat{p}_\varepsilon(x, y) = \#\{u : x \leq u \leq y, \Delta T_u \geq \varepsilon\}. \tag{4.2.82}$$

Asymptotic properties of an estimator $\hat{G}(\varepsilon; x, y)$ of $G(y) - G(x)$ are obtained as ε and σ tend to zero simultaneously, where

$$\hat{G}(\varepsilon; x, y) = (y - x)\sigma\left(\frac{2}{\pi\varepsilon}\right)^{1/2} - \sigma^2 \hat{p}_\varepsilon(x, y). \tag{4.2.83}$$

Theorem 4.2.9 *Suppose $a(\cdot)$ is continuously differentiable with derivative $a'(\cdot)$ and there exist constants $k_0 > 0$ and $k \geq 0$ such that*

$$a(u) \geq k_0, -k \leq a'(u) \leq 0$$

for all $u \in R$. Let $\varepsilon = \sigma^\delta$ with $\delta > 0$ and $\sigma \to 0$.

(i) If $\delta \le 2$ or $\delta \ge 6$, then $\hat{G}(\sigma^\delta; x, y)$ is not consistent.
(ii) If $2 < \delta < 6$, then $\hat{G}(\sigma^\delta; x, y)$ is consistent and the process

$$\sigma^{-\beta}\left(\hat{G}(\sigma_\delta; x, y) - \int_x^y a(u)\,du\right), \qquad x \le y \le \bar{\alpha},$$

converges in distribution on $D[x, \bar{\alpha}]$ as $\sigma \to 0$ to

(a) the nonrandom function $(-(2\pi)^{-1/2} \int_x^y a^2(u)\,du)$ if $2 < \delta < 10/3$ and $\beta = \delta/2 - 1$ (β increases from 0 to 2/3);
(b) the Gaussian process with the mean $(-(2\pi)^{-1/2} \int_x^y a^2(u)\,du)$ and the covariance function $R(y, z) \equiv (2/\pi)^{1/2}((y-x) \wedge (z-x))$ if $\delta = 10/3$ and $\beta = 3/2 - \delta/4 = 2/3$; and
(c) to the Gaussian process with mean 0 and covariance function $R(y, z) = (2/\pi)^{1/2}$ $((y - x) \wedge (z - x))$ if $10/3 < \delta < 6$ and $\beta = 3/2 - \delta/4 < 2/3$ (β decreases from 2/3 to 0).

(iii) The function $\varepsilon(\sigma) = \sigma^{10/3}$ (which yields the maximum rate of convergence $\alpha = 2/3$) corresponds to the minimax estimator within the class $\hat{G}(\varepsilon(\sigma); x, y)$.

We omit the proof. For complete details, see Genon-Catalot and Larédo (1986).

We now discuss some recent results of Kutoyants (1997a; 1997b) on efficient estimation of the density of the stationary distribution of an ergodic process.

4.2.4 Efficient Density Estimation

Let $\{X_t, t \ge 0\}$ be a stochastic process satisfying the stochastic differential equation

$$dX_t = a(X_t)\,dt + \sigma(X_t)\,dW_t, \qquad X_0 = X_0,\ t \ge 0. \tag{4.2.84}$$

Suppose the functions $a(\cdot)$ and $\sigma(\cdot)$ are such that the integrals

$$\int_0^y \frac{a(v)}{\sigma^2(v)}\,dv\,\mathrm{sgn}(y) \to -\infty \qquad \text{as } |y| \to \infty. \tag{4.2.85}$$

Then the process $\{X_t, t \ge 0\}$ is recurrent. Further assume that

$$G_* = G_*(a) \equiv \int_{-\infty}^{\infty} \frac{1}{\sigma^2(x)} \exp\left\{2\int_0^x \frac{a(y)}{\sigma^2(y)}\,dy\right\} dx < \infty. \tag{4.2.86}$$

This condition ensures the existence of a stationary distribution

$$F(z) = G_*^{-1} \int_{-\infty}^{z} \frac{1}{\sigma^2(x)} \exp\left\{2\int_0^x \frac{a(y)}{\sigma^2(y)}\,dy\right\} dx \tag{4.2.87}$$

for the diffusion process $\{X_t, t \ge 0\}$ satisfying (4.2.84). In particular, if $g(\cdot)$ is any continuous function such that

$$\int_{-\infty}^{\infty} |g(x)| F(dx) < \infty, \tag{4.2.88}$$

then

$$\lim_{T\to\infty} \frac{1}{T} \int_0^T g(X_t)\,dt = \int_{-\infty}^{\infty} g(x) F(dx) = Eg(\xi), \tag{4.2.89}$$

where ξ is a random variable with distribution function $F(\cdot)$. In other words, the process $\{X_t, t \geq 0\}$ possesses the ergodic property (Mandl 1968). The density function of the stationary distribution F is given by

$$f(x) = G_*^{-1} \frac{1}{\sigma^2(x)} \exp\left\{2 \int_0^x \frac{a(y)}{\sigma^2(y)} dy\right\}. \tag{4.2.90}$$

We assume that X_0 has the density f so that the process $\{X_t\}$ is stationary. We now consider the problem of efficient estimation of the density f.

Let $\mathcal{F}$ denote the class of functions $a(\cdot)$ such that (4.2.84) has a unique solution and condition (4.2.86) is satisfied. Let $\mathcal{F}_\delta$ be the class of functions from $\mathcal{F}$ such that conditon (4.2.86) is satisfied uniformly in some neighborhood U_δ (in the uniform metric) of each function $a(\cdot)$, that is, for some $\delta > 0$,

$$U_\delta = \{H(\cdot) : \sup_y |H(y)| \leq \delta\}$$

and

$$\sup_{a \in U_\delta G_*} (a) < \infty.$$

Fix $a(\cdot) \in \mathcal{F}_\delta$ and suppose the observed process is

$$dX_t = [a(X_t) + H(X_t)]\, dt + \sigma(X_t)\, dW_t, \qquad X_0 = X_0,\ 0 \leq t \leq T, \tag{4.2.91}$$

where the function $a(\cdot) + H(\cdot)$ is such that the solution of this equation exists and X_0 has the density function $f_H(\cdot)$ given by

$$f_H(x) = G_H^{-1} \frac{1}{\sigma^2(x)} \exp\left\{2 \int_0^x \frac{a(y) + H(y)}{\sigma^2(y)} dy\right\}. \tag{4.2.92}$$

Let $\{P_H^{(T)}, H(\cdot) \in U_\delta\}$ be the family of measures $P_H^{(T)}$ induced by the process $\{X_t, t \geq 0\}$ satisfying the equation on the space $C[0, T]$. Let E_H denote the expectation with respect to the probability measure P_H and

$$I_* = \left\{4 f^2(x) E\left(\frac{\chi_{\{\xi > x\}} - F(\xi)}{\sigma(\xi) f(\xi)}\right)^2\right\}^{-1}, \tag{4.2.93}$$

where χ_A is the indicator function of a set A and ξ is a random variable with the distribution function $F(\cdot)$.

Let $\ell(\cdot)$ denote a loss function satisfying the following conditions:

(i) $\ell(y, z) = \ell(y - z)$;

(ii) $\ell(\cdot) \geq 0$;

(iii) $\ell(0) = 0$, continuous at $z = 0$ but is not identically zero;

(iv) $\ell(\cdot)$ is symmetric on R and nondecreasing on R_+; and

(v) the function $\ell(z)$ grows as $z \to \infty$ more slowly than any function $\exp\{\varepsilon z^2\}$, $\varepsilon > 0$.

Let W_p denote the set of loss functions $\ell(y)$ satisfying conditions (i)–(v) with a polynomial majorant $C_p|y|^p$. The following theorem is proved in Kutoyants (1997a).

Theorem 4.2.10 *Let $a(\cdot) \in \mathcal{F}_\delta$ and $I_* > 0$. Then*

$$\underline{\lim}_{\delta\to 0}\ \underline{\lim}_{T\to\infty}\ \inf_{f_T}\ \sup_{H(\cdot)\in U_\delta} E_H \ell(T^{1/2}(f_T(x) - f_H(x))) \geq \frac{1}{\sqrt{2\pi}}\int_{-\infty}^{\infty} \ell(xI_*^{-1/2})e^{-x^2/2}\,dx, \tag{4.2.94}$$

where the infimum is taken over all possible estimators.

Remark. The proof of this inequality is based on the approach discussed in Ibragimov and Has'minskii (1981) and Prakasa Rao (1983a) for the estimation of a density function. Let $a(\cdot)$ be a fixed function. The supremum on U_δ is estimated from below on some parametric family with a special parameterization passing through the model with $a(\cdot)$. For this parametric model the Hájek–Le Cam inequality is applied and then one minimizes the lower bound so obtained to find the least favorable parametric family (with minimal Fisher information) and hence to obtain the lower bound as stated in the theorem.

Definition. An estimator $\hat{f}_T(x)$ is said to be asymptotically efficient for the loss function $\ell(\cdot)$ if

$$\lim_{\delta\to 0}\ \lim_{T\to\infty}\ \sup_{H(\cdot)\in U_\delta} E_H \ell(T^{1/2}(\hat{f}_T(x) - f_H(x))) = E\ell(I_*^{-1/2}\zeta) \tag{4.2.95}$$

for any function $a(\cdot) \in \mathcal{F}_\delta$ where ζ is $N(0,1)$. Such an estimator is also called a locally asymptotically minimax estimator.

Asymptotically efficient estimator Suppose $\sigma(\cdot)$ is a known function and positive. Consider the estimator

$$\tilde{f}_t(x) = \frac{1}{\sigma^2(x)T}\int_0^T \sigma(x - X_t)\,dX_t.$$

Suppose the following conditions hold:

(RR1)

$$P_H - \lim_{T\to\infty}\frac{1}{T}\int_0^T \left(\frac{I_{(X_t>x)} - F_H(X_t))}{\sigma(X_t)f_H(X_t)}\right)^2 dt = I_*(H)^{-1} \tag{4.2.96}$$

holds uniformly for $H(\cdot) \in U_\delta$, where

$$I_*(H) = \left\{4f_H^2(x)E_H\left(\frac{I_{(\xi>x)} - F_H(\xi)}{\sigma(\xi)f_H(\xi)}\right)^2\right\}^{-1}. \tag{4.2.97}$$

(RR2) For some $p_0 > 2$,

$$\sup H(\cdot) \in U_\delta E_H\left|\frac{I_{(\xi>x)} - F_H(\xi)}{\sigma(\xi)f_H(\xi)}\right|^{p_0} < \infty, \tag{4.2.98}$$

$$\sup H(\cdot) \in U_\delta E_H|\xi|^{p_0} < \infty \tag{4.2.99}$$

and

$$\sup H(\cdot) \in U_\delta E_H\left|\int_0^\xi \frac{I_{(v>x)} - F_H(v)}{\sigma(\xi)f_H(\xi)}\,dv\right|^{p_0} < \infty. \tag{4.2.100}$$

Theorem 4.2.11 *Suppose that conditions (RR1) and (RR2) hold and $I_*(H) > 0$ is a continuous function on U_δ. Then the estimator $\tilde{f}_T(x)$ is unbiased, uniformly consistent for $H(\cdot) \in U_\delta$, asymptotically normal, that is,*

$$T^{-1/2}(\tilde{f}_T(x) - f(x)) \xrightarrow{\mathcal{L}} N(0, I_*^{-1}) \tag{4.2.101}$$

and asymptotically efficient for the loss functions $\ell(\cdot) \in W_p$, $p < p_0$.

Remark. For a proof of Theorem 4.2.11, see Kutoyants (1997a). The estimator $\tilde{f}_T(x)$ can also be called the *local-time estimator* from the representation

$$\tilde{f}_T(x) = \frac{2\Lambda_T(x) - |X_T - x| + |X_0 - x|}{T\sigma^2(x)}, \tag{4.2.102}$$

where $\Lambda_T(x)$ is the local time of the diffusion process $\{X_t, t \geq 0\}$ defined as the limit

$$\lim \varepsilon \downarrow 0 \frac{\text{meas}\{t : |X_t - x| \leq \varepsilon,\ 0 \leq t \leq T\}}{4\varepsilon}, \qquad T \geq 0,\ x \in R. \tag{4.2.103}$$

Here meas(A) denotes the Lebegue measure of a set A (Karatzas and Shreve 1991). The representation (4.2.102) is a consequence of the Tanaka–Meyer formula

$$|X_T - x| = |X_0 - x| + \int_0^T \text{sgn}(X_t - x)\, dX_t + 2\Lambda_T(x) \tag{4.2.104}$$

(Karatzas and Shreve 1991, p. 220). For further discussion of asymptotically efficient estimators, see Kutoyants (1997a; 1997b).

4.3 Nonstationary Diffusion Model

Let us now consider the problem of estimating a function $\theta(t)$ on the basis of independent observations $X_j = \{X_j(t), 0 \leq t \leq T\}, 1 \leq j \leq n$, of the diffusion process given by the nonstationary diffusion model

$$dX(t) = \theta(t)X(t)\, dt + dW(t), \qquad X(0) = x_0,\ 0 \leq t \leq T. \tag{4.3.1}$$

4.3.1 Estimation of the Drift by the Kernel Method

Let us assume that the function $\theta(t)$ is continuous on $[0, t]$. Equation (4.3.1) can be written in the form

$$X(t) = x_0 + \int_0^T \theta(s)X(s)\, ds + W(t), \qquad 0 \leq t \leq T. \tag{4.3.2}$$

Let Θ_{L_α} be the set of functions $m(t)$ on $[0, T]$ which have k bounded derivatives in t and

$$|m^{(k)}(t_2) - m^{(k)}(t_1)| \leq L_\alpha |t_2 - t_1|^\alpha, \tag{4.3.3}$$

where $m^{(i)}(\cdot)$ denotes the ith derivative of $m(\cdot)$.

Consider the estimator $\tilde{\theta}_n(t)$ defined by

$$\tilde{\theta}_n(t) = \frac{1}{n\gamma_n} \sum_{j=1}^n \int_0^T G\left(\frac{t - \tau}{\gamma_n}\right) X_j(\tau)^{-1}\, dX_j(\tau) \tag{4.3.4}$$

as an estimator for $\theta(t)$, where G is a suitable kernel satisfying (4.2.56) and $\gamma_n \to 0$ as $n \to \infty$. However, the stochastic integral might not exist. Let us consider a modified version of that integral. Define

$$Y_{n,j}(t) = \begin{cases} X_j(t)^{-1}, & \text{if } |X_j(t)| \geq x_n, \\ x_n^{-1}, & \text{if } |X_j(t)| < x_n, \end{cases} \tag{4.3.5}$$

where $\{x_n\}$ is a positive sequence tending to zero to be chosen later. Let

$$\hat{\theta}_n(t) = \frac{1}{n\gamma_n} \sum_{j=1}^{n} \int_0^T G\left(\frac{t-\tau}{\gamma_n}\right) Y_{n,j}(\tau)\, dX_j(\tau). \tag{4.3.6}$$

It is easy to check that

$$\begin{aligned}\hat{\theta}_n(t) - \theta(t) = \gamma_n^{-1} \int G\left(\frac{t-\tau}{\gamma_n}\right) & \left\{ n^{-1} \sum_{j=1}^{n} I(|X_j(\tau)| \geq x_n) \right. \\ & \left. + n^{-1} \sum_{j=1}^{n} X_j(\tau) x_n^{-1} I(|X_j(\tau| < x_n) - 1 \right\} \theta(\tau)\, d\tau \\ & + (n\gamma_n)^{-1} \sum_{j=1}^{n} \int_0^T G\left(\frac{t-\tau}{\gamma_n}\right) Y_{n,j}(\tau)\, dW_j(\tau) \\ & + (\gamma_n)^{-1} \int_0^T G\left(\frac{t-\tau}{\gamma_n}\right) \theta(\tau)\, d\tau - \theta(t).\end{aligned}$$

It is obvious that $|Y_{n,j}(t)| \leq x_n^{-1}$ and hence $E(Y_{n,j}(t))^2 \leq x_n^{-2}$. Hence

$$\begin{aligned} E\left\{ (n\gamma_n)^{-1} \sum_{j=1}^{n} \int_0^T G\left(\frac{t-\tau}{\gamma_n}\right) Y_{n,j}(\tau)\, dW_j(\tau) \right\}^2 \\ = (n\gamma_n)^{-2} \sum_{j=1}^{n} \int_0^T G^2\left(\frac{t-\tau}{\gamma_n}\right) E(Y_{n,j}(\tau))^2\, d\tau \\ \leq (n\gamma_n x_n^2)^{-1} \int_{-\infty}^{\infty} G^2(u)\, du = c(n\gamma_n x_n^2)^{-1} \end{aligned}$$

for some constant $c > 0$. Furthermore,

$$\begin{aligned}\left| (\gamma_n)^{-1} \int_0^T G\left(\frac{t-\tau}{\gamma_n}\right) \theta(\tau)\, d\tau - \theta(t) \right| &= \left| \int_{-\infty}^{\infty} G(u)[\theta(t + u\gamma_n) - \theta(t)]\, du \right| \\ &\leq \frac{L_\alpha \gamma_n^{k+\alpha}}{k!} \int_{-\infty}^{\infty} |u^{k+\alpha} G(u)|\, du\end{aligned}$$

from the properties of the kernel G. It is easy to check that

$$E\left[n^{-1} \sum_{j=1}^{n} X_j(\tau) x_n^{-1} I\{|X_j(\tau) < x_n|\} \right]^2 \leq E[I(|X_1(\tau)| < x_n)]$$

$$
\begin{aligned}
&= P(|X_1(\tau)| < x_n) \\
&= \frac{1}{\sqrt{2\pi h^2}} \int_{-x_n}^{x_n} \exp\left\{-\frac{x-m}{2h^2}\right\} dx \\
&\leq \sqrt{\frac{2}{\pi}} \frac{x_n}{h} \leq c x_n,
\end{aligned}
$$

where

$$
m = x_0 \exp\left\{\int_0^\tau \theta(s)\, ds\right\}
$$

and

$$
h^2 = \int_0^\tau \exp\left\{2\int_s^\tau \theta(v)\, dv\right\} ds \geq (2L)^{-1}(1 - e^{-2Lc})
$$

for some positive constants L, c since $X_1(\tau)$ is a Gaussian random variable with parameters (m, h^2). This in turn follows from the representation

$$
X(t) = x_0 \exp\left\{\int_0^t \theta(s)\, ds\right\} + \int_0^t \exp\left\{\int_s^t \theta(v)\, dv\right\} dW(s), \qquad t \geq 0. \tag{4.3.7}
$$

Applying the above results, it can be checked that

$$
\sup_{c \leq t \leq d} E|\hat{\theta}_n(t) - \theta(t)|^2 \leq c_1 \gamma_n^{2\beta} + c_2 x_n + c_3 (n\gamma_n x_n^2)^{-1}. \tag{4.3.8}
$$

If we choose $x_n = n^{-2\beta/(2\beta+1)}$ and $\gamma_n = n^{-1/(2\beta+1)}$, then we obtain the following theorem.

Theorem 4.3.1 *Let $\theta(\cdot) \in \Theta_L$. Then, for any interval $(c, d) \subset [0, T]$,*

$$
\lim_{n\to\infty} \sup_{\theta(\cdot)\in\Theta_L} \sup_{c \leq t \leq d} E|\hat{\theta}_n(t) - \theta(t)|^2 n^{2\beta/(2\beta+1)} < \infty. \tag{4.3.9}
$$

Remark. Theorem 4.3.1 differs from that in Kutoyants (1985a) possibly due to some computational error. Kutoyants (1985b) proposed an alternate estimator $\tilde{\theta}_n(t)$ achieving the same rate (see Theorem 3.1 in Kutoyants 1985b). He further proved that this is the best rate possible under some conditions. We omit the details.

4.3.2 Estimation of the Drift by the Method of Sieves

Consider the stochastic differential equation

$$
dX(t) = \theta(t)X(t)\, dt + dW(t), \qquad X(0) = x,\ t \geq 0, \tag{4.3.10}
$$

where $\theta(\cdot) \in L^2([0, T])$. Suppose the process is observable on $[0, T]$ and $X_1, \ldots, X_n$ is a random sample of n independent observations of X on $[0, T]$. Then the log-likelihood function can be written in the form

$$
\begin{aligned}
\ell(X_1, \ldots, X_n : \theta) &\equiv \ell_n(\theta) \\
&= \sum_{k=1}^n \left[\int_0^T \theta(t) X_k(t)\, dX_k(t) - \frac{1}{2}\int_0^T \theta^2(t) X_k^2(t)\, dt\right].
\end{aligned} \tag{4.3.11}
$$

Let $\{V_n\}$ be an increasing sequence of subspaces of finite dimension d_n such that $\cup_{n\geq 1} V_n$ is dense in $L^2([0,T])$. Let $\{e_i\}$ be a set of independent vectors in $L^2([0,T])$ such that $(e_1,\ldots,e_{d_n})$ is a basis for V_n for every n. On V_n,

$$\ell_n(f) = n(-\tfrac{1}{2}\boldsymbol{\theta}^{(n)\prime}\boldsymbol{A}^{(n)}\boldsymbol{\theta}^{(n)} + \boldsymbol{B}^{(n)}), \tag{4.3.12}$$

where

$$\boldsymbol{\theta}^{(n)\prime} = (\theta_1,\ldots,\theta_{d_n}), \qquad \theta_i = \int_0^T \theta(x)e_i(x)\,dx,$$

$$\boldsymbol{B}^{(n)\prime} = (B_1^{(n)},\ldots,B_{d_n}^{(n)}), \qquad \boldsymbol{A}^{(n)} = ((A_{ij}^{(n)})),$$

with

$$B_j^{(n)} = n^{-1}\sum_{k=1}^{n}\int_0^T e_j(t)X_k(t)\,dX_k(t), \tag{4.3.13}$$

$$A_{ij}^{(n)} = n^{-1}\sum_{k=1}^{n}\int_0^T e_i(t)e_j(t)X_k^2(t)\,dt. \tag{4.3.14}$$

It can easily be checked that the maximum likelihood estimator of $\theta(\cdot)$ restricted to V_n is

$$\hat{\theta}^{(n)}(\cdot) = \sum_{j=1}^{d_n}\hat{\theta}_j^{(n)}e_j(\cdot), \tag{4.3.15}$$

with

$$\hat{\boldsymbol{\theta}}^{(n)\prime} = (\hat{\theta}_1^{(n)},\ldots,\hat{\theta}_{d_n}^{(n)})$$

satisfying

$$\boldsymbol{A}^{(n)}\hat{\boldsymbol{\theta}}^{(n)} = \boldsymbol{B}^{(n)}. \tag{4.3.16}$$

Observe that

$$A_{ij}^{(n)} \overset{\text{a.s.}}{\to} \int_0^T e_i(t)e_j(t)EX^2(t)\,dt \qquad \text{as } n\to\infty. \tag{4.3.17}$$

Let $(h_1,\ldots,h_{d_n})$ be the orthonormal basis for V_n with respect to the inner product

$$(p,q) = \int_0^T p(t)q(t)\,dm(t),\ dm(t) = E(X^2(t))\,dt. \tag{4.3.18}$$

Let $\hat{g}_1^{(n)},\ldots,\hat{g}_{d_n}^{(n)}$ be the coordinates of $\hat{\theta}^{(n)}(\cdot)$ with respect to the new basis and

$$\theta^{(n)}(\cdot) = \sum_{j=1}^{d_n} g_j^{(n)}h_j(\cdot) \tag{4.3.19}$$

be the orthogonal projection of $\theta(\cdot)$ on V_n with respect to the inner product on $L^2(dm(t))$. Then $\hat{\boldsymbol{g}}^{(n)\prime} = (\hat{g}_1^{(n)},\ldots,\hat{g}_{d_n}^{(n)})$ is a solution of

$$\boldsymbol{a}^{(n)}\hat{\boldsymbol{g}}^{(n)} = \boldsymbol{b}^{(n)}, \tag{4.3.20}$$

where

$$a_{ij}^{(n)} = n^{-1} \sum_{k=1}^{n} \int_0^T h_i(t) h_j(t) X_k^2(t)\, dt, \qquad \boldsymbol{a}^{(n)} = ((a_{ij}^{(n)})), \tag{4.3.21}$$

and

$$b_i^{(n)} = n^{-1} \sum_{k=1}^{n} \int_0^T h_i(t) X_k(t)\, dX_k(t), \qquad \boldsymbol{b}^{(n)\prime} = (b_1^{(n)}, \ldots, b_{d_n}^{(n)}). \tag{4.3.22}$$

Furthermore,

$$\boldsymbol{a}^{(n)}(\hat{\boldsymbol{g}}^{(n)} - \boldsymbol{g}^{(n)}) = \boldsymbol{c}^{(n)}, \tag{4.3.23}$$

where $\boldsymbol{g}^{(n)\prime} = (g_1^{(n)}, \ldots, g_{d_n}^{(n)})$ and $\boldsymbol{c}^{(n)\prime} = (c_1^{(n)}, \ldots, c_{d_n}^{(n)})$ with

$$c_j^{(n)} = n^{-1} \sum_{k=1}^{n} \left\{ \int_0^T h_j(t)[X_k^2(t) - E(X^2(t))][\theta(t) - \hat{\theta}^{(n)}(t)]\, dt \right.$$
$$\left. + \int_0^T h_j(t) X_k(t)\, dW_k(t) \right\}. \tag{4.3.24}$$

We have essentially changed the coordinate system to facilitate derivation of the asymptotic properties of the estimator $\hat{\theta}^{(n)}(t)$.

Using the above reparameterization, Nguyen and Pham Dinh Tuan (1982) proved the following result:

Theorem 4.3.2 *If $d_n \to \infty$ and $d_n^2/n \to 0$ as $n \to \infty$, then*

$$\lim_{n\to\infty} \int_0^T [\hat{\theta}^{(n)}(t) - \theta(t)] E(X^2(t))\, dt = 0 \tag{4.3.25}$$

in probability. Furthermore, if

$$\frac{d_n^3}{n} \to 0 \qquad \text{as } n \to \infty, \tag{4.3.26}$$

$$\int_0^T \frac{q^2(t)}{EX^2(t)}\, dt < \infty \qquad \text{for some } q(\cdot) \in L^2([0, T]) \tag{4.3.27}$$

and

$$\sqrt{n} \int_0^T q(t)[\theta^{(n)}(t) - \theta(t)]\, dt \to 0 \qquad \text{as } n \to \infty, \tag{4.3.28}$$

then

$$\sqrt{n} \int_0^T q(t)[\hat{\theta}^{(n)}(t) - \theta(t)]\, dt \xrightarrow{\mathcal{L}} N\left(0, \int_0^T \frac{q^2(t)}{EX^2(t)}\, dt\right) \qquad \text{as } n \to \infty. \tag{4.3.29}$$

Applying a different type of sieve, Leskow and Rozanski (1989) considered the problem of estimation of drift $\theta(t)$ in a stochastic differential equation of the type

$$dX(t) = \theta(t) a(t, X(t))\, dt + dW(t), \qquad X(0) = 0,\ t \geq 0, \tag{4.3.30}$$

where $a(t, x)$ is a known function and $\theta(\cdot) \in L^1([0, T])$. We briefly discuss their results.

Let P_λ be the probability measure generated by the process X when $\theta = \lambda$ in the above stochastic differential equation. Let P_W be the measure corresponding to the Wiener process. Assume that

$$\int_0^T a^2(t, X(t))\theta^2(t)\,dt < \infty \text{ a.s.} \tag{4.3.31}$$

Then $P_\lambda \ll P_W$ and

$$\frac{dP_\lambda}{dP_W} = \exp\left\{\int_0^T \theta(s)a(s, X(s))\,dX(s) - \frac{1}{2}\int_0^T \theta^2(s)a^2(s, X(s))\,ds\right\}. \tag{4.3.32}$$

Let $\{X_k(t), t \in [0, T]\}$, $1 \le k \le n$, be independent realizations of the process X on $[0, T]$. Then the log-likelihood function is given by

$$\ell_n(\theta) = \sum_{k=1}^{n}\int_0^T \theta(s)a(s, X_k(s))\,dX_k(s) - \frac{1}{2}\sum_{k=1}^{n}\int_0^T \theta^2(s)a^2(s, X_k(s))\,ds. \tag{4.3.33}$$

Let

$$S(n) = \left\{\theta(\cdot) \in L^1([0, T]) : \theta(s) = \sum_{j=1}^{m_n} x_j I_{A_{j,m_n}}(s)\right\}, \tag{4.3.34}$$

where

$$A_{j,m_n} = \left(\frac{j-1}{m_n}T, \frac{j}{m_n}T\right], \qquad 1 \le j \le m_n, \tag{4.3.35}$$

and

$$A_{0,m_n} = \left[0, \frac{1}{m_n}T\right] \tag{4.3.36}$$

and x_j, $1 \le j \le m_n$, are real numbers not all zero. The sequence of sets $S(n)$ is increasing in n and $\cup_n S(n)$ is dense in $L^1([0, T])$. Let $\hat{\theta}_n$ denote the maximum likelihood estimator of θ based on the sieve $S(n)$, that is,

$$\ell_n(\hat{\theta}_n) = \max_{\theta \in S(n)} \ell_n(\theta). \tag{4.3.37}$$

It is easy to check that

$$\hat{\theta}_n(s) = \sum_{j=1}^{m_n} \frac{\sum_{k=1}^{n}\int_{A_{j,m_n}} a(t, X_k(t))\,dX_k(t)}{\sum_{k=1}^{n}\int_{A_{j,m_n}} a^2(t, X_k(t))\,dt} I_{A_{j,m_n}}(s). \tag{4.3.38}$$

For any $s \in [0, T]$, choose $l(n, s) \in \{1, \ldots, m_n\}$ such that $s \in A_{l(n,s),m_n}$. Define $B_n^s = A_{l(n,s),m_n}$. Then

$$\hat{\theta}_n(s) = \frac{\sum_{k=1}^{n}\int_{B_n^s} a(t, X_k(t))\,dX_k(t)}{\sum_{k=1}^{n}\int_{B_n^s} a^2(t, X_k(t))\,dt}. \tag{4.3.39}$$

Suppose the following conditions hold:

(SS1) $\theta(\cdot)$ is Lipschitzian continuous of order $\alpha \ge \frac{1}{2}$, that is, there exists $C > 0$ such that

$$|\theta(t) - \theta(t)| \le C|t - s|^\alpha, \qquad t, s \in [0, T].$$

(SS2) (i) $\sup_{t \in [0,T]} E[a^4(t, X(t))] < \infty$;

(ii) $E[a^2(t, X(t))]$, $E[a^4(t, X(t))]$ are continuous in t; and

(iii) $\inf_{t\in[0,T]} |a(t, X(t))| > \delta > 0$ a.s.

Leskow and Rozanski (1989) proved the following result.

Theorem 4.3.3 *Suppose that conditions (SS1) and (SS2) hold and $m_n = n^{1/2}$. Then*

(i) $\hat{\theta}_n(s) \to \theta(s)$ a.s. as $n \to \infty$ for every $s \in [0, T]$;

(ii) $\int_0^T |\hat{\theta}_n(s) - \theta(s)|\, ds \to 0$ a.s. as $n \to \infty$;

(iii) $n^{1/4}(\hat{\theta}_n(s_i) - \theta(s_i), 1 \le i \le k)$ is asymptotically normal with mean zero and covariance matrix the diagonal matrix with the ith diagonal entry

$$\sigma_i = (Ea^2(s_i, X(s_i)))^{-1}$$

provided $Ea^2(s_i, X(s_i)) \neq 0$, $1 \le i \le k$.

Remarks. In general, if m_n is the dimension of the sieve subspace $S(n)$, then the normalizing factor in the above theorem is $(\frac{n}{m_n})^{1/2}$.

4.4 Linear Stochastic Systems

We now discuss some techniques for the estimation of unknown additive drifts (trends) present in the state and measurement processes of a Kalman–Bucy linear system, following McKeague and Tofoni (1991).

Consider a linear stochastic system of the following type due to Kalman and Bucy. A p-dimensional 'state' process $\boldsymbol{X}$ and a q-dimensional 'measurement' process $\boldsymbol{Z}$ are given by the stochastic differential system

$$d\boldsymbol{X}(t) = \boldsymbol{A}(t)\boldsymbol{X}(t) + \boldsymbol{B}(t)\boldsymbol{u}(t)\,dt + d\boldsymbol{W}(t), \tag{4.4.1}$$

$$d\boldsymbol{Z}(t) = \boldsymbol{C}(t)\boldsymbol{X}(t) + d\boldsymbol{V}(t), \tag{4.4.2}$$

$0 \le t \le T$, where $\boldsymbol{W}$ and $\boldsymbol{V}$ are independent p- and q-dimensional Wiener processes, $\boldsymbol{u}(\cdot)$ is a known deterministic input, $\boldsymbol{A}, \boldsymbol{B}, \boldsymbol{C}$ are known nonrandom time-dependent matricies of suitable dimensions and $\boldsymbol{X}(0)$ is independent of $\boldsymbol{W}$ and $\boldsymbol{V}$. Suppose the mean vector $E(\boldsymbol{X}(0)) = \boldsymbol{m}$ and the covariance matrix of $\boldsymbol{X}(0)$ are known and $\boldsymbol{Z}(0) = 0$. Further suppose that the process $\{\boldsymbol{Z}(s), 0 \le s \le t\}$ is observable, whereas $\{\boldsymbol{X}(s), 0 \le s \le t\}$ is not.

The Kalman filtering theory provides a recursive formula for the conditional expectation $\hat{\boldsymbol{X}}(t) = E(\boldsymbol{X}(t)|\mathcal{F}_t^{\boldsymbol{Z}})$ which is the optimal mean square estimator of the state $\boldsymbol{X}(t)$ given the past $\mathcal{F}_t^{\boldsymbol{Z}} = \sigma\{\boldsymbol{Z}(s), 0 \le s \le t\}$ of the measurement process (Liptser and Shiryayev 1978; Kallianpur 1980).

In applications to real problems, it is often the case that the unknown additive trends are present in both the state and the measurement processes, that is, the state process is given by

$$d\boldsymbol{X}(t) = \boldsymbol{f}(t)\,dt + \boldsymbol{A}(t)\boldsymbol{X}(t)\,dt + \boldsymbol{B}(t)\boldsymbol{u}(t)\,dt + d\boldsymbol{W}(t) \tag{4.4.3}$$

and, instead of observing $\boldsymbol{Z}$, we observe the process $\boldsymbol{Y}$ given by

$$d\boldsymbol{Y}(t) = \boldsymbol{g}(t)\,dt + d\boldsymbol{Z}(t), \qquad \boldsymbol{Y}(0) = 0, \tag{4.4.4}$$

where $\boldsymbol{Z}$ is as defined by (4.4.2) and $\boldsymbol{f}$ and $\boldsymbol{g}$ are *unknown* drift functions.

The problem considered here is to estimate $\boldsymbol{f}$ and $\boldsymbol{g}$ and use them in the Kalman filter $\hat{\boldsymbol{X}}(t) = E[\boldsymbol{X}(t) \mid \mathcal{F}_t^{\boldsymbol{Y}})$. We consider two types of sampling schemes: (I) observations of n

realizations $\{Y_i(t), 0 \le t \le T\}, 1 \le i \le n$, of the process Y satisfying (4.4.3) and (4.4.4) with the corresponding system realizations having independent noise processes W_i and V_i, $1 \le i \le n$; (II) observation of one trajectory of Y over $[0, nT]$ where the functions f, g, A, B and C are assumed to be periodic with period T.

Observation scheme II is relevant when there is a 'periodic' or 'seasonal' effect present in the model, for instance, in the analysis of the circadian rhythm in biology or in the study of cyclic systems in control engineering (Bittani and Guardabassi 1986; Prakasa Rao 1997a).

We are interested in the asymptotic properties of the estimators $\hat{f}$ and $\hat{g}$ as $n \to \infty$. For simplicity, we consider the case $p = q = 1$. Under the conditions that f, g, A, B, C and u are smooth and that $C(t)$ is nonzero on $[0, t]$, the equations for the Kalman filter are

$$d\hat{X}(t) = [f(t) + A(t)\hat{X}(t) + B(t)u(t)]\,dt + D(t)\,dv(t), \tag{4.4.5}$$

$$dv(t) = dY(t) - [g(t) + C(t)\hat{X}(t)]\,dt, \tag{4.4.6}$$

where $\hat{X}(0) = m$ and $v(\cdot)$ is the so called *innovation process* (Kallianpur 1980, Section 4.3). The process v is again a standard Wiener process and the function D is called the *Kalman gain*. In the present case, the function D does not depend on f and g. In fact $D(t) = C(t)P(t)$, where P is the unique positive solution of the *Ricatti differential equation*

$$P'(t) = 2A(t)P(t) - C^2(t)P^2(t) + 1, \tag{4.4.7}$$

with initial condition $P(0) = \text{var}(X(0))$. In view of (4.4.5) and (4.4.6), we have

$$d\hat{X}(t) = [A(t) - D(t)C(t)]\hat{X}(t)\,dt + [f(t) + B(t)u(t) - D(t)g(t)]\,dt + D(t)\,dY(t). \tag{4.4.8}$$

Solving this system of equations for $\hat{X}(t)$ (Davis 1977, Theorem 4.2.4) and substituting the same in (4.4.6), we obtain the innovation representation for Y:

$$Y(t) = \int_0^t [h(s) + U(s)]\,ds + v(t), \tag{4.4.9}$$

where

$$h(t) = g(t) + C(t)\int_0^t \Psi(t,s)[f(s) - D(s)g(s)]\,ds, \tag{4.4.10}$$

and $\Psi(t, s)$ is a solution of the linear time-varying system

$$\frac{\partial \Psi(t,s)}{\partial t} = [A(t) - D(t)C(t)] \qquad \Psi(t,s),\ \Psi(s,s) = 1. \tag{4.4.11}$$

Furthermore, U is given by

$$U(t) = C(t)\{\Psi(t,0)\,m + \int_0^t \Psi(t,s)[B(s)u(s)\,ds + D(s)\,dY(s)]\}. \tag{4.4.12}$$

We observe from (4.4.9) that the function h is identifiable given the observations of Y and U. However f and g are identifiable only if they are uniquely determined through h given by (4.4.10). The functions h and g are not simultaneously identifiable from the observations on Y. If $g \equiv 0$, then (4.4.10) reduces to

$$h(t) = \int_0^t \Phi(t,s) f(s)\,ds, \tag{4.4.13}$$

where $\Phi(t, s) = C(t)\Psi(t, s)$. If $f \equiv 0$, then (4.4.10) reduces to

$$h(t) = g(t) + \int_0^t \Gamma(t, s)g(s)\,ds, \tag{4.4.14}$$

where $\Gamma(t, s) = -C(t)\Psi(t, s)D(s)$.

As equations (4.4.13) and (4.4.14) are linear Volterra integral equations of the first kind and second kind respectively, (4.4.13) has a unique solution for f when $g \equiv 0$ and (4.4.14) has a unique solution for g when $f \equiv 0$ provided $C(t)$ does not vanish on $[0, T]$ (Linz 1985). Since h is identifiable, the drift f is identifiable when $g \equiv 0$ and g is identifiable when $f \equiv 0$.

Let μ_h be the probability measure induced by the process Y on $C[0, T]$. Let μ_W be be the Wiener measure. In view of the representation (4.4.9), it follows that $\mu_h \ll \mu_W$ and

$$\ell(h) \equiv \log \frac{d\mu_h}{d\mu_W} = \int_0^T \pi(s)\,dY(s) - \frac{1}{2}\int_0^T \pi^2(s)\,ds, \tag{4.4.15}$$

where

$$\pi(s) = h(s) + U(s), \qquad 0 \le s \le T. \tag{4.4.16}$$

Under observation scheme I, suppose the processes associated with the ith realization are indexed by i, as in ν_i, U_i, $\ell_i(h)$, etc. Even though the observed processes $\{Y_i, 1 \le i \le n\}$ are independent, they are not necessarily identically distributed since the inputs u_i are not assumed to be identical. However, the innovation processes ν_i are i.i.d. Wiener processes. We have

$$Y_i(t) = \int_0^t [h(s) + U_i(s)]\,ds + \nu_i(t), \tag{4.4.17}$$

where

$$U_i(t) = C(t)\{\Psi(t, 0)m + \int_0^t \Psi(t, s)[B(s)u_i(s)\,ds + D(s)\,dY_i(s)]\}. \tag{4.4.18}$$

In this case, the log-likelihood function $\ell_I^{(n)}(h)$ is given by

$$\ell_I^{(n)}(h) = \sum_{i=1}^n \ell_i(h), \tag{4.4.19}$$

where

$$\ell_i(h) = \int_0^T [h(s) + U_i(s)]\,dY_i(s) - \tfrac{1}{2}\int_0^T [h(s) + U_i(s)]^2\,ds.$$

Under observation scheme II, let h_i, U_i, Y_i and ν_i be the restrictions of h, U, Y and ν to the ith period:

$$\begin{aligned} h_i(t) &= h(iT + t), \\ U_i(t) &= U(iT + t), \\ Y_i(t) &= Y(iT + t) - Y(iT), \\ \nu_i(t) &= \nu(iT + t) - \nu(iT), \end{aligned}$$

for $0 \le t \le T$. Then

$$Y_i(t) = \int_0^t [h_i(s) + U_i(s)]\,ds + \nu_i(t), \qquad 0 \le t \le T. \tag{4.4.20}$$

Since ν is a Wiener process(which has stationary independent increments), it follows that ν_i, $1 \leq i \leq n$, are i.i.d. Wiener processes on $[0, T]$. The log-likelihood function $\ell_{II}^{(n)}(h)$ is given by (4.4.15) with T replaced by nT and it can be written in the form

$$\ell_{II}^{(n)}(h) = \sum_{i=1}^{n} \ell_i(h_i). \tag{4.4.21}$$

Note that $h_i \neq h$ since the function h is not perodic.

4.4.1 Estimation under Observation Scheme I

In this subsection, we consider the problem of estimation under observation scheme I (when $f \equiv 0$ – i.i.d. case).

Let $\hat{h}$ be an estimator for h. In view of (4.4.14), if $\hat{g}$ is an estimator for g, then $\hat{h}$ and $\hat{g}$ should be related by the Volterra integral equation

$$\hat{h}(t) = \hat{g}(t) + \int_0^t \Gamma(t, s)\hat{g}(s)\, ds. \tag{4.4.22}$$

Following Davis (1977, p. 125), given $\hat{h} \in L^2[0, T]$, there exists a unique solution $\hat{g} \in L^2[0, T]$ such that (4.4.22) holds. Let

$$\bar{h} = \hat{h} - h, \qquad \bar{g} = \hat{g} - g. \tag{4.4.23}$$

Let $M = \sup_{0 \leq t, s \leq T} |\Gamma(t, s)| < \infty$. Then it follows from (4.4.14) and (4.4.22) that

$$\bar{g}(t) = \bar{h}(t) - \int_0^t \Gamma(t, s)\bar{g}(s)\, ds \tag{4.4.24}$$

and hence

$$|\bar{g}(t)|^2 \leq 2TM^2 \int_0^t |\bar{g}(s)|^2\, ds + 2|\bar{h}(t)|^2.$$

An application of Gronwall's inequality (Kallianpur 1980, p. 94) implies that

$$\begin{aligned} |\bar{g}(t)|^2 &\leq 4TM^2 \int_0^T |\bar{h}(s)|^2 \exp\{2TM^2(t - s)\}\, ds + 2|\bar{h}(t)|^2 \\ &\leq 4TM^2 \exp\{2TM^2\} \int_0^T |\bar{h}(s)|^2\, ds + 2|\bar{h}(t)|^2. \end{aligned}$$

Integrating over $[0, T]$, we obtain that

$$\|\bar{g}\|_2^2 \leq (2 + 4TM^2 \exp\{2TM^2\})\|\bar{h}\|_2^2, \tag{4.4.25}$$

which shows that

$$\|\hat{g} - g\|_2^2 \leq C_T \|\hat{h} - h\|_2^2. \tag{4.4.26}$$

Therefore, if $\hat{h}$ is strongly L^2-consistent for h, then $\hat{g}$ is strongly L^2-consistent for g where $\hat{g}$ is defined by (4.4.22).

We now give some methods for construction of a strongly consistent estimator $\hat{h}$ for h.

Estimation by the method of sieves

Let $\{\psi_r, 1 \le r \le d_n\}$ span S_n and $\{\psi_r, r \ge 1\}$ be a complete orthonormal basis for $L^2[0,T]$ with $d_n \to \infty$ as $n \to \infty$. Let

$$h_r = \int_0^T h(t)\psi_r(t)\,dt, \qquad r \ge 1, \tag{4.4.27}$$

and $\boldsymbol{h}^{(n)} = (h_1, \ldots, h_{d_n})'$. It is easy to see that the log-likelihood function $\ell^{(n)}(h)$, omitting the terms not involving h, for $h \in S_n$, is

$$\ell^{(n)}(h) = \boldsymbol{h}^{(n)\prime}(\boldsymbol{Q}^{(n)} - \boldsymbol{P}^{(n)}) - \frac{n}{2}\boldsymbol{h}^{(n)\prime}\boldsymbol{h}^{(n)}, \tag{4.4.28}$$

where $\boldsymbol{Q}^{(n)}$ and $\boldsymbol{P}^{(n)}$ are $d_n \times 1$ vectors with components

$$Q_r^{(n)} = \sum_{i=1}^n \int_0^T \psi_r(t)\,dY_i(t) \tag{4.4.29}$$

and

$$P_r^{(n)} = \sum_{i=1}^n \int_0^T \psi_r(t)U_i(t)\,dt. \tag{4.4.30}$$

Maximizing the function $\ell^{(n)}(h)$ defined by (4.4.28) with respect to $\boldsymbol{h}^{(n)}$, we obtain

$$\hat{h}(t) = \sum_{r=1}^{d_n} \hat{h}_r\psi_r(t), \tag{4.4.31}$$

where

$$\hat{\boldsymbol{h}}^{(n)} = (\hat{h}_1, \ldots, \hat{h}_{d_n})'$$

is given by

$$\hat{\boldsymbol{h}}^{(n)} = \frac{1}{n}(\boldsymbol{Q}^{(n)} - \boldsymbol{P}^{(n)}). \tag{4.4.32}$$

Relations (4.4.17), (4.4.27) and (4.4.32) show that

$$(\hat{\boldsymbol{h}}^{(n)} - \boldsymbol{h}^{(n)})_r = n^{-1/2}\varepsilon_r^{(n)}, \tag{4.4.33}$$

where

$$\varepsilon_r^{(n)} = n^{-1/2}\sum_{i=1}^n \int_0^T \psi_r(t)\,dv_i(t), \qquad 1 \le r \le d_n. \tag{4.4.34}$$

In particular,

$$\|\hat{\boldsymbol{h}}^{(n)} - \boldsymbol{h}^{(n)}\|_2^2 = \frac{1}{n}\sum_{r=1}^{d_n}(\varepsilon_r^{(n)})^2, \tag{4.4.35}$$

where $\varepsilon_r^{(n)}$, $1 \le r \le d_n$, are i.i.d. $N(0,1)$ random variables. In other words $n\|\hat{\boldsymbol{h}}^{(n)} - \boldsymbol{h}^{(n)}\|_2^2$ has a χ^2 distribution with d_n degrees of freedom. It can be checked that if $d_n \to 0$ as $n \to \infty$, then $\|\hat{\boldsymbol{h}}^{(n)} - \boldsymbol{h}^{(n)}\|_2 \overset{\text{a.s.}}{\to} 0$ as $n \to \infty$. Note that

$$\|\hat{\boldsymbol{h}}^{(n)} - \boldsymbol{h}^{(n)}\|_2^2 = \sum_{1 \le i \le d_n}(\hat{h}_i^{(n)} - h_i)^2 + \sum_{i > d_n} h_i^2. \tag{4.4.36}$$

The second term goes to zero as $n \to \infty$ since $h \in L_2[0,T]$. The first term goes to zero by an application of the Borel–Cantelli lemma (Beder 1987). This shows that the sieve estimator $\hat{h}$ defined by (4.4.31) is strongly L^2-consistent for h and we have the following theorem.

Theorem 4.4.1 *The estimator $\hat{h}_n$ defined by (4.4.31) is strongly L^2-consistent for h if $d_n \to \infty$ and $d_n/n \to 0$ as $n \to \infty$.*

Estimation by the kernel method

Let K be a bounded kernel with support $[-1, 1]$, and integrating to unity. Let $b_n > 0$ be a bandwidth sequence. Define

$$\tilde{h}(t) = \frac{1}{b_n}\int_0^T K\left(\frac{t-s}{b_n}\right) d\tilde{H}(s), \tag{4.4.37}$$

with

$$\tilde{H}(t) = \frac{1}{n}\sum_{i=1}^{n}\left\{Y_i(t) - \int_0^t U_i(s)\,ds\right\}. \tag{4.4.38}$$

Note that $\tilde{H}(t)$ can be considered as an estimator for

$$H(t) = \int_0^t h(s)\,ds. \tag{4.4.39}$$

Theorem 4.4.2 *Suppose that $b_n \to 0$ such that $b_n n^{1-\delta} \to \infty$ as $n \to \infty$ for some $0 < \delta < 1$. Then $\tilde{h}$ defined by (4.4.37) is strongly L^2-consistent for h.*

Proof. Let

$$h^{(n)}(t) = \frac{1}{b_n}\int_0^T K\left(\frac{t-s}{b_n}\right) h(s)\,ds. \tag{4.4.40}$$

It is easy to see that $\|h^{(n)} - h\|_2 \to 0$ as $n \to \infty$ since h is continuous (Prakasa Rao 1983a). It is sufficient to prove that $\|\tilde{h} - h^{(n)}\|_2 \to 0$ as $n \to \infty$. It is easy to see that

$$\tilde{h} - h^{(n)} = (nb_n)^{-1/2}\varepsilon^{(n)}, \tag{4.4.41}$$

where

$$\varepsilon^{(n)}(t) = \frac{1}{\sqrt{b_n}}\int_0^T K\left(\frac{t-s}{b_n}\right) dW^{(n)}(s) \tag{4.4.42}$$

and $W^{(n)} = \sqrt{n}(\tilde{H} - H)$. It follows from (4.4.9) and (4.4.38) that $W^{(n)}$ is a standard Wiener process. Hence $\varepsilon^{(n)}(t)$ is a Gaussian random variable with mean zero and variance

$$\frac{1}{b_n}\int_0^T K^2\left(\frac{t-s}{b_n}\right) ds \le \int_{-1}^1 K^2(u)\,du.$$

Fix $\eta > 0$ and let $k > \frac{1}{\delta}$. By Hölder's inequality and Fubini's theorem, and observing that the $2k$th moment of $\varepsilon^{(n)}(t)$ is uniformly bounded in n and t, we obtain that

$$E\|\tilde{h} - h^{(n)}\|_2^{2k} \le \frac{1}{(nb_n)^k} T^{k-1}\int_0^T E(\varepsilon^{(n)}(t))^{2k}\,dt = O((nb_n)^{-k}) = O(n^{-k\delta}). \tag{4.4.43}$$

By Chebyshev's inequality, for any $\eta > 0$,

$$P(\|\tilde{h} - h^{(n)}\|_2 > \eta) \le \eta^{-2k} E\|\tilde{h} - h^{(n)}\|^{2k} = O(n^{-k\delta}).$$

Since $k\delta > 1$, it follows that

$$\sum_{n=1}^{\infty} P(\|\tilde{h} - h^{(n)}\|_2 > \eta) < \infty$$

for all $\eta > 0$. An application of the Borel–Cantelli lemma proves the result.

Asymptotic distribution for an estimator of g

Let $\gamma(t, s)$ be the resolvent kernel for $\Gamma(t, S)$. Then the unique solution g of (4.4.14) for a given h is given by

$$g(t) = h(t) + \int_0^T \gamma(t, s)h(s)\,ds \tag{4.4.44}$$

(Linz 1985, Theorem 3.3). Let the estimator $\tilde{g}$ be defined by

$$\tilde{g}(t) = \tilde{h}(t) + \int_0^T \gamma(t, s)\tilde{h}(s)\,ds \tag{4.4.45}$$

from the estimator $\tilde{h}(t)$ constructed earlier.

Theorem 4.4.3 *If $nb_n \to \infty$ such that $nb_n^3 \to 0$, then, for $0 < t < T$,*

$$(nb_n)^{1/2}(\tilde{g}(t) - g(t)) \xrightarrow{\mathcal{L}} N(0, \sigma^2) \qquad \text{as } n \to \infty, \tag{4.4.46}$$

where

$$\sigma^2 = \int_{-1}^{1} K^2(u)\,du. \tag{4.4.47}$$

Proof. It is easy to see from (4.4.44) and (4.4.45) and the proof of Theorem 4.4.2 that

$$\begin{aligned}(nb_n)^{1/2}(\tilde{g}(t) - g(t)) &= \varepsilon^{(n)}(t) + (nb_n)^{1/2}(h^{(n)}(t) - h(t)) \\ &\quad + b_n^{1/2}\eta^{(n)}(t) + (nb_n)^{1/2}\int_0^t \gamma(t, s)(h^{(n)}(s) - h(s))\,ds,\end{aligned} \tag{4.4.48}$$

where

$$\eta^{(n)}(t) = b_n^{-1}\int_0^t\int_0^T \gamma(t, s)K\left(\frac{s-u}{b_n}\right)dW^{(n)}(u)\,ds. \tag{4.4.49}$$

The first term on the right-hand side of (4.4.48) is Gaussian with mean zero and variance

$$b_n^{-1}\int_0^t K^2\left(\frac{t-s}{b_n}\right)ds,$$

which tends to σ^2 as $n \to \infty$ so that $\varepsilon^{(n)}(t) \xrightarrow{\mathcal{L}} N(0, \sigma^2)$. The remaining terms tend to zero in probability. For, using a Fubini type theorem for stochastic integrals (Liptser and Shiryayev 1977, Theorem 5.15), we have

$$\eta^{(n)}(t) = b_n^{-1}\int_0^T\int_0^t \gamma(t, s)K\left(\frac{s-u}{b_n}\right)ds\,dW^{(n)}(u),$$

which implies that $\eta^{(n)}(t)$ is Gaussian with mean zero and variance

$$\int_0^T \left\{ b_n^{-1} \int_0^t \gamma(t,s) K\left(\frac{s-u}{b_n}\right) ds \right\}^2 du \to \int_0^t \gamma^2(t,s)\, ds.$$

Hence the third term is of order $O_p(\sqrt{b_n})$. Since h is Lipschitzian, the second and the fourth terms are of order $O(\sqrt{nb_n^3})$. This proves the theorem.

Remark. (i) An alternate estimator for g is

$$g^*(t) = \tilde{h}(t) + \int_0^t \gamma(t,s)\, d\tilde{H}(s) \tag{4.4.50}$$

where $\tilde{h}$ and $\tilde{H}$ are as defined above. It can be checked that g^* and $\tilde{g}$ have the same asymptotic distribution.

(ii) For the corresponding results under the observation scheme II in the case when $f \equiv 0$, see McKeague and Tofoni (1991).

4.4.2 Estimation under Observation Scheme II

We conclude this section by briefly examining estimation under observation scheme II (when $g \equiv 0$) – i.i.d. case).

In order to estimate f, let us consider equation (4.4.13), namely,

$$h(t) = \int_0^t \Phi(t,s) f(s)\, ds, \tag{4.4.51}$$

which is a Volterra integral equation of the first kind. Here $\Phi(t,s) = C(t)\Psi(t,s)$. This can be solved by converting it into a Volterra integral equation of the second kind by differentiation (Linz 1985, p. 67). Since $C(t)$ does not vanish on $[0,T]$ by assumption, we have

$$f(t) = \frac{h'(t)}{C(t)} + F(t)h(t), \tag{4.4.52}$$

where

$$F(t) = D(t) - \frac{C'(t)}{C^2(t)} - \frac{A(t)}{C(t)}. \tag{4.4.53}$$

In order to estimate f, we have to estimate h and h' and construct estimators following (4.4.52).

Estimation by the kernel method

Let $K(\cdot)$ be a kernel function as defined earlier but assume in addition that K is differentiable. Let c_n be a bandwidth parameter different from b_n. Define

$$\tilde{h'}(t) = c_n^{-2} \int_0^T K'\left(\frac{t-s}{c_n}\right) d\tilde{H}(s) \tag{4.4.54}$$

and

$$\tilde{f}(t) = \frac{\tilde{h'}(t)}{C(t)} + F(t)\tilde{h}(t), \tag{4.4.55}$$

where $\tilde{h}(t)$ is as given by (4.4.37). It can be checked that if $c_n \to 0$ and $c_n n^{1/3-\delta} \to \infty$ where $0 < \delta < \frac{1}{3}$, then $\tilde{h}'(t)$ is strongly L^2-consistent. Under these conditions and the conditions stated in Theorem 4.4.2, it follows that $\tilde{f}$ is strongly L^2-consistent. Furthermore, if $nb_n \to \infty$, $nb_n^3 \to 0$, $nc_n^3 \to \infty$, $nc_n^5 \to 0$ and $c_n = o(b_n^{1/3})$, then, for $0 < t < T$,

$$(nc_n^3)^{1/2}(\tilde{f}(t) - f(t)) \xrightarrow{\mathcal{L}} N(0, \sigma^2(t)) \quad \text{as } n \to \infty,$$

where

$$\sigma^2(t) = \frac{\int_{-1}^{1} K'(u)^2\, du}{(C(t))^2}.$$

For the proofs of all these results, see McKeague and Tofoni (1991).

Remarks. In the multidimensional case, f is not identifiable if $p > q$, where p is the dimension of the X_t process (state process) and q is the dimension of the Z_t process (measurement process). Suppose $p \leq q$. Under the condition that the matrix $C(t)$ is invertible, the results obtained for the estimation of f earlier can be extended to the multidimensional case. If $f \equiv 0$, then g can be estimated for all p and q by the method of sieves or by the method of kernels discussed earlier.

5

Nonparametric Inference for Diffusion Type Processes from Sampled Data

5.1 Introduction

The importance of the problem of statistical inference from sampled data for stochastic processes was explained in Chapter 3. A general review of the methods of statistical inference from sampled data for stochastic processes was given in Prakasa Rao (1988b). The special case of nonparametric density estimation for stochastic processes from sampled data was discussed in Prakasa Rao (1990a). Before we discuss the problem of nonparametric inference for the special class of diffusion type processes from sampled data, we shall discuss some general results of nonparametric inference from sampled data for stochastic processes in the next section. As mentioned in Chapter 3, the problem of estimation of the drift coefficient as well as the diffusion coefficient from sampled data is of interest in view of their relevance in general for stochastic modeling; for instance, the problem of estimation of the diffusion coefficient from sampled data is of importance in mathematical finance.

5.2 Nonparametric Inference from Sampled Data for General Stochastic Processes

Let $(\Omega, \mathcal{F}, P)$ be a complete probability space and $\{\mathcal{F}_t, t \geq 0\}$, be a right continuous complete filtration defined on it. Suppose $X = \{X_t, t \geq 0\}$ is a stationary process adapted to $\{\mathcal{F}_t\}$. Suppose that $\mathcal{F}_0 = \{\phi, \Omega\}$. Let $N = \{N_t, t \geq 0\}$, $N_0 = 0$ be a point process adapted to $\{\mathcal{F}_t\}$ with stochastic intensity $\{\lambda_t\}$ (Bremaud 1981). Let $\{\tau_i, i \geq 1\}$ be the jump times for the process N with $\tau_0 = 0$. Then τ_i denotes the jump time for the ith jump of the process N. Note that

$$N_t = \sum_{i \geq 1} I[\tau_i \leq t], \tag{5.2.1}$$

where $I(A)$ is the indicator function of the set A. Suppose the process X is observed at the times $\tau_i, 0 \leq i \leq n$. In other words, the observations $X_{\tau_0}, \ldots, X_{\tau_n}$ are recorded. Note that

$$X_\tau(\omega) = X_{\tau(\omega)}(\omega) \tag{5.2.2}$$

by the definition of the stopping time τ. Suppose that $\{X_t\}$ is a progressively measurable process (Meyer 1966). It is easy to check that X_τ is $\mathcal{F}_\tau$-measurable, where $\mathcal{F}_\tau$ is the σ-algebra generated by $A \in \mathcal{F}$ such that $A \cap [\tau \leq t] \in \mathcal{F}_t$ for every $t \geq 0$.

Suppose X_0 has density f. By the stationarity of the process, f is the one-dimensional marginal density of the process $\{X_t\}$. Let us consider the problem of estimation of f from the data $\{X_{\tau_i}, 0 \leq i \leq n\}$.

Let $\{\delta_t(x), t \geq 0\}$ be a family of functions of delta type (Chapter 4). Let us recall the following definition.

A family of functions $\{\delta_t(x), t \geq 0\}$ is said to be a *family of delta type* if

$$
\begin{aligned}
&\int_{-\infty}^{\infty} |\delta_t(x)|\, dx \leq A < \infty, && t \geq 0;\\
&\int_{-\infty}^{\infty} \delta_t(x)\, dx = 1, && t \geq 0;\\
&\delta_t(x) \to 0 && \text{as } t \to \infty \text{ uniformly in } |x| > \lambda \text{ for any } \lambda > 0;\\
&\int_{|x|\geq\lambda} |\delta_t(x)|\, dx \to 0 && \text{as } t \to \infty \text{ for any } \lambda > 0.
\end{aligned}
\tag{5.2.3}
$$

An example of such a family is

$$\delta_t(x) = h_t^{-1} K(x h_t^{-1}), \qquad t \geq 0, \tag{5.2.4}$$

where $K(\cdot)$ is a bounded probability density function such that

$$|x|K(x) \to 0 \text{ as } |x| \to \infty, \qquad h_t > 0 \text{ and } h_t \to 0 \text{ as } t \to \infty$$

(Prakasa Rao 1983a; 1998).

Suppose the family $\{\delta_t(\cdot)\}$ is such that

(TT0) the process $\{\delta_T(x - X_t), 0 \leq t \leq T\}$ is $\mathcal{F}_t$-predictable for every x and every $T \geq 0$.

Let

$$\hat{f}_T(x) = \frac{1}{T}\int_0^T \delta_T(x - X_s)\, dN_s = \frac{1}{T}\sum_{i:[\tau_i \leq T]} \delta_T(x - X_{\tau_i}). \tag{5.2.5}$$

The stochastic integral on the right-hand side of (5.2.5) is well defined since $\delta_T(x - X_t)$ is $\mathcal{F}_t$-predictable and $\{N_t\}$ is $\mathcal{F}_t$-adapted (Bremaud 1981; Prakasa Rao 1987).

In order to study the asymptotic behavior of the estimator $\hat{f}_T(x)$ as an estimator of $f(x)$, define

$$f_T^*(x) = \frac{1}{T}\int_0^T \delta_T(x - x_s)\lambda_s\, ds. \tag{5.2.6}$$

Observe that

$$M_t = N_t - \int_0^t \lambda_s\, ds, \qquad t \geq 0, \tag{5.2.7}$$

is an $\mathcal{F}_t$-local martingale in general and is a zero-mean $\mathcal{F}_t$-martingale under the assumption $E(N_t) < \infty$. Furthermore, for any $\mathcal{F}_t$-predictable process $\{Z_t, t \geq 0\}$ such that

$$E\left[\int_0^t |Z_s|\lambda_s\, ds\right] < \infty, \qquad t \geq 0, \tag{5.2.8}$$

the process

$$\int_0^t Z_s\, dM_s, \qquad t \geq 0, \tag{5.2.9}$$

is an $\mathcal{F}_t$-martingale (Bremaud 1981).

The following central limit theorem, due to Kutoyants (1984a), will be useful. We omit the proof.

Theorem 5.2.1 *Suppose that $Z_T = \{Z_T(t), 0 \le t \le T\}$ is $\mathcal{F}_t$-predictable for every $T \ge 0$ and*

$$I_T = \int_0^T Z_T(t)\, dM_t, \qquad T \ge 0, \tag{5.2.10}$$

where M_t is as defined by (5.2.7). Suppose that

$$\int_0^T Z_T^2(t)\lambda_t\, dt \xrightarrow{p} \sigma^2 \qquad \text{as } T \to \infty, \tag{5.2.11}$$

and, for every $\varepsilon > 0$,

$$\int_0^T E[Z_T^2(s)\lambda_s I(|Z_T(s)| > \varepsilon)]\, ds \to 0 \qquad \text{as } T \to \infty. \tag{5.2.12}$$

Then

$$\int_0^T Z_T(s)\, dM_s \xrightarrow{\mathcal{L}} N(0, \sigma^2) \qquad \text{as } T \to \infty. \tag{5.2.13}$$

In addition to condition (TT0), we assume that:

(TT1) $E\left(\int_0^T \delta_T^2(x - X_s)\lambda_s\, ds\right) < \infty;$

(TT2) $E(N_T) = E\left(\int_0^T \lambda_s\, ds\right) < \infty.$

Under these conditions, it is easy to see that

$$E[\hat{f}_T(x)] = E[f_T^*(x)] \tag{5.2.14}$$

and the following result can be obtained by using Theorem 5.2.1.

Theorem 5.2.2 *(i) If*

$$\frac{1}{T^2}\int_0^T E[\delta_T^2(x - X_s)\lambda_s]\, ds \to 0 \qquad \text{as } T \to \infty, \tag{5.2.15}$$

then

$$\hat{f}_T(x) - f_T^*(x) \xrightarrow{\text{q.m.}} 0 \qquad \text{as } T \to \infty. \tag{5.2.16}$$

(ii) Suppose there exists $\nu_T > 0$ such that

$$\frac{\nu_T^2}{T^2}\int_0^T \delta_T^2(x - X_s)\lambda_s\, ds \xrightarrow{p} \sigma^2 > 0 \qquad \text{as } T \to \infty \tag{5.2.17}$$

and, for every $\varepsilon > 0$,

$$\frac{\nu_T^2}{T^2}\int_0^T \delta_T^2(x - X_s)\lambda_s I\{|\delta_T(x - X_s)| > \varepsilon T \nu_T^{-1}\}\, ds \to 0 \qquad \text{as } T \to \infty. \tag{5.2.18}$$

Then

$$\nu_T(\hat{f}_T(x) - f_T^*(x)) \xrightarrow{\mathcal{L}} N(0, \sigma^2) \qquad \text{as } T \to \infty.$$

Remark. Depending on whether the intensity process $\{\lambda_s\}$ is nonrandom or random but independent of the process $\{X_t\}$, a more detailed analysis can be made. We refer the reader to Prakasa Rao (1988b; 1990a) for further details.

5.3 Estimation of the Drift Coefficient

Consider the stochastic differential equation

$$dX_t = a(X_t)\,dt + dW_t, \qquad X_0 = x, t \geq 0.$$

Suppose the process is sampled at the time instants $t_i, 1 \leq i \leq n$. Let $X_i = X_{t_i}, 1 \leq i \leq n$. An estimator for $a(x)$ is

$$a_n(x) = \left[\frac{1}{2}\sum_{i=1}^{n}\frac{1}{h_i}K_2^{(1)}\left(\frac{x - X_i}{h_i}\right)\right]\left[\varepsilon + \sum_{i=1}^{n}K_1\left(\frac{x - X_i}{h_i}\right)\right]^{-1}, \tag{5.3.1}$$

where the kernels $K_1(\cdot)$ and $K_2(\cdot)$ are bounded densities on R with a bounded derivative $K_2^{(1)}$ almost everywhere for K_2. The sequence $\{h_i, i \geq 1\}$ is a decreasing sequence of positive numbers and ε is a fixed positive number. This estimator is motivated by the fact that

$$a(x) = \frac{1}{2}\frac{f^{(1)}(x)}{f(x)}, \tag{5.3.2}$$

where f is the stationary initial density of the process, assuming that it exists, and $f^{(1)}(x)$ is the derivative of $f(x)$. Banon (1977) investigated the weak consistency and asymptotic normality of such estimators. For simulation purposes, one can use,

$$t_i = iT, \quad T > 0, \qquad h_i = h_1 i^{-\beta}, \quad \text{where } h_1 > 0, \beta = \tfrac{1}{5},$$

$$K_1(x) = \begin{cases} x & \text{for } -1 \leq x \leq 1, \\ 0 & \text{otherwise,} \end{cases}$$

and

$$K_2(x) = \begin{cases} 2(1 - |x|) & \text{for } -1 \leq x \leq 1, \\ 0 & \text{otherwise.} \end{cases}$$

Let $a_{n1}(x)$ be the estimator so obtained.

Since the drift $a(x)$ can also be obtained as a limit of the conditional expectation, namely,

$$a(x) = \lim_{t \to 0}\frac{1}{t}E(X_t - X_0 | X_0 = x), \tag{5.3.3}$$

another estimator for $a(x)$, at x for which $f(x) > 0$, can be chosen, using a nonparametric method for estimating a regression function, as

$$\hat{a}_n(x) = \left[\sum_{i=1}^{n}(X_{i+1} - X_i)K\left(\frac{x - X_i}{h_i}\right)\right]\left[\varepsilon + \sum_{i=1}^{n}d_i K\left(\frac{x - X_i}{h_i}\right)\right]^{-1}, \tag{5.3.4}$$

where K is a square-integrable kernel with bounded support, $0 < d_i \downarrow 0$ as $i \to \infty$, $t_i = \sum_{j=1}^{i} d_j$, $h_i = h(t_i)$, where $h : R^+ \to R^+$ with $h(t) \downarrow 0$ as $t \to \infty$, and $\varepsilon > 0$. This estimator was suggested by Nguyen and Pham Dinh Tuan (1982). They showed that this estimator is asymptotically normal under some conditions. For simulation purposes, let us chose

$$d_i = d_1 i^{-\alpha}, \qquad h(t) = h_1 t^{-\beta},$$

where $\alpha = 2/5$ and $\beta = 1/3$. Let $a_{n2}(x)$ be the corresponding estimator with these specific values for α and β.

Another estimator based on the nonparametric estimation of the regression function

$$E(X_{t+\Delta}|X_t = x)$$

given a Markov sequence sampled from the stationary Markov process $\{X_t\}$ was suggested by Pham Dinh Tuan (1981). We do not discuss the details here.

Observing again that $a(x)$ is the limit of a conditional expectation, Geman (1979) suggested the estimator

$$\tilde{a}_n(x) = \frac{1}{n}\sum_{i=1}^{n}\frac{1}{d_i}(X_i - x), \tag{5.3.5}$$

where $d_i \downarrow 0$ as $i \to \infty$ and $X_i = X_{\tau_i + d_i}$, with $\{\tau_i\}$ a sequence of stopping times such that

$$\tau_1 = \inf\{t \geq 0,\, X_t = x\}, \qquad \tau_{i+1} = \inf\{t \geq \tau_i + d_i,\, X_t = x\}. \tag{5.3.6}$$

He has proved that $\tilde{a}_n(x)$ is consistent and asymptotically normal under some conditions. Let $d_i = d_1 i^{-\alpha}$. The optimal choice of α turns out to be $\alpha = 1/3$ in the sense of minimizing the mean square error. Let $a_{n3}(x)$ be the estimator so obtained with $\alpha = 1/3$.

Banon and Nguyen (1981a; 1981b) observed, through simulation studies, that the estimator $a_{n1}(x)$ gives the 'best' results among $a_{ni}(x)$, $1 \leq i \leq 3$. If $\sigma(\cdot) = \sigma$ is unknown but constant in the stochastic differential equation

$$dX_t = a(X_t)\,dt + \sigma(X_t)\,dW_t, \qquad X_0 = x,\, t \geq 0, \tag{5.3.7}$$

then one can use either $a_{n2}(x)$ or $a_{n3}(x)$ but the computation of $a_{n3}(x)$ is expensive. For further details about simulation studies, see Banon and Nguyen (1981a; 1981b).

Let us now consider the following multidimensional version of the above problem. Consider the stochastic differential equation

$$d\boldsymbol{X}(t) = \boldsymbol{a}(\boldsymbol{X}(t))\,dt + \boldsymbol{\sigma}(\boldsymbol{X}(t))\,d\boldsymbol{W}(t), \qquad t \geq 0, \boldsymbol{X}(0) = 0, \tag{5.3.8}$$

where $\boldsymbol{a} : R^d \to R^d$, $\boldsymbol{\sigma} : R^d \to R^d \times R^d$ are nonrandom functions and $\boldsymbol{W}$ is a d-dimensional vector of independent standard Wiener process components. We at first assume that $\sigma(\boldsymbol{x}) \equiv \boldsymbol{I}$, the identity matrix of order $d \times d$.

It is known that, if $P_{\boldsymbol{X}}^T$ and $P_{\boldsymbol{W}}^T$ denote the probability measures induced by $\boldsymbol{X}$ and $\boldsymbol{W}$ respectively on $((C[0, T])^d, \mathcal{B}_T)$, then

$$\log \frac{dP_{\boldsymbol{X}}^T}{dP_{\boldsymbol{W}}^T} = \int_0^T \langle \boldsymbol{a}(\boldsymbol{X}(s)), d\boldsymbol{X}(s)\rangle - \frac{1}{2}\int_0^T \|\boldsymbol{a}(\boldsymbol{X}(s))\|^2\,ds, \tag{5.3.9}$$

where $\langle \cdot, \cdot \rangle$ denotes the inner product and $\|\cdot\|$ denotes the norm in R^d under some conditions (Basawa and Prakasa Rao 1980). We assume that sufficient conditions hold for the existence of $dP_{\boldsymbol{X}}^T / dP_{\boldsymbol{W}}^T$.

Suppose the process $\boldsymbol{X}$ is observed at the times

$$0 \leq t_1 < t_2 < \cdots < t_n = T.$$

An obvious approximation for (5.3.9) based on the discrete set of observations $\boldsymbol{X}(t_i) = \boldsymbol{X}_i$, $1 \leq i \leq n$, is

$$\begin{aligned}\log \frac{dP_{\boldsymbol{X}}^T}{dP_{\boldsymbol{W}}^T} &\simeq \sum_{i=1}^{n} \langle \boldsymbol{a}(\boldsymbol{X}_{i-1}), \boldsymbol{X}_i - \boldsymbol{X}_{i-1}\rangle - \frac{1}{2}\sum_{i=1}^{n} \|\boldsymbol{a}(\boldsymbol{X}_{i-1})\|^2 (t_i - t_{i-1}) \\ &= \sum_{i=1}^{n} \langle \boldsymbol{a}(\boldsymbol{X}_{i-1}), \boldsymbol{X}_i - \boldsymbol{X}_{i-1} - \frac{1}{2}\boldsymbol{a}(\boldsymbol{X}_{i-1})(t_i - t_{i-1})\rangle. \end{aligned} \tag{5.3.10}$$

A smoothed version of the approximation (5.3.10) can be obtained by using a family of kernels $K_\lambda : R^d \to R^d \otimes R^d, \lambda > 0$. Let $K_\lambda(Y) = \{K_\lambda(Y)_{ij} : 1 \le i, j \le d\}$ be such that

$$\int_{R^d} K_\lambda(Y)_{ij}\, dY = 1, \qquad 1 \le i,\ j \le d, \tag{5.3.11a}$$

$$\lim_{\lambda\to\infty} \int_{R^d - B(0,\delta)} K_\lambda(Y)_{ij}\, dY = 0, \qquad 1 \le i,\ j \le d,\ \delta > 0, \tag{5.3.11b}$$

where $B(0, \delta)$ denotes the closed sphere with centre 0 and radius δ in R^d, and

$$K_\lambda(Y) \text{ is nonnegative definite for } Y \in R^d. \tag{5.3.11c}$$

It can be checked that the conditions (5.3.11a) and (5.3.11b) imply that

$$\lim_{\lambda\to\infty} \int_{R^d} K_\lambda(y - x) f(x)\, dx = f(y), \qquad x \in R^d \tag{5.3.12}$$

(Prakasa Rao 1983a) when $f(\cdot)$ is continuous at y. A smoothed version of the approximation given by (5.3.10) is

$$\log \frac{dP_X^T}{dP_W^T} \simeq \int_{R^d} \left\langle a(x), \sum_{j=1}^{n} K_\lambda(x - X_j)(X_j - X_{j-1}) \right\rangle dx$$
$$-\frac{1}{2} \int_{R^d} \left\langle a(x), \sum_{j=1}^{n} K_\lambda(x - X_j) a(x)(t_j - t_{j-1}) \right\rangle dx. \tag{5.3.13}$$

Minimization of this function with respect to $a(\cdot)$ leads to the equation

$$\left[\sum_{j=1}^{n} K_\lambda(x - X_j)(X_j - X_{j-1})\right] = \left[\sum_{j=1}^{n} K_\lambda(x - X_j)(t_j - t_{j-1})\right] \hat{a}(x). \tag{5.3.14}$$

Let

$$K_{x,\lambda}(x) = \sum_{j=1}^{n} K_\lambda(x - X_j)(X_j - X_{j-1}) \tag{5.3.15}$$

and

$$K_{\lambda,t}(x) = \sum_{j=1}^{n} K_\lambda(x - X_j)(t_j - t_{j-1}).$$

Then (5.3.14) can be written in the form

$$K_{\lambda,t}(x)\hat{a}(x) = K_{x,\lambda}(x). \tag{5.3.16}$$

Hence

$$\hat{a}(x) = K_{\lambda,t}(x)^{-} K_{x,\lambda}(x), \tag{5.3.17}$$

where $K_{\lambda,t}(x)^{-}$ is a generalized inverse of $K_{\lambda,t}(x)$. Eplett (1987) called $\hat{a}(x)$ defined by (5.3.17) as the *kernel maximum likelihood estimator* of $a(x)$.

Suppose $K_\lambda(x) = \lambda K(\lambda x) I$, where I is the identity matrix of order $d \times d$ and $K(\cdot)$ is a probability density over R^d. Further suppose that $t_i = i\Delta, 1 \le i \le n$, where $\Delta > 0$. Then $K_{\lambda,t}(x) = K_{\lambda,\Delta}(x) I$, where

$$K_{\lambda,\Delta}(x) = \Delta \sum_{j=1}^{n} K(\lambda(x - X_j)).$$

Then (5.3.17) reduces to

$$\hat{a}(x) = \begin{cases} (K_{\lambda,\Delta}(x))^{-1} K_{x,\lambda}(x), & \text{if } K_{\lambda,\Delta}(x) > 0, \\ 0, & \text{if } K_{\lambda,\Delta}(x) = 0. \end{cases} \tag{5.3.18}$$

A more general version of this estimator can be obtained in the following manner. For each $\lambda > 0$, let $f_\lambda : \{1, \ldots, d\} \times R^d \to R$ be such that

$$\begin{aligned} &f_\lambda(i, \cdot) \in C_{2,b}(R^d), && 1 \le i \le d, \\ &f_\lambda(i, x) = x_i, && 1 \le i \le d, x \in B(0, \beta_\lambda) \end{aligned} \tag{5.3.19}$$

where $\beta_\lambda \uparrow \infty$ as $\lambda \to \infty$.

Note that $f_\lambda(\cdot, x) \to x$ as $\lambda \to \infty$. Here $C_{2,b}(R^d)$ denotes the class of bounded functions continuously differentiable up to the second order. A smoothed version of the estimator (5.3.18) can be chosen in the form

$$\begin{aligned} \hat{a}(x)_j = &\left\{ \sum_{i=1}^{n} K(\lambda(x - X_{t_{i-1}}))(f_\lambda(j, X_{t_i} - x) - f_\lambda(j, X_{t_{i-1}} - x)) \right\} \\ &\times \left\{ \sum_{i=1}^{n} K(\lambda(x - X_{t_{i-1}}))(t_i - t_{i-1}) \right\}^{-1} \end{aligned} \tag{5.3.20}$$

where $\hat{a}(x)_j$ denotes the jth component of $\hat{a}(x)$.

Suppose $t_i = i\Delta(T), 1 \le i \le n \equiv T(\Delta) = [T/\Delta(T)]$. Let $T \to \infty$ such that $\lambda(T) \to \infty$ and $\Delta(T) \to 0$. Let

$$\hat{a}_T(x) = a(x; \lambda, \Delta, T) \tag{5.3.21}$$

as defined by (5.3.20). Eplett (1987) obtained the strong consistency of the estimator (5.3.21) under some conditions on f_λ defined by (5.3.19) using the following martingale characterization (Stroock and Varadhan 1979).

It is known that $X(t)$ is a solution of (5.3.8) if and only if, for every $f \in C_{2,K}(R^d)$,

$$\left\{ f(X(t)) - f(X(0)) - \int_0^t (\zeta f)(X(s))\, ds, \mathcal{F}_t, t \ge 0 \right\} \tag{5.3.22}$$

is a martingale with respect to $\mathcal{F}_t = \sigma\{X(s) : 0 \le s \le t\}$, where ζ is the infinitesimal generator of $\{X_t\}$ acting on $f \in C_{2,K}(R^d)$ defined by

$$(\zeta f)(x) = \sum_{i=1}^{d} a(x)_i \frac{\partial f}{\partial x_i}(x) + \frac{1}{2} \sum_{i=1}^{d} \sum_{j=1}^{d} \sigma(x)_{i,j} \frac{\partial^2 f}{\partial x_i \partial x_j}(x), \quad x \in R^d, \tag{5.3.23}$$

and $K \subset R^d$ is a compact set with the property

$$P(X(t) \in K, 0 \le t \le T) = 1. \tag{5.3.24}$$

Consider now the general case when the diffusion coefficient $\boldsymbol{\sigma}(\cdot)$ is unknown.

For estimating $\boldsymbol{\Sigma} = \boldsymbol{\sigma}\boldsymbol{\sigma}'$, where $\boldsymbol{\sigma}$ is the diffusion coefficient, Eplett (1987) suggested the estimator

$$\hat{\sum}(\boldsymbol{x}) = [K_{\lambda,T}(\boldsymbol{x})]^{-}\langle K_\lambda\rangle(\boldsymbol{x}), \qquad \boldsymbol{x} \in R^d, \tag{5.3.25}$$

where

$$\langle K_\lambda\rangle(\boldsymbol{x}) = \sum_{i=1}^{n} K_\lambda(\boldsymbol{x} - \boldsymbol{X}_{t_{i-1}})(\boldsymbol{X}_{t_i} - \boldsymbol{X}_{t_{i-1}})' \tag{5.3.26}$$

(here prime denotes the transpose of a vector).

The motivation for choosing (5.3.25) as an estimator for $\hat{\boldsymbol{\Sigma}}(\boldsymbol{x})$ is that

$$\lim_{n\to\infty} \sum_{j=1}^{n} (\boldsymbol{X}_{t_j} - \boldsymbol{X}_{t_{j-1}})(\boldsymbol{X}_{t_j} - \boldsymbol{X}_{t_{j-1}})' = \int_0^T \boldsymbol{\Sigma}(\boldsymbol{X}_s)\,ds \text{ a.s.} \tag{5.3.27}$$

for $0 < T < \infty$. Eplett (1987) obtained the strong consistency of the estimator $\hat{\boldsymbol{\Sigma}}(\boldsymbol{x})$ as $T \to \infty$ such that $\Delta(T) \to 0$ and $\lambda(T) \to \infty$. We omit the details.

5.4 Estimation of the Diffusion Coefficient

Consider the stochastic differential equation

$$dX_t = a(X_t)\,dt + \sigma(X_t)\,dW_t, \qquad t \geq 0, \tag{5.4.1}$$

with $X_0 = x$. We now study the problem of estimation of diffusion coefficient $\sigma(\cdot)$ from a discrete set of observations of the process $\{X_t\}$. Under some regularity conditions, it is known that $\{X_t\}$ is a stationary Markov process and

$$\sigma^2(x) = \lim_{t\to 0} \frac{(E(X_{t+s} - X_t)^2 | X_s = x)}{t}, \qquad x \in R, s \geq 0. \tag{5.4.2}$$

If $\sigma(\cdot)$ is a constant σ, then σ^2 can be estimated by

$$\sigma_n^2 = \frac{1}{n}\sum_{i=1}^{n} \frac{1}{\tau_i}(X_{t_i+\tau_i} - X_{t_i})^2, \tag{5.4.3}$$

where $\{\tau_i\}$ is a bounded sequence of positive numbers tending to zero and $\{t_i\}$ is a sequence such that $t_i + \tau_i \leq t_{i+1}$. It can be shown that

$$\sigma_n^2 \xrightarrow{\text{q.m.}} \sigma^2 \qquad \text{as } n \to \infty \tag{5.4.4}$$

if $E(X^4) < \infty$ by methods in Wong and Zakai (1965), and this estimator is also recursive.

The problem of estimating the diffusion coefficient has recently been of major interest in view of the application of diffusion processes to the modeling of financial data (see Black and Scholes 1973). In option pricing theory (Black and Scholes 1973) or in modeling interest rates (Vasicek 1977; Schaefer and Schwartz 1984; Cox *et al.* 1985), an important point is that the time-dependent diffusion coefficient is unknown. Its estimation is of importance for decision making in the financial markets. We have discussed some results, due to Dacunha-Castelle and Florens-Zmirou (1986) and Dohnal (1987), in this connection in Chapter 4.

5.4.1 Method of Nearest Neighbors

Consider the process $\{X_t\}$ defined by

$$dX_t = a(X_t) + \sigma(X_t)\,dW_t, \qquad t \in [0, 1], X_0 = x_0, \tag{5.4.5}$$

where $a(\cdot)$ and $\sigma(\cdot)$ are unknown. Suppose $a(\cdot)$ is twice differentiable with bounded derivatives and $\sigma(\cdot)$ is unknown with three continuous and bounded derivatives such that $0 < k \leq \sigma(x) \leq K < \infty$, Suppose the process is observed at the times $t_i = i/n, 1 \leq i \leq n, n \geq 1$. Let P^{x_0} denote the probability measure of the process $\{X_t\}$ and $I(A)$ denote the indicator function of a set A.

Formally, let

$$L_t(x) = \lim_{\delta \to 0} \frac{1}{2\delta} \int_0^t I(|X_s - x| < \delta)\,ds, \tag{5.4.6}$$

assuming that the limit exists in the sense of almost sure convergence. Then $L_t(x)$ defines the *local time* of X in x during $[0, t]$ and, for any positive sequence $\{h_n\}$,

$$L_t^{(n)}(x) = \frac{1}{2nh_n} \sum_{i=0}^{[nt]-1} I(|X_{(i/n)} - x| < h_n) \tag{5.4.7}$$

is a discrete approximation to $L_t(x)$. Let $h_n \to 0$ and

$$S_n(x) = n \frac{\sum_{i=1}^{n} I(|X_{(i/n)} - x| < h_n)[X_{((i+1)/n)} - X_{(i/n)}]^2}{\sum_{i=1}^{n-1} I(|X_{(i/n)} - x| < h_n)}. \tag{5.4.8}$$

The function $S_n(x)$ can be considered as estimator for $\sigma^2(x)$ and it can be shown that, conditionally on the event that 'x is visited',

$$\sqrt{nh_n}\left(\frac{S_n(x)}{\sigma^2(x)} - 1\right) \xrightarrow{\mathcal{L}} [L(x)]^{-1/2} Z \qquad \text{as } n \to \infty, \tag{5.4.9}$$

where Z is a standard normal random variable independent of the local time $L(x) \equiv L_1(x)$. The limiting distribution is mixed normal. We now sketch a proof of this result. For full details, see Florens-Zmirou (1993).

Lemma 5.4.1 *If $nh_n^4 \to 0$ as $n \to \infty$, then $L_t^{(n)}(x) \xrightarrow{\text{q.m.}} L_t(x)$, the local time under P^{x_0}.*

Lemma 5.4.2 *If $(nh_n^2)^{-1} \log n \to 0$ as $n \to \infty$, then $L_t^{(n)}(x) \xrightarrow{\text{a.s.}} L_t(x)$ as $n \to \infty$ under P^{x_0}.*

Lemma 5.4.3 *If $nh_n^4 \to 0$ as $n \to \infty$, then*

$$V_t^{(n)}(x) \xrightarrow{\text{q.m.}} \sigma^2(x) L_t(x)$$

under P^{x_0}, where

$$V_t^{(n)}(x) = \frac{1}{2h_n} \sum_{i=1}^{[nt]-1} I(|X_{(i/n)} - x| < h_n)[X_{((i+1)/n)} - X_{(i/n)}]^2.$$

For proofs of Lemmas 5.4.1–5.4.3, see Florens-Zmirou (1993).

Applying Lemmas 5.4.1–5.4.3, the following main results can be proved.

Theorem 5.4.4 *If $h_n \to 0$ and $nh_n^4 \to 0$ as $n \to \infty$, then $S_n(x)$ is a consistent estimator of $\sigma^2(x)$. Further, if $nh_n^3 \to 0$, then*

$$\sqrt{nh_n}\left(\frac{S_n(x)}{\sigma^2(x)} - 1\right) \xrightarrow{\mathcal{L}} L(x)^{-1/2}Z \qquad \text{as } n \to \infty$$

under P^{x_0}, where Z is $N(0, 1)$ independent of $L(x)$.

Theorem 5.4.4 is a consequence of the following result. Let

$$M_t^{(n)} = \sqrt{\frac{n}{2h_n}} \sum_{i=0}^{[nt]-1} I(|X_{(i/n)} - x| < h_n)\left[(X_{((i+1)/n)} - X_{(i/n)})^2 - \frac{\sigma^2(x)}{n}\right]$$
$$= \sum_{i=0}^{[nt]-1} m_i \quad \text{(say)}. \tag{5.4.10}$$

Let

$$W_t^{(n)} = \sum_{i=0}^{[nt]-1} w_i, \quad w_i = W_{((i+1)/n)} - W_{(i/n)}. \tag{5.4.11}$$

Theorem 5.4.5 *If $nh_n^3 \to 0$ as $n \to \infty$, then the sequence of processes $(M_t^{(n)}, W_t^{(n)})_{0\le t\le 1}$ defined above converges in distribution to the process $(B_{2\sigma^4 L_t(x)}, W_t)_{0\le t\le 1}$, where $\{B_t, 0 \le t \le 1\}$ and $\{W_t, 0 \le t \le 1\}$ are independent standard Wiener processes. In particular*

$$M_1^{(n)} \xrightarrow{\mathcal{L}} \sqrt{2}\sigma^2(x)\sqrt{L_1(x)}Z \qquad \text{as } n \to \infty \tag{5.4.12}$$

under P^{x_0}, where Z is $N(0, 1)$ independent of $L_1(x)$.

Observe that

$$\sqrt{2nh_n}\left(\frac{S_n(x)}{\sigma^2(x)} - 1\right) = \frac{M_1^{(n)}}{\sigma^2(x)L_1^{(n)}(x)}. \tag{5.4.13}$$

The results stated above make use of the work on local times due to Yor (1978) and Knight (1971).

Brugière (1993) extended the above results to the multidimensional diffusion processes. We briefly sketch these results.

Consider the diffusion process $\boldsymbol{X} = \{\boldsymbol{X}_s, 0 \le s \le 1\}$ in R^d satisfying the stochastic differential equation

$$d\boldsymbol{X}_t = \boldsymbol{a}(\boldsymbol{X}_t)\,dt + \boldsymbol{\sigma}(\boldsymbol{X}_t)\,d\boldsymbol{W}_t, \qquad \boldsymbol{X}_0 = \boldsymbol{x}_0,\ 0 \le t \le 1. \tag{5.4.14}$$

Let $P^{\boldsymbol{x}_0}$ be the probability measure generated by the process $\boldsymbol{X}$ with $\boldsymbol{X}_0 = \boldsymbol{x}_0$ and suppose the process is observed at the times i/n, $0 \le i \le n$. Define

$$\boldsymbol{\gamma}_n(\boldsymbol{x}) = \frac{n\sum_{i=1}^{n-1} I\{\|\boldsymbol{X}_{(i/n)} - \boldsymbol{x}\| < h_n\}(\boldsymbol{X}_{((i+1)/n)} - \boldsymbol{X}_{(i/n)})(\boldsymbol{X}_{((i+1)/n)} - \boldsymbol{X}_{(i/n)})'}{\sum_{i=1}^{n-1} I\{\|\boldsymbol{X}_{(i/n)} - \boldsymbol{x}\| < h_n\}}, \tag{5.4.15}$$

where $0 < h_n \to 0$ as $n \to \infty$. We study the properties of $\boldsymbol{\gamma}_n(\boldsymbol{x})$ as an estimator of $\boldsymbol{\gamma}(\boldsymbol{x}) = \boldsymbol{\sigma}(\boldsymbol{x})\boldsymbol{\sigma}'(\boldsymbol{x})$, conditional on the event that '$\boldsymbol{x}$ is visited'. Suppose $\boldsymbol{a}$ and $\boldsymbol{\sigma}$ are bounded and Lipschitz. Further suppose that $\boldsymbol{a}$ and $\boldsymbol{\sigma}$ are C^∞ functions. In addition, assume that $\boldsymbol{\gamma}(\boldsymbol{x}) = \boldsymbol{\sigma}(\boldsymbol{x})'\boldsymbol{\sigma}(\boldsymbol{x})$ is invertible and the family $\{\boldsymbol{\gamma}^{-1}(\boldsymbol{x}), \boldsymbol{x} \in R^d\}$ is norm bounded. Let

$$\begin{aligned}(\Delta_n \boldsymbol{X})_{(i/n)} &= \boldsymbol{X}_{(i+1/n)} - \boldsymbol{X}_{(i/n)},\\ Z^{\boldsymbol{X}}_{n,(i/n)}(\boldsymbol{x}) &= \boldsymbol{\gamma}^{-1}(\boldsymbol{x})[(\Delta_n \boldsymbol{X})'_{(i/n)}(\Delta_n \boldsymbol{X})_{(i/n)} - \boldsymbol{\gamma}(\boldsymbol{x})],\\ \chi^{\boldsymbol{X}}_{n,s}(\boldsymbol{x}) &= I\{\|\boldsymbol{X}_s - \boldsymbol{x}\| < h_n\},\\ N^{\boldsymbol{X}}_{n,t}(\boldsymbol{x}) &= \sum_{i=1}^{[nt]-1} \chi^{\boldsymbol{X}}_{n,\left(\frac{i}{n}\right)}(\boldsymbol{x}).\end{aligned} \tag{5.4.16}$$

Define

$$\begin{aligned}\alpha_n^{(1)} &= 2h_n \qquad \text{for } d = 1,\\ \alpha_n^{(2)} &= h_n^2 \log\left(\frac{1}{h_n^2}\right) \qquad \text{for } d = 2,\\ \alpha_n^{(d)} &= h_n^2 \qquad \text{for } d \geq 3.\end{aligned} \tag{5.4.17}$$

Define

$$\begin{aligned}L^{\boldsymbol{X}}_{n,t}(\boldsymbol{x}) &= \frac{1}{n\alpha_n^{(d)}} N^{\boldsymbol{X}}_{n,t}(\boldsymbol{x}), \text{ and}\\ M^{\boldsymbol{X}}_{n,t}(\boldsymbol{x}) &= \frac{1}{\sqrt{n\alpha_n^{(d)}}} \sum_{i=1}^{[nt]-1} \chi^{\boldsymbol{X}}_{n,(i/n)}(\boldsymbol{x}) Z^{\boldsymbol{X}}_{n,(i/n)}(\boldsymbol{x}).\end{aligned} \tag{5.4.18}$$

If $t = 1$, we omit the index t and $\boldsymbol{X}$ throughout the above notation.

The following theorem is due to Brugière (1993).

Theorem 5.4.6 *Suppose $nh_n^2 \to \infty$ and $n\alpha_n^{(d)} h_n^2 \to 0$ as $n \to \infty$. Then, for $d \geq 2$,*

$$\sqrt{N_n(\boldsymbol{x}_0)}[\boldsymbol{\gamma}^{-1}(\boldsymbol{x}_0)\boldsymbol{\gamma}_n(\boldsymbol{x}_0) - \boldsymbol{I}] \xrightarrow{\mathcal{L}} \sqrt{2}\boldsymbol{W} \qquad \text{as } n \to \infty \tag{5.4.19}$$

under $P^{\boldsymbol{x}_0}$, where $\boldsymbol{W}$ is $N(0, \boldsymbol{I})$. Furthermore,

$$(L_n(\boldsymbol{x}_0), M_n(\boldsymbol{x}_0)) \xrightarrow{\mathcal{L}} \left(L^{(d)}(\boldsymbol{x}_0), \sqrt{2L^{(d)}(\boldsymbol{x}_0)}\boldsymbol{W}^*\right) \qquad \text{as } n \to \infty \tag{5.4.20}$$

under $P^{\boldsymbol{x}_0}$, where

$$L^{(2)}(\boldsymbol{x}_0) = \frac{\varepsilon}{2(\det[\boldsymbol{\gamma}(\boldsymbol{x}_0)])^{1/2}}, \tag{5.4.21a}$$

with ε exponential with mean one, and

$$L^{(d)}(\boldsymbol{x}_0) = \int_0^\infty I[\|\boldsymbol{\sigma}(\boldsymbol{x}_0)\boldsymbol{W}_s^{(d)}\| < 1]\, ds, \qquad \text{for } d \geq 3, \tag{5.4.21b}$$

under $P^{\boldsymbol{x}_0}$. Here $\boldsymbol{W}^*$ is $N(0, \boldsymbol{I})$ independent of $L^{(d)}(\boldsymbol{x}_0)$ and $\boldsymbol{W}^{(d)}$ is the standard Wiener process on R^d.

Remarks. Brugière (1993) utilizes the estimates on Markov transition density function from Friedman (1964, p. 257) and a generalized version of the local time in the multidimensional case. For $d = 2$, the tools from Azencott (1984) and Ikeda and Watanabe (1981) are utilized; and for $d \geq 3$, the local L^2 approximation for paths of the diffusion processes are used. For details, see Brugière (1993). Brugière (1991) proved the consistency of the estimator without using the notion of local time (which does not exist for $d > 1$) and by using the results on the modulus of continuity of the diffusion processes. Theorem 5.4.6 gives asymptotic properties of $\boldsymbol{\gamma}_n(\boldsymbol{x})$ at $\boldsymbol{x} = \boldsymbol{x}_0$ where $\boldsymbol{x}_0$ is the initial state. If $\boldsymbol{x} \neq \boldsymbol{x}_0$, Brugière indicates that the results can be extended by using the Markov property of the diffusion process.

5.4.2 Method of Wavelets

Consider the stochastic differential equation

$$dX_t = a(t, X_t)\,dt + \theta(t)\,dW_t, \qquad X_0 = x_0, 0 \le t \le T \tag{5.4.22}$$

where $\{W_t\}$ is the standard Wiener process, $x_0 \in R$ is known, and the functions $a(t,u)$ and $\theta(t)$ are unknown. We assume that sufficient conditions on the coefficients $a(t,u)$ and $\theta(t)$ hold so that there exists a unique solution for (5.4.22). Suppose that a sample path of the process $\{X(t), 0 \le t \le T\}$ is observed at the time instants $t_i = i2^{-n}$, $i = 0, 1, \ldots, [2^n T]$ in the interval $[0, T]$. We will now construct an estimator of $\theta(t)$ based on the method of wavelets and study its asymptotic properties. For a brief discussion of wavelets, see Appendix C.

Assume that the following conditions hold:

(UU1) The function $a(t,u)$ belong to $C^1([0,\infty) \times R)$.

(UU2) For all $T > 0$, there exists a constant K_T such that for all $t, 0 \le t \le T$,

$$|a(t,(u)| \le K_T(1 + |u|), \qquad u \in R.$$

(UU3) The function $\theta(t)$ belongs to $C^m([0,\infty))$ for $m \ge 1$ and $\theta(t)$ is positive for all $t \ge 0$.

Let $\{V_j, -\infty < j < \infty\}$ be an increasing sequence of closed subspaces of $L^2(R)$ and suppose that $\{V_j, -\infty < j < \infty\}$ is an r-regular multiresolution analysis of $L^2(R)$ such that the associated scale function ϕ and the wavelet function ψ are compactly supported and belong to $C^r(R)$ (Daubechies 1988; Meyer 1990).

Let the subspace W_j be defined by

$$V_{j+1} = V_j \oplus W_j, \tag{5.4.23}$$

where $\oplus$ denotes the direct sum. Define

$$\phi_{j,k}(x) = 2^{j/2}\phi(2^j x - k), \qquad -\infty < j, k < \infty, \tag{5.4.24}$$

and

$$\psi_{j,k}(x) = 2^{j/2}\psi(2^j x - k), \qquad -\infty < j, k < \infty. \tag{5.4.25}$$

It is known that (i) for $-\infty < j < \infty$, $\{\phi_{j,k}(x), -\infty < k < \infty\}$ is an orthonormal basis for V_j, and $\{\psi_{j,k}(x), -\infty < k < \infty\}$ is an orthonormal basis for W_j; and (ii) the family $\{\psi_{j,k}, -\infty < j, k < \infty\}$ is an orthonormal basis for $L^2(R)$. This basis of $L^2(R)$ is called the *wavelet basis*.

We associate a resolution $j(n)$ for each n such that $j(n) \to \infty$ as $n \to \infty$ and estimate the projection of $\theta^2(\cdot)$ on $V_{j(n)}$. Since the function $\theta^2(\cdot)$ does not necessarily belong to $L^2(R)$, we replace θ^2 by a function $\bar{\theta}^2 \in L^2(R)$ such that

$$\bar{\theta}(t) = \theta(t) \qquad \text{for all } t \in [0, T]. \tag{5.4.26}$$

We assume that $\bar{\theta}(\cdot) \in C^m(R)$ has a compact support contained in $[-\varepsilon, T+\varepsilon]$ for some $\varepsilon > 0$. It is easy to see that the diffusion process $\{\bar{X}_t\}$ satisfying

$$d\bar{X}_t = a(t, \bar{X}_t)\,dt + \bar{\theta}(t)\,dW_t, \qquad \bar{X}_0 = x_0, \tag{5.4.27}$$

coincides with $\{X_t\}$ satisfying (5.4.22) on $[0, T]$ from the uniqueness of the solution. Using the decomposition

$$L^2(R) = V_{j(n)} \oplus (\oplus_{j \ge j(n)} W_j), \tag{5.4.28}$$

the function $\bar{\theta}^2(\cdot)$ can be represented in the form

$$\bar{\theta}^2(t) = \sum_{k=-\infty}^{\infty} \alpha_{j(n),k}\phi_{j(n),k}(t) + \sum_{\substack{j\geq j(n)\\ k=-\infty}}^{\infty} \beta_{j,k}\psi_{j,k}(t), \tag{5.4.29}$$

where

$$\alpha_{j,k} = \int_{-\infty}^{\infty} \bar{\theta}^2(t)\phi_{j,k}(t)\,dt \tag{5.4.30}$$

and

$$\beta_{j,k} = \int_{-\infty}^{\infty} \bar{\theta}^2(t)\psi_{j,k}(t)\,dt. \tag{5.4.31}$$

Let us define an estimator $\hat{\alpha}_{j,k}$ of $\alpha_{j,k}$ by

$$\hat{\alpha}_{j,k} = \sum_{i=0}^{N-1} \phi_{j,k}(t_i)(X_{t_i+1} - X_{t_i})^2, \tag{5.4.32}$$

where

$$t_i = i2^{-n}, \qquad 0 \leq i \leq N = [2^n T]. \tag{5.4.33}$$

Even though the subspace V_j is not finite-dimensional, since $\bar{\theta}^2(\cdot)$ and $\phi(\cdot)$ are compactly supported, it follows from the definition of $\phi_{j,k}$ that the two sets $\{-\infty < k < \infty : \alpha_{j,k} \neq 0\}$ and $\{-\infty < k < \infty : \hat{\alpha}_{j,K} \neq 0\}$ are contained in a finite set with cardinality L_j for each j depending only on T and the support of ϕ. In fact $L_j = O(2^j)$. Define an estimator for $\theta^2(\cdot)$ by

$$\hat{\theta}^2(t) = \sum_{k\in L_{j(n)}} \hat{\alpha}_{j(n),k}\phi_{j(n),k}(t) = \sum_{k=-\infty}^{\infty} \hat{\alpha}_{j(n),k}\phi_{j(n),k}(t). \tag{5.4.34}$$

Observe that the right-hand side of (5.4.34) is a finite sum in view of the above remarks.

In addition to conditions (UU1)–(UU3), let $h(\cdot)$ be a function satisfying the following condition:

(UU4) $h(\cdot)$ is continuous on $[0, T]$ with compact support in $[0, T]$ and belongs to the Sobolev space $H^{m'}(R)$ with $m' > 1/2$.

If $s \geq 0$, the Sobolev space $H^s(R)$ is the set defined by

$$H^s(R) = \left\{ f \in L^2(R), \int_{-\infty}^{\infty} |\hat{f}(\xi)|^2(1+|\xi|^2)^s\,d\xi < \infty \right\}$$

and the norm in $H^s(R)$ is defined by $\|f\|_0 = (\int_{-\infty}^{\infty} |\hat{f}(\xi)|^2(1+|\xi|^2)^s\,d\xi)^{1/2}$. If $s < 0$, the space $H^s(R)$ is defined as the topological dual of $H^{-s}(R)$. Note that $\hat{f}(\xi) = \int_{-\infty}^{\infty} f(x)e^{-i\xi x}\,dx$.

Theorem 5.4.7 *Suppose that conditions (UU1)–(UU4) hold. Then*

$$2^{n/2} \int_{-\infty}^{\infty} h(t)(\hat{\theta}^2(t) - \theta^2(t))\,dt \xrightarrow{\mathcal{L}} N(0, D(\theta, h)) \qquad \text{as } n \to \infty, \tag{5.4.35}$$

where

$$D(\theta, h) = 2\int_0^T h^2(t)\theta^4(t)\,dt \tag{5.4.36}$$

provided $\min(r, m + r, m') > 2$, $j(n) = [qn]$ *with* $1/[2\min(r, m + r, m')] \le q < 1/4$.

Here r is the regularity of the multiresolution analysis, and m and m′ are the exponents of the Sobolev space for $\bar{\theta}^2$ *and h, respectively.*

The proof of Theorem 5.4.7 is based on the following result.

Lemma 5.4.8 *Under the assumptions of Theorem 5.4.7, the following decomposition holds:*

$$2^{n/2} \int_0^T h(t)(\hat{\theta}^2(t) - \theta^2(t))\,dt$$
$$= 2^{n/2} \sum_{i=0}^{n-1} h_{j(n)}(t_i)\left[\left(\int_{t_i}^{t_{i+1}} \theta(s)\,dW_s\right)^2 - \int_{t_i}^{t_{i+1}} \theta^2(s)\,ds\right] + o_p(1). \tag{5.4.37}$$

We now briefly sketch the proof of this representation. For complete details, see Genon-Catalot *et al.* (1992).

Let P_j denote the orthogonal projection on V_j and set $f_j = P_j f$ for any $f \in L^2(R)$. It is known that for any $f \in H^s(R)$,

$$\|f - P_j f\|_{L^2(R)} \le 2^{-j(s \wedge r)} \varepsilon_j, \tag{5.4.38}$$

where $\varepsilon_j \to 0$ as $j \to \infty$ (Meyer 1990, pp. 41–48). Here $s \wedge r$ denotes $\min(s, r)$. Using the decomposition (5.4.28), let $h(t)$ be represented in the form

$$h(t) = \sum_{k=-\infty}^{\infty} \mu_{j(n),k}\phi_{j(n),k}(t) + \sum_{\substack{k=-\infty \\ j \ge j(n)}}^{\infty} \nu_{j,k}\psi_{j,k}(t). \tag{5.4.39}$$

Since $\bar{\theta}(\cdot)$ and $\theta(\cdot)$ are equal on $[0, T]$ and the support of h is included in $[0, T]$, it follows that

$$\int_{-\infty}^{\infty} h(t)(\hat{\theta}^2(t) - \theta^2(t))\,dt = Z_n + T_n, \tag{5.4.40}$$

where

$$Z_n = \sum_{k=-\infty}^{\infty} \mu_{j(n),k}(\hat{\alpha}_{j(n),k} - \alpha_{j(n),k}) \tag{5.4.41}$$

and

$$T_n = -\sum_{\substack{k=-\infty \\ j \ge j(n)}}^{\infty} \nu_{j,k}\beta_{j,k}. \tag{5.4.42}$$

In view of (5.4.38), it follows that

$$\begin{aligned} |T_n| &\le \|h - h_j\|_2 \|\bar{\theta}^2 - P_j\bar{\theta}^2\|_2 \\ &\le 2^{-j(n)(\min(r,m+r,m'))}\varepsilon_{j(n)}, \end{aligned} \tag{5.4.43}$$

where $\varepsilon_{j(n)} \to 0$ as $n \to \infty$. Hence $\Delta_n^{-1/2} T_n \to 0$ for $\Delta_n = 2^{-n}$ provided

$$2^{(n/2) - j(n)\min(r,m+r,m')} = O(1). \tag{5.4.44}$$

Since ϕ and h are compactly supported, it can be checked that the support of $h_{j(n)}$ is included in $[0, T]$ for large n. Hence

$$\sum_{k=-\infty}^{\infty} \mu_{j(n),k}\alpha_{j(n),k} = \int_{-\infty}^{\infty} h_{j(n)}(t)\bar{\theta}^2(t)\,dt$$

$$= \int_0^T h_{j(n)}(t)\theta^2(t)\,dt. \tag{5.4.45}$$

Therefore

$$Z_n = \sum_{i=0}^{N-1} (X_{t_{i+1}} - X_{t_i})^2 h_{j(n)}(t_i) - \int_0^T h_{j(n)}(t)\theta^2(t)\,dt \tag{5.4.46}$$

from (5.4.32) and (5.4.41). Applying the Itô formula, we have

$$Z_n = 2\left(\sum_{i=0}^{N-1} \int_{t_i}^{t_{i+1}} (X_s - X_{t_i})\,dX_s\right) h_{j(n)}(t_i) + C_n, \tag{5.4.47}$$

where

$$C_n = \left(\sum_{i=0}^{N-1} \int_{t_i}^{t_{i+1}} \theta^2(t)\,dt\right) h_{j(n)}(t_i) - \int_0^T h_{j(n)}(t)\theta^2(t)\,dt. \tag{5.4.48}$$

Using the fact that ϕ is a C^1 function and compactly supported, it can be checked that

$$\Delta_n^{-1/2} C_n = o_p(1) \tag{5.4.49}$$

under the condition $2^{2j(n)-n/2} = o(1)$, which holds provided $j(n) - (n/4) \to -\infty$ as $n \to \infty$.

The term $Z_n - C_n$ can be written in the form

$$Z_n - C_n = \sum_{i=1}^{5} A_{in}, \tag{5.4.50}$$

where

$$\begin{aligned}
A_{1n} &= 2\sum_{i=0}^{N-1} h_{j(n)}(t_i) \int_{t_i}^{t_{i+1}} a(s, X_s)\left\{\int_{t_i}^{s} a(u, X_u)\,du\right\} ds,\\
A_{2n} &= 2\sum_{i=0}^{N-1} h_{j(n)}(t_i) \left\{\int_{t_i}^{t_{i+1}} \theta(u)(t_{i+1} - u)\,dW_u\right\} a(t_i, X_{t_i}),\\
A_{3n} &= 2\sum_{i=0}^{N-1} h_{j(n)}(t_i) \int_{t_i}^{t_{i+1}} (a(s, X_s) - a(t_i, X_{t_i}))\left\{\int_{t_i}^{s} \theta(u)\,dW_u\right\} ds,\\
A_{4n} &= 2\sum_{i=0}^{N-1} h_{j(n)}(t_i) \int_{t_i}^{t_{i+1}} \left(\int_{t_i}^{s} a(u, X_u)\,du\right)\theta(s)\,dW_s,\\
A_{5n} &= 2\sum_{i=0}^{N-1} h_{j(n)}(t_i) \int_{t_i}^{t_{i+1}} \left(\int_{t_i}^{s} \theta(u)\,dW_u\right)\theta(s)\,dW_s.
\end{aligned} \tag{5.4.51}$$

Integrating by parts of A_{5n} leads to the relation

$$A_{5n} = 2\sum_{i=0}^{N-1} h_{j(n)}(t_i)\left[\left(\int_{t_i}^{t_{i+1}} \theta(u)\,dW_u\right)^2 - \int_{t_i}^{t_{i+1}} \theta^2(u)\,du\right]. \tag{5.4.52}$$

In the light of earlier computations, Lemma 5.4.8 will be proved if $\Delta_n^{-1/2} \times \sum_{i=1}^{4} A_{in} = o_p(1)$. For detailed estimates leading to proof of this fact, see Genon-Catolot *et al.* (1992).

Theorem 5.4.7 is now a consequence of Lemma 5.4.8 provided that

$$\Delta_n^{-1/2} A_{5n} \xrightarrow{\mathcal{L}} N(0, D(\theta, h)) \qquad \text{as } n \to \infty, \tag{5.4.53}$$

where $D(\theta, h)$ is as defined in (5.4.36). This can be checked by noting that

$$\Delta_n^{-1/2} A_{5n} = \sum_{i=0}^{N-1} \xi_{in}, \tag{5.4.54}$$

where

$$\xi_{in} = \Delta_n^{-1/2} h_{j(n)}(t_i) \left[\left(\int_{t_i}^{t_{i+1}} \theta(u)\, dW_u \right)^2 - \int_{t_i}^{t_{i+1}} \theta^2(u)\, du \right] \tag{5.4.55}$$

with $\xi_{in}, 0 \le i \le N-1$, independent random variables with mean zero and variance given by

$$\operatorname{var}(\xi_{in}) = 2\Delta_n^{-1} h_{j(n)}^2(t_i) \left(\int_{t_i}^{t_{i+1}} \theta^2(u)\, du \right)^2, \qquad 0 \le i \le N-1. \tag{5.4.56}$$

Under the condition $\min(r, m+r, m') > 2$, $j(n) = [qn]$ with $1/[2\min(r, m+\gamma, m')] \le q < 1/4$, it can be shown that the family of random variables $\{\xi_{in}, 0 \le i \le N-1\}$ obey the Lyapunov condition of Lindeberg's theorem (Billingsley 1986) and (5.4.53) holds, completing the proof of Theorem 5.4.7.

Let

$$R_n = E\left[\int_{-\infty}^{\infty} (\hat{\theta}^2(t) - \theta^2(t))^2 \gamma(t)\, dt \right], \tag{5.4.57}$$

where $\gamma(t)$ is a nonnegative continuous function with the support contained in $(0, T)$. Note that R_n is the mean integrated squared error of the estimator $\hat{\theta}(t)$ corresponding to the weight function $\gamma(t)$. Note that

$$R_n = B_n^2 + M_n, \tag{5.4.58}$$

where

$$B_n^2 = \int_{-\infty}^{\infty} (E\hat{\theta}^2(t) - \theta^2(t))^2 \gamma(t)\, dt \tag{5.4.59}$$

and

$$M_n = E\left(\int_{-\infty}^{\infty} (\hat{\theta}^2(t) - E\hat{\theta}^2(t))^2 \gamma(t)\, dt \right). \tag{5.4.60}$$

Then

$$M_n \le c' D_n \tag{5.4.61}$$

where $c' = \sup_{t \in R} \gamma(t)$ and

$$D_n = E\left(\int_{-\infty}^{\infty} (\hat{\theta^2}(t) - E\hat{\theta}^2(t))^2\, dt \right). \tag{5.4.62}$$

The following result is due to Genon-Catalot *et al.* (1992). We omit the proof.

Theorem 5.4.9 *Suppose the conditions (UU1)–(UU3) hold and $j(n) - n/2 \to -\infty$ as $n \to \infty$. Then*

$$R_n = B_n^2 + M_n, \tag{5.4.63}$$

where

$$B_n^2 \le c(2^{4j(n)-2n} + 2^{-2j(n)\min(m,r)} + 2^{-n}) \tag{5.4.64}$$

and

$$D_n = 2^{j(n)-n}2\int_0^T \theta^4(t)\,dt + o(2^{j(n)-n}), \tag{5.4.65}$$

in which the constant c depends only on ϕ, γ and θ^2 and not on the drift function $a(t, u)$.

Remark. This method can also be used to construct an estimate $\hat{\theta}^2(t)$ of $\theta^2(t)$ for the stochastic differential equation

$$dX_t = a(t, X_t)\,dt + \theta(t)h(X_t)\,dW_t, \qquad t \ge 0,\ X_0 = x_0, \tag{5.4.66}$$

where h is a known positive function belonging to $C^2(R)$. This can be seen by using the transformation $Y_t = F(X_t)$, where $F(x) = \int_0^x du/h(u)$ and (5.4.66) is transformed into an equation of the form (5.4.22) under some conditions on a and h.

Note that the estimate $\hat{\theta}^2(t)$ given by (5.4.34) can also be written in the form

$$\hat{\theta}^2(t) = \sum_{i=0}^{N-1}(X_{t_{i+1}} - X_{t_i})^2 K_n(t_i, t), \tag{5.4.67}$$

where

$$K_n(s, t) = 2^{j(n)}K(2^{j(n)}s, 2^{j(n)}t) \tag{5.4.68}$$

with

$$K(s, t) = \sum_{k=-\infty}^{\infty}\phi(s-k)\phi(t-k). \tag{5.4.69}$$

Hence $\hat{\theta}^2(t)$ is similar to a kernel estimate with the kernel $K(s, t)$ and bandwidth $h_n = 2^{-j(n)}$. Nonparametric estimation of density, its derivatives and their functionals via wavelets are discussed in Masry (1994) and Prakasa Rao (1996; 19997b; 1997c).

5.4.3 Estimation by Matching the Drift and the Diffusion to the Marginal Density

Consider the stochastic differential equation

$$dX_t = a(X_t, \theta)\,dt + \sigma(X_t)\,dW_t, \qquad t \ge 0, \tag{5.4.70}$$

where $\{W_t, t \ge 0\}$ is the standard Wiener process and the functional form of $a(\cdot, \cdot)$ is known but $\sigma(\cdot)$ is unknown. Suppose the process $\{X_t\}$ satisfies the following conditions:

(VV1) (i) The drift function $a(\cdot, \theta)$ and the diffusion function $\sigma(\cdot)$ have continuous derivatives on $(0, \infty)$ of order $s \ge 2$.

(ii) $\sigma^2(\cdot) > 0$ on $(0, \infty)$.

(iii) The integral of

$$m(v) \equiv \left(\frac{1}{\sigma^2(v)}\right) \exp\left\{-\int_v^{\bar{\varepsilon}} (2a(u,\theta)\sigma^2(u))\, du\right\},$$

the speed measure, converges at both the boundaries of $(0, \infty)$.

(iv) The integral of

$$s(v) \equiv \exp\left\{\int_v^{\bar{\varepsilon}} \left(\frac{2a(u,\theta)}{\sigma^2(u)}\right) du\right\},$$

the scale measure, diverges at both the boundaries of $(0, \infty)$.

Here the choice of $\bar{\varepsilon}$ is fixed but arbitrary in $(0, \infty)$. Under conditions (VV1)(i) and (ii), the stochastic differential equation admits a unique stationary strong solution (Ait-Sahalia 1996b). By the mean-value theorem, condition (VV1)(i) implies the following local Lipschitz and growth conditions: for each parameter vector θ and each compact subset of $R_+ = [0, \infty)$ of the form $K = [\frac{1}{\alpha}, \alpha]$, $\alpha > 0$, there exist constants C_1 and C_2 such that for all x, y in K,

$$|a(x,\theta) - a(y,\theta)| + |\sigma(x) - \sigma(y)| \le C_1|x - y|,$$

and

$$|a(x,\theta)| + |\sigma(x)| \le C_2(| + |x|).$$

Under conditions (VV1)(iii) and (iv), the solution of (5.4.70) is a Markov process with time-homogeneous transition densities, the stationary density $\pi(\cdot)$ exists and is continuously differentiable on $(0, \infty)$ up to order $s \ge 2$.

Suppose the following condition holds:

(VV2) The stationary density $\pi(\cdot)$ is positive on $(0, \infty)$ and the initial random variable X_0 has density $\pi(\cdot)$.

Then the process $\{X_t, t \ge 0\}$ is a stationary Markov process. Let

$$\eta(x) = \int_0^x a(u,\theta)\pi(u)\, du \tag{5.4.71}$$

and $\theta(x,t)$ be the solution of the backward Kolmogorov equation

$$\frac{\partial\theta(x,t)}{\partial t} = A\theta(x,t), \qquad \theta(x,0) = x, \tag{5.4.72}$$

where A is the backward Kolmogorov operator

$$A\theta(x,t) = a(x,\theta)\frac{\partial\theta(x,t)}{\partial x} + \frac{1}{2}\sigma^2(x)\frac{\partial^2\theta(x,t)}{\partial x^2}. \tag{5.4.73}$$

Then

$$\begin{aligned}\pi(x)\frac{\partial\theta}{\partial t} &= a(x,\theta)\pi(x)\frac{\partial\theta}{\partial x} + \frac{1}{2}\sigma^2(x)\pi(x)\frac{\partial^2\theta}{\partial x^2} \\ &= \left\{\frac{\partial\eta}{\partial x}\right\}\left\{\frac{\partial\theta}{\partial x}\right\} + \eta(x)\left\{\frac{\partial^2\theta}{\partial x^2}\right\}.\end{aligned} \tag{5.4.74}$$

This can be seen from the following argument.

Let $p(\Delta, X_{t+\Delta}|X_t)$ be the transition density for $(X_t, X_{t+\Delta})$. Consider the forward Kolmogorov equation

$$\frac{\partial p(\Delta, X_{t+\Delta}|X_t)}{\partial \Delta} = -\frac{\partial}{\partial X_{t+\Delta}}(a(X_{t+\Delta}, \theta)p(\Delta, X_{t+\Delta}|X_t)) + \frac{1}{2}\frac{\partial^2}{\partial X_{t+\Delta}^2}(\sigma^2(X_{t+\Delta})p(\Delta, X_{t+\Delta}|X_t)). \quad (5.4.75)$$

In order to construct an estimator for σ^2, we characterize the diffusion coefficient $\sigma^2(\cdot)$ using (5.4.75). By the stationarity of the process,

$$\int_0^\infty p(\Delta, X_{t+\Delta}|X_t)\pi(X_t)\, dX_t = \pi(X_{t+\Delta}). \quad (5.4.76)$$

Multiplying (5.4.75) by $\pi(X_t)$ and integrating with respect to the conditioning variable X_t, we obtain the differential equation

$$\frac{d^2}{dx^2}(\sigma^2(x)\pi(x)) = 2\frac{d}{dx}(a(x, \theta)\pi(x)), \quad (5.4.77)$$

where $x = X_{t+\Delta}$. This relation must hold for all $0 < x < \infty$ and the true parameter θ. Integrating (5.4.77) twice, under the boundary condition $\pi(0) = 0$, gives the relation

$$\sigma^2(x) = \frac{2}{\pi(x)}\int_0^x a(u, \theta)\pi(u)\, du, \qquad 0 < x < \infty. \quad (5.4.78)$$

This formula proves that if the drift parameter vector θ can be estimated, then the diffusion coefficient can be estimated from (5.4.78). Relation (5.4.77) justifies equation (5.4.74).

Applying Dynkin's formula (Karlin and Taylor 1981, p. 310) (under assumptions (VV1) and (VV2)), it follows that

$$\theta(x, t) = E[X_t|X_0 = x]. \quad (5.4.79)$$

Solving this equation, one can obtain the parameter θ, which can be estimated consistently by a nonlinear least-square approach using (5.4.79).

Ait-Sahalia (1996b) proved that the diffusion coefficient $\sigma^2(\cdot)$ of the process $\{X_t\}$ can be identified from its joint and the marginal densities by (5.4.78) under assumptions (VV1) and (VV2).

Suppose the process is observed at equal time intervals $\Delta = 1$ and $\{X_1, \ldots, X_n\}$ are the observations on the process. The density $\pi(\cdot)$ can be estimated by a kernel type density estimator (Prakasa Rao 1983a)

$$\hat{\pi}(x) = \frac{1}{nh_n}\sum_{i=1}^{n} K\left(\frac{x - X_i}{h_n}\right).$$

For the estimation of $\sigma^2(\cdot)$ using (5.4.78), one can plug in the estimator $\hat{\pi}(\cdot)$ in the integrand as well as in the denominator on the left-hand side of (5.4.78) or only in the denominator with the density $\pi(\cdot)$ inside the integral replaced by an empirical density. Let $\hat{\theta}$ be a $\sqrt{n}$-consistent estimator for θ following the relation given by (5.4.79). Let $\hat{\sigma}^2(\cdot)$ be the estimator obtained from the above procedure.

Suppose the following additional conditions hold:

(VV3) The observed sequence $\{X_i, 1 \le i \le n\}$ is a strictly stationary β-mixing sequence satisfying $k^\delta \beta_k \to 0$ as $k \to \infty$ for some $\delta > 1$.

(VV4) (i) $K(\cdot)$ is an even function, continuously differentiable up to order r with $2 \le r \le s$, belongs to $L^2(R)$ and $\int_{-\infty}^{\infty} K(x)\,dx = 1$.

(ii) $K(\cdot)$ is of order r, that is,

$$\int_{-\infty}^{\infty} x^i K(x)\,dx = 0, \qquad 1 \le i \le r-1,$$

$$\int_{-\infty}^{\infty} x^r K(x)\,dx \ne 0, \qquad \int_{-\infty}^{\infty} |x|^r K(x)\,dx < \infty.$$

(VV5) As $n \to \infty$, $h_n \to 0$, and
(i) (to estimate π) $n^{1/2} h_n^{(2r+1)/2} \to 0$ and $n^{1/2} h_n^{1/2} \to \infty$;
(ii) (to estimate σ^2) $n^{1/2} h_n^{(2r+1)/2} \to 0$ and $n^{1/2} h_n^{3/2} \to \infty$.

Ait-Sahalia (1996b) proved the following theorem.

Theorem 5.4.10 *Suppose that conditions (VV1)–(VV5) hold. Then the estimator $\hat{\sigma}^2(\cdot)$ is consistent and asymptotically normal. In fact*

$$h_n^{1/2} n^{1/2} (\hat{\sigma}^2(x) - \sigma^2(x)) \xrightarrow{\mathcal{L}} N(0, V(x)) \qquad \text{as } n \to \infty,$$

where

$$V(x) = \left(\int_{-\infty}^{\infty} K^2(u)\,du \right) \sigma^4(x)/\pi(x).$$

The limiting variance $V(x)$ can be estimated consitently by

$$\hat{V}(x) = \left(\int_{-\infty}^{\infty} K^2(u)\,du \right) \hat{\sigma}^4(x)/\hat{\pi}(x).$$

Furthermore, $\hat{\sigma}^2(x)$ and $\hat{\sigma}^2(y)$ are asymptotically independent for $x \ne y$ in $(0, \infty)$.

Applications of these results to pricing of interest rate derivative securities are discussed in Ait-Sahalia (1996b).

Remark. In order for the discrete process $\{X_i, i \ge 1\}$ to satisfy (VV3), the stochastic process $\{X_t, t \ge 0\}$ has to satisfy a stronger continuous-time mixing condition as discussed in Prakasa Rao (1990b). However, a condition on the drift and diffusion coefficients that ensures that the discrete process $\{X_i, i \ge 1\}$ satisfies (A3) is the following:

(VV3)′ $\lim_{x \to 0 \text{ or } x \to \infty} \sigma(x)\pi(x) = 0$ and $\lim_{x \to 0 \text{ or } x \to \infty} \left| \frac{\sigma(x)}{2a(x,\theta) - \sigma(x)\sigma^{(1)}(x)} \right| < \infty$,

where $\sigma^{(1)}(x)$ is the first derivative of $\sigma(x)$. Condition (VV3)′ ensures geometric ergodicity (Hansen and Scheinkman 1995) and it follows that there exists λ, $0 < \lambda < 1$, such that

$$|E(f(X_i) g(X_{i+k}))| \le \|f\|_2 \|g\|_2 \lambda^k$$

for f and g measurable such that $E[f^2(X_1)] < \infty$ and $E[g^2(X_1)] < \infty$. Applying the inequality for $f = I_A$ and $g = I_B$, we obtain a bound which can be used for checking (VV3).

5.4.4 Estimation by Minimizing L_2-Distance

Consider again the stochastic differential equation of type

$$dX(t) = a(t, X(t))\,dt + \sigma(t)\,dW(t), \qquad 0 \le t \le 1,$$

where $a(\cdot,\cdot)$ and $\sigma(\cdot)$ are unknown functions. Further suppose that the process $\{X(t), 0 \le t \le 1\}$ is observed at the times $t_{iN} = i/N, 0 \le i \le N$. The problem is to estimate the function $\sigma^2(\cdot)$. Suppose the following condition holds:

(WW1) (i) $\sup_{0\le t\le 1} |\sigma(t)| < G < \infty$.

(ii) For any K, there exists a constant C_K such that

$$\sup_{\substack{|x|<K\\ 0\le t\le 1}} |a(t,x)| < C_K < \infty.$$

Let D be the class of functions $h(\cdot)$ on $[0, 1]$ with bounded variation. Define

$$B_N(X,h) = \sum_{j=1}^{N} h(t_{jN})(X(t_{jN}) - X(t_{j-1,N}))^2.$$

The following Baxter type result is proved in Maiboroda (1995).

Theorem 5.4.11 *Suppose that condition (WW1) holds. Then there exists a constant $C > 0$ such that*

$$\sup_{h\in D} \frac{|B_N(X,h) - \int_0^1 h(t)\sigma^2(t)\,dt|}{2\sup|h(\cdot)| + \mathrm{var}(h(\cdot)) + 1} \le C\varepsilon_N,$$

where $\mathrm{var}(h(\cdot))$ *denotes the total variation of* $h(\cdot)$ *on the interval* $[0,1]$ *and* $\varepsilon_N = N^{-1/2}\log N$.

In addition, suppose that:

(WW2) $\sigma^2(\cdot) \in \zeta$ such that there exist $A > 0$, $V > 0$ such that for all $h \in \zeta$, $\sup_{t\in[0,1]} |h(t)| \le A < \infty$ and $\mathrm{var}(h(\cdot)) \le V < \infty$.

For $h \in \zeta$, define

$$R_N^*(h) = \int_0^1 h^2(t)\,dt - 2B_N(X,h).$$

Theorem 5.4.11 implies that

$$R_N^*(h) \to R(h) = \int_0^1 (h(t) - \sigma^2(t))^2\,dt - \int_0^1 \sigma^4(t)\,dt \text{ a.s.} \qquad \text{as } N \to \infty.$$

It is clear that $h(t) = \sigma^2(t), 0 \le t \le 1$, is the minimizer of $R(h)$. Hence we can possibly choose a minimizer S_N^* of $R_N^*(h)$ as an estimator of σ^2. However, it is possible that $R_N^*(h)$ has no minimum on ζ. Hence, we consider a δ-minimizing sequence of the following form.

Let

$$y_\delta = \arg\min(R_N^*, \zeta, \delta)$$

if

$$R_N^*(y_\delta) < R_N^*(h) + \delta$$

for any $h \in \zeta$.

The following result is due to Maiboroda (1995).

Theorem 5.4.12 *Let $S_N^* = \arg\min(R_N^*, \zeta, \varepsilon_N)$. Suppose that conditions (WW1) and (WW2) hold. Then*

$$\int_0^1 (S_N^*(t) - \sigma^2(t))^2\, dt \le \Lambda\varepsilon_N,$$

where $P(\Lambda < \infty) = 1$.

If condition (WW2) does not hold, then the functional R_N^* can be modified by adding a 'stabilizing term' $\alpha_N U(h)$, where

$$\sup_{0\le t\le 1} |h(t)| + \mathrm{var}(h(\cdot)) \le \max(CU(h), C)$$

for some constant C and $\alpha_N \to 0$. In other words, consider

$$R_N^{**}(h, \alpha_N) = R_N^*(h) + \alpha_N U(h)$$

and choose S_N^{**} as an estimater of σ^2 defined by

$$S_N^{**} = \arg\min(R_N^{**}(\cdot, \alpha_N), \zeta, \delta_N).$$

Theorem 5.4.13 *If $\sigma^2 \in \zeta$, $\zeta \subseteq D$ and $\varepsilon_N = o(\alpha_N)$, then*

$$\int_0^1 (S_N^{**}(t) - \sigma^2(t))^2\, dt \le \Lambda(\alpha_N + \varepsilon_N + \delta_N)\ a.s.,$$

where $P(\Lambda < \infty) = 1$.

Remarks. For proofs of Theorems 5.4.11–5.4.13, see Maiboroda (1995). Vovk and Maiboroda (1993) study the problem of estimation of the change point θ for coefficients of the type $a(t, x) = dx$, and

$$\sigma(t) = \begin{cases} \sigma_1, & \text{if } t \le \theta, \\ \sigma_2, & \text{if } t > \theta. \end{cases}$$

An example of a stabilizing function is

$$U(h) = \int_0^1 \left(|h(t)|^2 + \left| \frac{d}{dt} h(t) \right|^2 \right) dt$$

and the class ζ may be chosen as the set $\{h : U(h) < \infty\}$. It is known that

$$\sup |h| + \mathrm{var}(h) \le C[U(h)]^{1/2},$$

and it follows from Theorem 5.4.13 that the estimator

$$S_N^{**} = \arg\min(R_N^{**}(\cdot, \alpha_N), \zeta, 0)$$

is a consistent estimator for σ^2. Let

$$h(t) = \sum_{k=-\infty}^{\infty} \beta_k e_k(t), \qquad 0 \le t \le 1$$

where $e_k(t) = \exp(2\pi i k t)$. Then

$$S_N^{**}(t) = \sum_{k=-\infty}^{\infty} \beta_k^{**} e_k(t),$$

where

$$\beta_k^{**} = \frac{B_N(X, e_k)}{(1 + \alpha_N((2\pi k)^2 + 1))}.$$

Since $|B_N(X, e_k)| \leq B_N(X, e_0)$, it follows that $U(S_N^{**}) < \infty$. By using the estimator

$$\tilde{S}_{\ell N}(t) = \sum_{k=-\ell}^{\ell} \beta_k^{**} e_k(t)$$

instead of $S_N^{**}(t)$ as an estimator for $\sigma^2(t)$, we obtain that

$$\int_0^1 (S_N^{**}(t) - \tilde{S}_{\ell N}(t))^2 \, dt \leq |B_N(X, e_0)|^2 \sum_{|k|>\ell} \frac{1}{((\alpha_n(2\pi k)^2 + 1) + 1)^2}.$$

5.4.5 Efficient Estimation under the L_p-Loss Function

Upper and lower bounds for the minimax risk

Consider the one-dimensional stochastic differential equation

$$dX_t = a(t, X_t)\, dt + \sigma(X_t)\, dW_t, \qquad 0 \leq t \leq 1,\ X_0 = x_0 \tag{5.4.80}$$

where $a(\cdot, \cdot)$ and $\sigma(\cdot)$ are unknown. Suppose the following hold.

(XX0) Equation (5.4.80) admits a strong solution and the coefficients $a(\cdot, \cdot)$ and $\sigma(\cdot)$ satisfy the following conditions.

(XX1) $\sigma^2(\cdot)$ is Lipschitz continuous and there exist constants M_0 and M_1 such that

$$0 < M_0 \leq \sigma^2(x) < M_1 < \infty, \qquad x \in R;$$

(XX2) The function $\sigma^2(\cdot)$ belongs to the sphere of radius M_2 of the Besov space $B_{sp\infty}$ restricted to the interval $[m, M]$ with $s > 1 + \frac{1}{p}$ and $1 \leq p < \infty$.

(XX3) The drift coefficient $a(\cdot, \cdot)$ is continuous and there exists a constant M_3 such that

$$a^2(t, x) \leq M_3^2(1 + x^2), \qquad (t, x) \in [0, 1] \times R.$$

Let us first recall the definition of a Besov space $B_{sp\infty}$.

For any function $f(\cdot)$, define $\tau_h f = f(\cdot + h)$ and let

$$\omega_p(t; f) = \sup_{|h| \leq t} \|\tau_h f - f\|_p$$

for any $f \in L_p(R)$, where $\|\cdot\|_p$ denotes the L_p norm. Then

$$f \in B_{sp\infty} \Leftrightarrow f \in L_p(R) \quad \text{and} \quad [\omega_p(t; f)/t^s] \in L_\infty(R_+).$$

For $s = 1$, we can replace $\tau_h f - f$ by $\tau_h f + \tau_{-h} f - 2f$ and for $s = N + \alpha$, where N is positive integer and $0 < \alpha \leq 1$,

$$f \in B_{sp\infty} \Leftrightarrow f \in L_p(R) \quad \text{and} \quad f^{(N)} \in B_{\alpha p\infty},$$

where $f^{(N)}$ is an Nth derivative of f.

The following theorem, due to Meyer (1990), characterizes the Besov space $B_{sp\infty}$.

Theorem 5.4.14 *Let $0 < s < N$, N a positive integer and $1 \le p \le \infty$. Then $f \in B_{sp\infty}$ if and only if $f = \sum_{n=0}^{\infty} u_n$ with $\|u_n\|_p \le \varepsilon_n 2^{-sn}$, $n \ge 1$, and $\|u_n^{(N)}\|_p \le 2^{nN}\varepsilon_n 2^{-sn}$, $n \ge 1$, with $\{\varepsilon_n, n \ge 1\} \in \ell_\infty(N)$. Here $u_n^{(N)}$ is the Nth derivative of u_n.*

Let V_{sp} denote the class of functions (σ^2, a) satisfying conditions (XX0)–(XX3). Let

$$L(x) = \lim_{\delta \to 0} \frac{1}{2\delta} \int_0^1 I(|X_s - x| < \delta)\, ds \text{ a.s.}$$

be the local time at x as in (5.4.6) and J_ν be the event defined by

$$J_\nu = \left[\inf_{x \in [m,M]} L(x) \ge \nu > 0 \right].$$

Define

$$R_{n,\nu}(\hat{\sigma}^2, V_{sp}) = \sup_{(\sigma^2, a) \in V_{sp}} E_{\sigma^2,a}\left(\int_m^M |\hat{\sigma}^2(x) - \sigma(x)|^p\, dx | J_\nu \right).$$

Theorem 5.4.15 *Let D be the set of all estimators of σ^2. Suppose that conditions (XX0)–(XX3) hold. Then there exist positive constants C_1 and C_2 depending on sp and M_i, $0 \le i \le 3$ such that*

$$C_1 n^{-sp/(1+2s)} \le \inf_{\hat{\sigma}^2 \in D} R_{n,\nu}(\hat{\sigma}^2, V_{sp}) \le C_2 n^{-sp/(1+2s)}.$$

The above theorem is due to Hoffman (1997). It gives an upper and a lower bound for the minimax risk under the L_p-loss function.

Construction of an optimum estimator

For simplicity, we assume that $a = 0$ and $M - m = 1$. Let

$$\begin{aligned} Y_{(i/n)} &= n(X_{((i+1)/n)} - X_{(i/n)})^2 \\ &= n \int_{i/n}^{(i+1)/n} \sigma^2(X_s)\, ds + \varepsilon_{(i/n)}, \qquad i = 1, \ldots, n, \end{aligned}$$

where

$$\varepsilon_{(i/n)} = n\left(\int_{i/n}^{(i+1)/n} \sigma(X_s)\, dW_s \right)^2 - n \int_{i/n}^{(i+1)/n} \sigma^2(X_s)\, ds.$$

Let us partition the interval $[m, M]$ into subintervals

$$C_\lambda = [m + (\lambda - 1)h_n, m + \lambda h_n], \qquad \lambda = 1, \ldots, h_n^{-1},$$

where h_n is to be chosen. Define

$$N_{(i/n)}^\lambda = \left\{ \sum_{j \le i} I(X_{(j/n)} \in C_\lambda \right\} \wedge \{nh_n\nu\}.$$

Let $T_1 = 0$ and, for $i \ge 2$, define

$$T_i = \inf\left\{ j/n > T_{i-1} : \sum_\lambda (N_{(j/n)}^\lambda - N_{T_{i-1}}^\lambda) \ge 1 \right\} \wedge 1.$$

The sequence $T_i, i = 1, \ldots, [nv]$, constitutes an increasing sequence of stopping times with respect to the filtration $\{\mathcal{F}_t\}$ generated by the process $\{X_t\}$. Let

$$Y_{T_i} = n(X_{T_i+(1/n)} - X_{T_i})^2.$$

Let $\phi(\cdot)$ be a compactly supported, r times differentiable function with $r > s$, and define

$$\phi_{j,k}(x) = 2^{j/2}\phi(2^j x - k)$$

and

$$\alpha_{j,k} = \int_{-\infty}^{\infty} \phi_{j,k}(x)\sigma^2(x)\, dx$$

(see Appendix C and (5.4.24)). Define

$$\hat{\sigma}^2(\cdot) = \sum_{k=-\infty}^{\infty} \hat{\alpha}_{j_n,k}\phi_{j_n,k},$$

with

$$\hat{\alpha}_{j_n,k} = \frac{1}{[nv]}\sum_{i=0}^{[nv]} Y_{T_i}\phi_{j_n,k}(x_{T_i}).$$

Here

$$x_{T_i} = m + (\lambda_{T_i} - 1)h_n + \ell_{T_i}/[nv],$$

where λ_{T_i} is the index of the set C_λ to which X_{T_i} belongs and $\ell_{T_i} = \#\{X_{T_j} \in C_{\lambda_{T_j}} : j \leq i\}$.

The estimator $\hat{\sigma}^2(\cdot)$ is a wavelet type estimator as discussed earlier in this chapter. Hoffman (1997) proved the following result establishing that the above estimator of $\hat{\sigma}^2(\cdot)$ achieves the optimal rate of convergence. We omit the details.

Theorem 5.4.16 *Let* $1 \leq p \leq \infty$. *Let* $\hat{\sigma}^2(\cdot)$ *be an estimator as defined above with 'v' replaced by 'v/2'. Suppose that conditions (XX0)–(XX3) hold. If* $2^{j_n} \simeq n^{1/(1+2s)}$ *and* $h_n \simeq n^{-s/(1+2s)}$, *then*

$$R_{n,v}(\hat{\sigma}^2, V_{sp}) \leq Cn^{-sp/(1+2s)},$$

where C is a constant depending on $\phi, s, p, M_i, 0 \leq i \leq 3$ *and* v.

6

Applications to Stochastic Modeling

6.1 Introduction

Several applications of diffusion processes for stochastic modeling can be found in Karlin and Taylor (1975; 1981). We will discuss some recent applications to mathematical finance, forest management, stochastic hydrology etc., with principal emphasis on the statistical inference for such models. For other applications to stochastic modeling by semimartingales, in particular of counting processes to survival analysis and reliability and of diffusion processes with jumps, see Andersen *et al.* (1993), Fleming and Harrington (1991), Karr (1991) and Prakasa Rao (1999).

6.2 Mathematical Finance

6.2.1 Models for Pricing, Interest Rates and Risk

Many models in financial economics such as option pricing models, discount bond pricing models and other types of asset pricing models have been developed using stochastic differential equations in continuous time. Merton (1971), Brennan and Schwartz (1979), Vasicek (1977), Dothan (1978), Cox *et al.* (1985), Longstaff (1989) and Longstaff and Schwartz (1992) developed models for discount bond pricing that are closely related to the term structure of interest rates. Shoji (1995) studied the problem of model selection among some alternate short-term interest rate models. Earlier work on comparison of alternate models of short-term interest rates is due to Chan *et al.* (1992). Shoji (1995) used the Akaike information criterion (AIC) for model selection. In most of the models used, it is assumed that the drift function is linear:

1. Merton: $dX(t) = \alpha\, dt + \sigma\, dW(t)$
2. Dothan: $dX(t) = X(t)\sigma\, dW(t)$
3. Geometric Brownian motion: $dX(t) = \beta X(t)\, dt + X(t)\sigma\, dW(t)$
4. Vasicek: $dX(t) = (\alpha + \beta X(t))\, dt + \sigma\, dW(t)$
5. CIR SR: $dX(t) = (\alpha + \beta X(t))\, dt + \sigma\sqrt{X(t)}\, dW(t)$
6. Brennan–Schwartz: $dX(t) = (\alpha + \beta X(t))\, dt + X(t)\sigma\, dW(t)$
7. CIR VR: $dX(t) = X(t)^{3/2}\, dW(t)$
8. CEV: $dX(t) = \beta X(t)\, dt + X(t)^{\gamma}\sigma\, dW(t)$
9. Unrestricted: $dX(t) = (\alpha + \beta X(t))\, dt + X(t)^{\gamma}\sigma\, dW(t)$.

Models 1–8 are used in modeling the term structure of interest rates and valuing bond options, futures, futures option and other types of contingent claims. Model 9 includes all other models as special cases. The alternate models given above can be compared using the AIC, defined by

$$AIC = 2(-\log L + N_p)$$

where $\log L$ is the natural log-likelihood function which is being maximized with respect to the parameters for parameter estimation and N_p is the number of parameters. The 'best' model is the one with the minimum AIC value (Akaike 1977; 1983). Shoji (1995) studied parameter estimation for these models based on the process $X(t)$ discretized by local linearization (see Chapter 2).

In order to study the possibility of nonlinearity of the drift, Shoji (1995) introduced the polynomial drift model

$$dX(t) = a(X_t)\,dt + \sigma X_t^{\gamma}\,dW_t,$$

where $a(x) = \sum_{i=1}^{m} \alpha_i x^i$ and σ is a constant, and studied its applicability for various real data with no definite conclusions. It was observed that the drift could be linear or nonlinear.

The models described above deal with interest rate structures within a country. These models do not take into account the interaction of interest rates between and among countries. In an open economy between two countries, the rise or fall of interest rate in one country affects the rates in another country. For instance, if the exchange rate is fixed in a narrow band, there is no effective way to balance the economies other than adjusting the level of interest rates of the two countries (Svensson 1991). Shoji (1995) suggests the following model for the study of such problems.

Suppose there is an economically independent country A with a strong influence on the economy of another country B. Suppose the factors determining the interest rate in A depend on its own economic conditions, whereas that of B depends on its own as well as on that of A. Let r_A and r_B be the short-term rates of A and B, respectively. Let $s = r_B - r_A$ be the interest rate differential. Suppose s and r_A satisfy the following

$$ds = \alpha(\beta - s)\,dt + \gamma\,dW_s^{(1)},$$
$$dr_A = k(\theta - r_A)\,dt + \sigma\,dW_{r_A}^{(2)},$$

where $W_s^{(1)}$ and $W_{r_A}^{(2)}$ are the standard Brownian motions with covariation

$$d[W_s^{(1)}, W_{r_A}^{(2)}] = \rho\,dt$$

and $\alpha, \beta, \gamma, k, \theta$ and σ are constants. Then the process r_B satisfies the stochastic differential equation

$$dr_B = \{k(\theta - r_A) + \alpha(\beta - s)\}dt + \sigma\,dW_{r_A}^{(2)} + \gamma\,dW_s^{(1)}.$$

Since r_A and s follow continuous-time Gaussian processes, the parameters involved can be estimated by the maximum likelihood method. The continuous-time Gaussian process can be discretized and the log-likelihood function of the discretized process, $\log L$, is obtained as follows: let $p(y_{t_1}, \ldots, y_{t_N})$ be the joint probability density function of the Y_{t_n}s, where Y_{t_n} is the nth observation of either r_A or s at time t_n and N is the number of observations. By the Markov property

$$p(Y_{t_1}, \ldots, Y_{t_N}) = \left\{\prod_{n=1}^{N-1} p(Y_{t_{n+1}} | Y_{t_n})\right\} p(Y_{t_1}),$$

and hence, by the normality of the conditional distribution and the marginals,

$$\log L = -\frac{1}{2}\sum_{n=1}^{N-1}\left\{\frac{(Y_{t_{n+1}} - E_n)^2}{V_n} + \log(2\pi V_n)\right\} + \log p(Y_{t_1}),$$

where

$$E_n = E_{t_n}[r_{t_{n+1}}] \qquad (\text{or } E_{t_n}[s_{t_{n+1}}])$$

and

$$V_n = \text{var}[r_{t_{n+1}}] \qquad (\text{or } \text{var}_{t_n}(s_{t_{n+1}})).$$

Note that

$$E_t[r_{A,T}] = \theta + (r_{A,t} - \theta)e^{-k(T-t)}$$

and

$$\text{var}_t[r_{A,T}] = \frac{\sigma^2(1 - e^{-2k(T-t)})}{2k}$$

for $t \le T$. Similar relations hold for $E_{t_n}[s_{t_{n+1}}]$ and $\text{var}_{t_n}[s_{t_{n+1}}]$. Here E_{t_n} and var_{t_n} denote the conditional expectation and the conditional variance respectively given the observation at the time t_n. The parameter estimation can be done by the maximum likelihood approach.

The management of the market risk and credit risk is the most important part of the financial economics. The market risk is the loss due to the market fluctuations from dealings of the financial institutions and the individual investors who have the financial assets. The credit risk is defined to be the loss by the default of the counter parties of those who have the contracts. The financial regulators propose the method of risk management and intend to control risks by its methods. The value at risk (VaR) is defined to be the maximum loss of an asset return. Most of the financial asset returns tend to be analyzed on the assumption of linearity and Gaussianity. However, it is now observed that the actual distributions involved are heavy-tailed and not normal (Kariya *et al.* 1995; Nagahara 1996). Fama (1965) and Mandelbrot (1963) suggested the use of a heavy-tailed distribution such as the stable Pareto distribution as a model for the daily return of stocks.

Estimation

Following Nagahara (1996), we now consider a stochastic differential equation whose stationary distribution is Pearson type VII or IV (Wong 1963). The Cauchy distribution and the t distribution (with low degrees of freedom) have heavy tails compared to the normal distribution. They belong to the Pearson type VII family. Suppose the density function $p(\cdot)$ satisfies the differential equation

$$\frac{-p'(x)}{p(x)} = \frac{\frac{2b}{\tau}\left(\frac{x-\mu}{\tau} - \delta\right)}{\left(\frac{x-\mu}{\tau}\right)^2 + 1},$$

which implies that

$$p(x; \mu, \tau, \delta, b) = \frac{C\exp\left\{2b\delta \arctan\left(\frac{x-\mu}{\tau}\right)\right\}}{\{(x-\mu)^2 + \tau^2\}^b}$$

for $b > 1/2$, $\tau > 0$, $-\infty < \delta < \infty$ and $-\infty < \mu < \infty$, where C is the normalizing constant. This type of density is known as the Pearson type IV family (Johnson and Kotz 1970). Nagahara (1996) studied the properties of this family of densities. Wong (1963) showed that there is a relationship between the Pearsonian system and a Markov diffusion process. He showed that for any density $p(x)$ which belongs to the Pearsonian system defined by

$$-\frac{dp(x)}{dx} = \frac{c_0 + c_1 x}{d_0 + d_1 x + d_2 x^2} p(x),$$

it is possible to derive a diffusion process whose marginal density is $p(x)$. Ozaki (1985) generalized this result. Ozaki proved that if the density $p(x)$ satisfies

$$\frac{dp(x)}{dx} = \frac{c(x)}{d(x)} p(x), \tag{6.2.1}$$

where $c(x)$ and $d(x)$ are analytic functions, then there is a Markov diffusion process such that its transition density function $q(x|x_0, t)$ satisfies the Fokker–Plank equation

$$\begin{aligned} \frac{\partial q(x|x_0,t)}{\partial t} &= -\frac{\partial}{\partial x}[\{c(x) + d'(x)\} q(x|x_0,t)] \\ &\quad + \frac{1}{2}\frac{\partial^2}{\partial x^2}[2d(x) q(x|x_0,t)]. \end{aligned} \tag{6.2.2}$$

Following Mortensen (1979), the diffusion process $X(t)$ defined by (6.2.2) satisfies the stochastic differential equation

$$dX(t) = a(X(t))\,dt + \sqrt{b(X(t))}\,dW(t), \tag{6.2.3}$$

where

$$a(x) = c(x) + d'(x), \qquad b(x) = 2d(x). \tag{6.2.4}$$

We now consider the diffusion processes corresponding to the equation

$$\begin{aligned} \frac{-p'(x)}{p(x)} &= \frac{\frac{2b}{\tau}\left(\frac{x-\mu}{\tau} - \delta\right)\frac{\sigma^2}{2}}{\left\{\left(\frac{x-\mu}{\tau}\right)^2 + 1\right\}\frac{\sigma^2}{2}} \\ &= \frac{2b(x - \mu - \tau\delta)\frac{\sigma^2}{2}}{\{(x-\mu)^2 + \tau^2\}\frac{\sigma^2}{2}} \end{aligned} \tag{6.2.5}$$

to represent the symmetric and asymmetric heavy-tailed distribution of returns. The corresponding diffusion process satisfies the stochastic differential equation

$$dX_t = \sigma^2\{(1-b)(X_t - \mu) + b\tau\delta\}\,dt + \sigma\sqrt{(X_t - \mu)^2 + \tau^2}\,dW_t. \tag{6.2.6}$$

Equation (6.2.3) can be transformed into an equation with a constant as the diffusion coefficient by the transformation

$$y = \int_{-\infty}^{x} \frac{1}{\sqrt{b(u)}}\,du. \tag{6.2.7}$$

It becomes

$$dY(t) = A(Y(t))\,dt + dW(t), \tag{6.2.8}$$

where

$$A(y) = \frac{a(x)}{\sqrt{b(x)}} - \frac{1}{4}\frac{b'(x)}{\sqrt{b(x)}}.$$

For the special case of (6.2.6), the transformation is given by

$$y = \int_{-\infty}^{x} \frac{1}{\sigma\sqrt{\left(\frac{\zeta-\mu}{\tau}\right)^2 + 1^2}}\, d\zeta$$

$$= \frac{\tau}{\sigma} \log\left\{\frac{(x-\mu)}{\tau} + \sqrt{\left(\frac{(x-\mu)}{\tau}\right)^2 + 1^2}\right\}$$

$$= \frac{\tau}{\sigma} \operatorname{arcsinh}\left(\frac{x-\mu}{\tau}\right). \tag{6.2.9}$$

Hence

$$x = \mu + \tau \sinh\left(\frac{\sigma y}{\tau}\right)$$

and

$$A(y) = \tau\sigma\left(\frac{1}{2} - b\right)\tan h\left(\frac{\sigma y}{\tau}\right) + \frac{\sigma b \tau \delta}{\cosh\left(\frac{\sigma y}{\tau}\right)}.$$

Then the transformed differential equation is

$$dY(t) = A(Y(t))\, dt + \tau\, dW(t)$$

and this stochastic differential equation has a constant term as its diffusion coefficient.

We now apply the methods described earlier to the estimation of the parameters. Let us first consider the extended local linearization method following Ozaki (1985) and apply it to the stochastic differential equation

$$dX(t) = A(X(t))\, dt + \sigma\, dW(t).$$

For $t > s$, we approximate $A(x_t)$ by the first-order Taylor expansion

$$A(x_t) \simeq A(x_s) + \frac{\partial A(x_s)}{\partial x_s}(x_t - x_s)$$

$$= L_s x_t + M_s,$$

where

$$L_s = \frac{\partial A(x_s)}{\partial x_s},$$

and

$$M_s = A(x_s) - \frac{\partial A(x_s)}{\partial x_s} x_s,$$

$$= A(x_s) - L_s x_s.$$

Then

$$dX(t) = A(X(t))\, dt + \sigma\, dW(t)$$

$$= (L_s X_t + M_s)\, dt + \sigma\, dW(t) \tag{6.2.10}$$

for $t \geq s$. We assume that L_s and M_s are constant over short intervals.

Consider the corresponding ordinary differential equation

$$\frac{dx_t}{du} = P(u)x_t + Q(u), \qquad t \geq s, \ x(s) = x_s. \tag{6.2.11}$$

The solution of this differential equation is

$$x_t = \exp\left(\int_s^t P(u)\,du\right)\left[\int_s^t Q(u)\exp\left(\int_s^u -P(v)\,dv\right)du + x_s\right], \qquad t \geq s. \tag{6.2.12}$$

It can be checked that

$$X_t = X_s + \frac{A(X_s)}{L_s}(e^{L_s(t-s)} - 1) + \sigma \int_s^t e^{L_s(t-u)}\,dW_u, \qquad t \geq s, \tag{6.2.13}$$

is the solution of the stochastic differential equation (6.2.10). Furthermore,

$$\operatorname{var}(X_t | X_s = x_s) = \frac{\{\exp(2L_s(t-s)) - 1\}\sigma^2}{2L_s}, \qquad t \geq s.$$

If $X_1, \ldots, X_T$ are the observations sampled at equal time intervals ΔT, then the log-likelihood function is given by

$$\begin{aligned}
\log[p(X_1, \ldots, X_T)] &= \log p(X_1) + \sum_{t=1}^{T-1} \log p(X_{t+1}|X_t) \\
&= \log p(X_1) - \sum_{t=1}^{T-1} \frac{1}{2V_t}\left\{X_{t+1} - X_t - \frac{A(X_t)}{L_t}(e^{L_t \Delta_t} - 1)\right\}^2 \\
&\quad - \frac{T-1}{2}\log(2\pi V_t),
\end{aligned} \tag{6.2.14}$$

where

$$V_t = \operatorname{var}[X_{t+1}|X_t].$$

If the transformation $y = \psi(x)$ is applied, then the likelihood functions are related by the equation

$$p(x_1, \ldots, x_T) = p(y_1, \ldots, y_T)\left|\frac{\partial(y_1, \ldots, y_T)}{\partial(x_1, \ldots, x_T)}\right|$$

and the log-likelihood is given by

$$\log p(x_1, \ldots, x_T) = \log p(y_1, \ldots, y_T) + \sum_{t=1}^{T-1} \log\left|\frac{d\psi}{dx}\right|_{x=X_t}.$$

For instance, for the transformation given by (6.2.9), the log-likelihood is given by

$$\begin{aligned}
\ell(\theta) = &- \sum_{t=1}^{T-1} \frac{1}{2V_t}\{Y_{t+1} - Y_t - \frac{A_t}{L_t}(\exp(L_t\Delta_t) - 1)\}^2 \\
&- \frac{T-1}{2}\log(2\pi V_t) + \sum_{t=1}^{T-1} \log \frac{1}{\sigma\sqrt{\left(\frac{X_t-\mu}{\tau}\right)^2 + 1}}.
\end{aligned} \tag{6.2.15}$$

Here $A_t = A(X_t)$,

$$L_t = \frac{\partial A_t}{\partial y} = \left\{\left(\frac{1}{2} - b\right) - b\delta \sinh\left(\frac{\sigma y}{\tau}\right)\right\} \frac{\sigma^2}{\cosh^2\left(\frac{\sigma y}{\tau}\right)} \tag{6.2.16}$$

and

$$V_t = \frac{\{\exp(2L_t \Delta t) - 1\}}{2L_t} \tau^2. \tag{6.2.17}$$

The parameters can be estimated by the maximum likelihood method.

Let us now apply the modified local linearization method due to Shoji and Ozaki to this problem. For convenience, we choose $\Delta t = 1$. Consider the stochastic differential equation

$$dX_t = a(X_t, t)\,dt + \sigma\,dW_t. \tag{6.2.18}$$

For $t > s$, the function $a(X_t, t)$ is approximated by the Itô formula,

$$a(X_t, t) = a(X_s, s) + \left(\frac{\sigma^2}{2}\frac{\partial^2 a(X_s, s)}{\partial X_s^2} + \frac{\partial a(X_s, s)}{\partial s}\right)(t - s)$$
$$+ \frac{\partial a(X_s, s)}{\partial X_s}(X_t - X_s) \tag{6.2.19}$$

and hence

$$a(X_t, t) = L_s X_t + M_s t + N_s, \tag{6.2.20}$$

where

$$L_s = \frac{\partial a(X_s, s)}{\partial X_s}, \tag{6.2.21}$$

$$M_s = \frac{\sigma^2}{2}\frac{\partial^2 a(X_s, s)}{\partial X_s^2} + \frac{\partial a(X_s, s)}{\partial s}, \tag{6.2.22}$$

$$N_s = a(X_s, s) - L_s X_s - M_s s. \tag{6.2.23}$$

Hence

$$\begin{aligned} dX_t &= a(X_t, t)\,dt + \sigma\,dW_t \\ &= (L_s X_t + M_s t + N_s)\,dt + \sigma\,dW_t, \qquad t \geq s. \end{aligned} \tag{6.2.24}$$

Suppose that L_s, M_s and N_s are constant over short intervals. By Girsanov's theorem,

$$dX_t = (L_s X_t)\,dt + \sigma\,d\tilde{W}_t, \tag{6.2.25}$$

where $\tilde{W}_t$ is the Wiener process defined by

$$\tilde{W}_t = W_t - \int_0^t \beta(u)\,du, \tag{6.2.26}$$

and

$$\beta(u) = -\frac{1}{\sigma}(M_s u + N_s). \tag{6.2.27}$$

Applying the transformation $Y_t = e^{-L_s t} X_t$, we have

$$dX_t = (L_s X_t)\,dt + e^{L_s t}\,dY_t, \qquad t \geq s. \tag{6.2.28}$$

Hence

$$dY_t = \sigma e^{-L_s t}\, d\tilde{W}_t \tag{6.2.29}$$

for $t \geq s$, which implies that

$$\begin{aligned} Y_t &= Y_s + \sigma \int_s^t e^{-L_s u}\, d\tilde{W}_u \\ &= Y_s + \int_s^t (M_s u + N_s) e^{-L_s u}\, du + \sigma \int_s^t e^{-L_s u}\, dW_u \end{aligned} \tag{6.2.30}$$

for $t \geq s$. The transformation $Y_t = e^{-L_s t} X_t$ implies that

$$\begin{aligned} X_t = X_s &+ \frac{a(X_s, s)}{L_s}\{e^{L_s(t-s)} - 1\} + \frac{M_s}{L_s^2}\{(e^{L_s \Delta t} - 1) - L_s \Delta t\} \\ &+ \sigma \int_s^t e^{L_s(t-u)}\, dW_u, \qquad t \geq s, \end{aligned} \tag{6.2.31}$$

where

$$\operatorname{var}_s(X_t) = \frac{\{\exp(2L_s(t-s)) - 1\}\sigma^2}{2L_s} \tag{6.2.32}$$

and $\operatorname{var}_s(X_t)$ is the conditional variance of X_t given X_s. If $X_1, \ldots, X_T$ are observed at equal time intervals Δt, then the log-likelihood function of $(X_1, \ldots, X_T)$ is

$$\begin{aligned} \log p(x_1, \ldots, x_T) &= \log p(x_1) + \sum_{t=1}^{T-1} \log p(x_{t+1} | x_t) \\ &= \log p(x_1) - \sum_{t=1}^{T-1} \frac{1}{2V_t}\Bigg[x_{t+1} - x_t - \frac{A_t}{L_t}\{\exp(L_t \Delta t) - 1\} \\ &\quad - \frac{M_t}{L_t^2}\{(\exp(L_t \Delta t) - 1) - L_t \Delta t\}\Bigg]^2 - \frac{T-1}{2}\log(2\pi V_t), \end{aligned} \tag{6.2.33}$$

where M_t, V_t are as given below. We have seen earlier that if $y = \psi(x)$ is the transformation used, then

$$\log p(x_1, \ldots, x_T) = \log p(y_1, \ldots, y_T) + \sum_{t=1}^{T-1} \log \left|\frac{d\psi}{dx}\right|_{x=x_t}. \tag{6.2.34}$$

In the present problem,

$$L = \frac{\partial a}{\partial y} = \left\{\left(\frac{1}{2} - b\right) - b\delta \sinh\left(\frac{\sigma y}{\tau}\right)\right\} \frac{\sigma^2}{\cosh^2\left(\frac{\sigma y}{\tau}\right)}, \tag{6.2.35}$$

$$\begin{aligned} \frac{\partial L}{\partial y} = \frac{\partial^2 a}{\partial y^2} &= \frac{\sigma^2}{\tau \cosh\left(\frac{\sigma y}{\tau}\right)} \\ &\times \left\{-b\delta + (2b - 1)\frac{\tanh\left(\frac{\sigma y}{\tau}\right)}{\cosh\left(\frac{\sigma y}{\tau}\right)} + 2b\delta \tanh^2\left(\frac{\sigma y}{\tau}\right)\right\}, \end{aligned} \tag{6.2.36}$$

$$M_t = \frac{\tau^2}{2}\frac{\partial L}{\partial y} = \frac{\tau^2}{2}\frac{\partial^2 a}{\partial y^2}, \tag{6.2.37}$$

$$V_t = \frac{\{\exp(2L_t\Delta t) - 1\}}{2L_t}\tau^2. \tag{6.2.38}$$

The log-likelihood function in terms of $x_1, \ldots, x_T$ is given by

$$\begin{aligned}\log p(x_1, \ldots x_T) = &-\sum_{t=1}^{T-1} \frac{1}{2V_t}\left[Y_{t+1} - Y_t - \frac{A_t}{L_t}(\exp L_t\Delta t - 1)\right. \\ &\left. - \frac{M_t}{L_t^2}\{(\exp L_t\Delta t - 1) - L_t\Delta t\}\right]^2 \\ &- \frac{T-1}{2}\log(2\pi V_t) + \sum_{t=1}^{T-1}\log\frac{1}{\sigma\sqrt{\left(\frac{x_t-\mu}{\tau}\right)^2 + 1}}.\end{aligned} \tag{6.2.39}$$

Here $A_t = a(X_t, t)$. The parameters are estimated by maximizing the log-likelihood function.

Nagahara (1996) compared the performance of the two methods discussed above as well as the maximum likelihood method using the transition probability function of the process $\{X_t\}$ which can be explicitly computed in this case.

6.3 Forest Management

In forest management the decision to plant trees and the decision to harvest the timber produced are of practical importance. Since people in the forest market bid and quote their prices for the logs and lumber depending on the various types of information they possess, the stochastic behavior of the prices has to be modeled to enable rational decision making. Such models were proposed by Brennan and Schwartz (1985), Clarke and Reed (1989; 1990), Bailey (1991), Thompson (1992), Reed and Ye (1994) and Yoshimoto and Shoji (1995). Geometric Brownian motion is commonly used to describe the stochastic behavior of the price. Shoji (1995) gave a comparative analysis of several models focusing on the uncertainty of lumber prices. The models considered satisfy the stochastic differential equation

$$dX_t = (\alpha + \beta X_t)\,dt + X_t^\gamma\,dW_t, \qquad t \geq 0, \tag{6.3.1}$$

where X_t is the state of a price at time t. If $\gamma = 0$, then the stochastic differential equation is reduced to a linear differential equation with a constant coefficient as the diffusion. Otherwise, the movement of the process depends on the state of the process. If $\alpha = 0$ and $\gamma = 1$, the above stochastic differential equation gives geometric Brownian motion. Different choices of the parameters (α, β, γ) give different models for use in forest science, financial economics and other areas. Since (6.3.1) has a state-dependent diffusion for $\gamma \neq 0$, the model can be transformed into another model with a constant coefficient of diffusion by a transformation $y = \phi(x)$ which satisfies the ordinary differential equation

$$\frac{d\phi}{dx} = x^{-\gamma}.$$

It is easy to see that

$$\phi(x) = \begin{cases} \log x, & \text{if } \gamma = 1, \\ x^{1-\gamma}/(1-\gamma), & \text{otherwise.} \end{cases} \tag{6.3.2}$$

Using this transformation, (6.3.1) is tranformed into

$$dY_t = A(Y_t)\,dt + \sigma\,dW_t, \qquad t \geq 0, \tag{6.3.3}$$

where $A(\cdot)$ is a nonlinear function induced by ϕ.

Suppose the process is observed at equally spaced times t_i with $t_1 < t_2 < \cdots < t_N$. Let $X_i = X_{t_i}$, $1 \leq i \leq N$, be the observed data and $\Delta t = t_i - t_{i-1}$.

Applying the local linearization method and using the Markov property of the discretized process, the joint density of $Y_i = \phi(X_i)$, $1 \leq i \leq N$, is

$$p(y_1, \ldots, y_N) = p(y_1) \prod_{i=1}^{N} p(y_i | y_{i-1}), \tag{6.3.4}$$

where

$$p(Y_i | y_{i-1}) = \frac{1}{\sqrt{2\pi V_{i-1}}} \exp\left(-\frac{1}{2} \frac{(Y_i - E_{i-1})^2}{V_{i-1}} \right), \tag{6.3.5}$$

$$E_i = y_i + \frac{A(y_i)}{J_i}(\exp(J_i \Delta t) - 1) + \frac{M_i}{J_i^2}(\exp(J_i \Delta t) - 1 - J_i \Delta t), \tag{6.3.6}$$

$$V_i = \sigma^2 \frac{\exp(2J_i \Delta t) - 1}{2J_i}, \tag{6.3.7}$$

$$J_i = A'(y_i), \tag{6.3.8}$$

$$M_i = \frac{\sigma^2}{2} A'(y_i). \tag{6.3.9}$$

Note that

$$p(x_0, \ldots, x_N) = p(y_0, \ldots y_N) \prod_{i=0}^{N} |\phi'(x_i)|. \tag{6.3.10}$$

The log-likelihood function $\ell(\alpha, \beta, \gamma, \sigma^2)$ of $\{X_i\}_{i=0}^N$ is

$$\ell(\alpha, \beta, \gamma, \sigma^2) = \log p(\phi(x_0), \ldots, \phi(x_N)) + \sum_{i=0}^{N} \log(|\phi'(x_i)|). \tag{6.3.11}$$

The parameters α, β, γ and σ^2 can be estimated by maximizing ℓ with respect to the parameters.

Model selection can be done by applying the AIC. The smaller the AIC is, the better a model fits. Note that

$$AIC = 2(-\ell_m + N_p),$$

where ℓ_m is the maximum log-likelihood and N_p is the number of free parameters.

6.4 Market Policy

Yoshimoto *et al.* (1996) proposed a stochastic model to analyze the management of rice production under stochastic rice prices. The prices are assumed to follow a stochastic process that evolves through time as a geometric Brownian motion given by

$$dX_t = \alpha X_t\,dt + X_t \sigma\,dW_t, \qquad t \geq 0, \tag{6.4.1}$$

where X_t is the price for a fixed weight (say, 100 kg) of rice produced at time t. The coefficient α represents a trend in the relative price and σ can be interpreted as the instantaneous standard deviation of the relative price change.

The maximum likelihood method is used for estimation of α and σ from a discretized set of observations on $\{X_t\}$. Let $Y_t = a(X_t) = \log X_t$. Then

$$a'(X_t) = \frac{1}{X_t}, \qquad a''(X_t) = -\frac{1}{X_t^2}.$$

Using the Itô formula, we have

$$\begin{aligned} dY_t &= a'(X_t)\,dX_t + \frac{1}{2}a''(X_t)(dX_t)^2 \\ &= \frac{1}{X_t}(\alpha X_t\,dt + dW_t) + \frac{1}{2}\left(-\frac{1}{X_t^2}\right)\sigma^2 X^2(t)\,dt \\ &= \alpha\,dt + \sigma\,dW_t - \frac{1}{2}\sigma^2\,dt \\ &= \left(\alpha - \frac{1}{2}\sigma^2\right)dt + \sigma\,dW_t. \end{aligned} \tag{6.4.2}$$

Discretizing dy_t by $(y_{n+1} - y_n)$, we have

$$y_{n+1} - y_n = (\alpha - \frac{1}{2}\sigma^2)(t_{n+1} - t_n) + \sigma(W_{t_{n+1}} - W_{t_n}) \tag{6.4.3}$$

or

$$y_{n+1} = y_n + (\alpha - \tfrac{1}{2}\sigma^2)(t_{n+1} - t_n) + \sigma\varepsilon_{n+1}, \tag{6.4.4}$$

where

$$\varepsilon_{n+1} = W_{t_{n+1}} - W_{t_n} \simeq N(0, t_{n+1} - t_n). \tag{6.4.5}$$

Since $\{\varepsilon_i\}$ are independent $N(0, t_{n+1} - t_n)$, the likelihood function of $\{y_i, 0 \le i \le N\}$ is

$$p(y_0, \ldots, y_N) = p(y_0)\prod_{n=0}^{N-1} p(y_{n+1}|y_n), \tag{6.4.6}$$

where

$$p(y_{n+1}|y_n) = \frac{1}{\sqrt{2\pi\sigma^2(t_{n+1} - t_n)}}\exp\left[-\frac{1}{2}\frac{\{(y_{n+1} - y_n - (\alpha - \frac{1}{2}\sigma^2)(t_{n+1} - t_n)\}^2}{\sigma^2(t_{n+1} - t_n)}\right]. \tag{6.4.7}$$

Hence

$$\begin{aligned} \log p(y_0, \ldots, y_n) &= \log p(y_0) + \sum_{n=0}^{N-1} \log p(y_{n+1}|y_n) \\ &= -\frac{1}{2}\sum_{n=0}^{N-1}\Bigg[\log\{2\pi\sigma^2(t_{n+1} - t_n)\} \\ &\quad + \frac{\{y_{n+1} - y_n - (\alpha - \frac{\sigma^2}{2})(t_{n+1} - t_n)\}^2}{\sigma^2(t_{n+1} - t_n)}\Bigg] + \log p(y_0). \end{aligned} \tag{6.4.8}$$

Hence the log-likelihood based on $\{x_0, x_1, \ldots, x_N\}$ is

$$\log p(x_0, \ldots, x_n) = \log p(y_0, \ldots, y_N) + \sum_{n=0}^{N} \log \left| \frac{1}{x_n} \right|$$

$$= \log p(y_0, \ldots, y_N) - \sum_{n=0}^{N} \log |x_n| \qquad (6.4.9)$$

Given the observations on the rice prices, α and σ of the model (6.4.1) can be estimated by maximizing the above likelihood. Using these estimates, one can study stochastic control models to search for an optimal rice production management scheme. For details, see Yoshimoto *et al.* (1996).

7

Numerical Approximation Methods for Stochastic Differential Equations

7.1 Introduction

As we have seen in the earlier chapters, the development of numerical methods for simulating the solution of a stochastic differential equation (SDE) is of extreme importance and interest. We now briefly discuss some results of this nature for SDEs of more general type than those which give rise to diffusion type processes, following Kloeden and Platen (1989). Such SDEs are used for stochastic modeling purposes when there are diffusion and jump components present in the phenomenon (Prakasa Rao 1999). For an extensive discussion on numerical approximation methods for simulation of the solution of an SDE, see Kloeden and Platen (1992) and Kloeden *et al.* (1994).

Let us consider a process $X = \{X_t, t \geq 0\}$ with diffusion and jump components of the form

$$X_t = X_0 + \int_0^t a(X_s)\,ds + \int_0^t b(X_s)\,dW_s + \int_0^t \int_U c(X_s, u)M(du, ds) \tag{7.1.1}$$

for $0 \leq t \leq T$. Here the initial value $X_0 = x_0$ may be random. In the previous chapter, we gave some applications of modeling, through such SDEs, of financial economics (cf. Merton 1971), policy planning for production management, forest management etc. Additional applications of SDEs with jump components for modeling the nervous system, engineering and economic systems etc. are discussed in Prakasa Rao (1999). Other applications include hydrology (cf. Unny 1984), structural engineering and seismology (Kozin 1977; Shinozuka and Sato 1967), biological waste treatment (Harris 1976), fatigue cracking (Sobcyzk 1986), turbulence (Bywater and Chung 1973; Haworth and Pope 1986; Yaglom 1980), satellite dynamics (Sagirow 1970), helicopter stability (Pardoux and Pignol 1984) and telecommunications (Viterbi 1966). They are also used in modeling in radio-astronomy (Legland 1981) and in physics and chemistry (Horsthemke and Lefevre 1984; Van Kempen 1981). Application to genetics (Kimura and Ohta 1971) and population dynamics (Gard 1988) is well known (see also Karlin and Taylor 1981). Application to social sciences such as experimental psychology (Schoener *et al.* 1986) is recent.

Explicit solutions of an SDE are in general infrequent. As we have seen in the earlier chapters, it is necessary to develop efficient time-discretized numerical methods so that a large number of sample realizations of the process can be simulated and the relevant parameters are estimated. An extensive investigation of methods for the simulation of the solutions of SDEs can be found in Kloeden and Platen (1992). Gard (1988) discussed numerical methods for solving SDEs. In statistical inference for such processes, it is essential that the discrete approximations are uniform in the parameter in some sense as the parameters are unknown a priori. We will discuss a result due to Prakasa Rao and Rubin (1998) in this direction later in this chapter.

Consider a d-dimensional stochastic process $\{X_t, 0 \le t \le T\}$ satisfying the SDE (7.1.1). We assume that the initial condition $X_0 = x_0$ (possibly random), $U = R^r - \{0\}$ for some $r \ge 1$, and $\boldsymbol{W} = \{\boldsymbol{W}_t, 0 \le t \le T\}$ is an m-dimensional Wiener process with independent components $\{W_t^j\}, 1 \le j \le m$. In addition, suppose that

$$M(d\boldsymbol{u}, dt) = p(d\boldsymbol{u}, dt) - \pi(d\boldsymbol{u})\, dt$$

is a Poisson martingale measure on $U \times [0, T]$ defined in the following way. For each Borel set $B \subset U$, the process $\{p(B, [0, t]), 0 \le t \le T\}$ is a Poisson process with finite mean

$$E\{p(B, [0, t])\} = \int_0^t \int_B \pi(d\boldsymbol{u})\, ds \le K < \infty$$

for all $0 \le t \le T$ and is independent of the Wiener process $\boldsymbol{W}$. Note that $\pi(d\boldsymbol{u})\, ds$ is the intensity measure which is chosen to be of the form

$$\pi(d\boldsymbol{u})\, ds = \frac{d\boldsymbol{u}\, ds}{\|\boldsymbol{u}\|^{r+1}}.$$

The drift coefficient $\boldsymbol{a}(x) = \{a^i(x)\}_{i=1}^d$ and the jump coefficient $\boldsymbol{c}(x, u) = \{c^i(x, u)\}_{i=1}^d$ are d-dimensional vectors and the diffusion coefficient $\boldsymbol{b}(x) = \{\boldsymbol{b}^j(x)\}_{j=1}^m$ is a $d \times m$ matrix with d-dimensional vectors $\boldsymbol{b}^j(x), 1 \le j \le m$. Time-dependent drift and diffusion coefficients can also be included in this formulation by considering the first component of $X_t = \{X_t^i\}_{i=1}^d$ as the time component $X_t^1 = t$. This requires $a^1 \equiv 1, c^1 \equiv 0$ and $b^{1,j} \equiv 0$ for $1 \le j \le m$. If there is no jump component, that is, $c^i \equiv 0, 1 \le i \le d$, then the process $\{X_t\}$ is a diffusion process. Gihman and Skorohod (1972) and Ikeda and Watanabe (1981) discuss conditions for the existence and uniqueness of a solution to SDEs of type (7.1.1). A sufficient condition for such a result is that the coefficients $\boldsymbol{a}$, $\boldsymbol{b}$ and $\boldsymbol{c}$ are Lipschitz continuous and satisfy a linear growth condition.

7.2 Approximation of a Stochastic Differential Equation

Consider the SDE given by (7.1.1) and an equidistant time discretization $\{t_i\}_\delta$ given by

$$0 = t_0 < t_1 < \cdots < t_n = T$$

of $[0, T]$ with step size $\delta = T/n$. Let

$$Y_{i+1} = Y_i + \boldsymbol{a}(Y_i)\Delta_i + \boldsymbol{b}(Y_i)\Delta \boldsymbol{W}_i + \int_{t_i}^{t_{i+1}} \int_U \boldsymbol{c}(Y_i, \boldsymbol{u}) p(d\boldsymbol{u}, ds)$$
$$- \int_U \boldsymbol{c}(Y_i, \boldsymbol{u})\pi(d\boldsymbol{u})\, \Delta_i \tag{7.2.1}$$

for $0 \le i \le n-1$, with the intitial value $Y_0 = x_0$, $\Delta_i = t_{i+1} - t_i$ and $\Delta \boldsymbol{W}_i = \boldsymbol{W}_{t_{i+1}} - \boldsymbol{W}_{t_i}$ for $0 \le i \le n-1$. This is the *Euler–Maruyama approximation*. This approximation gives values at the discretization times $\{t_i\}$. If the values are required at intermediate points, one can use either piecewise constant values from the preceeding discretization points or a linear interpolation from the two values at the two discretization times on either side.

The random variables $\Delta W_i = W_{t_{i+1}} - W_{t_i}$ are independent m-dimensional random vectors $\Delta W_i = \{\Delta W_i^j\}_{j=1}^m$ with independent $N(0, \Delta_i)$ components. One can obtain such a random sample by standard procedures starting from uniformly distributed samples on [0, 1] by using either the Box–Muller method or the Polar–Marsagalia method.

The Poisson jump measure p can be generated in the following way. The process $\{p(U, [0, t]), 0 \leq t \leq T\}$ is a Poisson process with intensity $\pi(U) < \infty$. Since the Poisson process has independent increments, the time intervals between the successive jumps are exponentially distributed with mean $\pi(U)^{-1}$ and are independent of each other. An exponentially distributed random variable B_μ with mean μ can be obtained from a uniformly distributed random varible U_1 on [0, 1] by the transformation $B_\mu = -\mu \log U_1$. By this method, one can generate a sequence of jump points (times) $\sigma_1, \sigma_2, \ldots, \sigma_k$ of the Poisson process p. The random marks $\{u_1, u_2, \ldots, u_k\}$ associated with these jump times need to be generated. Note that $\{u_1, u_2, \ldots, u_k\}$ are i.i.d. with distribution function

$$F_{u_k}(B) = P(u_k \in B) = \pi(B)/\pi(U) \tag{7.2.2}$$

for all Borel sets $B \subset U$. Suppose $r = 1$. If the set of marks U is discrete, then one can generate $u_1, u_2, \ldots, u_k$ from random variables which are uniform on [0, 1]. If the distribution is continuous and strictly increasing, then we can use the inverse transformation method and obtain an F_{u_k}-distributed random variables u^* from a random variable U_1 uniformly distributed on [0, 1] by $u^* = F_{u_k}^{-1}(U_1)$. If the distribution F_{u_k} has both discrete and continuous components, then we combine both the methods. Note that

$$\int_{t_k}^{t_{k+1}} \int_U c(Y_i, u)p(du, ds) = \sum_{i=1}^{\infty} c(Y_i, u_k) I_{[\sigma_k \in (t_i, t_{i+1})]}, \tag{7.2.3}$$

where I_A is the indicator function of set A.

The approximation (7.2.1), with the noise generated by the above procedure, gives a recursive algorithm for calculating approximations for the values of the process $\{Y_t\}$ at the specified discretization points.

More general discretization schemes are given in Platen (1981; 1982a) and Mikulevicius and Platen (1988). Here the discretization jump times might include jump times of the Poisson measure, and they need not be equidistant but a maximum step size may be fixed. Maruyama (1955) proved the mean square convergence of the Euler approximation for diffusion processes. Gihman and Skorohod (1979) proved a similar result for processes with jumps. Results concerning the weak convergence of the approximation (convergence in distribution) are studied by Jacod and Shiryayev (1987), Platen and Rebolledo (1985) and others.

Let us consider the case $d = 1$. For simulation of the trajectories of a process, it is essential to know how close the approximations are to the true trajectories. Let $A_\delta = E|X_T - Y_n|$, where Y_n is the Euler–Maruyama approximation to the value of the process X_t on $[0, T]$ at the terminal instant $t = T$. The approximation is said to be in the *strong sense* of order γ if $E|X_T - Y_n| \leq K\delta^\gamma$ for some constant $K > 0$, where δ is the maximum step size of true discretization. Note that $0 < \delta < 1$. It is known that the Euler–Maruyama approximation is of strong order $\gamma = 1/2$ for diffusion processes (Gihman and Skorohod 1979; Maruyama 1955). Higher-order time-discrete strong approximations for the diffusion processes have been studied in Chang (1985; 1987), Clark (1978), Clark and Cameron (1980), Jannsen (1982; 1984), Milstein (1974), Newton (1986), Nikitin and Rasevig (1978), Platen (1980), Rao *et al.* (1974),

Shimuzu and Kawachi (1984), Talay (1982) and Wagner and Platen (1978). Strong approximations for processes with jumps are studied in Dsagnidse and Tschitashvili (1975), Platen (1982a; 1982b) and Wright (1980).

In many practical problems, one may be intrested in approximation of the moments and in general of the expection $Eg(X_T)$ for some function $g(\cdot)$ which we term as an approximation in the *weak sense*. Such results have been discussed in the literature. For a list of references, see Kloeden and Platen (1989).

Wagner and Platen (1978) obtained a stochastic Taylor formula by repeated application of Itô's lemma. This was generalized by Azencott (1982), Platen (1982a) and Platen and Wagner (1982). The stochastic Taylor formula allows a function of the process X_t, that is $f(X_t)$, to be expanded about $f(X_{t_0})$ in terms of the multiple stochastic integrals weighted by the coefficients evaluated at X_{t_0}. In the one-dimensional case, a stochastic Taylor expansion for $f(X_t)$ about $f(X_{t_0})$ for $t \in [t_0, T]$ is of the form

$$\begin{aligned} f(X_t) = {} & f(X_{t_0}) + [a(X_{t_0}) f^{(1)}(X_{t_0}) + \tfrac{1}{2} b^2(X_{t_0}) f^{(2)}(X_{t_0})] \int_{t_0}^{t} ds \\ & + \int_U [f(X_{t_0} + c(X_{t_0}, u)) - f(X_{t_0}) - c(X_{t_0}, u) f^{(1)}(X_{t_0})] \pi(du) \int_{t_0}^{t} ds \\ & + b(X_{t_0}) f^{(1)}(X_{t_0}) \int_{t_0}^{t} dW_s \\ & + \int_{t_0+}^{t} \int_U \{f(X_{t_0} + c(X_{t_0}, u)) - f(X_{t_0})\} M(du, ds) \\ & + b(X_{t_0}) \{b(X_{t_0}) f^{(2)}(X_{t_0}) + b^{(1)}(X_{t_0}) f^{(1)}(X_{t_0})\} \int_{t_0}^{t} \int_{t_0}^{s_2} dW_{s_1}\, dW_{s_2} \\ & + R, \end{aligned} \tag{7.2.4}$$

where R is the remainder term consisting of stohastic integrals of higher multiplicity than those appearing in the expansion in the earlier part of the formula on the right-hand side of (7.2.4). Here $f^{(i)}$ denotes the ith derivative of f. This formula can be considered as a generalization of both the Itô formula and the deterministic Taylor formula. By truncating the stochastic Taylor expansions about successive discretization points, one can form a discrete Taylor approximation as a function of a Wiener process and these approximations may be taken as the basic numerical schemes for simulation.

The simplest strong Taylor approximation for a diffusion process is the Euler scheme. In the one-dimensional case, it has the form

$$Y_{i+1} = Y_i + a\Delta_i + b\Delta W_i, \qquad 0 \le i \le n-1,\ Y_0 = x_0. \tag{7.2.5}$$

Here $\Delta_i = t_{i+1} - t_i$ and $\Delta W_i = W_{t_{i+1}} - W_{t_i}$. In the multidimensional case, the approximation is of the form

$$Y_{i+1}^{(k)} = Y_i^{(k)} + a^{(k)}\Delta_i + \sum_{j=1}^{m} b^{(k,j)} \Delta W_i^j, \qquad 0 \le i \le n-1, \tag{7.2.6}$$

where k denotes the kth component for $1 \le k \le d$. Under the Lipschitz and the linear growth conditions, Gihman and Skorohod (1972) proved that this approximation is of strong order $\gamma = 1/2$.

If an additional term from the Taylor expansion is included in the one-dimensional case, then one has

$$Y_{i+1} = Y_i + a\Delta_i + b\Delta W_i + \tfrac{1}{2}bb^{(1)}\{(\Delta W_i)^2 - \Delta_i\} \tag{7.2.7}$$

for $0 \le i \le n-1$ (cf. Milstein 1974). This additional term is from the double Wiener integral

$$\int_{t_i}^{t_{i+1}} \int_{t_i}^{s_2} dW_{s_1}\, dW_{s_2} = \tfrac{1}{2}\{(\Delta W_i)^2 - \Delta_i\} \tag{7.2.8}$$

in (7.2.4). In the multidimensional case, the kth component of the Milstein scheme is

$$Y_{i+1}^{(k)} = Y_i^{(k)} + a^{(k)}\Delta_i + \sum_{j=1}^{m} b^{(k,j)}\Delta W_i^j$$
$$+ \frac{1}{2}\sum_{j_1,j_2=1}^{m}\sum_{i=1}^{d} b^{(i,j_1)}\frac{\partial b^{(k,j_2)}}{\partial x^i}\int_{t_i}^{t_{i+1}}\int_{t_i}^{s_2} dW_{s_1}^{j_1}\, dW_{s_2}^{j_2} \tag{7.2.9}$$

for $i = 0, 1, \ldots, n-1$ and $k = 1, 2, \ldots, d$. For $m \ge 2$, this involves multiple Wiener integrals of the form

$$I_{(j_1,j_2)} = \int_{t_i}^{t_{i+1}}\int_{t_i}^{s_2} dW_{s_1}^{j_1}\, dW_{s_2}^{j_2} \tag{7.2.10}$$

for $j_1 \ne j_2$ which cannot be expressed in terms of $\Delta W_i^{j_1}$ and $\Delta W_i^{j_2}$ as in (7.2.8). Kloeden and Platen (1989) discuss approximation of such stochastic integrals. They also study the strong and the weak approximations by the stochastic Taylor formula for diffusion processes as well as for processes with diffusion and jump components. For details, see Kloeden and Platen (1989) and additional references given there. Computational complexity and additional problems of a practical nature in the simulation of SDEs are discussed in Kloeden *et al.* (1994).

7.3 Effects of Discretization in Estimation for Diffusion Processes

The need for parametric as well as nonparametric inference based on sampled data from a diffusion process or in general from any stochastic process has been discussed earlier in this book. Following Kloeden *et al.* (1996), we now discuss the effects of the discretization on parameter estimation for diffusion processes.

Consider the solution $\{X_t\}$ of the SDE

$$dX_t = a(X_t)\,dt + \sigma(X_t)\,dW_t, \qquad 0 \le t \le T,\ X_0 = x_0. \tag{7.3.1}$$

We would like to simulate the sample paths of the process $\{X_t\}$. Since the solution $\{X_t\}$ is generally unknown, we would approximate $\{X_t\}$ by $\{Y_t^{\Delta}\}$ which can be simulated and which has the property that Y^{Δ} converges to X as $\Delta \to 0$ in a suitable sense. As was mentioned in Section 7.2, a discrete time approximation Y^{Δ} is said to converge with strong order $\gamma > 0$ at time T if there exist positive contants K and $\delta_0 < T$ such that, for all $\Delta \in (0, \delta_0)$,

$$E(|X_T - Y_T^{\Delta}|) \le K\Delta^{\gamma}, \tag{7.3.2}$$

where K does not depend on the time step size Δ. We restrict attention to the equidistant time discretization of $[0, T]$ in the following discussion, with

$$t_n = n\Delta, \qquad n = 1, \ldots, n_T, \tag{7.3.3}$$

and

$$\Delta = T/n_T$$

for some integer n_T. The Euler scheme in this case is

$$Y_{n+1} = Y_n + a(Y_n)\Delta + \sigma(Y_n)\Delta W_n \tag{7.3.4}$$

for $n \geq 0$ with $Y_0^\Delta = X_0$ and $\Delta W_n = W_{t_{n+1}} - W_{t_n}$. This scheme provides a recursive algorithm for simulating an approximate solution of (7.3.1). Under the usual Lipschitz and growth conditions on $a(\cdot)$ and $\sigma(\cdot)$, it is known that the Euler approximation converges with strong order $\gamma = 1/2$. If we include an extra term in (7.3.4) and consider

$$Y_{n+1} = Y_n + a(Y_n)\Delta + \sigma(Y_n)\Delta W_n + \tfrac{1}{2}\sigma(Y_n)\sigma'(Y_n)\{(\Delta W_n)^2 - \Delta\}, \tag{7.3.5}$$

where $\sigma'(\cdot)$ denotes the derivative of $\sigma(\cdot)$, then it can be shown that the approximation is of strong order $\gamma = 1$ (Milstein 1974). Further additional terms in (7.3.5) lead to an approximation which is of strong order $\gamma = 3/2$. This is given by

$$\begin{aligned} Y_{n+1} = Y_n &+ a\Delta + \sigma\Delta W_n + \tfrac{1}{2}\sigma\sigma'\{\Delta W_n)^2 - \Delta\} \\ &+\sigma a'\Delta Z_n + \tfrac{1}{2}\{aa' + \tfrac{1}{2}\sigma^2 a''\}\Delta^2 \\ &+\{a\sigma' + \tfrac{1}{2}\sigma^2\sigma''\}\{\Delta W_n\Delta - \Delta Z_n\} \\ &+\tfrac{1}{2}\sigma(\sigma\sigma')'\{\tfrac{1}{3}(\Delta W_n)^2 - \Delta\}\Delta W_n, \end{aligned} \tag{7.3.6}$$

where the coefficients $a, \sigma, \ldots$ are calculated at Y_n. Here prime denotes the derivative, and

$$\Delta Z_n = \int_{t_n}^{t_{n+1}} \int_{t_n}^{s_2} dW_{s_1}\, ds_2 \tag{7.3.7}$$

which is Gaussian with mean zero and variance $\frac{1}{3}\Delta^3$, with $E(\Delta Z_n \Delta W_n) = \frac{1}{2}\Delta^2$. The pairs $(\Delta W_n, \Delta Z_n)$ are independent for $n = 1, 2, \ldots$.

Kloeden *et al.* (1996) considered the scheme

$$Y_{n+1} = Y_n + \tilde{a}(Y_n)\Delta + \tfrac{1}{2}\{\sigma(Y_n + \tilde{a}(Y_n)\Delta + \sigma(Y_n)\Delta W_n) + \sigma(Y_n)\}\Delta W_n, \tag{7.3.8}$$

where

$$\tilde{a} = a - \tfrac{1}{2}\sigma\sigma' \tag{7.3.9}$$

which is of strong order $\gamma = 1$ (see also Kloeden and Platen 1992).

Example 7.3.1 *(Ornstein–Uhlenbeck process)*
Consider the SDE

$$dX_t = (\theta_1 X_t + \theta_2)\, dt + \sigma\, dW_t, \qquad t \geq 0, \tag{7.3.10}$$

where σ is known. The maximum likelihood estimators based on the continuous paths of $\{X_t\}$ over $[0, T]$ are

$$\begin{aligned} \hat{\theta}_{1,T} &= \left\{T \int_0^T X_t\, dX_t - (X_T - X_0)\int_0^T X_t\, dt\right\} \Big/ N_T \\ &= \left\{\tfrac{1}{2}T(X_T^2 - X_0^2 - \sigma^2 T) - (X_T - X_0)\int_0^T X_t\, dt\right\} \Big/ N_T \end{aligned} \tag{7.3.11}$$

and

$$\hat{\theta}_{2,T} = \left\{(X_T - X_0)\int_0^T X_t^2\,dt - \int_0^T X_t\,dX_t \int_0^T X_t\,dt\right\}/N_T$$
$$= \left\{(X_T - X_0)\int_0^T X_t^2\,dt - \tfrac{1}{2}(X_T^2 - X_0^2 - \sigma_T^2)\int_0^T X_t\,dt\right\}\Big/N_T, \quad (7.3.12)$$

where

$$N_T = T\int_0^T X_t^2\,dt - \left(\int_0^T X_t\,dt\right)^2. \quad (7.3.13)$$

One can obtain approximations to $\hat{\theta}_{1,T}$ and $\hat{\theta}_{2,T}$ by using the trapezoidal formula. The approximate estimators are

$$\hat{\theta}_{1,T}^{(\Delta)} = \frac{1}{2N_T^{(\Delta)}}\left\{T(X_T^2 - X_0^2 - \sigma^2 T) - (X_T - X_0)\sum_{n=0}^{n_T-1}(X_{t_{n+1}} + X_{t_n})\Delta\right\} \quad (7.3.14)$$

and

$$\hat{\theta}_{2,T}^{(\Delta)} = \frac{1}{2N_T^{(\Delta)}}\left\{X_T - X_0\sum_{n=0}^{n_T-1}(X_{t_{n+1}}^2 + X_{t_n}^2)\Delta\right.$$
$$\left. -\tfrac{1}{2}(X_T^2 - X_0^2 - \sigma^2 T)\sum_{n=0}^{n_T-1}(X_{t_{n+1}}X_{t_n})\Delta\right\}, \quad (7.3.15)$$

where

$$N_T^{(\Delta)} = \frac{T}{2}\sum_{n=0}^{n_T-1}(X_{t_{n+1}}^2 + X_{t_n}^2)\Delta - \left(\frac{1}{2}\sum_{n=0}^{n_T-1}(X_{t_{n+1}} + X_{t_n})\Delta\right)^2. \quad (7.3.16)$$

Kloeden *et al.*(1996) used an approximation of the form

$$Y_{n+1} = Y_n + \{\theta_1 Y_n + \theta_2\}\Delta + \tfrac{1}{2}\theta_1(\theta_1 Y_n + \theta_2)\Delta^2 + \sigma\Delta W_n + \sigma\theta_1\Delta Z_n \quad (7.3.17)$$

which is of strong order $\gamma = 2$ with step size $\Delta = 2^{-4}$ for simulation of the paths of the process.

Example 7.3.2 (*Duffing–Van der Pol oscillator*)
The velocity $X_t^{(2)}$ of a Duffing–Van der Pol oscillator is determined by the SDE

$$dX_t^{(2)} = [\alpha X_t^{(1)} - (X_t^{(2)} + X_t^{(1)3})]\,dt + \sigma X_t^{(1)}\,dW_t, \quad (7.3.18)$$

where

$$X_t^{(1)} = X_0^{(1)} + \int_0^t X_s^{(2)}\,ds \quad (7.3.19)$$

is the position of the oscillator. We assume that σ is known. Then the maximum likelihood estimator for α is

$$\hat{\alpha}_T = T^{-1}\left\{\int_0^T X_t^{(1)+}\,d\tilde{X}_t + \int_0^T X_t^{(1)^2}\,dt\right\}, \quad (7.3.20)$$

where

$$\tilde{X}_t = X_t^{(2)} + \int_0^t X_s^{(2)}\, ds \tag{7.3.21}$$

and

$$X_t^{(1)+} = \begin{cases} X_t^{(1)^{-1}}, & \text{if } X_t^{(1)} \neq 0, \\ 0, & \text{otherwise.} \end{cases} \tag{7.3.22}$$

It is easy to see that

$$\hat{\alpha}_T = \alpha + T^{-1}\sigma W_T \tag{7.3.23}$$

and hence $\hat{\alpha}_T$ is normal with mean α and variance $T^{-1}\sigma^2$. Note that, for trajectories where $X_t^{(1)}$ does not cross the level zero in $[0, T]$,

$$\int_0^T X_t^{(1)^{-1}}\, dX_t^{(2)} = \frac{X_T^{(2)}}{X_T^{(1)}} - \frac{X_0^{(2)}}{X_0^{(1)}} + \int_0^T \left[\frac{X_t^{(2)}}{X_t^{(1)}}\right]^2 dt \tag{7.3.24}$$

and hence

$$\hat{\alpha}_T = T^{-1}\left\{\frac{X_T^{(2)}}{X_T^{(1)}} - \frac{X_0^{(2)}}{X_0^{(1)}} + \int_0^T \left(\left[\frac{X_t^{(2)}}{X_t^{(1)}}\right]^2 + \frac{X_t^{(2)}}{X_t^{(1)}} + X_t^{(1)^2}\right) dt\right\}. \tag{7.3.25}$$

Kloeden *et al.* (1996) used the scheme

$$Y_{n+1}^{(2)} = Y_n^{(2)} + \{\theta Y_n^{(1)} - (Y_n^{(2)} + Y_n^{(1)^3})\}\Delta + \sigma Y_n^{(1)} \Delta W_n \tag{7.3.26}$$

using $Y_{n+1}^{(1)} = Y_n^{(1)} + Y_n^{(2)}\Delta$ for simulating the paths of the process. The time step size is $\Delta = 10^{-2}$. It is known that this scheme is of strong order $\gamma = 1$.

An approximation for the estimator $\hat{\alpha}_T$ is

$$\begin{aligned} \hat{\alpha}_T^{(\Delta)} = T^{-1}\Bigg\{ & \frac{X_T^{(2)}}{X_T^{(1)}} - \frac{X_0^{(2)}}{X_0^{(1)}} + \frac{1}{2}\sum_{n=0}^{n_T-1}\Bigg(\left(\frac{X_{t_{n+1}}^{(2)}}{X_{t_{n+1}}^{(1)}}\right)^2 + \left(\frac{X_{t_n}^{(2)}}{X_{t_n}^{(1)}}\right)^2 \\ & + \frac{X_{t_{n+1}}^{(2)}}{X_{t_{n+1}}^{(1)}} + \frac{X_{t_n}^{(2)}}{X_{t_n}^{(1)}} + (X_{t_{n+1}}^{(1)})^2 + (X_{t_n}^{(1)})^2\Bigg)\Delta\Bigg\}. \end{aligned} \tag{7.3.27}$$

This estimator can be used when $X_{t_n} \neq 0$ for $n = 0, 1, \ldots, n_T$.

Example 7.3.3 (*Population model*)
Consider a logistic population model of the diffusion branching type

$$dX_t = \alpha X_t(K - X_t)\, dt + \sigma\sqrt{X_t}\, dW_t, \qquad t \geq 0, \tag{7.3.28}$$

where K is the carrying capacity of the environment. Let $\theta_1 = \alpha K$ and $\theta_2 = -\alpha$. We assume that σ is known. Then (7.3.28) can be written in the form

$$dX_t = (\theta_1 X_t + \theta_2 X_t^2)\, dt + \sigma\sqrt{X_t}\, dW_t. \tag{7.3.29}$$

The maximum likelihood estimators of θ_1 and θ_2 based on $\{X_t, 0 \le t \le T\}$ are

$$\hat{\theta}_{1,T} = \left\{(X_T - X_0)\int_0^T X_t^3\,dt - \int_0^T X_t\,dX_t \int_0^T X_t^2\,dt\right\} \Big/ N_T$$
$$= \left\{(X_t - X_0)\int_0^T X_t^3\,dt - \tfrac{1}{2}\left(X_T^2 - X_0^2 - \sigma^2\int_0^T X_t\,dt\right)\int_0^T X_t^2\,dt\right\} \Big/ N_T \tag{7.3.30}$$

and

$$\hat{\theta}_{2,T} = \left\{\int_0^T X_t\,dX_t \int_0^T X_t\,dt - (X_T - X_0)\int_0^T X_t^2\,dt\right\} \Big/ N_T$$
$$= \left\{\tfrac{1}{2}\left(X_T^2 - X_0^2 - \sigma^2\int_0^T X_t\,dt\right)\int_0^T X_t\,dt - (X_T - X_0)\int_0^T X_t^2\,dt\right\} \Big/ N_T, \tag{7.3.31}$$

where

$$N_T = \left(\int_0^T X_t\,dt \int_0^T X_t^3\,dt\right) - \left(\int_0^T X_t^2\,dt\right)^2. \tag{7.3.32}$$

Kloeden *et al.* (1996) used the scheme

$$Y_{n+1} = Y_n + \{\theta_1 Y_n + \theta_2 Y_n^2 - \tfrac{1}{4}\sigma^2\}\Delta$$
$$+ \tfrac{1}{2}\sigma\{[(Y_n + (\theta_1 Y_n + \theta_2 Y_n^2 - \tfrac{1}{4}\sigma^2)\Delta + \sigma\sqrt{Y_n^+}\,\Delta W_n)^+]^{1/2}$$
$$\sqrt{Y_n^+}\}\Delta W_n \tag{7.3.33}$$

with time step size $\Delta = 2^{-6}$ for simulation purposes. This scheme is of strong order $\gamma = 1.0$. Here x^+ denotes the positive part of x. The parameters θ_1 and θ_2 are estimated by discretizing the integrals in (7.3.30)–(7.3.32) by the trapezoidal formula. The discretized estmators $\hat{\theta}_{1,T}^{(\Delta)}$ and $\hat{\theta}_{2,T}^{(\Delta)}$ are given by

$$\hat{\theta}_{1,T}^{(\Delta)} = \frac{1}{2N_T^{(\Delta)}}\left\{(X_T - X_0)\sum_{n=0}^{n_T-1}(X_{t_{n+1}}^3 + X_{t_n}^3)\Delta\right.$$
$$\left. -\frac{1}{2}\left[X_T^2 - X_0^2 - \frac{\sigma^2}{2}\sum_{n=0}^{n_T-1}(X_{t_{n+1}} + X_{t_n})\Delta\right]\sum_{n=0}^{n_T-1}(X_{t_{n+1}}^2 + X_{t_n}^2)\Delta\right\} \tag{7.3.34}$$

and

$$\hat{\theta}_{2,T}^{(\Delta)} = \frac{1}{2N_T^{(\Delta)}}\left\{\frac{1}{2}\left[X_T^2 - X_0^2 - \frac{\sigma^2}{2}\sum_{n=0}^{n_T-1}(X_{t_{n+1}} + X_{t_n})\Delta\right]\sum_{n=0}^{n_T-1}(X_{t_{n+1}} + X_{t_n})\Delta\right.$$
$$\left. -(X_T - X_0)\sum_{n=0}^{n_T-1}(X_{t_{n+1}}^2 + X_{t_n}^2)\Delta\right\}, \tag{7.3.35}$$

where

$$N_T^{(\Delta)} = \frac{1}{4} \sum_{n=0}^{n_T-1} (X_{t_{n+1}} + X_{t_n})\Delta \sum_{n=0}^{n_T-1} (X_{t_{n+1}}^3 + X_{t_n}^3)\Delta - \left\{ \frac{1}{2} \sum_{n=0}^{n_T-1} (X_{t_{n+1}}^2 + X_{t_n}^2)\Delta \right\}^2. \tag{7.3.36}$$

For typical simulated trajectories for all three examples discussed here, see Kloeden *et al.* (1996). A comparative study of estimation methods for continuous-time stochastic processes by simulation techniques is given in Shoji and Ozaki (1997).

7.4 Uniform Approximation for Stochastic Integrals of Functions of a Solution of a Stochastic Differential Equation

Suppose $\{X(t), t \geq 0\}$ is the unique stationary solution of an SDE

$$dX(t) = a(X(t), \theta)\, dt + dW(t), \qquad 0 \leq t \leq T,\ X(0) = X_0, \tag{7.4.1}$$

where $\theta \in \Theta$ is unknown. In problems of statistical inference, one encounters stochastic integrals of the form

$$\int_0^T f(X(t), \theta)\, dW(t), \qquad \theta \in \Theta, \tag{7.4.2}$$

where θ is the unknown parameter in a compact set $\Theta \subset R$. The problem of interest is to find a uniform approximation to the above stochastic integral uniform in the parameter θ. In addition to the Lipschitzian and the linear growth conditions for the existence of a unique solution $\{X(t)\}$ for fixed $\theta \in \Theta$, we assume the following conditions:

(YY1) $f(x, \theta)$ is differentiable with respect to x and θ with partial derivatives $f_{x\theta}$ and $f_{xx\theta}$. Furthermore, $f_{x\theta}$ is Lipschitzian in θ of order $\alpha > 1/2$ and $f_{xx\theta}$ is Lipschitzian in θ of order α uniformly in x.

(YY2) $E(X_0^8) < \infty$.

Let $I(\theta)$ and $S(\theta)$ denote the integral in (7.4.2) defined *sensu* Itô and Rubin–Fisk–Stratonovich, respectively. Prakasa Rao and Rubin (1998) show that Rubin–Fisk–Stratonovich approximation $S_n(\theta)$ to $S(\theta)$ converges at a faster rate than the Itô approximation $I_n(\theta)$ to $I(\theta)$. In fact

$$E[\sup_\theta |S_n(\theta) - S(\theta)|^2] = O(n^{-2}), \qquad E[\sup_\theta |I_n(\theta) - I(\theta)|^2] = O(n^{-1}). \tag{7.4.3}$$

Specifically, the Rubin–Fisk–Stratonovich type approximation obtained from equal size partitions converges faster to the corresponding Rubin–Fisk–Stratonovich stochastic integral than the approximation of Itô type for such partitions converges to the corresponding Itô integral. The rate of convergence is uniform over the parameter θ in a compact parameter space Θ.

Appendix A

Uniform Ergodic Theorem

The following result is due to Prakasa Rao (1982).

Theorem A.1 *Let $\{X_t, t \geq 0\}$ be a stationary ergodic process under $P_\theta, \theta \in \Theta$ open $\subset R$. Suppose that $g(s, x)$ is continuous in s a.s. $[P_\theta]$ and $|g(s, x)| \leq h(x)$ for all $s \in I(\theta)$, where $I(\theta)$ is a closed interval containing θ. Furthermore, suppose that $E_\theta[g(s, X_0)] = 0$ for all $s \in I(\theta)$ and $E_\theta[h(X_0)] < \infty$. Then*

$$\sup_{s\in I(\theta)} \left| \frac{1}{T} \int_0^T g(s, X_t)\, dt \right| \to 0 \text{ a.s. } [P_\theta] \qquad \text{as } T \to \infty. \tag{A.1}$$

Proof. Without loss of generality, assume that $g(s, x)$ is continuous in s for all $x \in R$. Let $I(s, \rho)$ be an open interval of length 2ρ centered at s and define

$$U(x; s, \rho) = \inf\{g(s, x) : s \in I(\theta) \cap \overline{I(s, \rho)}\}. \tag{A.2}$$

Since $g(s, x)$ is continuous and $I(\theta) \cap \overline{I(s, \rho)}$ is closed, it follow that $U(x; s, \rho) \uparrow g(s, x)$ as $\rho \downarrow 0$ a.s. $[P_\theta]$. Note that $|E_\theta(U(X_0; s, \rho))| < \infty$ since $E_\theta[h(X_0)] < \infty$. Hence, by the monotone convergence theorem, it follows that

$$E_\theta[U(X_0; s, \rho)] \to 0 \text{ as } \rho \downarrow 0. \tag{A.3}$$

Therefore, given $\varepsilon > 0$, there exists $\rho_1 = \rho_1(s, \varepsilon) > 0$ such that

$$E_\theta[U(X_0; s, \rho_1] > -\varepsilon. \tag{A.4}$$

Let $\rho_2 = \frac{1}{2}\rho_1(s, \varepsilon)$ and consider the closed interval $\overline{I(s, \rho_2)}$ centered at s and of length $2\rho_2$. Note that for any $s' \in I(\theta) \cap \overline{I(s, \rho_2)}$,

$$U(x; s', \rho_2) \geq U(x; s', \rho_1), \qquad x \in R,$$

since $I(s', \rho_2) \subset I(s, \rho_1)$. Therefore, given $s \in I(\theta)$,

$$E_\theta[U(X_0; s', \rho_2)] \geq -\varepsilon \tag{A.5}$$

for all $s' \in I(\theta) \cap \overline{I(s, \rho_2)}$.

Observe that $\{I(s, \rho_2) : s \in I(\theta)\}$ is an open covering for $I(\theta)$. Since $I(\theta)$ is compact, there exist a finite number of open intervals $I(s_i, \rho_{2i})$, $1 \le i \le k$, covering $I(\theta)$. It is clear that

$$\min_{1\le i\le k} \{E_\theta(U(X_0; s_i, \rho_{2i}))\} \ge -\varepsilon. \tag{A.6}$$

Given $s \in I(\theta)$, there exists an s_i such that $s \in I(s_i, \rho_{2i})$ for some $1 \le i \le k$ and

$$g(s, x) \ge U(x; s_i, \rho_{2i}), \qquad x \in R, \tag{A.7}$$

and hence

$$\frac{1}{T}\int_0^T g(s, X_t)\, dt \ge \frac{1}{T}\int_0^T U(X_t; s_i, \rho_{2i})\, dt \tag{A.8}$$

for all $T \ge 0$. Therefore

$$\inf_{s\in I(\theta)} \frac{1}{T}\int_0^T g(s, X_t)\, dt \ge \min_{1\le i\le k} \left\{\frac{1}{T}\int_0^T U(X_t; s_i, \rho_{2i})\, dt\right\}. \tag{A.9}$$

However, by the ergodic theorem,

$$\frac{1}{T}\int_0^T U(X_t; s_i, \rho_{2i})\, dt \to E_\theta[U(X_0; s_i, \rho_{2i})] \tag{A.10}$$

as $T \to \infty$ a.s. $[P_\theta]$ for $1 \le i \le k$. Hence, by (A.6), we have

$$\liminf_{T\to\infty} \inf_{s\in I(\theta)} \left\{\frac{1}{T}\int_0^T g(s, X_t)\, dt\right\} \ge -\varepsilon \text{ a.s. } [P_\theta]. \tag{A.11}$$

Similarly we obtain that

$$\limsup_{T\to\infty} \sup_{s\in I(\theta)} \left\{\frac{1}{T}\int_0^T g(s, X_t)\, dt\right\} \le \varepsilon \text{ a.s. } [P_\theta]. \tag{A.12}$$

Relations (A.11) and (A.12) together prove the theorem.

Appendix B

Stochastic Integration and Limit Theorems for Stochastic Integrals

B.1 Stochastic Integrals with Respect to a Wiener Process

Definition. Let $(\Omega, \mathcal{F}, P)$ be a probability space and $\{\mathcal{F}_t, t \geq 0\}$ be a filtration. A stochastic process $W = \{W_t, \mathcal{F}_t, t \geq 0\}$ is called a *standard Wiener Process* relative to the filtration $\{\mathcal{F}_t, t \geq 0\}$ if:

(i) the trajectories $W_t(\omega), t \geq 0$ are continuous (P-a.s.) in t;

(ii) $W = \{W_t, \mathcal{F}_t, t \geq 0\}$ is a square-integrable martingale with $W_0 = 0$; and

(ii) $E(W_t - W_s)^2 | \mathcal{F}_s] = t - s, t \geq s$.

Definition. A stochastic process $B = \{B_t, t \geq 0\}$ on a probability space $(\Omega, \mathcal{F}, P)$ is called *standard Brownian motion* if

(i) $B_0 = 0$ a.s. $[P]$;

(ii) B is a process with stationary independent increments; and

(iii) $B_t - B_s \simeq N(0, |t - s|)$, where $N(0, \sigma^2)$ denotes the Gaussian distribution with mean 0 and variance σ^2.

Theorem B.1 *Any Wiener process is a Brownian motion.*

Theorem B.2 *Let* $0 \equiv t_0^{(n)} < t_1^{(n)} < \cdots < t_n^{(n)} \equiv T$ *be a subdivision of* $[0, T]$ *such that* $\max_i |t_{i+1}^{(n)} - t_i^{(n)}| \to 0$ *as* $n \to \infty$. *Then*

$$\lim_{n\to\infty} \sum_{i=0}^{n-1} [W_{t_{i+1}^{(n)}} - W_{t_i^{(n)}}]^2 = T \text{ a.s.}$$

and

$$\underset{n\to\infty}{\text{l.i.m.}} \sum_{i=0}^{n-1} [W_{t_{i+1}^{(n)}} - W_{t_i^{(n)}}]^2 = T,$$

where l.i.m. *denotes convergence in the quadratic mean.*

Proof. It is clear that

$$E\left[\sum_{i=0}^{n-1} \{W_{t_{i+1}^{(n)}} - W_{t_i^{(n)}}\}^2\right] = T,$$

and hence it is sufficient to prove that

$$\operatorname{var}\left[\sum_{i=0}^{n-1}\{W_{t_{i+1}^{(n)}} - W_{t_i^{(n)}}\}^2\right] \to 0.$$

This follows from the fact that the process W has independent increments with $\operatorname{var}[W_t - W_s] = |t - s|$. The almost sure convergence property can be proved by the Borel–Cantelli lemma in the special case $t_i^{(n)} = \frac{i}{n}T$.

Definition. A stochastic process $\{f(t, \omega), t \geq 0, \omega \in \Omega\}$ is called *nonanticipative* with respect to the filtration $\{\mathcal{F}_t\}$, if for each t, $f(t, \cdot)$ is $\mathcal{F}_t$-measurable.

Definition. A stochastic process $f(t, \omega)$ is called *simple nonanticipative* if there exists a finite subdivision $0 = t_0 < t_1 < \cdots < t_n = T$ of the interval $[0, T]$ where

$$\begin{aligned} f(t,\omega) &= \alpha_0(\omega)\ (\mathcal{F}_0\text{-measurable}) \qquad &&\text{for } t = t_0 = 0 \\ &= \alpha_i(\omega)\ (\mathcal{F}_{t_i}\text{-measurable}) \qquad &&\text{for } t_i < t \leq t_{i+1}, 1 \leq i \leq \mathcal{F}_{n-1} \end{aligned}$$

and $E[\int_0^T f^2(t, \omega)\, dt] < \infty$.

For f simple, define

$$I_T(f) \equiv \int_0^T f\, dW \equiv \int_0^T f(t, \omega)\, dW(t, \omega) = \sum_{i=0}^{n-1} \alpha_i[W(t_{i+1}) - W(t_i)].$$

Lemma B.3 *Let f be nonanticipative and $E[\int_0^T f^2(t, \omega)\, dt] < \infty$, that is, $f \in \mathcal{M}_T$. Then there exists a sequence f_n of simple functions such that*

$$E\left\{\int_0^T [f - f_n]^2\, dt\right\} \to 0 \qquad \textit{as } n \to \infty.$$

For $f \in \mathcal{M}_T$, we define

$$I_T(f) = \underset{n\to\infty}{\text{l.i.m.}}\ I_T(f_n).$$

Let f be nonanticipative such that

$$P\left\{\int_0^T f^2(t, \omega)\, dt < \infty\right\} = 1.$$

Denote this class by $\mathcal{P}_T$.

Lemma B.4 *Let $f \in \mathcal{P}_T$. Then there exist $f_n \in \mathcal{M}_T$ such that $\int_0^T [f_n - f]^2\, dt \to 0$ in probability as $n \to \infty$.*

For $f \in \mathcal{P}_T$, define

$$I_T(f) = \int_0^T f\, dW = \lim_n -p \int_0^T f_n\, dW,$$

where $\{f_n\}$ is as given in Lemma B.4 and $\lim_n -p$ denotes the limit in probability. It can be shown that the limit $I_T(f)$ does not depend on the choice of the sequence $\{f_n\}$ in Lemma B.4. This integral is called the *Itô stochastic integral* of f with respect to the Wiener process.

Properties of an Itô stochastic integral

(i) If $f \in \mathcal{M}_T$, then

$$E\left[\int_0^T f\,dW\right] = 0,$$

and

$$E\left[\int_0^T f\,dW\right]^2 = E\left[\int_0^T f^2(t)\,dt\right].$$

(ii) If f_1 and $f_2 \in \mathcal{M}_T$, then

$$E\left[\int_0^T f_1\,dW \int_0^T f_2\,dW\right] = E\left[\int_0^T f_1 f_2\,dt\right].$$

(iii) If $f \in \mathcal{P}_T$, then, for any $\varepsilon > 0$ and $\delta > 0$,

$$P\left\{\left|\int_0^T f(t)\,dW(t)\right| > e\right\} \le P\left\{\int_0^T f^2(t)\,dt > \delta\right\} + \frac{\delta}{\varepsilon^2}.$$

All the above properties are first proved for simple functions and then extended by a limiting argument.

Remark. Note that

$$\int_0^T W\,dW = \frac{W^2(T) - T}{2}.$$

The calculus of an Itô integral is not the same as ordinary calculus. Stratonovich (1996) defined a symmetic version of a stochastic integral which adapts to the ordinary calculus.

Further properties of an Itô stochastic integral

Let

$$I_t(f) = \int_0^t f\,dW, \qquad f \in \mathcal{P}_T,\ 0 \le t \le T.$$

Then $\{I_t(f), \mathcal{F}_t, 0 \le t \le T\}$ is a martingale with continuous sample paths almost surely. Further

$$P\left\{\sup_{0\le t\le T}\left|\int_0^t f(s)\,dW(s)\right| > \varepsilon\right\} \le P\left\{\int_o^T f^2(t)\,dt > \delta\right\} + \frac{\delta}{\varepsilon^2}.$$

In particular, if $f \in \mathcal{M}_T$, then

$$P\left\{\sup_{0\le t\le T}\left|\int_o^t f(s)\,dW(s)\right| > \varepsilon\right\} \le \frac{1}{\varepsilon^2}\int_o^T Ef^2(t)\,dt.$$

Suppose τ is a stopping time with respect to $\{\mathcal{F}_t\}$. Then

$$\int_0^\tau f\,dW$$

stands for the process $I_t(f)$ stopped at time τ. If $\tau_1 \le \tau_2$ are stopping times, then

$$\int_{\tau 1}^{\tau_2} f\,dW = \int_0^{\tau_2} f\,dW - \int_0^{\tau_1} f\,dW.$$

If $f \in \mathcal{M}_T$, then

(i) $E\left[\int_{\tau_1}^{\tau_2} f\,dW\right] = 0$,

(ii) $E\left[\int_{\tau_1}^{\tau_2} f\,dW\right]^2 = E\left[\int_{\tau_1}^{\tau_2} f^2\,dt\right]$,

(iii) $E\left[\int_{\tau_1}^{\tau_2} f\,dW|\mathcal{F}_{\tau_1}\right] = 0$ a.s., and

(iv) $E\left[\left\{\int_{\tau_1}^{\tau_2} f\,dW\right\}^2|\mathcal{F}_{\tau_1}\right] = E\left[\int_{\tau_1}^{\tau_2} f^2\,dt|\mathcal{F}_{\tau_1}\right]$ a.s.

Lemma B.5 *For any $f \in \mathcal{M}_1$ and $\alpha > 0$, $\beta > 0$,*

$$P\left[\sup_{0\le t\le 1}\left\{\int_0^t f(s)\,dW(s) - \frac{\alpha}{2}\int_0^t f^2(s)\,ds\right\} > \beta\right] \le e^{-\alpha\beta}.$$

This lemma follows from the fact that $\mathcal{Z}(t) = \exp\{\int_0^t f(s)\,dW(s) - \frac{1}{2}\int_0^t f^2(s)\,ds\}$, $0 \le t \le 1$, is a supermartingale.

Remark. As a consequence of the above properties, it follows that if $W = \{W_t, \mathcal{F}_t\}$ is a Wiener process, then for any stopping time τ adapted to $\{\mathcal{F}_t\}$ with $E\tau < \infty$, $EW_\tau = 0$ and $EW_\tau^2 = E\tau$.

Definition. A random process $X = \{X_t, \mathcal{F}_t, 0 \le t \le T\}$ is called an *Itô process* (relative to the Wiener process $W = \{W_t, \mathcal{F}_t, 0 \le t\}$) if there exist nonanticipative processes $a = \{a_t, \mathcal{F}_t, 0 \le t \le T\}$ and $b = \{b_t, \mathcal{F}_t, 0 \le t \le T\}$ such that

$$P\left\{\int_0^T |a_t|\,dt < \infty\right\} = 1,$$

$$P\left\{\int_0^T b_t^2\,dt < \infty\right\} = 1,$$

and

$$X_t = X_0 + \int_0^t a(s,\omega)\,ds + \int_0^t b(s,\omega)\,dW(s) \text{ a.s.}$$

for all $0 \le t \le T$. Then X is said to have the stochastic differential

$$dX_t = a(t,\omega)\,dt + b(t,\omega)\,dW(t), \qquad 0 \le t \le T.$$

Remark. If $a(s,\cdot)$ and $b(s,\cdot)$ are measurable with respect to the σ-algebra generated by $X(t)$, $0 \le t \le s$, then the process X is said to be of diffusion type. If $a(s,\omega)$ and $b(s,\omega)$ are of the form $A(s, X(s,\omega))$ and $B(s, X(s,\omega))$, then X is said to be a diffusion.

Suppose an Itô process X has the stochastic differential

$$dX_t = a(t)\,dt + b(t)\,dW(t), \qquad 0 \le t \le T.$$

Let $F(t,x)$ be a continuous function on $[0,T] \times R$. The following result gives the stochastic differential for the process $F(t, X(t))$ under some smoothness assumptions.

Lemma B.6 (*Itô*) *Let $F(t,x)$ be a continuous function on $[0,T] \times R$ with continuous derivatives $F_t(t,x)$, $F_x(t,x)$, $F_{xx}(t,x)$, and let X be an Itô process with stochastic differential*

$$dX_t = a(t)\,dt + b(t)\,dW(t), \qquad 0 \le t \le T.$$

Then the process $Z(t) = F(t, X(t))$ is an Itô process with stochastic differential

$$dZ(t) = [F_t(t, X(t)) + F_x(t, X(t))a(t) + \tfrac{1}{2}F_{xx}(t, X(t))b^2(t)]\,dt \\ +F_x(t, X(t))b(t)\,dW(t),$$

where $Z(0) = F(0, X(0))$.

For a more general version, see Kunita (1990).
We now obtain a maximal inequality.

Theorem B.7 *Let a random process $\{f(u,t), \mathcal{F}_t, 0 \le t \le T\}$ be continuously differentiable with respect to u with probability one and $\left|\frac{\partial f(u,t)}{\partial u}\right| < c_1 < \infty$. Let*

$$\zeta(u) = \int_0^T f(u,t)\,dW(t).$$

Then there exists a constant $C > 0$ such that, for any $N > 0$,

$$P\left(\sup_{A<u<B} |\zeta(u) - \zeta(A)| > N\right) \le 8K_2 \exp\left\{-\frac{N}{2K_1(B-A)}\right\} \tag{B.1}$$

where $K_1 = C\sqrt{T}$ and $K_2 = 1 + \exp\{-\frac{1}{2}K_1^2\}$.

Proof. Note that

$$\begin{aligned}
&P\{\zeta(u+h) - \zeta(u) > c\} \\
&= P\left\{\int_0^T [f(u+h,t) - f(u,t)]\,dW(t) > c\right\} \\
&\le P\left\{\int_0^T \frac{\Delta f}{h}\,dW - \frac{1}{2}\int_0^T \left(\frac{\Delta f}{h}\right)^2 dt > \frac{c}{2h}\right\} \\
&\quad + P\left\{\frac{1}{2}\int_0^T \left(\frac{\Delta f}{h}\right)^2 dt > \frac{c}{2h}\right\} \\
&< e^{-\frac{c}{2h}}\left(1 + E\exp\left\{\frac{1}{2}\int_0^T \left(\frac{\Delta f}{h}\right)^2 dt\right\}\right) \\
&\le K_2 e^{-\frac{c}{2h}},
\end{aligned}$$

where $\Delta f = f(u+h,t) - f(u,t)$ by Chebyshev's inequality and the inequality

$$E\left[\exp\left\{\int_0^T g(t)\,dW(t) - \frac{1}{2}\int_0^T g^2(t)\,dt\right\}\right] \le 1$$

(cf. Liptser and Shiryayev 1977; or Basawa and Prakasa Rao 1980). See Lemma B.5. Let a_n be a sequence such that

$$\sum_{n=1}^{\infty} a_n 2^{-n} = N' = \frac{N}{K_1(B-A)}. \tag{B.2}$$

Then, a lemma due to Burnashev (see Gihman and Skorohod 1977, p. 240) shows that

$$P\{\sup_{A<u<B} [\zeta(u) - \zeta(A)] > N\} \le \frac{K_2}{2} \sum_{n=1}^{\infty} 2^n e^{-a_n}. \tag{B.3}$$

Minimizing the quantity on the right-hand side of inequality (B.3) with respect to $\{a_n\}$, under (B.2), by the method of Lagrange multipliers, we obtain that $a_n = N' - 4\log 2 - 2n\log 2$, and, substituting this value for a_n into (B.3), we arrive at inequality (B.1).

Theorem B.8 *Let $f(\cdot)$ be a nonanticipative process with $E\left[\exp\left\{\frac{1}{2}\|f\|^2\right\}\right] < \infty$. Then*

$$E\left[\exp\left\{i\int_0^T f(t)\,dW(t) + \tfrac{1}{2}\|f\|^2\right\}\right] = 1.$$

Proof. Let

$$V(T) = \exp\left\{i\int_0^T f(t)\,dW(t) + \frac{1}{2}\int_0^T f^2(t)\,dt\right\}.$$

Apply Itô's lemma. Then

$$V(t) = 1 + i\int_0^T V(t)f(t)\,dW(t).$$

Note that $E[V(T)] = 1$ provided

$$E\left[\int_0^T V(t)f(t)\,dW(t)\right] = 0.$$

It can be checked that

$$E\left[\int_0^T |V(t)|^2|f(t)|^2\,dt\right] < \infty,$$

under the condition

$$E[\exp(\tfrac{1}{2}\|f\|^2)] < \infty.$$

Hence, it follows that

$$E\left[\int_0^T V(t)f(t)\,dW(t)\right] = 0 \quad \text{and} \quad E[V(t)] = 1.$$

□

Corollary B.9 *Let $\tilde{f}(\cdot) \in \mathcal{P}_T$ and suppose that $P\{\int_0^T \tilde{f}^2(t)\,dt \ge \sigma^2\} = 1$. Define*

$$\tau = \inf\left\{t : \int_0^t \tilde{f}^2(s)\,ds = \sigma^2\right\}, \quad \textit{and} \quad \zeta = \int_0^\tau \tilde{f}(t)\,dW(t).$$

Then ζ is $N(0, \sigma^2)$.

This result follows from the above lemma: choose

$$f(t) = \lambda \tilde{f}(t) I[t \leq \tau].$$

Then

$$E\left[\exp\left\{i\lambda \int_0^\tau \tilde{f}(t)\, I[t \leq \tau]\, dW(t) + \tfrac{1}{2}\lambda^2\sigma^2\right\}\right] = 1,$$

that is,

$$E[e^{i\lambda\zeta}] = e^{-\frac{1}{2}\lambda^2\sigma^2}.$$

Hence ζ is $N(0, \sigma^2)$.

Theorem B.10 (*Central Limit Theorem; Kutoyants (1984a)*) *If $f_\varepsilon(\cdot) \in \mathcal{P}_{T_\varepsilon}$ and*

$$\int_0^{T_\varepsilon} f_\varepsilon^2(t)\, dt \xrightarrow{p} \sigma^2 \qquad \text{as } \varepsilon \to 0,$$

then

$$\int_0^{T_\varepsilon} f_\varepsilon(t)\, dW_\varepsilon(t) \xrightarrow{\mathcal{L}} N(0, \sigma^2) \qquad \text{as } \varepsilon \to 0,$$

where $\{W_\varepsilon(t), 0 \leq t \leq T_\varepsilon\}$ is a standard Wiener process.

Proof. Define

$$g_\varepsilon(t) = \begin{cases} f_\varepsilon(t) & \text{for } 0 \leq t \leq T_\varepsilon \\ \sigma & \text{for } T_\varepsilon < t \leq T_\varepsilon + 1, \end{cases}$$

$$\tau_\varepsilon = \inf\left\{t : \int_0^t g_\varepsilon^2(s)\, ds = \sigma^2\right\},$$

and

$$\zeta_\varepsilon = \int_0^{\tau_\varepsilon} g_\varepsilon(t)\, dW_\varepsilon(t).$$

Let

$$F_\varepsilon = \int_0^{T_\varepsilon} f_\varepsilon(t)\, dW_\varepsilon(t).$$

Note that $\pounds(\zeta_\varepsilon)$ is $N(0, \sigma^2)$ by Corollary B.9. It is sufficient to prove that

$$\zeta_\varepsilon - F_\varepsilon \xrightarrow{p} 0 \qquad \text{as } \varepsilon \to 0.$$

For $a > 0$ and $b > 0$,

$$\begin{aligned}
&P(|\zeta_\varepsilon - F_\varepsilon| > a) \\
&= P\left\{\left|\int_0^{\tau_\varepsilon} f_\varepsilon(t)\, dW_\varepsilon(t) - \int_0^{T_\varepsilon} g_\varepsilon(t)\, dW_\varepsilon(t)\right| > a\right\} \\
&= P\left\{\left|\int_0^{T_\varepsilon+1} g_\varepsilon(t)[I_{\{t \leq T_\varepsilon\}} - I_{\{t \leq \tau_\varepsilon\}}] dW_\varepsilon(t)\right| > a\right\} \\
&\leq \frac{b}{a^2} + P\left\{\int_0^{T_\varepsilon+1} g_\varepsilon^2(t) |I_{\{t \leq T_\varepsilon\}} - I_{\{t \leq \tau_\varepsilon\}}|\, dt > b\right\}.
\end{aligned}$$

Let $A = \{\omega : \int_0^{T_\varepsilon+1} g_\varepsilon^2(t)|I_{\{t\le T_\varepsilon\}} - I_{\{t\le\tau_\varepsilon\}}|\,dt > b\}$. Note that $P\{A \cap [\tau_\varepsilon = T_\varepsilon]\} = 0$, and

$$P(A \cap [\tau_\varepsilon < T_\varepsilon]) = P\left\{\int_0^{T_\varepsilon+1} g_\varepsilon^2(t)|I_{\{t\le T_\varepsilon\}} - I_{\{t\le\tau_\varepsilon\}}|\,dt > b, \tau_\varepsilon < T_\varepsilon\right\}$$
$$= P\{\|f_\varepsilon\|^2 - \sigma^2 > b, \tau_\varepsilon < T_\varepsilon\}.$$

Similarly,

$$P(A \cap [\tau_\varepsilon > T_\varepsilon]) = P\{\sigma^2 - \|f_\varepsilon\|^2 > b, \tau_\varepsilon > T_\varepsilon\}.$$

Hence,

$$P(A) \le P\{|\|f_\varepsilon\|^2 - \sigma^2| > b\} \to 0 \qquad \text{as } \varepsilon \to 0.$$

Hence the theorem is proved.

The following result leading to a stochastic Taylor expansion of a diffusion process $X^{(\varepsilon)}$ satisfying the stochastic differential equation

$$dX_t^{(\varepsilon)} = \mu(\varepsilon, X_t^{(\varepsilon)})\,dt + \varepsilon\sigma(X_t^{(\varepsilon)})\,dW_t, \qquad X_0^{(\varepsilon)} = x \in (l, r)$$

is due to Azencott (1982, pp. 239–285). Suppose $\mu(0, u) > 0$ for all $u \in (l, r)$. Here $\{W_t\}$ is the standard Wiener process.

Theorem B.11 *If $\mu(\varepsilon, u) \in C^{N+1}([0, \infty) \times (l, r))$ and $\sigma(u)$ is $C^{N+1}((l, r))$, then there exist real-valued semimartingales $g_j = (g_j(t))$, $1 \le j \le N$, with $g_j(0) = 0$, continuous on $[0, \zeta^0)$ and such that*

$$X_t^{(\varepsilon)} = x(t) + \sum_{j=1}^{N} \varepsilon^j g_j(t) + \varepsilon^{(N+1)} R_{N+1}^{(\varepsilon)}(t) \qquad \text{if } t < \zeta^{(0)} \wedge \zeta^{(\varepsilon)},$$

$$R_{N+1}^{(\varepsilon)}(t) = \infty \qquad \text{if } t \ge \zeta^{(0)} \wedge \zeta^{(\varepsilon)},$$

and for all $T \in (0, \zeta^0)$,

$$\lim_{\varepsilon\to 0,\, r\to+\infty} P\left(\sup_{0\le s\le T} |R_{N+1}^{(\varepsilon)}(s)| \ge r\right) = 0.$$

Moreover,

$$\lim_{\varepsilon\to 0} P(\zeta^{(\varepsilon)} < \zeta^{(0)}) = 0.$$

Here $x(t)$ is the solution of the ordinary differential equation

$$dx(t) = \mu(0, x(t))\,dt, \qquad x(0) = x,$$

$$\zeta^{(\varepsilon)} = \inf\{t \ge 0 \;:\; X_t^{(\varepsilon)} \notin (l, r)\},$$

and $\zeta^{(0)}$ is the deterministic explosion time of $x(t)$, that is, $\zeta^{(0)} = t(r-)$ where

$$t(a) = \int_x^a \frac{du}{\mu(u, 0)} = x^{-1}(a).$$

Remark. It is known that the semimartingales $g_j, 1 \leq j \leq N$, in the above result are uniquely determined as the solutions of a system of stochastic differential equations. The process g_1 is defined for $t < \zeta^0$ by

$$dg_1(t) = [\mu'_\varepsilon(0, X(t)) + \mu'_u(0, X(t))g_1(t)]\, dt + \sigma(X(t))\, dW_t, \qquad g_1(0) = 0,$$

where μ'_ε and μ'_u are the partial derivaives of $\mu(\varepsilon, u)$ with respect to ε and u, respectively and $\{W_t\}$ is the standard Wiener process. In fact

$$g_1(t) = \mu(0, X(t)) \left[\int_0^t \frac{\mu'_\varepsilon(0, X(s))}{\mu(0, X(s))}\, ds + \int_0^t \frac{\sigma(X(s))}{\mu(0, X(s))}\, dW_s \right],$$

and hence $\{g_1(t), t < \zeta^0\}$ is a Gaussian process.

Let $\{W_t, t \geq 0\}$ be a standard Wiener process and $\{\mathcal{F}_t\}$ be an increasing family of σ-algebras such that $\sigma\{W_s; 0 \leq s \leq t\} \subset \mathcal{F}_t$ and $\mathcal{F}_t$ is independent of $\{W_s - W_t : s \geq t\}$. Let $\{\zeta_t, t \geq 0\}$ be a stationary process adapted to $\{\mathcal{F}_t\}$. Let $h(\theta, x)$ be differentiable with respect to θ for $\theta \in [-1, 1]$ and suppose that $h'(\theta, x)$ is Lipschitzian in θ with Lipschitzian function $C(x)$. Further assume that

$$E[C(\zeta_0)]^2 < \infty, \tag{B.4}$$

$$h(-1, x) = h(1, x) = 0 \tag{B.5}$$

and

$$h'(-1, x) = h'(1, x) = 0. \tag{B.6}$$

It is easy to see that

$$\int_0^T h(\theta, \zeta_t)\, dW_t$$

exists as an Itô integral for every $\theta \in [-1, 1]$. It was proved in Lemma 4.2 of Prakasa Rao and Rubin (1981) that given $\gamma > \frac{1}{2}$, there exists a constant $H > 0$ such that

$$\limsup_{T\to\infty} \frac{\sup_\theta |\int_0^T h(\theta, \zeta_t)\, dW_t|}{T^{1/2}(\log T)^\gamma} \leq H < \infty \text{ a.s.} \tag{B.7}$$

Hence

$$\sup_\theta \left| \frac{1}{T} \int_0^T h(\theta, \zeta_t)\, dW_t \right| \leq \frac{2HT^{1/2}(\log T)^\gamma}{T}$$

with probability one for large T. Let $T \to \infty$. Then

$$\sup_\theta \left| \frac{1}{T} \int_0^T h(\theta, \zeta_t)\, dW_t \right| \to 0 \text{ a.s.} \tag{B.8}$$

In general, let $g(\theta, x)$ be differentiable with respect to θ in $[-1, 1]$. Then there exists a cubic polynomial $P(\theta, x)$ in θ such that

$$g(-1, x) = P(-1, x); \qquad g(1, x) = P(1, x), \tag{B.9}$$

$$g'(-1, x) = P'(-1, x); \qquad g'(1, x) = P'(1, x), \tag{B.10}$$

and the coefficients of $P(\theta, x)$ are linear functions of $g(-1, x), g(1, x), g'(-1, x)$ and $g'(1, x)$. Define

$$h(\theta, x) = g(\theta, x) - P(\theta, x). \tag{B.11}$$

The $h(\theta, x)$ satisfies the conditions (5) and (6). Suppose that $g'(\theta, x)$ is Lipschitzian in θ with Lipschitzian function $C(x)$, satisfying

$$E[g'(1, \zeta_0)]^2 < \infty \tag{B.12}$$

and

$$E[C(\zeta_0)]^2 < \infty. \tag{B.13}$$

It can be checked that the conditions (9)–(13) imply that the function $h(\theta, x)$ satisfies the conditions (4)–(7) and hence

$$\sup_\theta \left| \frac{1}{T} \int_0^T h(\theta, \zeta_t)\, dW_t \right| \to 0 \qquad \text{as } T \to \infty \text{ a.s.}$$

It is obvious that

$$\sup_\theta \left| \frac{1}{T} \int_0^T P(\theta, \zeta_t)\, dW_t \right| \to 0 \qquad \text{as } T \to \infty \text{ a.s.}$$

since $\theta \in [-1, 1]$ and $P(\theta, x)$ is a polynomial in θ. Hence

$$\sup_\theta \left| \frac{1}{T} \int_0^T g(\theta, \zeta_t)\, dW_t \right| \to 0 \qquad \text{as } T \to \infty \text{ a.s.} \tag{B.14}$$

The restriction that $\theta \in [-1, 1]$ can be changed to θ ranging over a compact set by a suitable reparametrization.

Theorem B.12 (*Prakasa Rao 1982*) *Let $g(\theta, x)$ be differentiable with respect to θ in $[-1, 1]$. Further, suppose that $g'(\theta, x)$ is Lipschitzian in θ satisfying the conditions (12) and (13). Then property (14) holds.*

We now discuss some results on sufficient conditions for the interchangeability of stochastic integration and ordinary differentiation.

B.2 Interchanging Stochastic Integration and Ordinary Differentation

Let $(\Omega, \mathcal{F}, P)$ be a complete probability space and $\{\mathcal{F}_t, t \geq 0\}$ be an increasing family of sub-σ-algebras of $\mathcal{F}$ such that $\mathcal{F}_0$ contains all P-null sets in $\mathcal{F}$. Let $\{W_t, t \geq 0\}$ be an $\mathcal{F}_t$-adapted Wiener process. The following theorem (proved in Stroock 1982) generalizes Kolomogrov's theorem for Banach space valued random variables.

Theorem B.13 *Let B be a separable Banach space endowed with the norm $\|\cdot\|$ and let $\{Z(\theta) : \theta \in R^d\}$ be a family of B-valued random variables. Suppose there exist $0 < c < \infty$, $0 < \alpha \leq 1$ and $p \geq d + \alpha$ such that*

$$E\|Z(\theta_1) - Z(\theta_2)\|^p \leq c\, \|\theta_1 - \theta_2\|^{d+\alpha}, \qquad \theta_1, \theta_2 \in R^d.$$

Then there exists a continuous version $\{Z_1(\theta)\}$ of the process $\{Z(\theta)\}$, that is, $P(Z(\theta) = Z_1(\theta)) = 1$ for all θ and for all $\omega \in \Omega$, the map $\theta \to Z_1(\theta, \omega)$ is a continuous map from R^d into B.

Let $\mathcal{L}_2$ be the family of all progressively measurable processes f on $[0, 1]$ such that $E\{\int_0^1 f^2(t, \omega)\, dt\} < \infty$.

Theorem B.14 *Let $p \geq 2$. Then there exists a universal constant c_p such that*

$$E \sup_{0 \leq t \leq 1} \left| \int_0^t f(u, \cdot)\, dW(u) \right|^p \leq c_p E \int_0^1 |f(u, \cdot)|^p\, du.$$

See Stroock (1982) or Stroock and Varadhan (1979) for a proof.

Theorem B.15 *Suppose there exist constants $c > 0$ and $0 < \beta \leq 1$ such that*

ZZ1 $|g(\theta_1, t, \omega) - g(\theta_2, t, \omega)| \leq c\|\theta_1 - \theta_2\|^\beta$ *for all t, ω and θ_1, $\theta_2 \in R^d$. Further, suppose that $\{g(\theta, t, \cdot)\} \in \mathcal{L}_2$ for all $\theta \in R^d$. Then there exists a continuous version $Z(\theta, t, \omega)$ of the stochastic integral $\int_0^1 g(\theta, u, \cdot)\, dW(u)$, that is, for all $\omega \in \Omega$, the map $\theta \to Z(\theta, \cdot, \omega)$ is a continuous map from R^d into $C[0, 1]$ equipped with the supremum norm.*

Proof. Let $p \geq 2$ such that $p\beta > d$. Let

$$Z(\theta, t) = \int_0^t g(\theta, u)\, dW(u).$$

Then, by Theorem B.14, it follows that

$$\begin{aligned}
&E \sup_{0 \leq t \leq 1} |Z(\theta_1, t) - Z(\theta_2, t)|^p \\
&\leq c_p E \int_0^1 |g(\theta_1, u) - g(\theta_2, u)|^p\, du \\
&\leq c_p c \|\theta_1 - \theta_2\|^{p\beta}.
\end{aligned}$$

The result now follows from Theorem B.14.

The following theorem gives sufficient conditions for the differentiability under the stochastic integral.

Theorem B.16 *Let $\{f(\theta, \cdot, \cdot), \theta \in R\} \subset \mathcal{L}_2$ be such that for all t, ω, $\frac{d}{d\theta} f(\theta, t, \omega) = f'(\theta, t, \omega)$ exists and $\{f'(\theta, \cdot, \cdot), \theta \in R\} \subset \mathcal{L}_2$. Further, assume that there exist constants $0 < c < \infty$, $0 < \beta_i \leq 1$, $i = 1, 2$, such that*

(ZZ2)

$$\begin{cases} |f(\theta, t, \omega) - f(\theta_2, t, \omega)| \leq c|\theta_1 - \theta_2|^{\beta_1}, \\ |f'(\theta, t, \omega) - f'(\theta_2, t, \omega)| \leq c|\theta_1 - \theta_2|^{\beta_2}. \end{cases}$$

Then there exists a version $X(\theta, t, \omega)$ of $\int_0^t f(\theta, u)\, dW(u)$ such that for all $\omega, t, \theta \to X(\theta, t, \omega)$ is differentiable in θ and

$$\frac{d}{d\theta} X(\theta, t, \omega) = \int_0^t f'(\theta, t, \omega)\, dW(u).$$

Proof. Let us first choose a continuous version $X_1(\theta, t)$ of $\int_0^t f(\theta, u)\, dW(u)$. This is possible under the condition (ZZ2) since (ZZ2) implies (ZZ1) of Theorem B.15. Let

$$\begin{aligned} Y(\theta_1, \theta_2, t, \omega) &= \frac{X_1(\theta_1, t, \omega) - X_1(\theta_2, t, \omega)}{\theta_1 - \theta_2} \qquad \text{if } \theta_1 \neq \theta_2 \\ &= \int_0^t f'(\theta_1, u)\, dW(u) \qquad \text{if } \theta_1 = \theta_2. \end{aligned}$$

Note that Y is continuous on the set $[(\theta_1, \theta_2) : \theta_1 \neq \theta_2]$ and any two continuous versious agree outside a null set N. Hence $X_1(\theta, t, \omega)$ is a differentiable function of θ for all t outside a null set N. Define $X_1(\theta, t, \omega) = 0$ on the null set N and equal to $X_1(\theta, t, \omega)$ elsewhere.

For any θ_1 and θ_2,

$$Y(\theta_1, \theta_2, t, \cdot) = \int_0^t \left[\int_0^1 f'(\lambda\theta_1 + (1-\lambda)\theta_2, u, \cdot)\, d\lambda \right] dW(u).$$

Let

$$g(\theta_1, \theta_2, u, \omega) = \int_0^1 f'(\lambda\theta_1 + (1-\lambda)\theta_2, u, \omega)\, d\lambda.$$

Then

$$\begin{aligned} &|g(\theta_1, \theta_2, u, \omega) - g(\theta_3, \theta_4, u, \omega)| \\ &\leq \int_0^1 |f'(\lambda\theta_1 + (1-\lambda)\theta_2, u, \omega) - f'(\lambda\theta_3 + (1-\lambda)\theta_4, u, \omega)|\, d\lambda \\ &\leq c \int_0^1 |\lambda\theta_1 + (1-\lambda)\theta_2 - \lambda\theta_3 - (1-\lambda)\theta_4|^{\beta_2}\, d\lambda \\ &\leq c_1(|\theta_1 - \theta_3|^{\beta_2} + |\theta_2 - \theta_4|^{\beta_2}) \\ &\leq c_2(|\theta_1 - \theta_3|^2 + |\theta_2 - \theta_4|^2)^{\beta_2/2}, \end{aligned}$$

for some constants c_1 and c_2. An application of Theorem B.15 proves that Y has a continuous modification.

Remark. The above result can be extended to include processes on $[0, \infty)$ instead of $[0, 1]$. A more general result valid for semimartingales is as follows. Suppose $\{g(\theta, t, \omega), \theta \in R^d\}$ satisfies the following conditions:

(ZZ3) For all $n \geq 1$, there exist constants c_n, α_n, $0 < c_n < \infty$, $0 < \alpha_n \leq 1$ and stopping times T_n such that

- (i) $T_n \uparrow \infty$ a.s.,
- (ii) $|g(0, t, \omega)| \leq c_n$ on $[t < T_n(\omega)]$, and
- (iii) $|g(\theta_1, t, \omega) - g(\theta_2, t, \omega) \leq c_n|\theta_1 - \theta_2|^{\alpha_n}$ if $|\theta_1| \leq n$, $|\theta_2| \leq n$ and on $[t < T_n(\omega)]$.

Then the following result holds.

Theorem B.17 *Let S be a continuous semimartingale and $\{f(\theta, \cdot, \cdot), \theta \in R^d\}$ be a family of progressively measurable processes. The following properties hold:*

(i) If $f(\theta, \cdot, \cdot) \in R^d$ satisfies (ZZ3), then there exists a continuous version $X(\theta, t, \cdot)$ of $\int_0^t f(\theta, u), \cdot)\, dS(u)$.

(ii) If $g(\theta, t, \omega) = \frac{d}{d\lambda} f(\theta+\lambda e, t, \omega)|_{\lambda=0}$ *exists for all* θ *where* $e \in R^d$ *and* $\{g(\theta, \cdot, \cdot) \in R^d\}$ *satisfies (ZZ3), then there exists a version of* $X(\theta, t, \cdot)$ *of* $\int_0^t f(\theta, u)\, dS(u)$ *such that*

$$Y(\theta, t, \omega) = \frac{d}{d\lambda} X(\theta + \lambda e, t, \omega)\bigg|_{\lambda=0}$$

exists and

$$Y(\theta, t, \omega) = \int_0^t g(\theta, u, \cdot)\, dS(u).$$

Results given above are due to Karandikar (1983).

B.3 Fubini Type Theorem for Stochastic Integrals

Let $(\Omega, \mathcal{F}, P)$ and $(\tilde{\Omega}, \tilde{f}, \tilde{P})$ be two probability spaces. Define

$$(\bar{\Omega}, \bar{\mathcal{F}}, \bar{P}) = (\Omega \times \tilde{\Omega}, \mathcal{F} \times \tilde{\mathcal{F}}, P \times \tilde{P}).$$

Let $\{\mathcal{F}_t, 0 \leq t \leq 1\}$ and $\{\tilde{\mathcal{F}}_t, 0 \leq t \leq 1\}$ be filtrations in $\mathcal{F}$ and $\tilde{\mathcal{F}}$, respectively. Let $W = \{W_t, 0 \leq t \leq 1\}$ be a Wiener process on $(\Omega, \mathcal{F}, P)$ such that W_t is $\mathcal{F}_t$ measurable. Let $\mathcal{F}_t^W = \sigma\{W_s, s \leq t\}$.

Theorem B.18 *Let* $\{g_t(\omega, \tilde{\omega}), \mathcal{F}_t^W \times \tilde{\mathcal{F}}_t, 0 \leq t \leq 1\}$ *be a random process such that*

$$E\left[\int_0^1 g_t^2(\omega, \tilde{\omega})\, dt\right] < \infty,$$

where E denotes expectation with respect to $P \times \tilde{P}$*. Then, for all t,* $0 \leq t \leq 1$,

$$\int_{\tilde{\Omega}}\left[\int_0^t g_s(\omega, \tilde{\omega})\, dW_s(\omega)\right] d\tilde{P}(\tilde{\omega}) = \int_0^t \left[\int_{\tilde{\Omega}} g_s(\omega, \tilde{\omega})\, d\tilde{P}(\tilde{\omega})\right] dW_s(\omega).$$

A proof is given in Liptser and Shiryayev (1977, p. 187).

B.4 Sufficient Conditions for the Differentiability of an Itô Stochastic Integral

Theorem B.19 *Let a random process* $\{f(u, t), \mathcal{F}_t, 0 \leq t \leq T\}$ *be, with probability one,* $k+1$ *times continuously differentiable with respect to u, and let*

$$E\left[\int_0^T \left[\frac{\partial^j f(u, t)}{\partial u^j}\right]^2 dt\right] \leq D_j < \infty, \qquad j = 0, 1, \ldots, k+1.$$

Suppose the stochastic integral

$$\zeta(u) = \int_0^T f(u, t)\, dW(t) \tag{B.15}$$

exists with respect to a Wiener process $\{W(t), 0 \leq t \leq T\}$*. Then the stochastic integral has, with probability one, k continuous derivatives and*

$$\frac{\partial^j \zeta(u)}{\partial u^j} = \int_0^T \frac{\partial^j f(u, t)}{\partial u^j}\, dW(t), \qquad 1 \leq j \leq k. \tag{B.16}$$

Proof. For simplicity, we consider the case $k = 1$. The general case can be proved analogously. Define the process

$$\dot{\zeta}(u) = \int_0^T \dot{f}(u, t)\, dw(t),$$

where $\dot{f}(u, t) = \frac{\partial}{\partial u} f(u, t)$. It can be seen, from the continuous differentiability of $\dot{f}(u, t) = \frac{\partial}{\partial u} f(u, t)$ with respect to u and the properties of a stochastic integral, that

$$\begin{aligned} E|\dot{\zeta}(u+h) - \dot{\zeta}(u)|^2 &= E\left\{\int_0^T [\dot{f}(u+h) - \dot{f}(u)]^2\, dt\right\} \\ &\le h \int_0^h \left(E \int_0^T \ddot{f}(u+v, t)^2\, dt \right) dv \\ &\le D_2 h^2, \end{aligned} \tag{B.17}$$

where $\ddot{f}(u, t) = \frac{\partial^2}{\partial u^2} f(u, t)$. Hence the process $\{\zeta(u)\}$ is is continuous with probability one by Kolmogorov's theorem. Taking limit as $h \to 0$ in the expression, we have

$$\begin{aligned} \frac{\zeta(u+h) - \zeta(u)}{h} &= \frac{1}{h} \int_0^h \left(\int_0^T \dot{f}(u+v, t)\, dW(t) \right) dv \\ &\overset{a.s.}{\to} \int_0^T \dot{f}(u, t)\, dW(t). \end{aligned} \tag{B.18}$$

This proves (B.16).

Appendix C

Wavelets

We now present a brief review of the theory of wavelets relevant to our discussion in Chapter 5. An extensive discussion can be found in Meyer (1990) and Daubechies (1988; 1992). All the definitions and results given here for $L^2(R)$ are also valid for $L^2(R^n)$.

Definition C.1 (*Meyer 1990, p. 21*) A multiresolution analysis of $L^2(R)$ is an increasing sequence $\{V_j, -\infty < j < \infty\}$ of closed linear subspaces of $L^2(R)$ such that:

(i) for all $-\infty < j < \infty$, $V_j \subset V_{j+1}$ and $\cap_j V_j = \{0\}$ and $\cup_j V_j$ is dense in $L^2(R)$;
(ii) for all $-\infty < j < \infty$, $f(x) \in V_j$ if and only if $f(2x) \in V_{j+1}$;
(iii) for all $-\infty < k < \infty$, $f(x) \in V_0$ if and only if $f(x-k) \in V_0$; and
(iv) there exists a function g in V_0 such that the sequence $\{g(x-k), -\infty < k < \infty\}$ is a Riesz basis for V_0.

Let H be a Hilbert space with the norm $\|\cdot\|$. Suppose that $\{e_i, i \geq 0\}$ is a sequence in H such that the closure of the span of $\{e_i, i \geq 0\}$ is equal to H. The sequence $\{e_i, i \geq 0\}$ is a *Riesz basis* if there exist two constants C_1 and C_2 such that $0 < C_1 \leq C_2$ and for all real α_i, $i \leq 0$,

$$C_1\Big(\sum |\alpha_i|^2\Big)^{1/2} \leq \Big\|\sum \alpha_i e_i\Big\| \leq C_2\Big(\sum |\alpha_i|^2\Big)^{1/2}.$$

Definition C.2 (*Meyer (1990), p. 22*) A multiresolution analysis $\{V_j, -\infty < j < \infty\}$ of $L^2(R)$ is *r-regular* if the function g defined in Definition C.1(iv) can be chosen so that

$$|g^{(i)}(x)| \leq C_m(1+|x|)^{-m}$$

for some constants $C_m > 0$, for all $0 \leq i \leq r$ and for all $m \geq 1$.

Let $\hat{f}(\zeta)$ denote the Fourier transform of a function f, that is,

$$\hat{f}(\zeta) = \int f(x)\exp(-i\,\zeta x)\,dx.$$

Theorem C.1 (*Meyer (1990), pp. 27–29*) *Let $\{V_j, -\infty < j < \infty\}$ be an r-regular multiresolution analysis of $L^2(R)$ and let $\phi \in L^2(R)$ be defined by*

$$\hat{\phi}(\zeta) = \hat{g}(\zeta)\Bigg(\sum_{k=-\infty}^{\infty} |\hat{g}(\zeta + 2k\pi)|^2\Bigg)^{-1/2}.$$

Then $\{\phi(x-k), -\infty < k < \infty\}$ *is an orthonormal basis of* V_0 *which satisfies*

$$|\phi^{(i)}(x)| \le C_m(1+|x|)^{-m}$$

for all $0 \le i \le r$ *and for all* $m \ge 1$.

The function ϕ is called the *scale function* associated with $\{V_j, -\infty < j < \infty\}$. Consider the family of functions

$$\phi_{j,k}(x) = 2^{j/2}\phi(2^j x - k), \qquad -\infty < j, k < \infty.$$

Then, for all $-\infty < j < \infty$, $\{\phi_{j,k}, -\infty < k < \infty\}$ is an orthonormal basis for V_j.

For $s \ge 0$, let $H^s(R)$ be the Sobolev space defined by

$$H^s(R) = \left\{ f \in L^2(R) : \int |\hat{f}(\zeta)|^2 (1+|\zeta|^2)^s \, d\zeta < \infty \right\},$$

with the norm of $H^s(R)$ given by

$$\|f\|_s = \left(\int |\hat{f}(\zeta)|^2 (1+|\zeta|^2)^s \, d\zeta \right)^{1/2}, \qquad f \in H^s(R).$$

If $s < 0$, the Sobolev space $H^s(R)$ is defined as the topological dual of $H^{-s}(R)$. Let P_j be the projection operator from $L^2(R)$ to V_j, that is, the operator of orthogonal projection on V_j. The following theorem gives the quality of approximation of a function f obtained by using a multiresolution analysis.

Theorem C.2 (*Meyer (1990), p. 41*) *Let* $r \ge 1$ *and* $\{V_j, -\infty < j < \infty\}$ *be an r-regular multiresolution analysis of* $L^2(R)$. *If* $f \in H^s(R)$ *with* $-r \le s \le r$, *then* $P_j f$ *converges to* f *in* $H^s(R)$, *that is,*

$$\|f - P_j f\|_s \to 0 \qquad \text{as } j \to \infty.$$

Let W_j *be the orthogonal complement of* V_j *in* V_{j+1}. *Then*

$$V_{j+1} = V_j \oplus W_j.$$

Then the space $L^2(R)$ *can be represented as the direct sum*

$$L^2(R) = \bigoplus_{j=-\infty}^{\infty} W_j = V_{j_0} \oplus \left(\bigoplus_{j \ge j_0} W_j \right)$$

for all $-\infty < j_0 < \infty$. *Let* D_j *be the orthogonal projection on* W_j. *Then* $D_j = P_{j+1} - P_j$ *and* $f(x) \in W_0$ *if and only if* $f(2^j x) \in W_j$. *The Sobolev spaces can be characterized by using the above decomposition.*

Theorem C.3 ((*Meyer (1990), p. 48*) *If* $f \in H^{-s}(R)$, $\{V_j, -\infty < j < \infty\}$ *is an r-regular multiresolution analysis of* $L^2(R)$ *and if* $-r < s < r$, *then* $f \in H^s(R)$ *if and only if*

$$P_0 f \in L^2(R) \text{ and } \|D_j f\| = \varepsilon_j 2^{-js}, \qquad j \ge 1,$$

where $\{\varepsilon_j, j \ge 1\}$ *is a sequence in* $\ell^2(\mathbb{N})$, *that is,* $\sum_{j=1}^{\infty} \varepsilon_j^2 < \infty$. *Moreover, the* H^s*-norm of* f *is equivalent to*

$$\|P_0 f\|_{L^2} + \left(\sum_{j=1}^{\infty} \varepsilon_j^2 \right)^{1/2}.$$

We conclude this appendix by turning our attention to wavelets associated with a multiresolution analysis.

Theorem C.4 *(Meyer (1990), pp. 70–71) Let* $\{V_j, -\infty < j < \infty\}$ *be an r-regular multiresolution analysis of* $L^2(R)$. *(Recall that* $V_{j+1} = V_j \oplus W_j$, $L^2(R) = \oplus_{j=-\infty}^{\infty} W_j$). *Then there exists a function* ψ *in* W_0 *such that*

(i) $|\psi^{(i)}(x)| \leq C_m(1+|x|)^{-m}$ *for all integers* $i \leq r$ *and* $m \geq 1$;

(ii) *the sequence* $\{\psi(x-k), -\infty < k < \infty\}$ *is an orthonormal basis for* W_0.

Such a function ψ is called a *wavelet associated with the multiresolution analysis* $\{V_j, -\infty < j < \infty\}$. Let

$$\psi_{j,k}(x) = 2^{j/2}\psi(2^j x - k), \qquad -\infty < j, k < \infty.$$

The family $\{\psi_{j,k}(x), -\infty < j, k < \infty\}$ is an orthonormal basis of $L^2(R)$.

The function ψ can be obtained from ϕ in the following way. Note that $\phi(\frac{x}{2}) \in V_{-1}$ and $V_{-1} \subset V_0$. Hence

$$\frac{1}{2}\phi\left(\frac{x}{2}\right) = \sum_{k=-\infty}^{\infty} \alpha_k \phi(x+k),$$

where

$$\alpha_k = \frac{1}{2}\int \phi\left(\frac{x}{2}\right)\phi(x+k)\,dx.$$

Therefore

$$\hat{\phi}(2\zeta) = m(\zeta)\hat{\phi}(\zeta)$$

with

$$m(\zeta) = \sum_{k=-\infty}^{\infty} \alpha_k e^{ik\zeta}.$$

Define ψ by

$$\hat{\psi}(2\zeta) = e^{-i\zeta}\bar{m}(\zeta+\pi)\hat{\phi}(\zeta),$$

or equivalently,

$$\frac{1}{2}\psi\left(\frac{x+1}{2}\right) = \sum_{k=-\infty}^{\infty} \bar{\alpha}_k(-1)^k\phi(x-k).$$

Then the function ψ satisfies the condition of Theorem C.6. Furthermore,

$$\int_{-\infty}^{\infty} x^i\psi(x)\,dx = 0 \qquad \text{for } i \leq r.$$

Remark. Our discussions here is based on Genon-Catalot *et al.* (1992). They present two explicit examples of wavelets obtained with the multiresolution analyses due to Meyer (1990 p. 72).

Appendix D

Gronwall–Bellman Type Lemma

Theorem D.1 *If $v(\cdot)$ is a positive bounded function satisfying the inequality*

$$v(t) \leq c_0 + c_1 \int_0^t v(s)\, ds + c_2 \int_0^t \int_0^s v(\tau)\, d\tau\, ds, \qquad t \geq 0$$

then

$$v(t) \leq c_0 e^{(c_1+c_2)t}, \qquad t \geq 0.$$

In general, the following result holds.

Theorem D.2 *Let c_0, c_1 and c_2 be nonnegative constants, $u(t)$ be a nonnegative bounded function and $v(t)$ be a nonnegative integrable function on $[0, 1]$ such that*

$$u(t) \leq c_0 + c_1 \int_0^t v(s)u(s)\, ds + c_2 \int_0^t v(s) \left[\int_0^s u(t)\, dK(t) \right] ds$$

where $K(\cdot)$ is a nondecreasing right continuous function with $0 \leq K(s) \leq 1$. Then

$$u(t) \leq c_0 \exp\left\{ (c_1 + c_2) \int_0^t v(s)\, ds \right\}, \qquad 0 \leq t \leq 1.$$

For proofs, see Liptser and Shiryayev (1977, p. 130).

References

Ait-Sahalia, Y. (1996a) Testing continuous-time models of the spot interest rate, *Review of Financial Studies*, **9**, 385–426.

Ait-Sahalia, Y. (1996b) Nonparametric pricing of interest rate derivative securities, *Econometrica*, **64**, 527–560.

Ait-Sahalia, Y. (1997) Maximum likelihood estimation of discretely sampled diffusions: A closed form approach, Preprint, University of Chicago.

Akaike, H. (1977) On entropy maximization principle, in: P.R. Krishnaiah, ed., *Applications of Statistics*, North-Holland, Amsterdam, 27–41.

Akaike, H. (1983) Statistical inference and measurement of entropy, in: G.E.P. Box, T. Leonard and C.F.J. Wu, eds, *Scientific Inference, Data Analysis, and Robustness*, Academic Press, New York, 165–189.

Aldous, D. and Eagleson, G. (1978) On mixing and stability of limit theorems, *Annals of Probability*, **6**, 325–331.

Andersen, P.K., Borgan, Ø., Gill, R.D. and Keiding, N. (1993) *Statistical Models Based on Counting Processes*, Springer, New York.

Arato, M. (1982) *Linear Stochastic Systems with Constant Coefficients: A Statistical Approach*, Lecture Notes in Control and Information Sciences **45**, Springer, Berlin.

Arnold, L. and Kleimann, W. (1987) On unique ergodicity for degenerate diffusions, *Stochastics*, **21**, 41–61.

Azencott, R. (1982) Formule de Taylor stochastique et développement asymptotique d'intégrales de Feynman, in: J. Azema and M. Yor, eds, *Seminaire de Probabilités XVI*, Lecture Notes in Mathematics **921**, Springer, Berlin, 237–285.

Azencott, R. (1984) Densités des diffusions en temps petits: développements asymptotiques, in: J. Azema and M. Yor, eds, *Seminaire de Probabilités XVIII*, Lecture Notes in Mathematics **1059**, Springer, Berlin.

Bahadur, R.R., Zabell, S.J., and Gupta, J.C. (1980) Large deviations, tests and estimates, in: I.M. Chakravarthi, ed., *Asymptotic Theory of Statistical Tests*, Academic Press, New York, 33–64.

Bailey, W. (1991) Valuing agricultural firms, *Journal of Economic Dynamics and Control*, **15**, 771–791.

Banon, G. (1977) Estimation non parametrique de la densité de probabilité pour les processus de Markov, Thesis, Université Paul Sabatier de Toulouse, France.

Banon, G. (1978) Nonparametric identification for diffusion processes, *SIAM Journal on Control and Optimization*, **16**, 380–395.

Banon, G. and Nguyen, H.T. (1978) Sur l'estimation recurrente de la densite et de sa derivee pour un processus de Markov, *Comptes Rendus de l'Académie des Sciences A*, **286**, 691–694.

Banon, G. and Nguyen, H.T. (1981a) Recursive estimation in diffusion model, *SIAM, Journal on Control and Optimization*, **19**, 676–685.

Banon, G. and Nguyen, H.T. (1981b) A nonparametric estimation in diffusion model by discrete sampling, *Publications Statistiques de l'Université de Paris* **26**, 89–109.

Barndorff-Nielsen, O.E. (1977) Exponentially decreasing distributions for the logarithm of the particle size, *Proceedings of the Royal Society*, (London) *A*, **353**, 401–419.

Barndorff-Nielsen, O.E. and Sørensen, M. (1994) A review of some aspects of asymptotic likelihood theory for stochastic processes, *International Statistical Review*, **62**, 133–165.

Barndorff-Nielsen, O.E., Kent. J. and Sørensen, M. (1982) Normal variance-mean mixtures and *z* distributions, *International Statistical Review*, **50**, 145–159.

Basawa, I.V. and Prabhu, N.U. (1994) *Statistical Inference in Stochastic Processes*, Special issue of *Journal of Statistical Planning and Inference*, **39**.

Basawa, I.V. and Prakasa Rao, B.L.S. (1980) *Statistical Inference for Stochastic Processes*, Academic Press, London.

Basawa, I.V. and Scott, D.J. (1983) *Asymptotic Optimal Inference for Non-ergodic Models*, Lecture Notes in Statistics **17**, Springer, Heidelberg.

Basu, A. (1983a) Asymptotic theory of estimation in nonlinear stochastic differential equations for the multiparameter case, *Sankhyā A*, **45**, 56–65.

Basu, A. (1983b) The Bernstein–von Mises theorem for a certain class of diffusion processes, *Sankhyā A*, **45**, 150–160.

Beder, J.H. (1987) A sieve estimator for the mean of a Gaussian process, *Annals of Statistics*, **15**, 59–78.

Bibby, B.M. (1994) Optimal combination of martingale estimating functions for discretely observed diffusion processes, Research Report 264, Dept. of Theoretical Statistics, University of Aarhus.

Bibby, B.M. and Sørensen, M. (1995a) Martingale estimating functions for discretely observed diffusion processes, *Bernoulli*, **1**, 17–39.

Bibby, B.M. and Sørensen, M. (1995b) A hyperbolic diffusion model for stock prices, Research Report 331, Dept. of Theoretical Statistics, University of Aarhus.

Bibby, B.M. and Sørensen, M. (1995c) On estimation for discretely observed diffusions: A review, Research Report 334, Dept. of Theoretical Statistics, University of Aarhus.

Billingsley, P. (1961) *Statistical Inference for Markov Processes*, University of Chicago Press, Chicago.

Billingsley, P. (1968) *Convergence of Probability Measures*, Wiley, New York.

Billingsley, P. (1986) *Probability and Measure*, Wiley, New York.

Bishwal, J.P.N. (1996a) Rates of convergence of the maximum likelihood and Bayes estimators in the Ornstein–Uhlenbeck process, Preprint.

Bishwal, J.P.N. (1996b) Asymptotic theory of approximate Bayes estimator for diffusion processes from discrete observations, Preprint.

Bishwal, J.P.N. and Bose, A. (1995) Speed of convergence of the maximum likelihood estimator in the Ornstein–Uhlenbeck process, *Calcutta Statistical Association Bulletin*, **45**, 245–251.

Bittani, S. and Guardabassi, G. (1986) Optimal periodic control and periodic systems analysis: An overview, in: *Proceedings of the 25th Conference on Decision Control, IFAC, Athens*, 1417–1423.

Black, F. and Scholes, M. (1973) The pricing of options and corporate liabilities, *Journal of Political Economy*, **81**, 637–654.

Blum, J.R. and Boyles, R.A. (1981) Random sampling from a continuous parameter stochastic process, in: D. Dugue, E. Lukacs and V.K. Rohatgi, eds, *Analytical Methods in Probability Theory*, Lecture Notes in Mathematics **861**, Springer, Berlin, 15–24.

Blum, J.R. and Rosenblatt, J. (1964) On random sampling from a stochastic process, *Annals of Mathematical Statistics*, **35**, 1713–1717.

Borwanker, J.D., Kallianpur, G. and Prakasa Rao, B.L.S. (1971) The Bernstein–von Mises theorem for Markov processes, *Annals of Mathematical Statistics*, **42**, 1241–1253.

Bremaud, P. (1981) *Point Processes and Queues: Martingale Dynamics*, Springer, Berlin.

Brennan, M.J. and Schwartz, E.S. (1979) A continuous time approach to the pricing of bonds, *Journal of Banking and Finance*, **3**, 133–155.

Brennan, M.J. and Schwartz, E.S. (1985) Evaluating natural resource investments, *Journal of Business*, **58**, 135–157.

Brown, B.M. and Hewitt, J.T. (1975) Asymptotic likelihood theory for diffusion processes, *Journal of Applied Probability*, **12**, 228–238.

Broze, L., Scaillet, O. and Zakoian, J.-M. (1995) Quasi indirect inference for diffusion processes, CORE Discussion Paper No. 9505.

Brugière, P. (1991) Théorème de limite centrale pour un estimateur de la variance d'un processus de diffusion dans le cas multidimensionnel, *Comptes Rendus de l'Académie des Sciences I*, **313**, 943–947.

Brugière, P. (1993) Théorème de limite centrale pour un estimateur non paramétrique de la variance d'un processus de diffusion multidimensionnelle, *Annales de l'Institut Henri Poincaré, Probabilités et Statistiques*, **29**, 357–389.

Burnashev, M.V. (1977) Asymptotic expansion of estimates of a signal parameter in white Gaussian noise, *Matematik Sbornik*, **104**, 179–206 (in Russian).

Bywater, R.J. and Chung, P.M. (1973) Turbulent flow fields with two dynamically significant scales, *AIAA Papers*, **73-646**.

Chan, K.C., Karolyi, G.A., Longstaff, F.A. and Sanders, A.B. (1992) An empirical comparison of alternative models of short term interest rate, *Journal of Finance*, **47**, 1209–1227.

Chang, C.C. (1985) Numerical solution of stochastic differential equations, Ph.D Thesis, University of California, Berkeley.

Chang, C.C. (1987) Numerical solution of stochastic differential equation with constant diffusion coefficients, *Mathematics of Computation*, **49**, 523–542.

Clark, J.M.C. (1978) The design of robust approximation to the stochastic differential equations of nonlinear filtering, in: J.K. Skwirzynski, ed., *Communication Systems and Random Process Theory*, Sijthoff and Noordhoff, Alphen aan de Rijn, Netherlands, 721–734.

Clark, J.M.C. and Cameron, R.J.(1980) The maximum rate of convergence of discrete approximations for stochastic differential equations, in: B. Grigelionis, ed., *Stochastic Differential Systems,* Lecture Notes in Control and Information Sciences **25**, Springer, Berlin, 162–171.

Clarke, H.R. and Reed, W.J. (1989) The tree-cutting problem in a stochastic environment: The case of age-dependent growth, *Journal of Economic Dynamics and Control*, **13**, 569–595.

Clarke, H.R. and Reed, W.J. (1990) Land development and wilderness conservation policies under uncertainity: A synthesis, *Natural Resource Modeling*, **4**, 11–37.

Clement, E. (1997) Estimation of diffusion processes by simulated moment methods, *Scandinavian Journal of Statistics*, **24**, 353–369.

Conley, T.G., Hansen, L.P., Luttmer, E.G.J. and Scheinkman, J.A. (1997) Short term interst rates as subordinate diffusions, *Review of Financial Studies*, **10**, 525–578.

Courtadon, G. (1985) Une synthèse des modèles d'évaluation d'options sur obligations, *Finance*, **6**.

Cox, D.R. and Miller, H.D. (1965) *The Theory of Stochastic Processes*, Methuen, London.

Cox, J.C. (1996) 1975–1996: The constant elasticity of variance option pricing model, *Journal of Portfolio Management*, special issue.

Cox, J.C., Ingersoll, E. and Ross, S.A. (1985) A thoery of the term structure of interest rates, *Econometrica*, **53**, 385–407.

Dacunha-Castelle, D. and Dufflo, M. (1983) *Probabilités et Statistiques 2. Problèmes à Temps Mobile*, Masson, Paris.

Dacunha-Castelle, D. and Florens-Zmirou, D. (1986) Estimation of the coefficients of diffusion from discrete observations, *Stochastics*, **19**, 263–284.

Daubechies, I. (1988) Orthogonal bases of compactly supported wavelets, *Communications on Pure and Applied Mathematics*, **41**, 909–996.

Daubechies, I. (1992) *Ten Lectures on Wavelets*, SIAM, Philadelphia.

Davis, M.H.A. (1977) *Linear Systems and Control*, Chapman & Hall, London.

De Winne, R. (1995) The discretization bias for processes of the short-term interest rate: An empirical analysis, CORE Discussion Paper No. 9564.

Deheuvels, P. (1973) Sur l'estimation sequentielle de la densité, *Comptes Rendus de l'Académie des Sciences A*, **276**, 1119–1121.

Dietz, H.M. (1987) Explicit solutions for a class of linear functional stochastic differential equations, Preprint No. 150, Humboldt-Universität, Berlin.

Dietz, H.M. (1989) Asymptotic properties of maximum likelihood estimators in diffusion type models, Part 1: General statements, Preprint No. 228, Humboldt-Universität, Berlin.

Dietz, H.M. (1992) A non-Markovian relative of the Ornstein–Uhlenbeck process and some of its local statistical properties, *Scandinavian Journal of Statistics*, **19**, 363–379.

Dietz, H.M. and Kutoyants, Yu.A. (1997) A class of minimum-distance estimators for diffusion processes with ergodic properties *Statistics and Decisions*, **15**, 211–227.

Dohnal, G. (1987) On estimating the diffusion coefficient, *Journal of Applied Probability*, **24**, 105–114.

Dorogovcev, A.Ja. (1976) The consistency of an estimate of a parameter of stochastic differential equation, *Theory of Probability and Mathematical Statistics*, **10**, 73–82.

Dothan, M.U. (1978) On the term structure of interest rates, *Journal of Financial Economics*, **6**, 59–69.

Dsagnidse, A,A. and Tschitashvili, R.J. (1975) Approximate integration of stochastic differential equations, *Tbilisi State Univ., Inst. Appl. Math., Trudy*, **4**, 267–279 (In Russian).

Dufkova, V. (1977) On controlled one-dimensional diffusion process with unknown parameter, *Advances in Applied Probability*, **9**, 105–124.

Elliott, R.J. (1982) *Stochastic Calculus and Applications*, Springer, New York.

Engelbert, H.J. and Schmidt, W. (1981) On the behaviour of certain functionals of the Wiener process and applications to stochastic differential equations, in: M. Arato, D. Vermes and A.V. Balakrishnan, eds, *Stochastic Differential Systems*, Lecture Notes in Control and Information Sciences **36**, Springer, Berlin, 47–55.

Engelbert, H.J. and Schmidt, W. (1985) On solutions of one-dimensional stochastic differential equations without drift, *Zeitschrift der Wahrscheinlichkeitstheorie und Verwandter Gebiete*, **68**, 287–314.

Eplett, W.J.R. (1987) Functional parameter estimation and hypothesis testing for diffusions, Preprint.

Erlandsen, M. and Sørensen, M. (1984) Statistical analysis of the variation of the oxygen cocentration in a river by means of diffusion processes, in: L.S. Mortensen, ed., *Applied Statistics Symposium*, Aarhus: RECAU, 421–431.

Fama, E.F. (1965) The behavior of stock market prices, *Journal of Business*, **38**, 34–105.

Feigin, P. (1976) Maximum likelihood estimation for stochastic processes – A martingale approach, *Advances in Applied Probability*, **8**, 712–736.

Feigin, P. (1979) Some comments on a curious singularity, *Journal of Applied Probability*, **16**, 440–444.

Feigin, P.D. (1985) Stable convergence of semimartingales, *Stochastic Processes and their Applications*, **19**, 125–134.

Fleming, T.R. and Harrington, D.P (1991) *Counting Processes and Survival Analysis*, Wiley, New York.

Florens-Zmirou, D. (1984a) Théorème de limite centrale pour une des fonctionnelles de diffusion, *Comptes Rendus de l'Académie des Sciences I*, **299**, 595–598.

Florens-Zmirou, D. (1984b) Théorème de limite centrale pour une diffusion et pour sa discrétisée, *Comptes Rendus de l'Académie des Sciences I*, **299**, 995–998.

Florens-Zmirou, D. (1989) Approximate discrete-time schemes for statistics of diffusion processes, *Statistics*, **20**, 547–557.

Florens-Zmirou, D. (1993) On estimating the diffusion coefficient from discrete observations, *Journal of Applied Probability*, **30**, 790–804.

Friedman, A. (1964) *Partial Differential Equations of Parabolic Type*, Prentice Hall, Englewood Cliffs, NJ.

Friedman, A. (1975) *Stochastic Differential Equations and Applications*, Vol. 1, Academic Press, New York.

Fuller, W.A. (1976) *Introduction to Statistical Time Series*, Wiley, New York.

Gallant, A.R. and Long, J.R. (1997) Estimating stochastic differential equations efficiently by minimum chi-squared, *Biometrika*, **84**, 125–141.

Gallant, A.R. and Tauchen, G. (1996) Which moments to match?, *Econometric Theory*, **12**, 657–681.

Gallant, A,R. and White, H. (1988) *A Unified Theory of Estimation and Inference for Nonlinear Dynamic Models*, Basil Blackwell, Oxford.

Gard, T.C. (1988) *Introduction to Stochastic Differential Equations*, Marcel Dekker, New York.

Geman, S. (1979) On a commonsense estimator for the drift of a diffusion, *Pattern Analysis No. 79*, Division of Applied Mathematics, Brown University, Providence, RI.

Geman, S. and Hwang, C.R. (1982) Nonparametric maximum likelihood estimation by the method of sieves, *Annals of Statistics*, **10**, 401–414.

Genon-Catalot, V. (1987) Test sur la dérive d'un mouvement brownien arrêté au temps d'atteinte d'une sphère, *Comptes Rendus de l'Académie des Sciences I*, **304**.

Genon-Catalot, V. (1989) First hitting times and positions of concentric spheres for testing the drift of a diffusion process, *Probability and Mathematical Statistics*, **10**, 27–44.

Genon-Catalot, V. (1990) Maximum contrast estimation for diffusion processes from discrete observations, *Statistics*, **21**, 99–116.

Genon-Catalot, V. and Jacod, J. (1993) On the estimation of the diffusion coefficient for multi-dimensional diffusion processes, *Annales de l'Institut Henri Poincaré, Probabilités et Statistiques*, **29**, 119–151.

Genon-Catalot, V. and Jacod, J. (1994) On the estimation of the diffusion coefficient for diffusion processes: Random sampling, *Scandinavian Journal of Statistics*, **21**, 193–221.

Genon-Catalot, V. and Larédo, C. (1986) Non-parametric estimation for partially observed transient diffusion processes, *Stochastics*, **18**, 169–196.

Genon-Catalot, V. and Larédo, C. (1987) Limit theorems for the first hitting times process of a diffusion and statistical applications, *Scandinavian Journal of Statistics*, **14**, 143–160.

Genon-Catalot, V. and Larédo, C. (1990) An asymptotic sufficiency property of observations related to the first hitting times of a diffusion process, *Statistics and Probability Letters*, **9**, 203–208.

Genon-Catalot, V., Larédo, C. and Picard, D. (1992) Non-parametric estimation of the diffusion coefficient by the wavelet methods, *Scandinavian Journal of Statistics*, **19**, 317–335.

Gerencsér, L., Gyongy, I. and Michaletsky, G. (1984) Continuous time recursive maximum likelihood method. A new approach to Ljung's scheme, in: *9th IFAC Proc.*, **II**, 683–685.

Ghysels, E., Harvey, A. and Renault, E. (1995) Stochastic volatility, CORE Discussion Paper No. 9569.

Gihman, I.I. and Skorohod, A.V. (1972) *Stochastic Differential Equations*, Springer, Berlin.

Gihman, I.I. and Skorohod, A.V. (1977) *Introduction to the Theory of Random Processes*, Nauka, Moscow (in Russian).

Gihman, I.I. and Skorohod, A.V. (1979) *The Theory of Stochastic Processes*, Vols I–III, Springer, Berlin.

Godambe, V.P. and Heyde, C.C. (1987) Quasi-likelihood and optimal estimation, *International Statistical Review*, **55**, 231–244.

Gouriéroux, C. and Monfort, A. (1989) *Statistique et modèles économétriques*, Economica.

Gouriéroux, C. and Monfort, A. (1993) Simulation based inference, *Journal of Econometrics*, **59**, 5–33.

Gouriéroux, C, Monfort, A. and Renault, E. (1993) Indirect inference, *Journal of Applied Econometrics*, **8**, 85–118.

Gradshteyn, I.S. and Ryzhik, I.M. (1965) *Tables of Integrals, Series, and Products*, 4th edition, Academic Press, New York.

Grenander, U. (1981) *Abstract Inference*, Wiley, New York.

Hájek, J. (1972) Local asymptotic minimax and admissibility in estimation, in: L. Le Cam, J. Neyman and E.L. Scott, eds, *Proceedings of the Sixth Berkeley Symposium on Mathematical Statistics and Probability*, Vol. 1, University of California Press, Berkeley, 175–194.

Hall, P. and Heyde, C. (1980) *Martingale Limit Theory and Its Applications*, Academic Press, New York.

Hansen, L.P. (1982) Large sample properties of generalized method of moments estimators, *Econometrica*, **50**, 1029–1054.

Hansen, L.P. and Scheinkman, J.A. (1995) Back to the future: Generating moment implications for continuous time Markov processes, *Econometrica*, **63**, 767– 804.

Harris, C.J. (1976) Simulation of nonlinear stochastic equations with applications in modeling water pollution, in: C.A. Brebbi, ed., *Mathematical Models for Environment Problems*, Pentech Press, London, 269–282.

Haworth, D.C. and Pope, S.B. (1986) A second order Monte Carlo method for the solution of the Ito stochastic differential equation, *Stochastic Analysis and Applications*, **4**, 151–186.

Heyde, C.C. (1988) Fixed sample and asymptotic optimality for classes of estimating functions, in: *Statistical Inference from Stochastic Processes*, ed. N.U. Prabhu. *Contemporary mathematics*, **80**, 241–247.

Hoffman, M. (1997) Estimation non-paramétrique du coefficient de diffusion pour une perte L_p, *Comptes Rendus de l'Académie des Sciences I* **324**, 475–480.

Horsthemke, W. and Lefevre, R. (1984) *Noise-Induced Transitions*, Springer, Berlin.

Hutton, J.E. and Nelson, P.I. (1986) Quasi-likelihood estimation for semimartingales, *Stochastic Processes and their Applications*, **22**, 245–257.

Ibragimov, I.A. and Has'minskii, R.Z. (1981) *Statistical Estimation: Asymptotic Theory*, Springer, Berlin.

Ibragimov, I.A. and Linnik, Yu.V. (1971) *Independent and Stationary Sequences of Random Variables*, Wolters-Noordhoff, Groningen, Netherlands.

Ikeda, N. and Watanabe, S. (1981) *Stochastic Differential Equations and Diffusion Processes*, North-Holland, Amsterdam.

Jacod, J. (1993) Random sampling in estimation problems for continuous Gaussian processes with independent increments, *Stochastic Processes and their Applications* **44**, 181–204.

Jacod, J. and Shiryayev, A.N. (1987) *Limit Theorems for Stochastic Processes*, Springer, Heidelberg.

Jannsen, R. (1982) Difference-methods for stochastic differential equations, Preprint No. 50, Universität Kaiserslautern.

Jannsen, R. (1984) Discretization of the Wiener process in difference methods for stochastic differential equations, *Stochastic Processes and their Applications*, **18**, 361–369.

Jeganathan, P. (1982) On the asymptotic theory of estimation when the limit of the loglikelihood ratio is mixed normal, *Sankhyā A*, **44**, 173–212.

Johnson, N.L. and Kotz, S. (1970) *Continuous Univariate Distributions, 1*, Wiley, New York.

Jørgensen, B. (1982) *Statistical Properties of the Generalised Inverse Gaussian Distribution*, Lecture Notes in Statistics **9**, Springer, New York.

Kallianpur, G. (1980) *Stochastic Filtering Theory*, Springer, New York.

Karandikar, R.L. (1983) Interchanging the order of stochastic integration and ordinary differentiation, *Sankhyā A*, **45**, 120–124.

Karatzas, I. and Shreve, S.E. (1991) *Brownian Motion and Stochastic Calculus*, Springer, New York.

Kariya, T., Tsukuda, Y., Maru, J., Matsue, Y. and Omaki, K. (1995) An extensive analysis on the Japanese markets via S.Taylor's model, *Financial Engineering and the Japanese Markets*, **2**, 15–86.

Karlin, S. and Taylor, H. (1975) *A First Course in Stochastic Processes*, Academic Press, New York.

Karlin, S. and Taylor, H. (1981) *A Second Course in Stochastic Processes*, Academic Press, New York.

Karr, A.F. (1991) *Point Processes and their Statistical Inference*, Marcel Dekker, New York.

Kasonga, R. (1988) The consistency of a nonlinear least squares estimator for diffusion processes, *Stochastic Processes and their Applications*, **30**, 263–275.

Keller, G., Kersting, G. and Roesler, U. (1984) On the asymptotic behaviour of solutions of stochastic differential equations, *Zeitschrift der Wahrscheinlichkeitstheorie und Verwandter Gebiete*, **68**, 163–189.

Kessler, M. (1995a) Estimation des paramètres d'une diffusion par des contrastes corrigés, *Comptes Rendus de l'Académie des Sciences I*, **320**, 359–362.

Kessler, M. (1995b) Quasi-likelihood inference for a discrete Markov chain.

Kessler, M. (1995c) Simple estimating functions for a discretely observed diffusion process, Research Report No. 336, Dept. of Theoretical Statistics, University of Aarhus.

Kessler, M. (1997) Estimation of an ergodic diffusion from discrete observations, *Scandinavian Journal of Statistics*, **24**, 211–229.

Kessler, M. and Sørensen, M. (1995) Estimating functions based on eigenfunctions for a discretely observed diffusion process, Research Report No. 332, Dept. of Theoretical Statistics, University of Aarhus.

Kimura, M. and Ohta, T. (1971) *Theoretical Aspects of Population Genetics*, Princeton University Press, Princeton, NJ.

Kloeden, P.E. and Platen, E. (1989) A survey of numerical methods for stochastic differential equations, *Stochastic Hydrology and Hydraulics*, **3**, 155–178.

Kloeden, P.E. and Platen, E. (1992) *The Numerical Solution of Stochastic Differential Equations*, Springer, Berlin.

Kloeden, P.E., Platen, E. and Schurz, H. (1994) *Numerical Solution of Stochastic Differential Equations through Computer Experiments*, Springer, Berlin.

Kloeden, P.E., Platen, E., Schurz, H. and Sørensen, M. (1996) On effects of discretization on estimators of drift parameters for diffusion processes, *Journal of Applied Probability*, **33**, 1061–1076.

Knight, F. (1971) A reduction of continuous square integrable martingales to Brownian motion, in: H. Dinges, ed., *Martingales*, Lecture Notes in Mathematics **190**, Springer, Berlin.

Kozin, F. (1977) An approach to characterizing, modeling and analyzing earthquake excitation records, in: H. Parkus, ed., *Random Excitation of Structures by Earthquakes and Atmospheric Turbulence*, CISM Courses and Lectures **225**, Springer, Vienna, 77–109.

Kunita, H. (1990) *Stochastic Flows and Stochastic Differential Equations*, Cambridge University Press, Cambridge.

Kutoyants, Yu.A. (1982) Expansion of a maximum likelihood estimate by diffusion powers, *Theory of Probability and its Applications*, **29**, 465–477.

Kutoyants, Yu.A. (1984a) *Parameter Estimation for Stochastic Processes* (trans. and ed. B.L.S. Prakasa Rao), Heldermann, Berlin.

Kutoyants , Yu.A. (1984b) Parameter estimation for diffusion type processes of observation, *Mathematische Operationsforschung Statistick Serie Statistik*, **15**, 541–551.

Kutoyants, Yu.A. (1985a) On nonparametric estimation of trend coefficients in a diffusion process, in: N.V. Krylov *et al.*, eds, *Statistics and Control of Stochastic Processes*, Optimization Software, New York, 230–250.

Kutoyants, Yu.A. (1985b) Efficient nonparametric estimation of trend coefficients, Preprint.

Kutoyants, Yu.A. (1991) Minimum distance parameter estimation for diffusion type observation, *Comptes Rendus de l'Académie des Sciences I*, **312**, 637–642.

Kutoyants, Yu.A. (1997a) Efficient density estimation for ergodic diffusions, Preprint, Université du Maine, Le Mans.

Kutoyants, Yu.A. (1997b) Several problems of nonparametric estimation by observations of ergodic diffusion process, Preprint, Université du Maine, Le Mans.

Kutoyants, Yu.A. and Pilibossian, P. (1994a) On minimum l_1-norm estimate of the parameter of Ornstein–Uhlenbeck process, *Statistics and Probability Letters*, **20**, 117–123.

Kutoyants, Yu.A. and Pilibossian, P. (1994b) On minimum uniform metric estimate of parameters of diffusion-type processes, *Stochastic Processes and their Applications*, **51**, 259–267.

Lanska, V. (1979) Minimum contrast estimation in diffusion processes, *Journal of Applied Probability*, **16**, 65–75.

Larédo, C. (1990) A sufficient condition for asymptotic sufficiency of incomplete observations of a diffusion process, *Annals of Statistics*, **18**, 1158–1171.

Le Breton, A. (1976) On continuous and discrete sampling for parameter estimation in diffusion type processes, *Mathematical Programming Studies*, **5**, 124–144.

Le Cam, L. (1986) *Asymptotic Methods in Statistical Decision Theory*, Springer, New York.

Le Cam, L. and Yang, G.L. (1990) *Asymptotics in Statistics: Some Basic Concepts*, Springer, New York.

Legland, F. (1981) Estimation de paramètres dans les processus stochastiques en observation incomplète: Application à un problème de radio-astronomie, Doctoral thesis, Université de Paris IX (Dauphine).

Leskow, J. and Rozanski, R. (1989) Maximum likelihood estimation of the drift function for a diffusion process, *Statistics and Decisions*, **7**, 243–262.

Levanony, D. (1994) Conditional tail probabilities in continuous-time martingale LLN with application to parameter estimation in diffusions, *Stochastic Processes and their Applications*, **51**, 117–134.

Levanony, D., Schwartz, A. and Zeitouni, O. (1994) Recursive identification in continuous-time stochastic processes, *Stochastic Processes and their Applications*, **49**, 245–275.

Linkov, Yu.N. (1989) Local asymptotic mixed normality for distributions of diffusion type processes, *Theory of Probability and Mathematical Statistics*, **40**, 67–73 (in Russian).

Linz, P. (1985) *Analytical and Numerical Methods for Volterra Integral Equations*, SIAM, Philadelphia.

Liptser, R.S. and Shiryayev, A.N. (1977) *Statistics of Random Processes: General Theory*, Springer, New York.

Liptser, R.S. and Shiryayev, A.N. (1978) *Statistics of Random Processes: Applications*, Springer, New York.

Ljung, L., Mitter, S.K. and Moura, J.M.F. (1988) Optimal recursive maximum likelihood estimation, in: *10th IFAC Proc.*, **IX**, 241–242.

Lo, A.W. (1988) Maximum likelihood estimation of generalized Ito processes with discretely sampled data, *Econometric Theory*, **4**, 231–247.

Loeve, M. (1963) *Probability Theory*, Van Nostrand, Princeton, NJ.

Longstaff, F.A. (1989) A nonlinear general equilibrium model of the term structure of interest rates, *Journal of Financial Economics*, **23**, 195–224.

Longstaff, F.A. and Schwartz, E.S. (1992) Interest rate volatility and the term structure: A two-factor general equilibrium model, *Journal of Finance*, **47**, 1259–1282.

Maiboroda, R.E. (1995) Estimates of intense noise for inhomogeneous diffusion processes, *Ukrainian Mathematical Journal*, **47**, 1085–1091.

Mandelbrot, B. (1963) The variation of certain speculative prices, *Journal of Business*, **36**, 394–419.

Mandl, P. (1968) *Analytical Treatment of One-Dimensional Markov Processes*, Springer, Berlin.

Maruyama, G. (1955) Continuous Markov processes and stochastic equations, *Rendiconti del Circolo Matematico di Palemo*, **4**, 48–90.

Masry, E. (1994) Probability density estimation from dependent observations using wavelet orthonormal bases, *Statistics and Probability Letters*, **21**, 181–194.

McKeague, I.W. and Tofoni, T. (1991) Nonparametric estimation of trends in linear stochastic systems, in: N.U. Prabhu and I.V. Basawa, eds, *Statistical Inference in Stochastic Processes*, Marcel Dekker, New York, 143–166.

Merton, R.C. (1971) Optimum consumption and portfolio rules in a continuous-time model, *Journal of Economic Theory*, **3**, 373–413.

Meyer, P. (1966) *Probability and Potentials*, Blaisdell, Waltham, MA.

Meyer, Y. (1990) *Ondelettes et Opérateurs 1*, Hermann, Paris.

Mikulevicius, R. and Platen, E. (1988) Time discrete Taylor approximation for Ito processes with jump component, *Mathematische Nachrichten*, **138**, 93–104.

Milstein, G.N. (1974) Approximate integration of stochastic differential equations, *Theory of Probability and its Applications*, **19**, 557–562.

Milshtein, G. (1976) A method of second order accuracy integration of stochastic differential equations, *Theory of Probability and its Applications*, **23**, 396–401.

Mishra, M.N. (1989) The Bernstein–von Mises theorem for a class of nonhomogeneous diffusion processes, *Statistics and Decisions*, **7**, 153–165.

Mishra, M.N. and Bishwal, J.P.N. (1995) Approximate maximum likelihood estimation for diffusion processes from discrete observations, *Stochastics and Stochastics Reports*, **52**, 1–13.

Mishra, M.N. and Prakasa Rao, B.L.S. (1985a) Asymptotic study of the maximum likelihood estimator for nonhomogeneous diffusion processes, *Statistics and Decisions*, **3**, 193–203.

Mishra, M.N. and Prakasa Rao, B.L.S. (1985b) On the Berry–Esseen bound for maximum likelihood estimator for linear homogeneous diffusion processes, *Sankhyā A*, **47**, 392–398.

Mishra, M.N. and Prakasa Rao, B.L.S. (1987) Rate of convergence in the Bernstein–von Mises theorem for a class of diffusion processes, *Stochastics*, **22**, 50–75.

Mishra, M.N. and Prakasa Rao, B.L.S. (1991) Bounds for the equivalence of Bayes and maximum likelihood estimators for a class of diffusion processes, *Statistics*, **52**, 613–625.

Mortensen, R.E. (1979) Mathematical problems of modeling stochastic systems, *Journal of Statistical Physics*, **1**, 271–296.

Nagahara, Y. (1996) A study of non-Gaussian modeling for financial economics, Doctoral thesis, Institute of Statistical Mathematics, Tokyo.

Newton, N.J. (1986) An asymptotically efficient difference formula for solving stochastic differential equations, *Stochastics*, **19**, 175–206.

Nguyen, H.T. (1979) Density estimation in a continuous time stationary Markov process, *Annals of Statistics*, **7**, 341–348.

Nguyen, H.T. and Pham Dinh Tuan (1982) Identification of nonstationary diffusion model by the method of sieves, *SIAM Journal on Control and Optimization*, **20**, 603–611.

Nikitin, N.N. and Rasevig, V.D. (1978) Methods of computer simulation of stochastic differential equation, *Vytshical. Matem. i Matem. Fiski*, **18**, 106–117 (in Russian).

Novikov, A.A. (1972) Sequential estimation of the parameters of diffusion-type processes, *Mathematical Notes*, **12**, 812–818.

Ozaki, T. (1985) Non-linear time series models and dynamical systems, in: E.J. Hannan, ed., *Handbook of Statistics, Vol. 5*, North Holland, Amsterdam, 25–83.

Ozaki, T. (1992) A bridge between nonlinear time series models and nonlinear stochastic dynamical systems: A local linearization approach, *Statistica Sinica*, **2**, 113–135.

Ozaki, T. (1993) A local linearization approach to nonlinear filtering, *International Journal of Control*, **57**, 75–96.

Pardoux, E. and Pignol, M. (1984) Étude de la stabilité de la solution d'une EDS bilinéaire à coefficients periodiques: Application au movement d'une pale hélicoptère, in: A. Bensoussan and J.L. Lions, eds, *Analysis and Optimization of Systems. Proceedings of the Sixth International Conference on Analysis and Optimization of Systems, Nice, June 19–22, 1984, Part 2*, Lecture Notes in Control and Information Sciences **63**, Springer, Berlin, 92–103.

Pedersen, A.R. (1995a) Consistency and asymptotic normality of an approximate maximum likelihood estimator for discretely observed diffusion processes, *Bernoulli*, **1**, 257–279.

Pedersen, A.R. (1995b) A new approach to maximum likelihood estimation for stochastic differential equations based on discrete observations, *Scandinavian Journal of Statistics*, **22**, 55–71.

Penev, S.I. (1985) Parametric statistical inference for multivariate diffusion processes using discrete observations, in: *Proc. Fourteenth Spring Conf. Bulgar. Math.*, Bulgarian Academy of Sciences, Sofia, 501–510.

Penev, S.I. (1986) Asymptotically normal estimators in nonlinear stochastic differential equations using Gauss–Newton procedures, Preprint.

Pham Dinh Tuan (1981) Nonparametric estimation of the drift coefficient in the diffusion equation, *Mathematische Operationsforschung Statistik Serie Statistik*, **12**, 61–73.

Platen, E. (1980) Approximation of Ito integral equation, in: B. Grigelionis, ed., *Stochastic Differential Systems*, Lecture Notes in Control and Information Sciences **25**, Springer, Berlin, 172–176.

Platen, E. (1981) An approximation method for a class of Ito processes, *Lietuvos Matem. Rink.*, **21**, 121–133.

Platen, E. (1982a) A generalised Taylor formula for solutions of stochastic differential equations, *Sankhyā A*, **44**, 163–172.

Platen, E. (1982b) An approximation method for a class of Ito processes with jump components, *Lietuvos Matem. Rink.*, **21**, 121–133.

Platen, E. and Rebolledo, R. (1985) Weak convergence of semimartingales and discretization methods, *Stochastic Processes and their Applications*, **20**, 41–58.

Platen, E. and Wagner, W. (1982) On a Taylor formula for a class of Ito processes, *Probability and Mathematical Statistics*, **3**, 37–51.

Prabhu, N.U., ed. (1988) *Statistical Inference from Stochastic Processes*, (*Contemporary Mathematics*, Vol. **80**), American Mathematical Society, Providence, RI.

Prabhu, N.U. and Basawa, I.V. (1991) *Statistical Inference in Stochastic Processes*, Marcel Dekker, New York.

Prakasa Rao, B.L.S. (1972) Maximum likelihood estimation for Markov processes, *Annals of the Institute of Statistical Mathematics*, **24**, 333–345.

Prakasa Rao, B.L.S. (1978a) Sequential nonparametric estimation of density via delta sequences, *Sankhyā A*, **41**, 82–94.

Prakasa Rao, B.L.S. (1978b) Density estimation for Markov processes using delta sequences, *Annals of the Institute of Statistical Mathematics*, **30**, 321–328.

Prakasa Rao, B.L.S. (1979) Nonparametric estimation for continuous time processes via delta families, *Publications de l'Institut de Statistique de l'Université de Paris*, **24**, 79–97.

Prakasa Rao, B.L.S. (1981) The Bernstein–von Mises theorem for a class of diffusion processes, *Theory of Random Processes*, **9**, 95–101 (in Russian).

Prakasa Rao, B.L.S. (1982) Maximum probability estimation for diffusion processes, in: G. Kallianpur, P.R. Krishnaiah and J.K. Ghosh, eds, *Statistics and Probability: Essays in Honor of C.R. Rao*, North-Holland, Amsterdam, 579–590.

Prakasa Rao, B.L.S. (1983a) *Nonparametric Functional Estimation*, Academic Press, Orlando, FL.

Prakasa Rao, B.L.S. (1983b) Asymptotic theory for nonlinear least squares estimator for diffusion processes, *Mathematische Operationsforschung Statistik Serie Statistik*, **14**, 195–209.

Prakasa Rao, B.L.S. (1984) On Bayes estimation for diffusion fields, in: J.K. Ghosh and J. Roy, eds, *Statistics: Applications and New Directions*, Statistical Publishing Society, Calcutta, 504–511.

Prakasa Rao, B.L.S. (1985) Estimation of the drift for diffusion processes, *Statistics*, **16**, 263–275.

Prakasa Rao, B.L.S. (1987) *Asymptotic Theory of Statistical Inference*, Wiley, New York.

Prakasa Rao, B.L.S. (1988a) Law of iterated logarithm for fluctuations of posterior distributions for a class of diffusion processes and a sequential test of power one, *Theory of Probability and its Applications*, **33**, 269–275.

Prakasa Rao, B.L.S. (1988b) Statistical inference from sampled data for stochastic processes, in: N.U. Prabhu, ed., *Statistical Inference from Stochastic Processes (Contemporary Mathematics*, Vol. **80**), American Mathematical Society, Providence, RI, 249–284.

Prakasa Rao, B.L.S. (1990a) Nonparametric density estimation for stochastic processes from sampled data, *Publications de l'Institut de Statistique de l'Université de Paris*, **35**, 51–83.

Prakasa Rao, B.L.S. (1990b) On mixing for flows of σ-algebras, *Sankhyā A*, **52**, 1–15.

Prakasa Rao, B.L.S. (1995) *Statistics and its Applications*, Indian National Science Academy, New Delhi.

Prakasa Rao, B.L.S. (1996) Nonparametric estimation of the derivatives of a density by the method of wavelets, *Bulletin on Informatics and Cybernetics*, **28**, 91–100.

Prakasa Rao, B.L.S. (1997a) On distributions with periodic failure rate and related inference problems, in: N. Balakrishnan and S. Panchapakesan, eds, *Advances in Statistical Decision Theory*, Birkhäuser, Boston.

Prakasa Rao, B.L.S. (1997b) Estimation of integral of square of density by wavelets, *Publications de l'Institut de Statistique de l'Université de Paris*, **41**, 29–48.

Prakasa Rao, B.L.S. (1997c) Estimation of the integrated squared density derivative by wavelets, Preprint, Indian Statistical Institute, New Delhi.

Prakasa Rao, B.L.S. (1998) Nonparametric functional estimation: An overview, in: S. Ghosh, ed., *Asymptotics, Nonparametrics, and Time Series*, Marcel Dekker, New York.

Prakasa Rao, B.L.S. (1999) *Semimartingales and Their Statistical Inference*, CRC Press, Boca Raton, FL.

Prakasa Rao, B.L.S. and Bhat, B.R. (1996) *Stochastic Processes and Statistical Inference*, New Age International, New Delhi.

Prakasa Rao, B.L.S. and Rubin, H. (1981) Asymptotic theory of estimation in non-linear stochastic differential equations, *Sankhyā A*, **43**, 170–189.

Prakasa Rao, B.L.S. and Rubin, H. (1998) Uniform approximations for families of stochastic integrals, *Sankhyā A*, **60**, 57–73.

Rao, N.J., Borwanker, J.D. and Ramakrishna, D. (1974) Numerical solution of Ito integral equations, *SIAM Journal on Control*, **12**, 124–139.

Reed, W.J. and Ye, J.J. (1994) The role of stochastic monotonicity in the decision to conserve or harvest old-growth forest, *Natural Resource Modeling*, **8**, 47–79.

Rogers, L.C.G. and Williams, D. (1987) *Diffusions, Markov Processes, and Martingales, Volume 2: Ito Calculus*, Wiley, Chichester.

Rozanov, Yu.V. (1967) *Stationary Random Processes*, Holden-Day, San Francisco.

Sagirow, P. (1970) *Stochastic Methods in the Dynamics of Satellites, CISM Courses and Lecture* **57**, Springer, Berlin.

Schaefer, S.M. and Schwartz, E.S. (1984) A two-factor model of the term structure: An approximate analytical solution, *Journal of Financial and Quantitative Analysis*, **19**, 413–424.

Schmetterer, L. (1974) *Introduction to Mathematical Statistics*, Springer, Berlin.

Schoener, G., Haken, H. and Kelso, J.A.S. (1986) A stochastic theory of phase transitions in human hand movement, *Biology and Cybernetics*, **53**, 247–257.

Shimizu, A. and Kawachi, T. (1984) Approximate solutions of stochastic differential equations, *Bulletin of Nagoya Institute of Technology*, **36**, 105–108.

Shinozuka, M. and Sato, Y. (1967) Simulation of nonstationary random processes, *Journal of the Engineering Mechanics Division, American Society of Civil Engineers*, **93** (EM1), 11.

Shoji, I. (1995) Estimation and inference for continuous time stochastic models, Doctoral Thesis, Institute of Statistical Mathematics, Tokyo.

Shoji, I. and Ozaki, T. (1997) Comparative study of estimation methods for continuous time stochastic processes, *Journal of Time Series Analysis*, **18**, 485–506.

Skorokhod, A.V. (1989) *Asymptotic Methods in Theory of Stochastic Differential Equations*, American Mathematical Society, Providence, RI.

Smith, A. (1990) Three essays on the solution and estimation of dynamic macroeconomic models, Ph.D. Thesis, Duke University.

Sobcyzk, K. (1986) Modelling of random fatigue crack growth, *Engineering Fracture Mechanics*, **24**, 609–623.

Sørensen, M. (1983) On maximum likelihood estimation in randomly stopped diffusion-type processes, *International Statistical Review*, **51**, 95–110.

Sørensen, M. (1986) On sequential maximum likelihood estimation for exponential families of stochastic processes, *International Statistical Review*, **54**, 191–210.

Stefanov, V.T. (1985) On efficient stopping times, *Stochastic Processes and their Applications*, **19**, 305–314.

Strasser, H. (1985) *Mathematical Theory of Statistics*, Walter de Gruyter, Berlin.

Stratonovich, R.L. (1966) A new representation for stochastic integrals and equations, *SIAM Journal on Control*, **4**, 362–371.

Stroock, D.W. (1982) *Lectures on 'Topics in Stochastic Differential Equations'*, Narosa, New Delhi.

Stroock, D. and Varadhan, S.R.S. (1979) *Multidimensional Diffusion Processes*, Springer, Berlin.

Svensson, L.E.O. (1991) The term structure of interest rate differentials in a target zone, *Journal of Monetary Economics*, **28**, 87–116.

Talay, D. (1982) Convergence pour chaque trajectoire d'un schema d'approximation des EDS, *Comptes Rendus de l'Académie des Sciences I*, **295**, 249–252.

Tauchen, G. (1997) New minimum chi-square methods in empirical finance, in: D. Kreps and K. Wallis, eds, *Advances in Econometrics: Seventh World Congress*, Cambridge University Press, Cambridge, 279–317.

Taylor, S. (1986) *Modeling Financial Time Series*, Wiley, Chichester.

Thompson, T.A. (1992) Optimal forest rotation when stumpage prices follow a diffusion process, *Land Economics*, **68**, 329–342.

Unny, T.E. (1984) Numerical integration of stochastic differential equations in catchment modeling, *Water Resources Research*, **20**, 360–368.

Van Kempen, N.G. (1981) *Stochastic Processes in Physics and Chemistry*, North-Holland, Amsterdam.

Vasicek, O. (1977) An equilibrium characterization of the term structure, *Journal of Financial Economics*, **5**, 177–188.

Viterbi, A.J. (1966) *Principles of Coherent Communication*, McGraw-Hill, New York.

Vovk, L.B. and Maiboroda, R.E. (1993) Estimation of the discard time for a process of Ornstein–Uhlenbeck type, *Ukrainian Mathematical Journal*, **45**, 1198–1205.

Wagner, W. and Platen, E. (1978) Approximation of Ito integral equations, Preprint ZIMM, Akademie der Wissenschaften der DDR, Berlin.

Walter, G. and Blum, J.R. (1979) Probability density estimation using delta sequences, *Annals of Statistics*, **7**, 328–340.

Weiss, L and Wolfowitz, J. (1974) *Maximum Probability Estimation and Related Topics*, Springer, Berlin.

Wong, E. (1963) The construction of a class of stationary Markoff processes, *Proceedings of the American Mathematical Society Symposium on Applied Mathematics*, **16**, 264–276.

Wong, E. (1971) *Stochastic Processes in Information and Dynamical Systems*, McGraw-Hill, New York.

Wong, E. and Zakai, M. (1965) The oscillation of stochastic integrals, *Zeitschrift der Wahrscheinlichkeitstheorie und Verwandter Gebiete*, **4**, 103–112.

Wright, D.J. (1980) Digital simulation of Poisson stochastic differential equations, *International Journal of Systems Science*, **11**, 781–785.

Yaglom, A. (1980) Application of stochastic differential equations for the description of turbulent equations, in: B. Grigelionis, ed., *Stochastic Differential Systems*, Lecture Notes in Control and Information Sciences **25**, Springer, Berlin, 1–13.

Yor, M. (1978) Temps locaux, *Astérisque*, **52–53**, 17–35.

Yoshida, N. (1988) Robust M-estimators in diffusion processes, *Annals of the Institute of Statistical Mathematics*, **40**, 799–820.

Yoshida, N. (1990) Asymptotic behavior of M-estimator and related random field for diffusion process, *Annals of the Institute of Statistical Mathematics*, **42**, 221–251.

Yoshida, N. (1992) Estimation for diffusion processes from discrete observation, *Journal of Multivariate Analysis*, **41**, 220–242.

Yoshimoto, A. and Shoji, I. (1995) Forest land valuation under stochastic log prices, Report No.79, Institute of Statistical Mathematics, Tokyo, 55–63.

Yoshimoto, A., Shoji, I. and Yoshimoto, Y. (1996) Application of a stochastic control model to a free market policy of rice, *Statistical Analysis of Time Series*, Research Report No. 90, Institute of Statistical Mathematics, Tokyo.

Author Index

Subject Index

Printed and bound by CPI Group (UK) Ltd, Croydon, CR0 4YY
16/04/2025
14658497-0005